PHOTOMICROGRA

For frontispiece legend see page ix.

PHOTOMICROGRAPHY

DOUGLAS LAWSON

1972
ACADEMIC PRESS   LONDON AND NEW YORK

ACADEMIC PRESS INC. (LONDON) LTD.
24/28 Oval Road,
London NW1

United States Edition published by
ACADEMIC PRESS INC.
111 Fifth Avenue
New York, New York 10003

Copyright © 1972 by
ACADEMIC PRESS INC. (LONDON) LTD.

All Rights Reserved
No part of this book may be reproduced in any form by photostat, microfilm, or any other means, without written permission from the publishers

Library of Congress Catalog Card Number: 71-185209
ISBN: 0.12.439750.6

PRINTED IN GREAT BRITAIN BY W. S. COWELL LTD, IPSWICH, SUFFOLK

Preface

This volume constitutes a complete new work though it follows the general principles of my first book, 'The Technique of Photomicrography', published in 1960. I was gratified and encouraged by the eagerness with which that publication was received and by the high opinion entertained by the public as to its usefulness and on this account have been prompted to write this new volume, which I trust will be even more acceptable.

I have endeavoured to include in its pages the most recent discoveries and applications, though I must confess that it is almost impossible to keep up with the rapid pace of progress and new design. It has been my object throughout to guide the operator to a better way of recording through the microscope and so to place photomicrography on a higher plane than hitherto. I have endeavoured to provide a comprehensive survey of the compound microscope, to inform on the construction of microscope lenses and their image-forming qualities, the varied use of the (generally neglected) reflected light, the simple microscope. Chapters have been replaced by nearly 80 subject headings to make for easy reference and to avoid repetition; so characteristic in a book of this nature.

I hope the work will be a satisfactory guide for those anxious to get to grips with the subject and that it will be the means of helping them on their way to produce better pictures through the microscope.

A book of this nature cannot possibly be written without the generous help of many people, personal friends, microscope manufacturers or those whose products are essential in the production of a good photomicrograph. I am grateful to Ilfords for their encouragement over many years, to the Royal Photographic Society, of which I have been a member for some 25 years, whose continual interest I value, and I take this opportunity to acknowledge the honours accorded me in the presentation of the Society's Medal in 1961 and The Rodman Medal in 1969.

Last but not least I should like to thank my sister Dulcie for her readiness to help at all times, Ken Austin for making many pieces of optical equipment and my employers for their kind permission to include work carried out whilst in their employment. I extend my sincere thanks to Mrs. Christine Stansfield for her co-operation in typing the final script, and to Sheila Tesh for her interest in the animal studies.

August 1972

Douglas Lawson
12 *Park Road, Banstead*
Surrey, England

Acknowledgements

I express my grateful thanks to the following for permitting me to publish information, drawings and photographs relating to their apparatus and applications; more especially to E. Leitz, Wetzlar, Carl Zeiss, Oberkochen, Watson (the Optical Division of M.E.L. Equipment Co., Ltd.) and John Swain and Son for the loan of colour blocks; American Optical Company, Bausch and Lomb, Blazers (Filters) High Vacuum Ltd., Ealing Beck (R and J Beck), General Electric X-ray Dept., Wisconsin, Graticules Ltd., E. Leitz, Wetzlar, Philips Electrical Ltd., Pyser-Britex (Swift) Ltd., C. Reichert, Austria, The Projectina Co., Ltd. (Nikon), W. R. Prior and Co., Ltd., Laser Products TRG, New York, Schott and Gen. Mainz, Vickers Instruments Ltd., Watson (the Optical Division of M.E.L. Equipment), Wild, Heerbrugg, Carl Zeiss, Jena and Carl Zeiss, Oberkochen. I am also grateful to TRW Systems, Los Angeles, California, for sponsoring and arranging several exhibitions in the United States in which over half a million people viewed my photomicrographs.

Frontispiece

The frontispiece depicts butterfly scales and a few diatoms arranged as grasses and flowers in a vase. Butterflies hover nearby. The scales and diatoms (added to shape the vase) are stuck on a black base. The original arrangement measures 7·5 mm from the uppermost point to the lower edge of the vase base, seen here ×21. |—— scale/1 mm ——| The colour photomicrograph was made on the apparatus set out in Fig. 141 (minus eyepiece) incorporating a 75 mm Objective 0·9 N.A. and camera bellows extended 50 cm. Thus, an image size of ×81/2 was achieved. Recorded on Ektachrome type B sheet film, 32 ASA, by reflected light from two 6V 30W lamps. Such work can only be arranged scale by scale, and viewed under a dissecting microscope or a low power lens system giving at least ×20 magnification. Even so, only a few scales can be added at one sitting. The scale-line will assist in an appreciation of the size of the original arrangement.

Contents

	Page
Preface	vii
Acknowledgements	viii
Introduction	1
The eye and simple microscope	1
The simple microscope	4
The early compound microscope	6
The compound microscope	10
Objective	16
Brightness	27
Refraction—air glass air	28
Numerical aperture and resolving power	29
Reference	39
Diffraction	39
Depth of focus	42
Depth of field	43
Refractive index	50
Reference	54
Cover glass	55
Reference	61
Microscope slide	61
Working distance	61
Objectives	62
Condensers	68
Eyepiece	74
Magnification	84
Illuminating the specimen	92
References	106
Balphot metallurgic microscope	107
Specular highlights	108
Köhler illumination: incident light	108
Some independent lamps	110
Lamp adjustment	110
Centering the condenser for Köhler illumination in transmitted bright-field	112
Köhler illumination in transmitted bright field	114
Tilting stage	117

xi

Colour expression	132
Optical staining	132
References	138
Monochromatic light (filters)	138
Diffusion disc in the field plane	145
Colour of light	146
Stained specimen	151
Camera	152
Exposure determination	160
Optical bench	164
Line overlaps in photomicrography	165
"Wedge" method for recording particles in liquid	171
Photographing precipitation bands	172
References	176
Transmitted darkground effects	176
Reference	178
Photographing large bulk materials	178
Recording gas bubbles inside a bottle	180
Back lighting with "simple microscope"	181
Phase contrast	183
References	194
Polarization	194
Interference technique	203
References	212
The interferometer microscope	212
Reflecting microscope	216
References	222
Infra-red photomicrography	223
References	231
Ultra-violet photomicrography	231
References	239
In-line mirror monochromator	239
Biolaser system	247
X-ray microscope and camera	249
Reference	253
Fluorescence illumination	253
References	263
Darkground photomicrography	263
Tyndall effect	274
Brownian movement	275
Schlieren photomicrography	275
References	276

CONTENTS

Stereo-photomicrography	276
Negative material	286
Processing plate and paper	287
Printing	288
Exhibition prints	289
The aesthetic and pictorial application	290
Maintenance and care of the microscope	298
Cinémicrography	299
References	310
Some common faults in photomicrography	310
Hot stages	313
Graticules	314
Mounting and staining	326
Mounting	332
Additional stains and routine methods	344
Replica technique	346
Some terms used in photomicrography	347
List of plates	355
List of colour plates	359
Plates	361
Colour plates	459
Subject index	481

Dedicated to the memory of the late
John Farquharson, Ph.D., D.Sc., F.R.I.C.,
in recognition of his untiring interest
and continual encouragement in the early
days of my work in the field of photomicrography.

Introduction

Photomicrography is the reproduction of minute objects, or parts of a small object, magnified by the simple and compound microscope and viewed in the form of a photographic transparency or print. It is a branch of science which demands knowledge of microscopy and photography and both these fields require perfect control. Photomicrography is a vast subject and it is therefore proposed to deal with some of the more important points such as apparatus, optical systems and the application of the microscope as a working tool in the production of a photomicrograph. For the purpose of this book it is assumed that all types of compound microscopes, from the simplest to the most up-to-date streamlined models, are being used. In any case, whatever type of compound microscope is used the same principles are almost always applied.

From time to time the word "microphotography" is used instead of "photomicrography". In order to keep the distinctions clear, Fig. 1 has been prepared to demonstrate a *photomicrograph*. At the opposite end of the scale the word microphotography is used to indicate an object recorded on a minute negative†, as illustrated by Fig. 2. Microphotography is also an exacting branch of photography, and as appreciated, the micro-negatives must be of a very high standard. Plate 1 has been made from the negative illustrated in Fig. 2.

When the eye is functioning at its best it can separate fine detail as small as 0·1 mm and particle size can be reduced to 0·2 of a micron by employing the optical microscope with a light source. In recent years the electron microscope has enabled even smaller particles (Plate 2e–f) down to 0·3–0·4 nm to be seen. These remarkable limits are made possible by operating at 100 kV. The Cambridge Steroscan Electron microscope, for instance, records solid objects direct, at comparatively low magnifications, with far greater depth of field (depth of sharpness) being obtained than hitherto.

The compound microscope, like other industrially designed apparatus, has not always been as streamlined as it is today. So before studying its application let us examine the use and function of early models of the microscope.

The eye and simple microscope

The human eye is so made that it can only have distinct vision when the light rays falling upon it are near parallel, or slightly divergent. The retina on which the image impinges requires the intervention of the crystalline lens to bring

---

†"And I will show you the microfilm. It is hidden in my cufflinks." Ivan Zabarin, *Daily Mail*, 9th Oct. 1971.

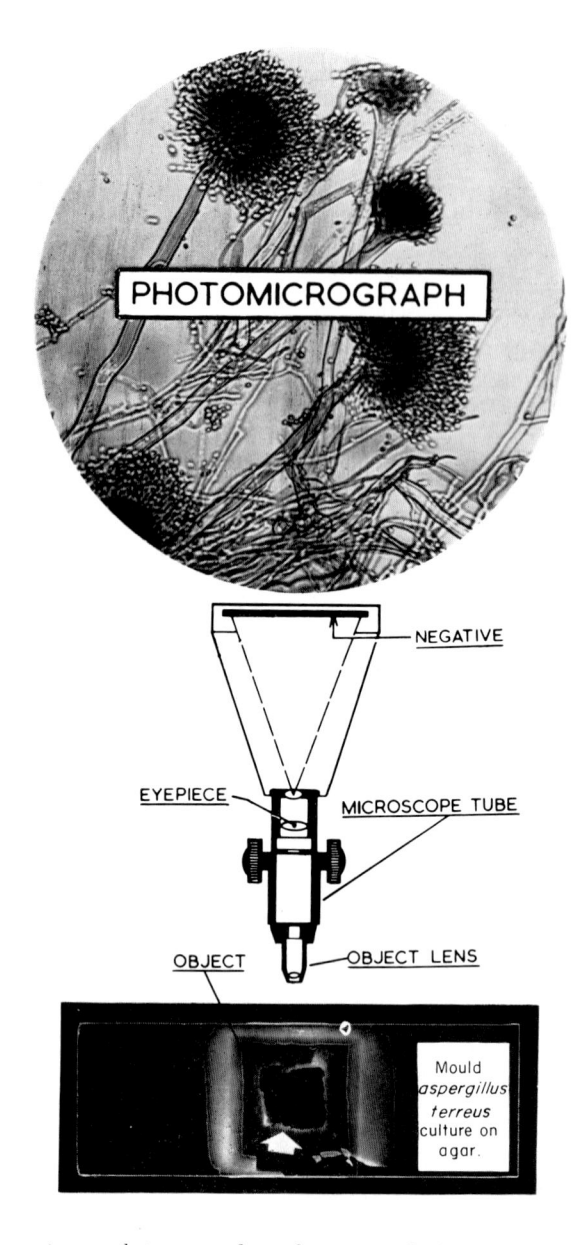

Fig. 1. A photomicrograph is a recording of a very small object such as illustrated here. The specimen area is indicated by the arrow on the microscope slide which, in this case, is of a mould. An image is made by the objective lens, passed on to the eyepiece and then on to the negative. Further enlargement can take place, as illustrated here. The final magnification is ×600.

THE EYE AND SIMPLE MICROSCOPE

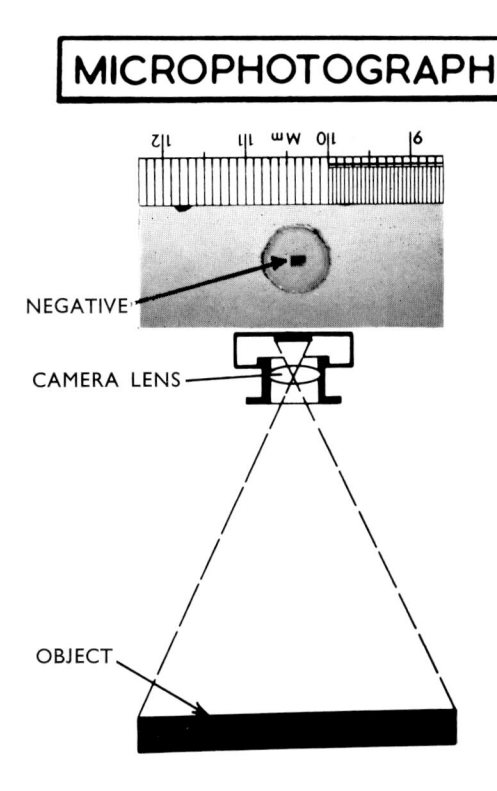

Fig. 2. A microphotograph is a minute negative or recording of a large object. The illustration shows a micro-negative actual size. The negative must be viewed by a magnifying glass or microscope. See Plate 1 which is a print from this negative.

some order and direct the light rays to an accurate point of focus upon its surface. The approximate unaided limit of near distinct vision is in the region of 250 mm (10 in.) and to most people with normal sight objects viewed nearer become indistinct. The apparent size of the object seen is due to the angle subtended to the eye or, as Fig. 3 demonstrates, the angle formed by two lines from the eye to the extremity of the object. When viewing minute objects the lines would be near parallel.

It will be seen that the lines from the eye to *A A* present an angle which, when the distance is short, is nearly twice as great as the angle *B B*, which in turn is twice the distance from the eye. This is the angle of vision, or the visual angle of the eye. The angle subtended at the eye by a near object can be increased by bringing it to a point closer to the eye. If the object is moved still nearer, the angle is also increased until clarity is lost. By changing the point and the distance of the object, a point is reached when an image of increased magnification is formed by a lens, as in Fig. 4. Thus a virtual image is formed at the near point. Such a lens, as illustrated, can be used as a simple microscope.

A converging lens placed between a near object and the eye produces divergence of light rays from the object so that they enter the eye as if they were

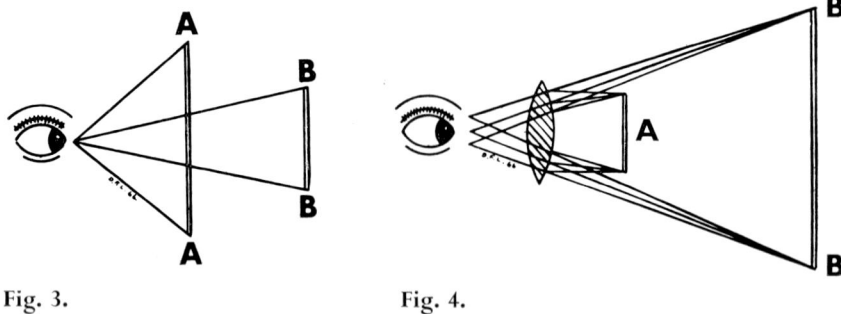

Fig. 3.

Fig. 4.

from an object beyond the normal limit of near distinct vision. The eye accepts the image as if it were from line B of Fig. 4, placed at a distance for normal near distinct vision. The difference between the real image and the imaginary line is known as the magnifying power of the lens. The diverging image seen in Fig. 5 has been brought about because the lens was not held parallel to the object. For visual purposes this error is not noticed: the eye is most accommodating.

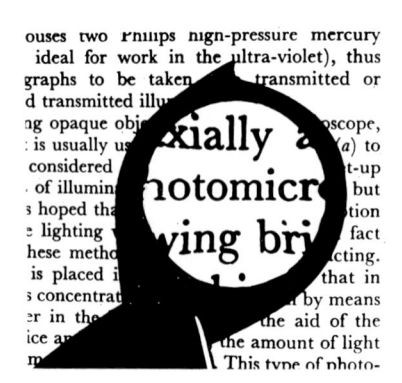

Fig. 5.

The simple microscope

Often a large diameter lens is expected to produce a high magnification of the specimen under investigation and at the same time project a wide field of view. In addition, the simple microscope is expected to have a long working distance between the object and lens. Unfortunately the laws of geometric optics do not

THE SIMPLE MICROSCOPE

permit this (see Fig. 6), as listed below:

(a) The higher the magnifying power of a lens the shorter is its working distance, as set out in the Figure. The working distance is approximately the same as the focal length of the lens. It does, however, differ when a simple microscope is made up of two or more elements cemented together.

(b) The higher the power of a lens the smaller is the field of view.

(c) The higher the power of a lens the smaller the lens becomes.

(d) The smaller the diameter of a lens the lower the light intensity and the greater the lens errors can be. Thus there is a limit set on the useful magnification

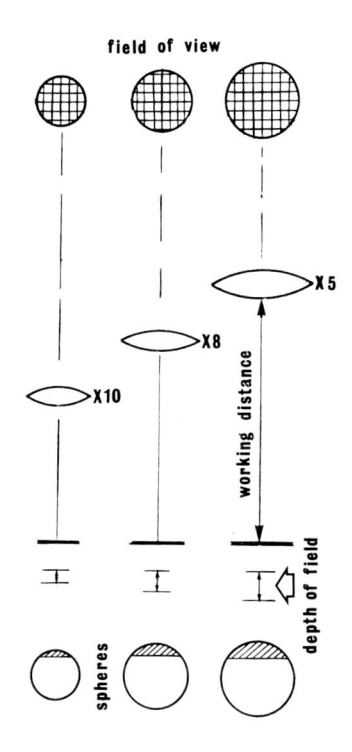

Fig. 6.

produced by such a lens which can be as high as ×20 and, in exceptional cases, even higher. If a simple microscope were to produce high magnifications, it would have a lens diameter of 2 mm or thereabouts.

(e) Lens' surfaces can be termed as sphere sections. The higher the power the sharper must be the curvature of the lens (see Fig. 6).

(f) The higher the power of a lens the shorter is the depth of field: i.e. depth of sharpness. Depth of field is essential when examining and photographing rough surfaces.

(g) To determine the approximate magnifying power of a lens, divide its focal length into 254 mm (10 in.), 5 in. f/l 10 in. = 2 power; or 50 mm f/l 254 mm = 5 power. A lens with a 5 in. focal length would be a ×2 magnifier. A ×10 single lens held close to the eye and focused on a near object will magnify that object 10 times its size, and it would be seen 10 in. from the eye. When using a reading glass, or any lens, at a distance from the eye other factors enter, which make it more difficult to compute the actual magnification.

Crude single lenses, hand held for reading, and spectacle glasses, have been in use since the fifteenth century. These were followed by more refined, powerful lenses, such as those contained in the simple microscope (Plate 3). Special techniques of grinding and polishing eliminated surface defects and other irregularities. Leeuwenhoek (1673) made some extraordinary observations with simple lenses held in brass collars and one of these is now preserved in Utrecht University Museum. Some simple microscopes have been shown to consist of two and three lenses placed close together as if one. Single and simple lenses still serve a useful purpose enabling the operator to examine objects prior to further study through the compound microscope. (See Plates 50-54 and 60).

The early compound microscope

Although the compound microscope was in use as early as 1609, its immense scientific possibilities were not generally appreciated until much later. Nevertheless, a number of interesting observations were made around that time and a few engravings of Francesco Stelluti's original drawings still exist. The invention of the compound microscope is claimed by many, yet there is little doubt that it was Galileo, then professor at Padua, who introduced the microscope in 1609. Like others, his first microscope was merely an inversion of his telescope. Kepler (1610), with a better instrument, aroused great interest by showing enlarged images of fleas, flies and itch-mites. It is not known whether these were drawings or projections. Thus the simple and compound microscope became known as flea-glasses or fly-glasses. The use of comparatively large specimens proves that Kepler's microscope was suited to lower power.

Galileo described on one occasion how, with his perspicillum (telescope), he observed the eye of an insect to be covered by a thick membrane and that the spectrum was perforated with holes like a warrior's vizor. Galileo was, of course, referring to the compound eye of an insect, commonly known to produce mosaic vision.

On September 23rd, 1624, Galileo wrote from Florence to Federigo Cesi, a wealthy nobleman who was at that time fully occupied in the study of the microscope and its applications, as follows:

Herewith an *occhialino* for examining minute things at near distance, I hope that from it you will have as much use and enjoyment as I. Its despatch has been delayed because I had not reduced it to perfection. The object should be attached to the mobile stage at the

THE EARLY COMPOUND MICROSCOPE 7

base and so moved as to bring the whole thing into view, since only a small part can be seen at a time in the *occhiale*. [High power objectives were obviously in use.] That there may be just the right distance between lens and object, the glass (objective) must be advanced or withdrawn, the little tube made movable on its base and adjusted as desired. With infinite wonder I have examined very many minute creatures, among which the most horrible are fleas, and the most beautiful are ants and moths. With delight I have seen how flies and other little animals can walk on mirrors and even upside down. You have now vast opportunity to observe thousands and thousands of details and I beg you to send me news of the most curious of them.

For many years after its invention, the compound microscope aroused surprisingly little interest among scientists until in 1667 Dr. Hooke published *Micrographia*, in which he mentions his compound microscope. It was 3 in. in diameter, 7 in. in length, and furnished with a draw-tube and three glasses: an object-glass, middle-glass and eye-glass. Until the early twentieth century, compound microscopes were crude and clumsy, and varied in style and pattern. Figure 7 shows a typical small clinical compound microscope. It has a fixed objective, screw-in type of eyepiece, a manually operated microscope slide and

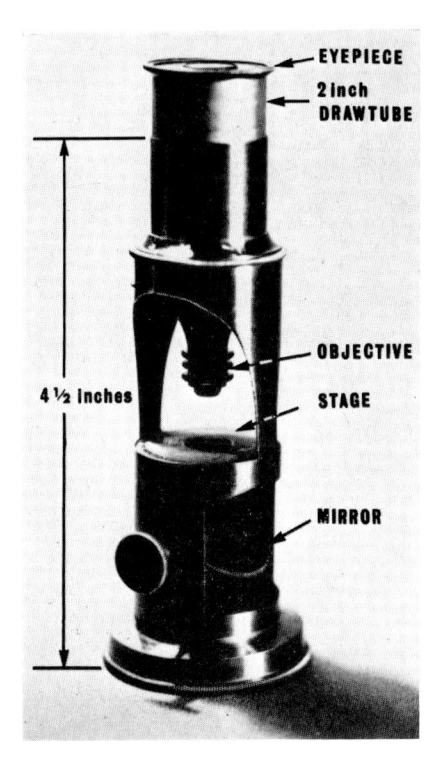

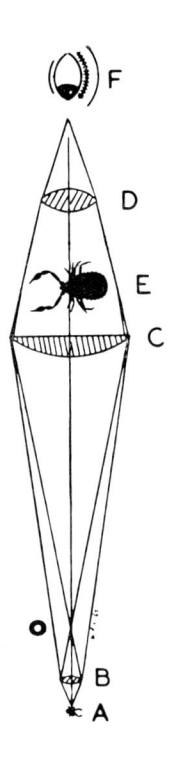

Fig. 7.

Fig. 8.

is illuminated by transmitted light from the very small mirror. This unique early compound microscope was purchased for the trivial sum of 2/– (10p) and the original price, still written on it was a modest £3.17s.0d (£3.85).

Figure 8 demonstrates the optical system of one of the earliest forms of compound microscope, fitted with a small simple object-glass B, a field lens C, and eye lens D. The specimen, placed in front of the object-glass at A its principal focal distance, made the image rays converge (at O) to a focus and form an inverted image at E. The light rays are emitted from the eye-lens at near parallel, cross, and are accepted by the eye F.

Up to about 1840 the compound microscope was used solely for visual and drawing purposes and incorporated the camera lucida, an eyepiece that projects an image from which drawings are made. Those shown in Figs. 9 and 10 were very popular. The micrographs (drawings) made with the aid of these instruments were

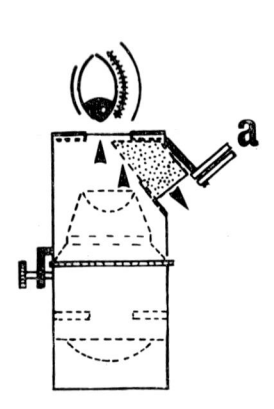

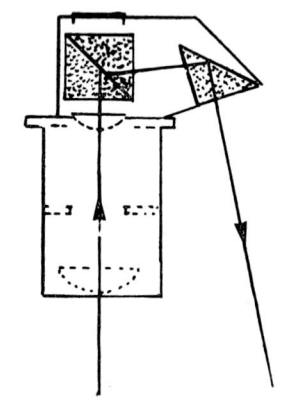

Fig. 9. Old Watson Eyepiece for drawing purposes. a: Filters.

Fig. 10. Old Swift-Ives camera lucida.

often works of art and many are still in existence. The eyepiece pre-directed the image by means of a prism and, if the light was too bright, neutral density filters were available (Fig. 9a). Camera lucida is still made, or at least the Leitz Micro Promar microprojecter is, and is used in schools, universities and industry today. Micrographs (drawings) can still be seen in a wide range of textbooks, sometimes with photomicrographs, thus making visual aids which are readily absorbed. The creative art of making delicate drawings (micrographs) has given place to photomicrographs. One of the writer's micrographs is seen in Fig. 11; by way of contrast Plate 4 demonstrates the value of an annotated photomicrograph.

Dancer (1840) produced photographs of microscopic objects taken with the aid of gas illumination. Various images were made on silvered plates and images of sections of wood and fossils were reproduced by means of a solar microscope† on sensitive or treated paper and on glass plates. A year later Richard Hodgson and the Rev. J. B. Reade obtained excellent daguerrotypes of microscopic objects. Dr. Donne (1840) of Paris presented copies of various magnified microscopic objects on daguerrotype plates to the Academy of Science. His photomicrographic apparatus has been described as consisting of several achromatic

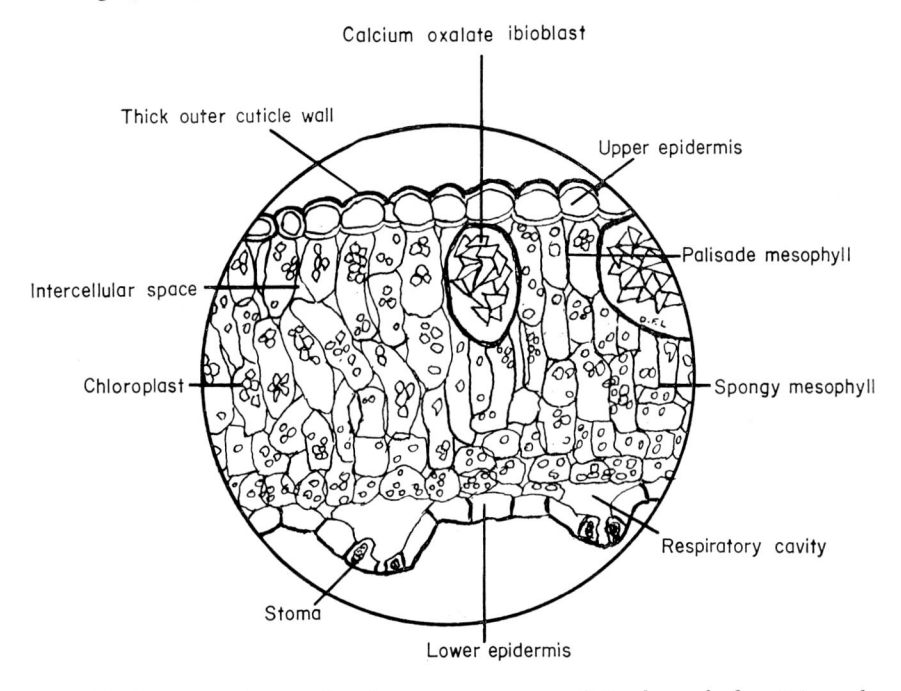

Fig. 11. Diagram (micrograph) of transverse section of Raspberry leaf ×400, made by projection, illustrated in Fig. 10. A 4 mm objective and substage condenser was also used.

objectives, which were always used without an eyepiece, but with a diverging lens to overcome the distances at which the image was formed. Five years later, M. Leon Foucault (1845) published an "Atlas of Microscopic Anatomy", in which the engravings were printed from daguerrotype plates which had been chemically etched after exposure and development. Delve (1852) produced some specimens of prints from collodion negatives which have been referred to as having "extreme details, beauty, texture and sharp delineation of the objects".

As early as 1855, Mr. Wenham referred to "Microscopic Photography" in the Quarterly Journal of Microscopical Science. Plainly, photography was applied to the compound microscope as soon as light-sensitive materials became available.

† Sunlight being the light source.

Prints from negatives marked an important step forward in the history of photomicrography when Dr. R. L. Maddox, in 1867, presented a paper to the Royal Society on his methods of taking photomicrographs. At the turn of the century the Royal Photographic Society awarded him the Progress Medal, the Society's highest award. Arthur Gill, writing in the October 1967 issue of The Photographic Journal, suggests that it would be of value to see work by such people as Maddox published in the photographic press. In addition, the illustration of award-winning work devised by special techniques, especially that of the last twenty years, would most certainly set and encourage high standards.

The compound microscope

This consists of six main units. Three optical: substage condenser, objective and eyepiece; and three mechanical: the microscope, illumination system and the camera (as in Fig. 12). The latest models are complete units (Fig. 13) and

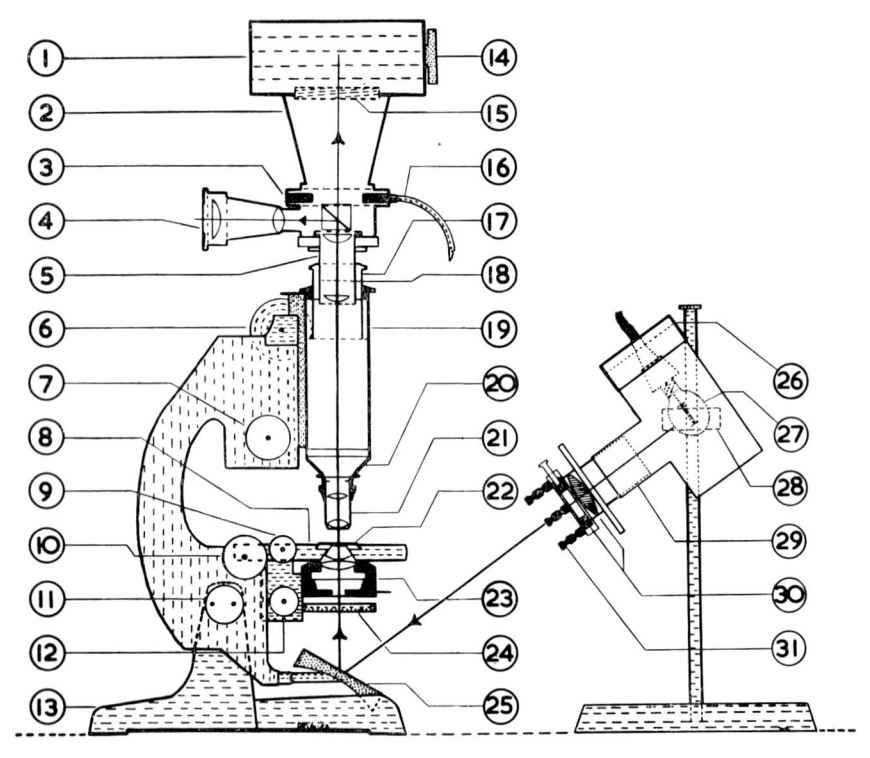

Fig. 12. In the case of a fixed body tube, such as the Leitz Ortholux microscope, the coarse and fine adjustment operate the microscope stage, moving the specimen to and from the objective in line with the optical axis. 1. Leica camera. 2. Micro-Ibso attach-

THE COMPOUND MICROSCOPE 11

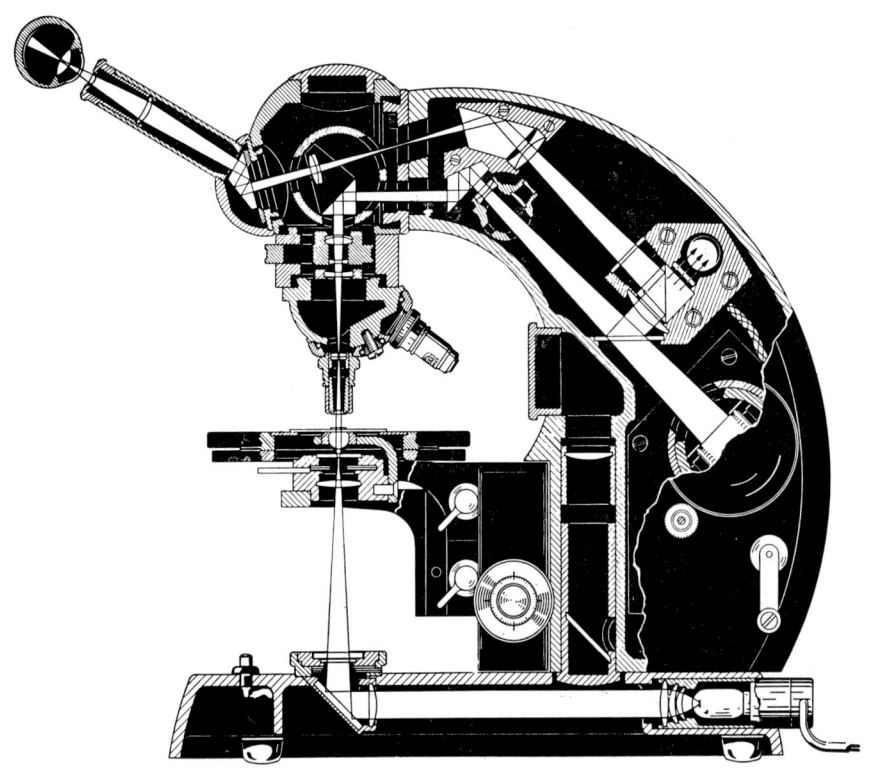

Fig. 13. Paths of rays in the Zeiss Photomicroscope II with transmitted illumination. The light of the low-voltage lamp mounted in the foot of the instrument is conducted across a mirror into the illuminating apparatus of the microscope. Above the objective are mounted the magnification changer Optovar, deflecting prisms as well as a projective light beam, switchable into two stages. A semi-permeable mirror then splits the course of the rays. One part of the light is directed to the film, the remainder to a photocell and into the viewing eyepiece.

in some respects resemble models of the old bench type of microscope. The illumination system is sometimes built within the apparatus and the camera is streamlined to match the appearance of the microscope. Instant photomicrographs can be made through the medium of the Polaroid Land camera. Bench

---

ment. 3. Diaphragm. 4. Viewing eyepiece. 5. Eyepiece. 6. Coarse adjustment. 7. Fine adjustment. 8. Stage. 9. Stage adjustment (lateral). 10. Stage adjustment (forward). 11. Joint (hinged). 12. Substage focusing adjustment. 13. Foot. 14. Film wind (free movement). 15. Attachment thread. 16. Cable release. 17. Draw tube. 18. Primary image. 19. Body. 20. Nosepiece. 21. Objective. 22. Specimen. 23. Substage, substage condenser and iris. 24. Filter carrier. 25. Mirror. 26. Bulb adjustment sleeve. 27. Coiled filament bulb. 28. Tilt adjustment. 29. Focusing sleeve. 30. Condenser and iris diaphragm. 31. Filter carrier.

B

models (Plate 5), of course, are still being made and are in fact very much to the fore. Some of these operate with independent light sources and detachable camera.

From Fig. 12 the eyepiece (5) is seen to be the uppermost part of the microscope and that it fits into the head of the draw-tube (17), as do most small portable cameras (1). The draw-tube is a smooth sliding fit in the body tube in which it is retained without any slackness. The body tube (19) carrying the draw-tube is operated by a rack and pinion. This is one of the few items which has retained its old form. It provides a coarse and fine (6 and 7) adjustment for focusing purposes. At the base of the body is fitted an objective holder, commonly known as a nose piece (20).

The stand. Whatever model of microscope is used, the baseplate must be heavy and free from bench tremors and able to withstand vibration caused by, say, running machinery; or nearby heavy traffic and trains. It is useless paying large sums of money for the very best outfit if it is not able to withstand all forms of vibration.

The limb. This is the main support and the depth through the optical axis of the lenses should allow a large specimen to be placed on the stage (Fig. 12, 8). It is therefore necessary to have considerable up and down movement. The horizontal optical bench demands other arrangements, according to the model. The stage and body are attached to the limb. The body moves on a sliding device, is tension loaded and operated by a thumb screw. The stage should always be at a true right angle to the tube housing and objective. Both limb and body tube are made to near perfection, guaranteeing parallelism between the mechanical and optical axis.

The stage. This (Fig. 12, 8) carries the specimen slide (22) at right angles to the optical axis and is a fixture, or slides mechanically moved by a thumb screw (9 and 10). Some stages are moved by coarse and fine adjustment in the line of the optical axis of the microscope. In addition, there is a stage which can be rotated for work by polarized light. The projected image can also be rotated to fit into the negative format. Such centering stages are usually fitted with a graduated circular scale and vernier. A complete selection of stages is available for the new Spencer Microscope. This includes graduated and ungraduated and allows complete coverage of a 51 mm by 77 mm (2 in. × 3 in.) slide.

Foot. This is attached to the limb (Fig. 12, 13), hinged, and provides any angle at which the microscope can be operated. It should therefore be sufficiently heavy to allow the instrument to operate without having to be clamped down. The mechanical adjustments are few in number, but nevertheless they are all very important in their particular application and provide a delicate control over the optical system.

Coarse adjustment. The coarse adjustment (Fig. 12, 6) provides an up and down movement of several inches, in the line of the optical axis. Some microscopes

THE COMPOUND MICROSCOPE

offer a working distance of over eight inches when a long focal length objective is in use. The coarse adjustment responds to the delicate touch of thumb and finger, allowing the body to be moved as little as one thousandth of a millimetre. Some microscopes of recent design have an adjustable stage in place of the tube (Fig. 13).

Fine adjustment. The fine focusing adjustment (Fig. 12, 7) carries on where the coarse focusing adjustment left off and, with the aid of a magnifying lens (Fig. 5), extremely fine adjustment of the image on the ground glass is ensured. The displacement of the up and down movement of tube or stage can be measured against the scale provided, and this is usually in units of 0·001 mm.

Nosepiece. The nosepiece (Fig. 12, 20) houses the objective. The up-to-date model of the rotating nosepiece can carry from two to six objectives, of varying focal length, according to the particular design of the microscope. With a slight turn of the friction loaded nosepiece an objective of a different focal length can be brought into operation, ensuring a quick changeover in magnification. Many designs now offer a par-focal lens system, the objective lens automatically coming into focus when placed under the tube. These objectives are approximately aligned and also image the same centre point of the specimen. However, it is advisable to check on the critical point of focus before making an exposure. Odd objectives cannot be used to effect as they do not offer the same refinements as the original set; neither are they computed to operate with the par-focal eyepiece.

Tube diameter. The internal diameter of the eyepiece tube has been set at 23·2 mm, and all tubes are based on this standard.

Mechanical tube length. This is the distance between the objective screw-in surface and the eyepiece rest surface, as indicated in Fig. 14. Objectives are constructed by particular makers to give their best performances at a given distance of mechanical tube length, when operating with transmitted light. When using the recommended cover-glass thickness, the distance is either 170 mm for Leitz optical systems and 160 mm for Zeiss and some other makes of the microscope. Objectives used with incident or reflected light often image the specimen at "infinity", and are corrected accordingly, which could be beyond the distance given.

Optical tube length. This comes into being from the optical centre of the objective, through the upper focal plane to the lower focal plane of the eyepiece (Fig. 14). Objectives have been computed to work at a set distance from the eyepiece, and if this is not observed the two lenses will not function at their best as one unit resulting in spherical errors and an inferior image. From this it is clear that objectives and eyepieces of different manufacturers also function as one unit. (But they should be used with caution. While certain standards of optical tube length are generally accepted, some microscopes, designed for metallurgical purposes, have a tube length ranging from 190 mm to 215 mm.) The mechanical

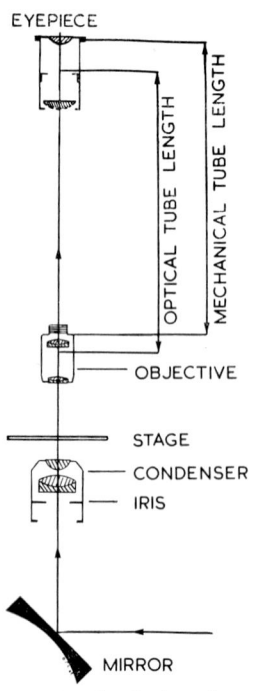

Fig. 14. The relationship between optical tube length and mechanical tube length.

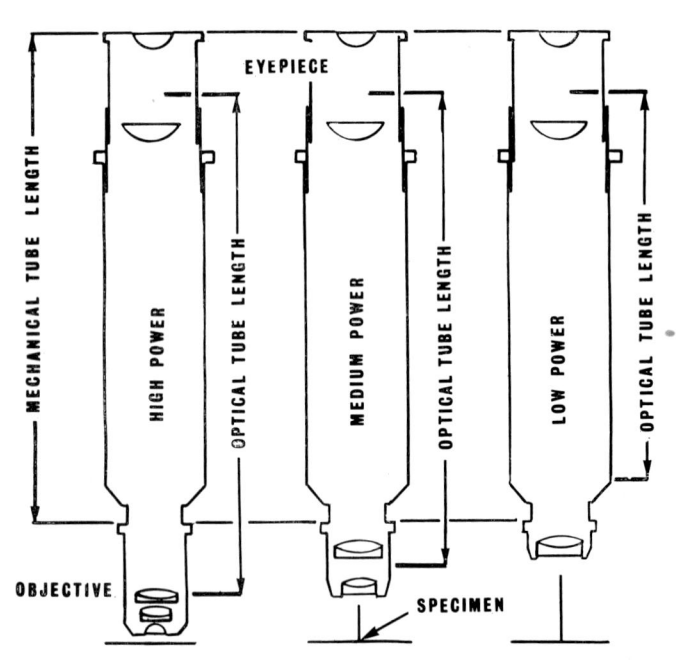

Fig. 15. Optical tube length for high, medium and low power, compared with the mechanical tube length.

tube length is constant for a given type of microscope, whereas the optical tube length may vary considerably, especially when working with low power objectives (Fig. 15). The photomicrographer should be prepared to find the best practical tube length above or below the theoretical tube length if he is to obtain the best results, particularly when photographing through a "thick" or "thin" cover glass. We will come to this in greater detail. Because an objective has been manufactured to work with an eyepiece, it in no way excludes it from being used on its own. The method used in the production of a photomicrograph has certain latitudes and it calls for the precautions learned through practical experience.

The following story concerns this particular point. When applying for the Fellowship of the Institute of British Photographers (now Institute of Incorporated Photographers) my thesis described practical use of the objective without an eyepiece. Supporting evidence amounted to many photomicrographs. The examiners refused to accept this method and wrote: "two optical systems, namely objective and eyepiece, were computed to operate together and in no way must be separated in order to produce an image". Thus the concern for a quality image was not even considered. My application failed, but it is encouraging that certain manufacturers now state that their low power objectives can be used without an eyepiece.

When making photomicrographs without the aid of the eyepiece it is advisable to remove the draw-tube. This will then give greater clearance and provide a larger projected image field. If reflections or a flare spot occur from within the body tube it should be lined with thin black paper or, better still, use a 50 mm body tube.

Substage. The substage houses the condenser (Fig. 12, 23) and allows free movement in or along the line of the optical axis for focusing. Some substages have a centering device in order to ensure the exact positioning of the condenser's optical system. This centering is essential when employing objectives such as the apochromatic and achromatic, especially in the medium and high power range. It is the view of the writer that all substage condensers should be fitted with centering adjustment screws; without these the lens can become eccentric. Those who operate the instrument should be familiar with the operation and image effect by changing the position of the optical axis. Nearly all substages have a built-in iris diaphragm which functions as an aperture diaphragm and must at all times be operated with caution. As the microscope is more convenient to use in the vertical position, a mirror (Fig. 12, 25) is placed on the free moving tailpiece immediately below the optical system. The mirror redirects light from a source in the near horizontal position some distance away. The mechanical rotating stage can be fitted to any microscope and often an electrically heated warm stage is necessary for some work. Cold stage is also available.

Mirror and condenser. The mirror has a concave side (Fig. 16) and a flat side (Fig.

17). The curved side of the mirror has a focal length of approximately 9 cm ($3\frac{1}{2}$ in.). This side of the mirror is rarely used because it causes some deterioration of the source, can mar the projected image and, of course, the final expression. If a condenser is used in conjunction with the light source then the flat side of the mirror should always be used. The converging rays from the concave

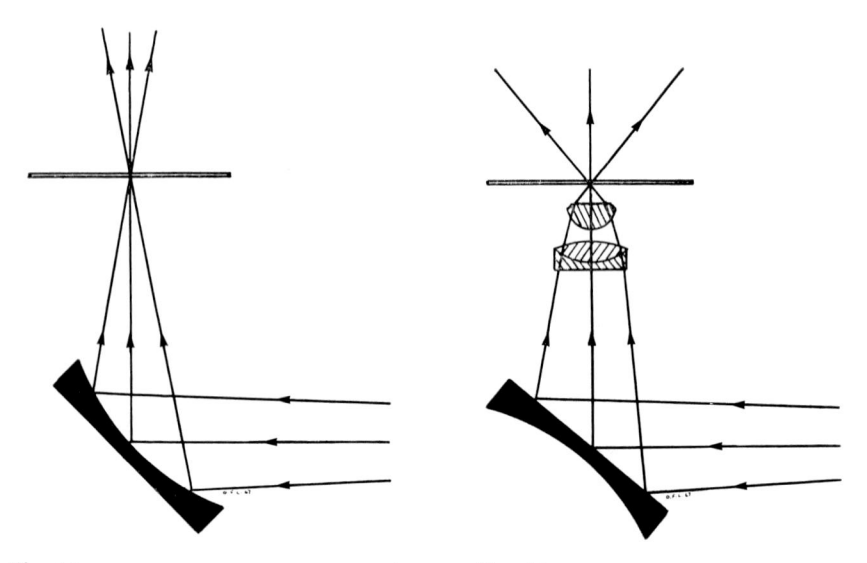

Fig. 16.   **Fig. 17.**

mirror (Fig. 16) will produce difficulties and prevent the condenser from fulfilling its prime function. Such curved surfaces are condensers in themselves, projecting a narrow converging cone of light and are only effective when a long focus (low power) optical system is in use. It is as well to bear in mind that such a cone of light is most unsuitable for objectives with a numerical aperture greater than 0·25, such as some 25 mm and 16 mm objectives.

Objective

Having looked at the construction of the metal parts of the microscope, we will now turn our attention to the basic lens arrangements within the instrument. The chief function of an objective is to form and project a high quality magnified image of a minute object placed in front of it and at close proximity to the front lens. The primary image or magnification is a real inverted image of the specimen and is at a distance called the optical tube length away (Fig. 15). This is the distance from the back focal plane of the objective in use to the anterior focal plane of the eyepiece.

The optical components of an objective are made up of various types of glass, ground into different shapes and cemented together in an effort to correct or lessen defects such as aberrations, which otherwise would occur. The objective consists of a combination of centred spherical surfaces, modelled around those shown in Fig. 18. The theoretical possibilities of such cannot come into practice unless the centres of curvature of the surfaces of the lenses are all coaxial and positioned correctly along the principal optical axis (Fig. 26). This applies equally with condensers and eyepieces and in particular with immersion objectives.

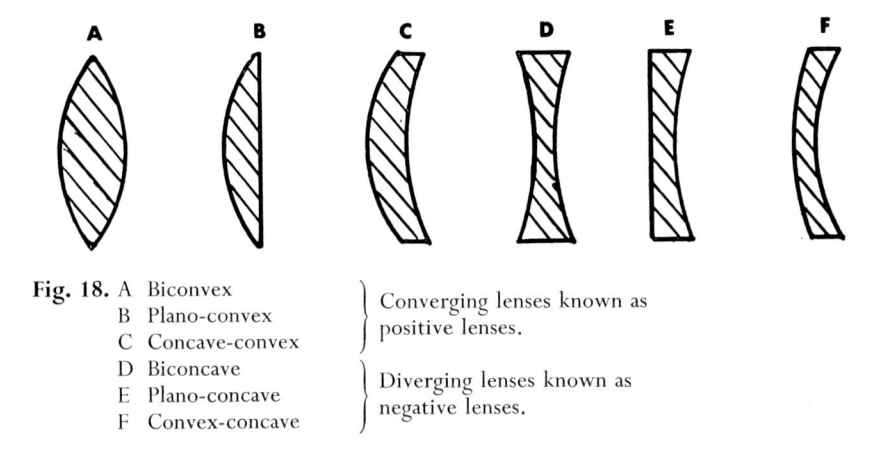

Fig. 18. A Biconvex
       B Plano-convex   } Converging lenses known as
       C Concave-convex   } positive lenses.
       D Biconcave
       E Plano-concave   } Diverging lenses known as
       F Convex-concave } negative lenses.

In addition to residual aberrations due to the structure, considerable disturbance is further caused by extra-axial image errors. An image error correction is brought about by using several lens surfaces at various angles. This in itself produces many problems for the manufacturer. One of these is the miniaturization of the many optical components. Shaping and polishing is essential and the production of high power objective lenses is more than exacting. The two most commonly used lenses are the negative lens, perhaps better known as diverging lens; and the positive lens, or converging lens.

1. *Diverging lens.* The two images on the right of Fig. 19 were produced by diverging lenses in the manner indicated in Fig. 20. These lenses spread the incident beam of light so that rays moving near parallel to the optical axis emerge as if they had originated from a distant point. The point can be found by continuing along the rays that emerge, to the point of intersection (Fig. 21). This is sometimes referred to as the rear focal point. As we can see from the figure, the point "o" is only apparent; therefore it derives the term "virtual", or virtual image of the real image. Rays travelling to the front focal point emerge parallel to the optical axis. Figure 22 illustrates a converging and diverging lens used as one unit. It will be seen that the diverging lens intercepts rays from the converging

lens. The illustration demonstrates the position of the real object (*RO*), virtual image (*VI*) and the real image (*RI*).

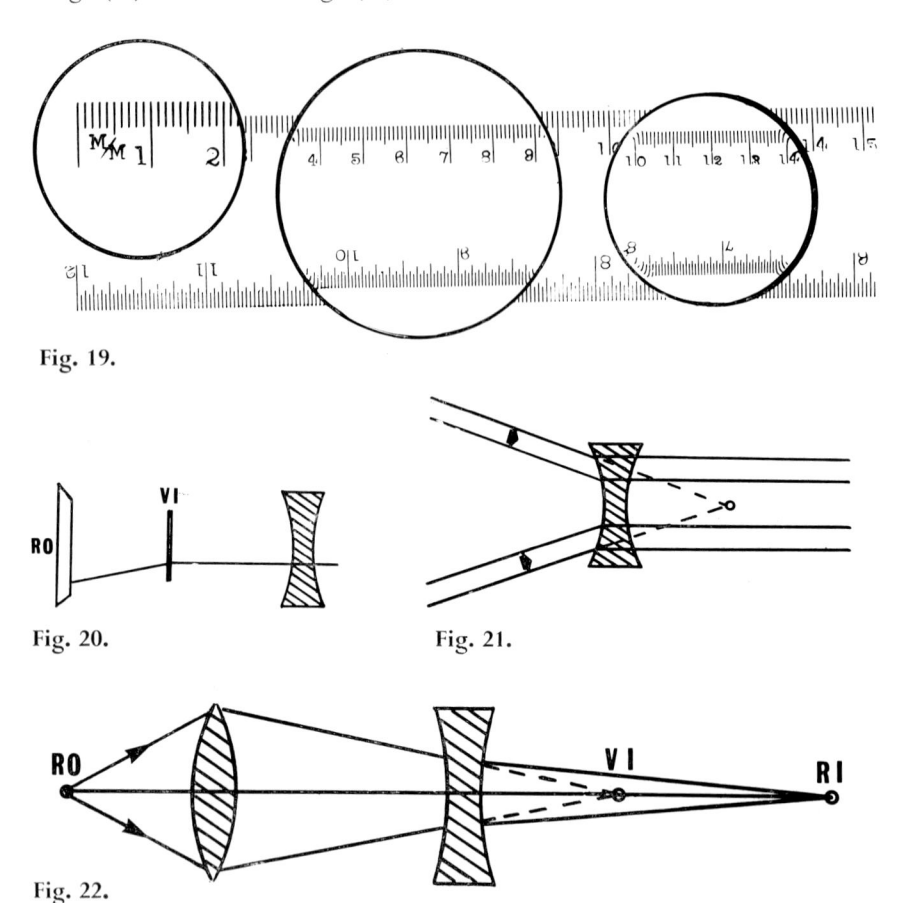

Fig. 19.

Fig. 20.

Fig. 21.

Fig. 22.

2. *Converging lens.* The image on the left of Fig. 19 was formed by a converging lens (Fig. 23). Compare the images with the other two seen in Fig. 19. Rays from an object some distance away travel on a route parallel to the optical axis and to the "focal point", when the object forms a real image.

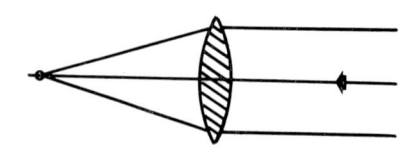

Fig. 23.

OBJECTIVE

Objective and primary magnification. How does a lens produce a magnified image of a near object? The diagram (Fig. 24) shows a positive lens with two infinite numbers of rays leaving point B of the specimen A, which at a distance from the front lens is greater than the focal length. One of the rays leaves B and passes parallel with the optical axis through the lens to B^1, is then refracted or bent, and passes through the focus F to B^2. The image of B is also brought to this point of interception by the undeviated ray C, which passes through the centre of the

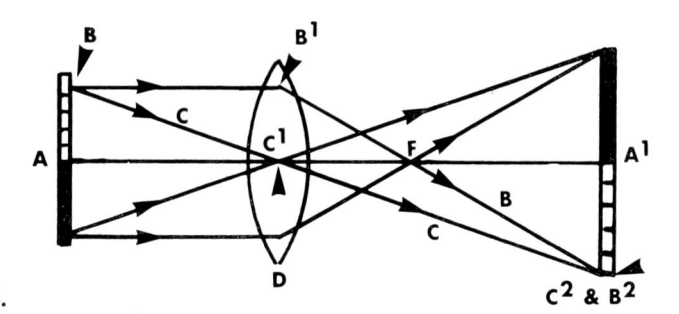

Fig. 24.

lens C^1. B is now imaged (see illustration), and forms a real inverted image of B. In the compound microscope, all paths from the specimen passing through the objective lens are approximately equivalent, and the effect of deviation of the illuminating pencils on the image is negligible.

The following (Fig. 25) is a similar illustration to that of the previous, but

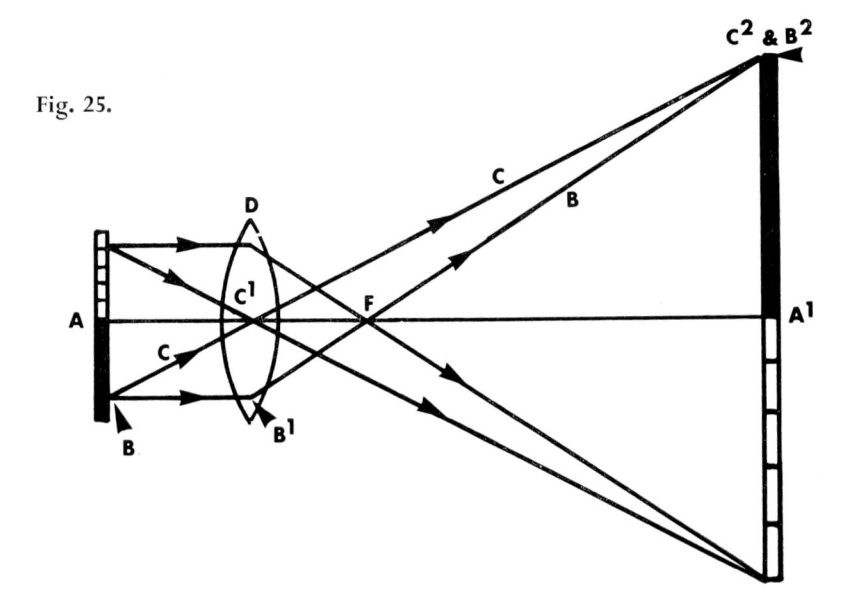

Fig. 25.

two changes have in fact been made. The specimen is placed nearer to the lens, thus introducing a further change in the performance—lens to point of focus. Here an inverted image of A is formed at C^2 and B^2, but further from the lens. From this it can be seen that as the distance from the specimen to the lens is lessened, the size of the projected image is increased. Other points on the specimen A can be formed and traced through the lens, forming an inverted image of the specimen.

The magnification of the projected image continues to increase as the specimen is brought nearer to the lens, until a stage is reached when it is at a distance equal to the focal length. At this point the two rays do not converge, but remain parallel and do not form a real image. When the specimen is at a distance equal to the focal length, an erect but virtual image is formed at infinity.

From Figs. 24 and 25, it can be seen that the magnification size of the image, divided by the size of the object, is given by

$$\frac{B^2A^1}{BA} \text{ or } \frac{B^2A^1}{C^1B^1}$$

The triangles FA^1B^2, FC^1B^1, are similar and so their corresponding sides are proportional, thus

$$\frac{B^2A^1}{C^1B^1} = \frac{FA^1}{FC^1}$$

FA^1 is the optical tube length (Fig. 15) and FC^1 is the focal length of the lens; therefore the magnification is the tube length divided by the focal length. Magnification is as a rule expressed in diameters, a certain number of time linear.

Now let us go a stage further than that seen in Fig. 25. We will now include an eyepiece in the illustration (Fig. 26) and observe the change brought about in

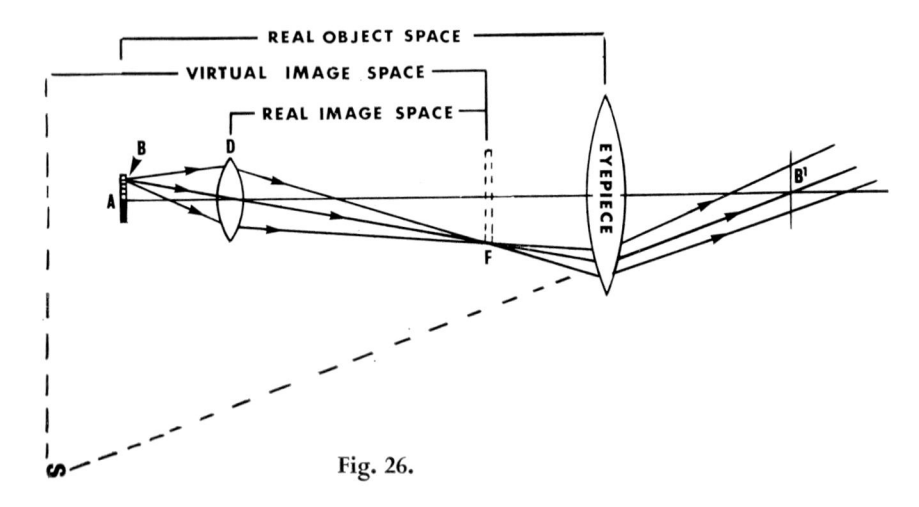

Fig. 26.

the rays. The objective D forms the real image of B at F. F has been used as object to construct the projected image by the eyepiece. B^1 is the image of B and is formed by the eyepiece, which is known as the real intermediate image of the objective, and all rays which emerge from the objective should go through the eyepiece. The real intermediate image produced by the objective will lie in the area of the focal plane of the eyepiece (Fig. 27 E), and be seen at infinity. The image focal plane lies in the area above the exit pupil of the microscope and,

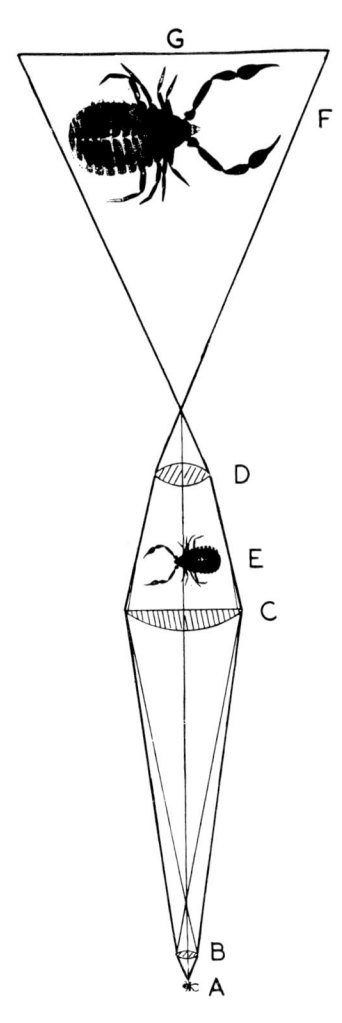

Fig. 27. Image formation from specimen to film. A. Specimen. B. Objective. C. The image plane within a Huygen eyepiece lies between the field-lens C and eye lens D, in the plane of the diaphragm. F. Camera. G. Image of A recorded at G, the focal plane.

when operating with a bellows camera, the projected image can be seen as indicated in Fig. 27 G. There is a considerable limit on the distance the image can be projected and by removing the ground glass screen an image can be formed on the ceiling or wall many feet away.

When studying fine details the microscope tends to become an extraneous instrument interposed between the specimen and the viewer as he looks through the eyepiece and sees the specimen clearly delineated against its background. The distance from the eye to the specimen does not concern him because he does not see the instrument through which he is viewing, to him it becomes an extraneous body. The real image is seen as though it were at S of Fig. 26, a comfortable distance away, approximately 250 mm (10 in.). The photomicrographer accepts the projected image and builds up a mental picture of what is seen before him, this being replaced later by an emulsion. When viewing the specimen, the image of which passes through several lenses, there appears to be a sense of homogeneity, a oneness, an uninterrupted sense of nearness, even touching distance between the specimen and the eye; irrespective of the smallness of the specimen and the distance, prisms and optical systems which really separate it from the eye. But when viewing the image on the ground glass screen the sense of homogeneity is completely lost as the operator becomes aware of the instrument, its complications and its vastness, which now separate him from the specimen.

Lens aberrations. There are five monochromatic aberrations which can affect an optical system and ultimately the photographic image. They are spherical aberration, chromatic aberration, coma, astigmatism and field curvature. In order to fulfil the requirements of a photomicrograph an objective must image a crisp, sharp recording of the specimen.

The illustrations are only valid for simple lenses placed in a narrow beam of light. When a thin specimen is placed before a wide angle lens the rays passing through the lens do not all arrive at a common point of focus, and aberrations occur. In order to obviate any shortcomings in a lens, manufacturers have designed lenses of special shapes. These consist of various types of glass (dispersion power) and by the insertion of an iris diaphragm in the appropriate place, rays are brought to a common point of focus. This is what happens in theory, and the potential correction is sufficient to render the aberration harmless, if the lenses are used as intended. Mind you, in taking a photomicrograph, we must try to use the microscope as it is intended, setting out to give the image a clear, distinct, and faithful representation at whatever magnification is demanded.

Maximum image quality can only be obtained when all the components are correctly adjusted and operating at their very best performance. Less than this reduces such a standard. It is possible to record the point of interest as registered on the emulsion including, for instance, foreign matter, scratches, out-of-focus colours etc. When this occurs the operator accepts only what he wishes to see.

Such workers are perhaps not yet alive to the differences in negative and print quality and are merely content that an image has been recorded.

The word system is used here because the projected image does not really depend upon any one particular component. We know all too well that in addition to the optical systems there follows the exposure, and that the after-exposure techniques too, affect the overall projected image. Good definition depends upon the extent to which these aberrations are rendered harmless.

Spherical aberration. Spherical aberration is brought about by the inability of a lens to bring marginal and inner rays to one common point of focus. Therefore, image points are not sharply defined on one single plane but the image or images extend along the optical axis as illustrated in Fig. 28. The general result of spherical aberration is a blurred image instead of sharpness and an overall flat image.

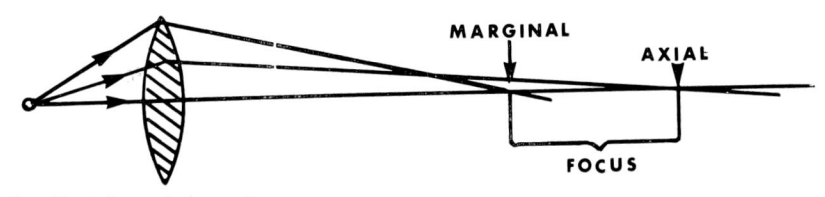

Fig. 28. Spherical aberration.

The correction demands an additional lens, as opposed to the last figure. Here in Fig. 29 a plano–convex doublet-achromatic lens has been added and is positioned with the flat side facing the specimen. Rays are directed from A to A^1 and are free from spherical errors. Points such as A^1 are said to be "aplanatic points". Van Deijl (1807) used a similar lens system in his objective, and a few years later

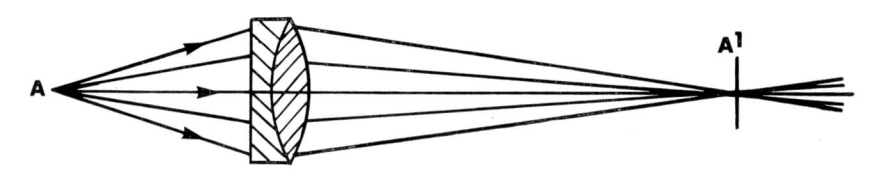

Fig. 29. Plano–convex doublet–achromatic lens diverging rays from A are brought together at A^1.

Lister improved the image quality by adding a further lens to his objective (Fig. 30). This high power objective included two achromatic lenses in line. Thus B gave aplanatic refraction, producing a virtual image at A^1 of the figure. A^1 served as one of the aplanatic points in the other lens combination, at the same time the conjugate aplanatic point of this second lens system B^1, lies at A^2, the position of the real image. The lenses represented in Fig. 29 and Fig. 30 are said to be free from spherical aberration. The improved spherical correction enables increased resolving power and image sharpness but no improvement in curvature of field.

Selligue-Chevalier (1824) discovered that image quality was greatly improved by incorporating four pairs of lens combinations; but unlike the lens by Van Deijl and Lister, the convex lens of each pair was facing the specimen with an air space between the convex and concave lens.

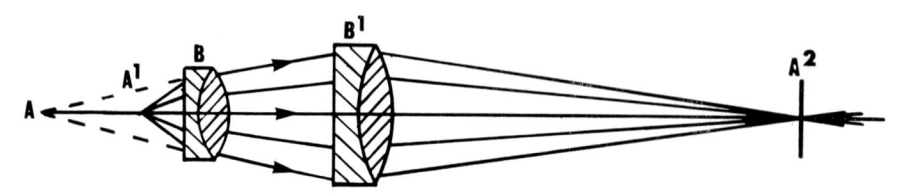

Fig. 30. Lister type of objective, two achromatics in train virtual image A, rays from A^1 diverge and are brought to a point A^2.

Chromatic aberration. Unlike spherical aberration, this is caused when various light wavelengths (colours) are used in the formation of the projected image (Fig. 31). This is particularly evident if unfiltered light is used to photograph a colourless specimen. This fault is also more pronounced when recording on colour film. In such an instance unwanted colour zones occur around the images when in fact such colours do not exist in the specimen. This effect can stem from the refractive index of an optical media, when it is greater for short wavelengths than for long wavelengths. From Fig. 31 it will be seen that a simple lens has a shorter focal length for blue than for red. If focusing on the yellow-green band appears clearly

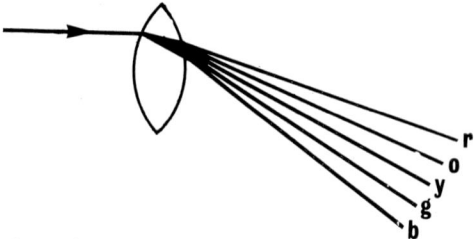

Fig. 31. Chromatic aberration.

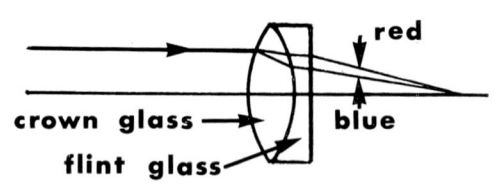

Fig. 32. Chromatic correction. A compound lens, using flint glass for one element and crown for the other. These lenses are usually cemented together and are said to be achromatic.

defined to the eye then out-of-focus colours will also be imaged. The shorter the wavelength of light, the stronger the refraction. Consequently, the blue rays are brought to a focus closer to the lens than the red rays. Also, when focusing on the blue image the fringe objects are surrounded by red zones. Likewise, when focusing on the red image one finds that blue is apparent round the edges, and so on. The higher the magnification the worse the fault becomes. Figure 32 B demonstrates an improvement brought about by a change in optical design.

In 1722 an Englishman, Charles M. Hall, discovered that flint glass differed very little in refractive power from crown glass but that the dispersive power of the former was double that of the latter. This remarkable discovery opened the way to the manufacture of lens units consisting of one or two strong crown lenses with a weaker negative lens of flint glass. The flint glass corrects the chromatic in the crown lenses but, unfortunately, reduces the magnification by approximately one half. In 1807 Deijl marketed his achromatic objective and some years later a better achromatic system comprised of various lenses was produced by the Optical Glass Works, Jena (now Zeiss, Jena).

By using an achromatic objective, with a good filter, yellow-green or green, placed in the light train either in front of the source or in the filter carrier, extremely good results can be obtained. The obvious improvement is an overall sharp picture, containing greater detail in the finer areas.

In addition to the achromatic objective there is the apochromatic objective. This is characterized by the inclusion of mineral fluorite lenses which disperse any residual colour that appears in the achromatic lens. Added components have achieved much, both in the production of the photographic image and in the cost of the production of the lens system. The correction in these lenses is so high that further filtering of the light is often unnecessary. However, filters are used when photographing multi-stained specimens which incorporate emulsions such as the Ilford chromatic type. The semi-apochromatic objective is less expensive, but does not produce quite the same high quality image as does the apochromatic.

Coma. This is the inability of a lens, or lens system, to bring the marginal and

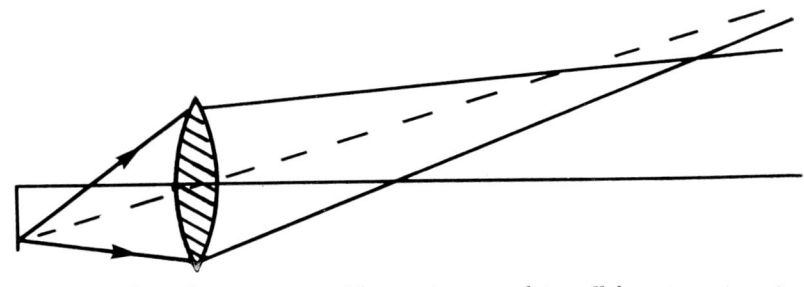

Fig. 33. A condition known as coma. The specimen area lying off the axis causes an image which will be a radial line.

axial rays together in such a way that the ratio of the subject–lens distance to lens–image distance is constant for all rays (Fig. 33). This is yet another problem which has been satisfactorily dealt with by the designers of high quality objectives. Sometimes coma will produce a one-sided hazy halo which borders the image proper. An added image, known as the optical membrane, may also appear off the optical axis. Coma can be brought into a corrected lens system merely by the misplacement of an optical element lying at an angle. A tilting specimen can also produce the effects described above.

Astigmatism. This is perhaps the most common defect in human vision and the commonest fault in optical systems. On the one hand the eye may need some assistance to see distant objects, whilst on the other it may need assistance in order to see near objects. The solution is for a converging lens to be placed in the near proximity of the eye, to change the image distance, and at the same time bringing rays to a common point of focus.

A narrow pencil beam of light striking a lens at an oblique angle fails to meet at a given point but, instead, at two focal lines. Any ray which passes through a lens strikes at least one surface at an oblique angle and so spherical aberration is also associated with astigmatism. By rotating the lens and watching the ground glass image, it will be seen that if this fault is present it will image two points, B and B^1 of Fig. 34. One focal line will be vertical and one horizontal and so,

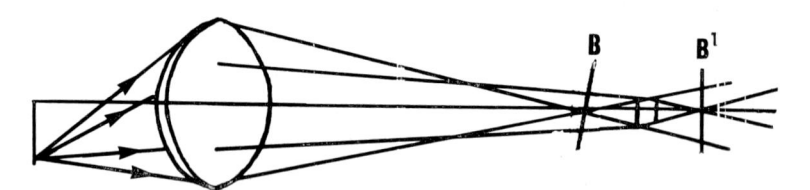

Fig. 34. A fault known as astigmatism. Instead of a single focal point, image point, the oblique pencil rays produce two focal "lines," image B and B^1 in different planes and directions.

when photographing an object, vertical lines appear sharp at B and the horizontal lines sharp at B^1. Somewhere between the two images, ignoring B and B^1, will be the accepted sharp image. These two lines in place of one are astigmatic. Sometimes spurious effects in the specimen image, such as small globular particles positioned near the edge of the field, may also appear to be rods. Images less sharp among sharp are the odd effects caused by this aberration.

Curvature of the image. Astigmatism is corrected by a stop when the two astigmatic images are seen as a combined image, but then the image will be curved, producing curvature of field. An object lying in a flat plane produces an image lying in a curved plane (Fig. 35 and Fig. 36). If a lens exhibits this particular fault it will be impossible to image a wide flat field. However, it is common knowledge that it is now possible to purchase corrected objectives for this aberration. If the aperture stop is placed on the specimen side of an uncorrected lens it will direct axial rays through the centre of the lens, and those on its edge will skirt the edge of the lens. The projected image will then be barrel-shaped. If the aperture

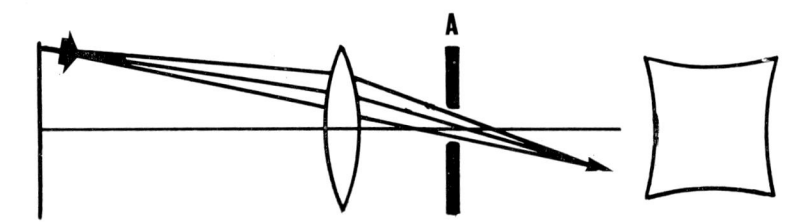

Fig. 35. Pin-cushion distortion.

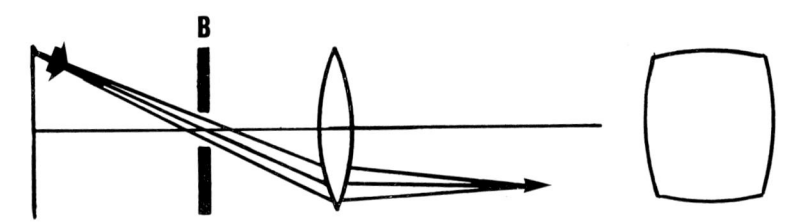

Fig. 36. Barrel distortion.

stop is moved to the image side of the lens, the rays from the edge of the specimen will be less than from central rays, and will image an increase in magnification on the edges, giving cushion-shaped distortion. By using two lenses, one an aspheric, and placing a stop between them, the distortions are cancelled out.

Brightness

The illumination, or brightness, of the image, recorded on the exposure meter, will determine the apparent brightness of a given object with a given meter. This brightness cannot exceed that of the object unless, of course, that object is very small, and then, through other factors too, it takes upon it an added brightness. If the sensitivity of the emulsion is constant, and the actual brightness of the transmitted or reflected light is constant, any change of brightness will depend upon the object and its character.

Let us pause for a moment and look at Plate 6. Each of the five steps have been cut from half plate prints. The subject is, from left to right, the highly polished surface of a cigarette lighter. Two light sources, objective and eyepiece, and distance from eyepiece to film, were constant throughout the series, as was the emulsion. Exposures varied considerably: (a) of the figure was given an exposure of 6 sec, (c) for 60 sec, while (d) had 20 min and (e) 30 min. The image brightness of (e) was such that it was impossible to focus on the projected image on the ground glass screen of the camera. Focusing was carried out by placing white powder on the surface of the metal, this being brushed off before exposure. As we have seen in this series of industrial application to photomicrography, the image brightness can change. Excess brightness can be controlled through the use of N.D. filters, mentioned later on. When an aperture is used to intercept some of the light it makes no difference to the brightness of the image but it does affect the illumination qualities of an image projected on a screen or emulsion. Reducing the aperture reduces the cone of light, not its strength. We will return to this when considering the condenser.

Refraction – air glass air

Before passing on to the microscope slide, cover-glass and mounting medium, let us look at a ray of light passing obliquely from one transparent substance to another. The first thing to be noticed is that the ray is refracted at the surface, the amount of refraction being proportional to the difference in optical density between the two substances. If a ray of light is perpendicular to the surface of the medium, its course would be unchanged. An oblique ray passing from a less dense to the more dense substance is refracted or bent towards the normal. The relationship between the ray, in glass and in air, is described by Snell's Law, and set out as an equation

$$\frac{\sin i}{\sin r} = n$$

where i and r are the angles made by the incident and refracted rays with the perpendicular at the point of entry, and n is the refractive index of the glass, Fig. 37.

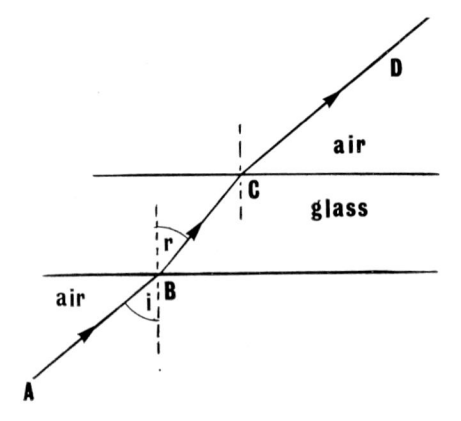

Fig. 37.

Now what happens when a ray meets the surface from the more dense side, taking the case of glass and air as illustrated in the figure? A ray such as A–B moves at an oblique angle and passes from one substance to the other, but is refracted away from the normal—the angle of incidence is less than the angle of refraction. The angle of incidence for which the angle of refraction is 90° is of particular importance, and is known as the critical angle. When the angle of incidence is smaller, light passes to the rarer medium; when it is greater, no light emerges but is reflected from the surface back into the more dense medium. If two rays impinge on such a surface, with angles of incidence less or greater than the critical angle, one will pass to the rarer medium, the other will be totally reflected at the surface. The one succeeds in passing the surface, the other fails.

Figure 38 demonstrates the case of a double refracting substance when the entering ray E–F divides, one ray emerges because d is less than the critical angle; another ray, K, is totally rejected at J, because B is greater than the critical angle.

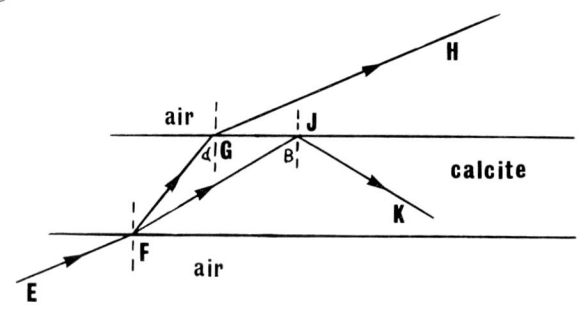

Fig. 38.

Numerical aperture and resolving power

The prime function of an objective is to separate fine detail at a determined magnification, without side effects, with the aim of seeing more and more structural detail. The higher the magnification the greater is the struggle to separate one line, or one dot, from another (Plate 7). In other words, the resolving power of a lens may be defined as the smallest distance between two points which can be rendered as separate images.

The resolving power of an objective is also conditioned by the magnification and the aperture, known as the numerical aperture or N.A. We have seen that the resolving power is the least distance between two lines an objective of a given N.A. and given wavelength will record as two separate lines, and can be expressed by the relation

$$\frac{0 \cdot 5 \lambda}{\text{N.A.}}$$

where λ is the wavelength of the light used, and N.A. the numerical aperture of the objective. The wavelength of light in use also incorporates the use of filters, whether the lens is used dry or immersed in oil or water. The refractive index (the medium occupying the space between specimen, cover-glass and the front lens of the objective) is the semi-vertical angle of the maximum cone of projected light accepted by the objective from a point in the specimen. The N.A. is an expression of the fraction of a wave-front admitted by an objective. The term was originally introduced by Abbe to express the sine of one half the angular aperture, multiplied by the refractive index of the specimen medium, with which the objective is designed to operate. Thus

N.A. = n sine U

where N.A. is of course the numerical aperture, and n is the refractive index of the medium. U represents one half the angular aperture, the corresponding angles that the marginal rays make with the principal axis, Fig. 39.

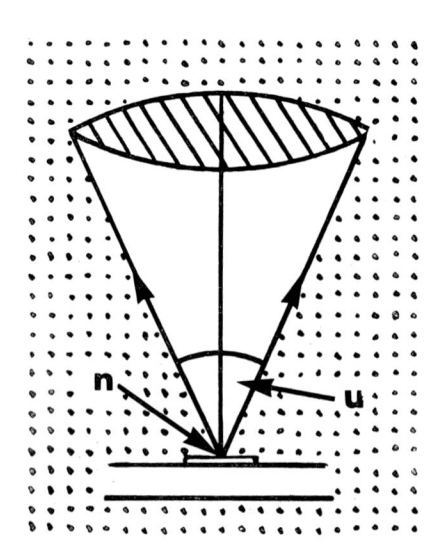

Fig. 39.

Let us consider the numerical aperture of an objective. Figure 40 (after Leitz) demonstrates the maximum aperture values which may be achieved with media air, water and oil between the front lens and the object. It is assumed that transmitted light rays emanate in various directions from a point below the cover-glass. Figure 40(a) illustrates the cover-glass and front lens of a dry objective, separated by air, the refractive index of which is known to be $n = 1 \cdot 000$. The rays leaving the top surface of the cover-glass enter the air within an angle of total reflection, this being $41 \cdot 5°$, any ray beyond this figure is reflected back from the top surface into the glass. With an angle of 90° a theoretical maximum of 1·0 would be obtained for the N.A. of a dry objective which, however, cannot be reached in practice since it would necessitate the use of the front lens of the objective in contact with the top of the cover-glass. Due to the necessary separation of the front lens and cover-glass, and the usual working distance, plus the small nature of the lens, an angle of 72° is the highest that can be used. This results in a maximum N.A. of 0·95 for a dry objective.

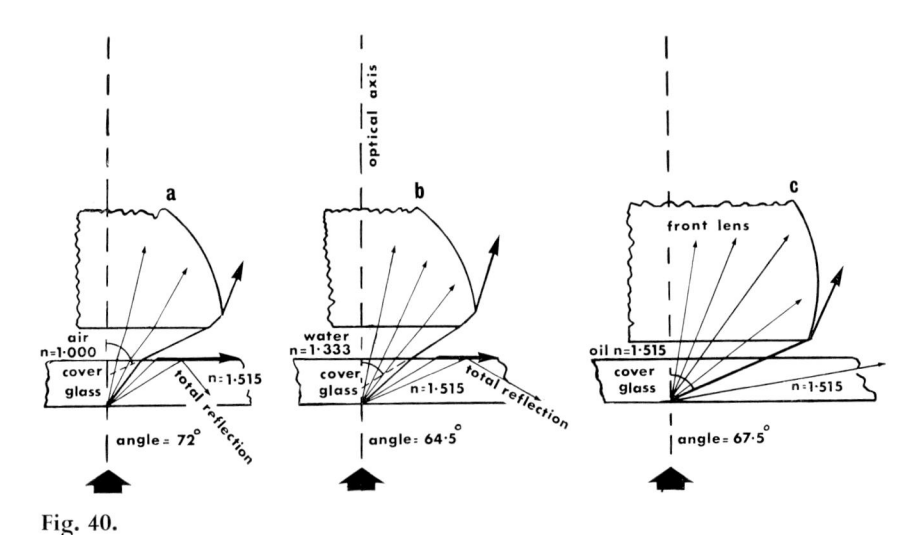

Fig. 40.

What has been said of the dry objective does not apply to the water immersion objective. From Fig. 40(b) it will be seen that the top of the cover-glass and the front lens of the objective are separated by water, $n = 1·333$ (see page 53). Total reflection occurs only for rays at a larger angle than 61·5° so that a theoretical maximum of 1·33 is obtained for the N.A. For similar technical reasons as with the dry lens, an angle of only 64·5° can be used bringing the N.A. value down to N.A. 1·25. From (c) of the figure it can be seen that a change has taken place in the rays leaving the top surface of the cover-glass, for here no total reflection occurs, due to the fact that the cover-glass and immersion oil have approximately the same refractive indices, therefore $n = 1·515$ is reached. All the light within an angle of 90° would reach the front lens if it could be made to accept such rays. As the lens is so very small this is impossible. The short focal length of the oil immersion objective however, results in a minute front lens of less than 1 mm in diameter: no mean achievement. Such a small lens and short working distance of the lens covers only an angle of only 67·5° at its maximum. The result is that the upper limit of N.A. is 1·40.

These three cases demonstrate the maximum apertures obtainable, and the rays of importance are outlined in heavy ink. Between the top surface of the cover-glass and objective these rays have almost the same inclination, but within the cover-glass the same rays form a considerably larger angle, that of 67·5° in connection with oil immersion, less than with dry objectives 39°, and water immersion system which is 52·5°. These angles also allow for determining the N.A. by multiplying the sine ratio of the angle by the refractive index of the cover-glass. These considerations show quite clearly how the N.A., and consequently the resolving power, increases as the beam of rays coming from the specimen and reaching the objective increase in angle.

It also follows that the higher the N.A., the less working distance there is, Hence, the more than hemispherical front lenses of the immersion objectives are held in position by a very narrow metal rim. These optical systems demand careful treatment for undue pressure on the front lens may dislodge it and even damage the second lens (see Figs. 57–59).

It is true to say that the main properties of the microscope depend upon the quality of the objective. The first requirement is a sufficiently large aperture of an angle 2μ of the bundle of rays reflected from the various points of the specimen. The size of the aperture is fixed and is the first requirement on account of the better analysing powers of the objective. When photographing "large" specimens maximum *angular* aperture (high N.A.) has to be sacrificed for area covered when a narrow *angular* aperture objective takes over. It is not possible to record a wide field *and* at the same time achieve high resolution, but it is possible to get maximum definition. It is therefore necessary to have a range of objectives, ranging from low-power, with narrow *angular* aperture (low N.A.) to the wide *angular* aperture (high N.A.), such as dry and immersion systems with an aperture of more than unity. The following will illustrate a range of objectives involving these points.

Table 1.

Type of objective	Designation power/N.A.	Focal length	Working distance	Initial magnification
Achromatic	0·07 narrow *angular* aperture	32 mm	27·0 mm	×3·5
	0·12 and wide *actual* aperture	32 mm	30·0 mm	×5
	0·25	25 mm	15·0 mm	×6
	0·25	16 mm	5·0 mm	×10
	0·17	8 mm	13·5 mm	×10
	0·54	8 mm	1·4 mm	×20
	0·54	4 mm	1·6 mm	×20
	0·65	4 mm	0·50 mm	×45
	0·65 wide *angular* aperture	2 mm	0·30 mm	×45
	1·00 and narrow *actual* aperture		0·40 mm	×100

NUMERICAL APERTURE AND RESOLVING POWER

To some it may seem odd that an objective having a wide *actual* aperture may have but a narrow *angular* aperture and that the lenses of widest *angular* aperture may be those of the narrow *actual* aperture. In Fig. 41 it will be seen that the angle of aperture of *a*, *b* and *c*, depends on the *actual* aperture of the objective lens, and also the distance of the objective when in focus from its front lens. Two objectives may have the same *actual* aperture (Fig. 42), and yet one may have a much wider *angular* aperture than the other because the focal distance of the specimen is less. On the other hand two objectives may have the same *angular* aperture and yet the *actual* aperture of one is much greater than that of the other, the focal distance of the specimen being greater. On pages 34 and 38 we will see how the *angular* aperture affects the resolving power of the two objectives with the same *actual* aperture. Further, compare notes from the above table. Here it will be seen that objectives of low-power or long-focal distance have the widest *actual* apertures and those of high-power or short focus group have the widest

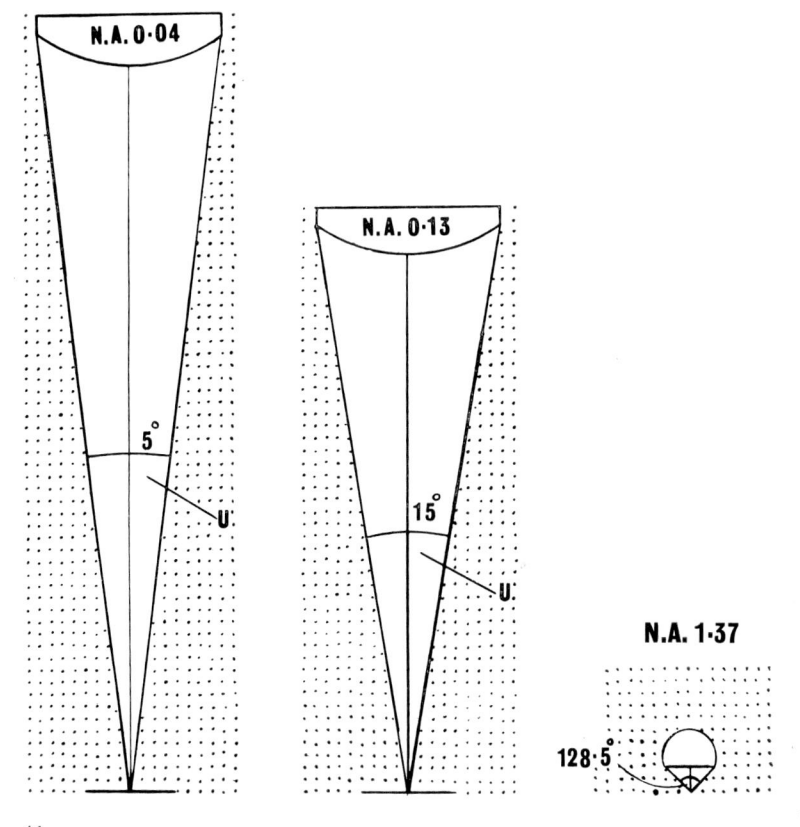

Fig. 41.

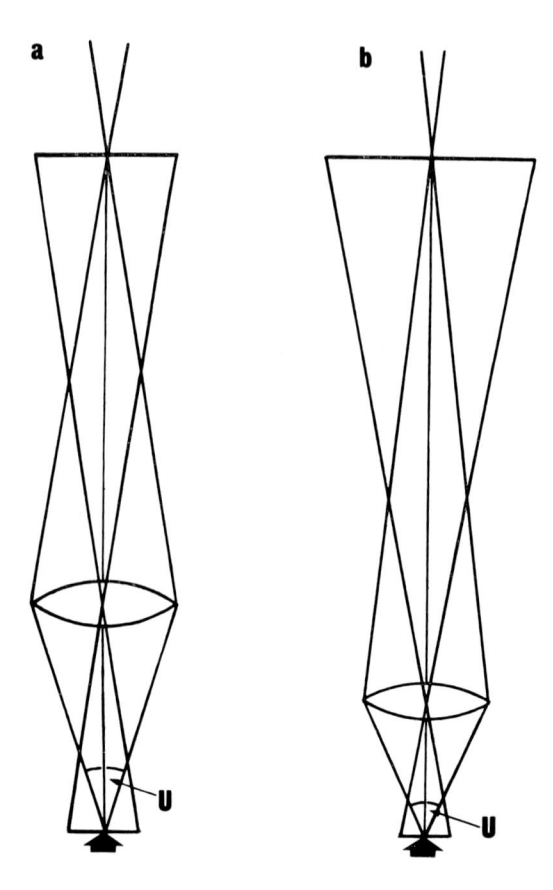

Fig. 42. Objectives of different focal length but equal N.A. The angle U is made by the marginal rays with the axis. (a) 25 mm N.A. 0·30. (b) 16 mm N.A. 0·30.

angular apertures. If the focal distance is constant, the *angular* aperture will increase or decrease with the *actual* aperture. If the *actual* aperture is constant, the *angular* aperture will increase with the shortening of the focal distance, and will decrease with its added distance.

The image definition produced by an objective mainly depends upon the degree of correction both in spherical and chromatic aberration, a condition essential to the performance of all objectives. Good definition may be more easily recognised in objectives of narrow or medium *angular* aperture. The defining properties of an objective will be evident by the quality of the image which is recorded of almost any specimen. Any imperfection in definition is amplified by the use of a high × number eyepiece. Sometimes a change in tube-length will improve definition. It is important to acquaint oneself in eyepiece-

NUMERICAL APERTURE AND RESOLVING POWER

to-objective distance, as clear definition throughout the depth of the specimen or, at least, of the principal features in essential.

Thus the *angular* aperture is an all important feature in the production of a good photomicrograph. By introducing a small substage aperture the angle of the cone of rays emerging from the substage lens is reduced, resulting in an inferior image, and the diameter of the slightly out-of-focus image becomes more easily visible. If such images appear on a photographic emulsion they are detrimental to the image proper, as well as being a nuisance (Fig. 48).

Depth of field is inversely proportional to the N.A. A point source of light may be considered to emit a continuous series of spherical wave-fronts which expand away from it into the surrounding space. Emerging rays which converge to form a point image of a point object, without meeting, produce a bright spot of light surrounded by diffraction rings or small disc, known as the Airy (1834) disc (Plate 8). The radius of the Airy disc is

$$r = \frac{0.61\lambda}{\sin U}$$

where λ is the wavelength of the light concerned, and U is the angle marginal rays make with the axis. Figure 43 demonstrates the light distribution curve

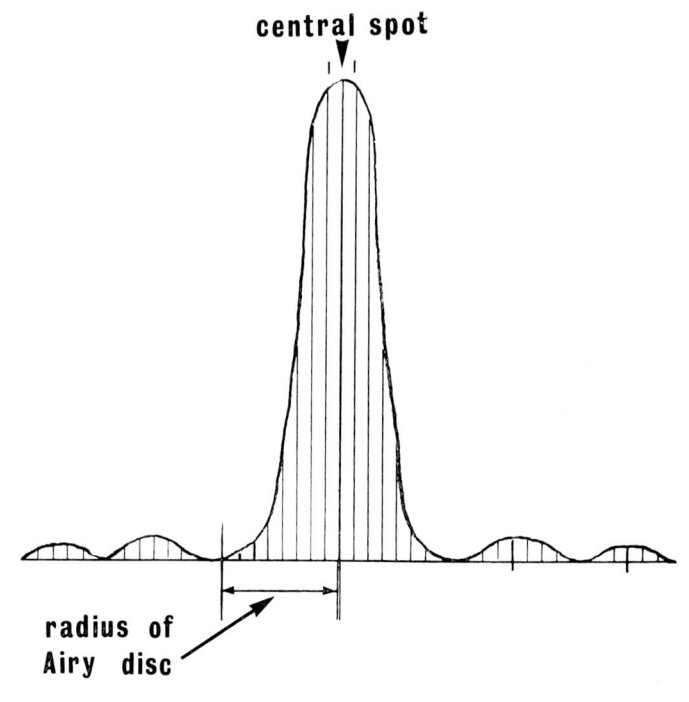

Fig. 43.

in an Airy disc, likely to be encountered when operating with an aplanatically corrected objective. Allen's sine condition for the Airy disc is as follows

$$\frac{\sin \sigma}{\sin \sigma} = k$$

Two rays will be separated by a darker region only when their centres are at a distance at least approximately equal to the radius of the Airy (1834) disc. The sine condition states that the sine of different aperture angles, pair in the object space, and must be constant. k corresponds to the image scale reproduced from the objective. Wavelength should be equal from image point 1 to 2. Figure 44 illustrates that the aperture can be reduced until the diameter of the Airy disc approaches the agreed value of the circle of confusion. Figure 45 demonstrates

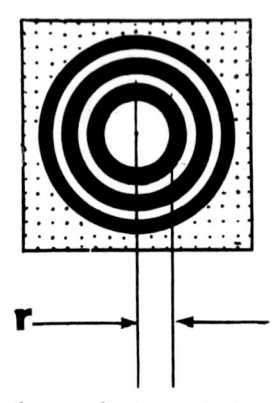

Fig. 44. Airy Disc. Instead of a lens producing a point image clearly defined, it produces a bright spot of light surrounded by diffraction rings. See also Fig. 48.

the circle of confusion in the out-of-focus images: A is a point image focused short of the screen; B an image of a point object focused in the plane of the ground glass screen, FP; C a point image formed above the ground glass screen. $CC =$ diameter of circle of confusion. It can be seen that a point object S in front of a convex lens produces a point image on a screen placed at the point of a focus, but any point in front or behind this point gives a "fuzzy" circular image and is known as the circle of confusion.

The acceptable diameter of the "circle of confusion" for normal photographic work is 0·254 mm (0·01 in.), but this is inadequate when viewing a photomicrograph, where a considerably higher standard is called for. These factors must be borne in mind when enlargements are made from a negative.

The diameter of the circle of confusion is increased directly with the degree of enlargement. If a circle of confusion of 0·0130 cm (0·005 in.) is to be tolerated in the final print at $\times 10$ it must be 0·00130 cm (0·0005 in.) on the negative.

This can impose quite severe restrictions on enlargement and can cause disappointment when a satisfactory result at, say, quarter plate size cannot be adequately enlarged to 10 in. by 8 in., or even higher, or a 25·4 cm by 20·4 cm (10 in. by 8 in.) to 76·2 cm by 101·6 cm (30 in. by 40 in.). This is the reverse

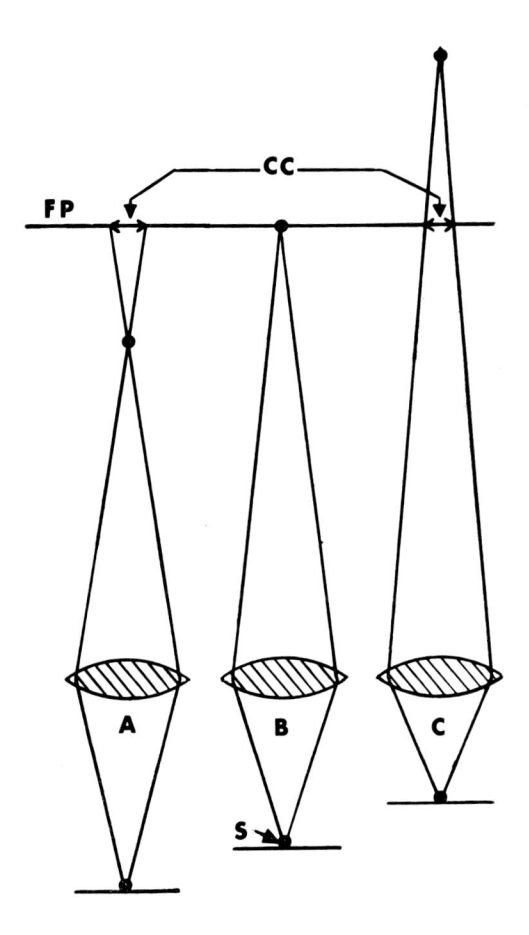

Fig. 45. Demonstrating the circle of confusion in out-of-focus images. A point image formed beneath the focal plane and C a point image formed above the focused point.

of photography, where an enlargement usually enhances the subject originally imaged on a small negative. However, the degree of penetration (depth of field), flatness of field, and working distance (lens to specimen) decreases as the N.A. increases (Fig. 42). In addition, the field covered decreases as the N.A. increases: as set out in the following table.

Table 2. *Fields of view*

Apochromatic objectives			Field of view (in.)		
Focal length (mm)	Numerical aperture	×6	Compensating eyepieces		
			×8	×17	×25
40	0·11	0·25	0·18	0·11	0·09
16	0·35	0·075	0·06	0·035	0·03
8	0·65	0·04	0·03	0·017	0·014
4	0·95	0·02	0·015	0·010	0·007
3[1]	1·2	0·015	0·010	0·009	0·005
2[1]	1·3	0·009	0·007	0·004	0·003
Fluorite	Objectives				
3·75[1]	0·95	0·02	0·014	0·009	0·007

R. and J. Beck Ltd.   [1] Oil immersion

Two objectives of different focal lengths can have the same N.A., for instance an objective of 25 mm focal length may have as large N.A. as a narrow focal length, a 16 mm, which will project a larger image of a given point, covering less field of view (Fig. 42). As the N.A. controls the resolving power it would be, in this case, an advantage to use the longer focal length objective and take advantage of the inclusion of a greater coverage of field.

Some controversy surrounds the precise manner in which the resolving power of a lens should be calculated when it is used to examine a non-luminous specimen by added light. For practical purposes, however, it is usually sufficient to assume that the resolving power of an objective under these conditions is given by

$$r = \frac{\lambda}{\text{N.A.}}$$

The minimum resolvable distance between two self-luminous points may be shown to be

$$\frac{0 \cdot 61 \lambda}{\text{N.A.}}$$

where λ is the wavelength, and N.A. is the numerical aperture of the microscope objective in use. 0·61 is the Airy constant. The resolving power also increases with a decrease in the wavelength of light, for example, by the use of blue light in place of red: and falls with a decrease in the numerical aperture of the objective. The following table sets out some comparisons in resolving power in various objectives.

When the instrument is adjusted to maximum resolving power, the maximum useful magnification which may be attained with an objective used in the visible

Table 3.

Focal length	Primary mag. at 160 mm tube length	N.A.	Lines/mm resolved	Useful magnification
50 mm	×2·75	0·15	435	50
25 mm	×6	0·21	588	150
16 mm	×10	0·30	820	200
8 mm	×20	0·65	1333	350
4 mm	×50	0·75	2222	550
2 mm (water)	×100	1·2–35	3570	1000
2 mm (oil)	×100	1·37	4166	1900

region of the spectrum is of the order of 1,000 × its N.A. (Fig. 87b). It cannot be emphasised too strongly that the whole question of resolving power is controversial and the operator must, when producing an image, take care that diffraction phenomena do not invalidate the true interpretation of the shape and texture of the subject (Plate 9). Similarly, it is most important that the N.A. of the condenser is properly matched with that of the objective. The upper limit of useful magnification which may be attained in a compound microscope employing visible light is, therefore, in the region of 1,500 to 2,000 diameters, but even this can be greatly increased by using ultra-violet (high resolving power) microscopy, and of course the electron microscope, as seen in Plate 10(b, d, f).

Reference

Airy, G. B. (1834) *Camb. Phil. Trans.* **5**, 238.

Diffraction

One difficulty in producing a point image of an object is that a bright spot of light surrounded by diffraction rings is produced instead. This also happens when recording the edge of an object (Plate 11(b), 1–6). By way of contrast Plate 10 (a, c, e) has been included.

The laws of geometrical optics are not absolutely true. Even if the lens were perfect, a point image of a point object would never be formed but the image would be of a finite size. Diffraction can be defined as a scattering of light rays (Fig. 46), occurring at the side of a light beam, caused by the interaction between minute structural detail in the specimen and the wave-front of the incident light. When the light passes across an edge of a specimen the light is deflected into a shadow which has a sharp image, bordered by further rings of lesser density as indicated in the figure. This defect is most apparent when the pencil beam of light traverses a narrow aperture. Excessive reduction of the substage iris diaphragm will emphasise this phenomenon. The light rays will fan out when emerging from the narrow aperture and produce a rotation-symmetrical diffraction

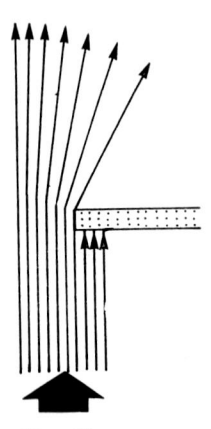

Fig. 46.

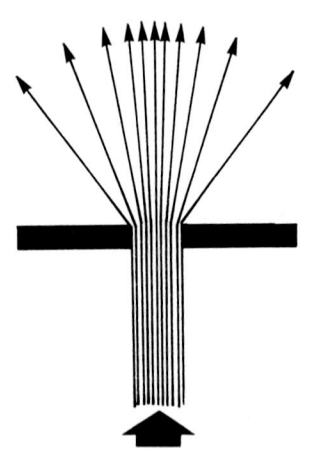

Fig. 47.

as in Figs 47 and 48. It will be seen from the figures that the diffraction rings vary in density away from the first ring, the beam of light has a rhythmic variation of brightness and long waves fanning out farther than short. 1–6 of the Plate indicates that the light and dark intervals are always of variable breadth; the light and dark striae show a gradually increasing and decreasing brightness. Without great care diffraction phenomenon will appear when photographing most structures, whatever the composition. The image of the specimen can become that of diffraction structure. However, it will be discovered by using a short wavelength such as a blue filter, that diffraction rings around a margin become less distinguishable, whereas with the use of a red filter the diffraction rings will mask the subject. From this it will be noticed that the bending can worsen by longer wavelength and be deflected by shorter ones.

DEPTH OF FOCUS 41

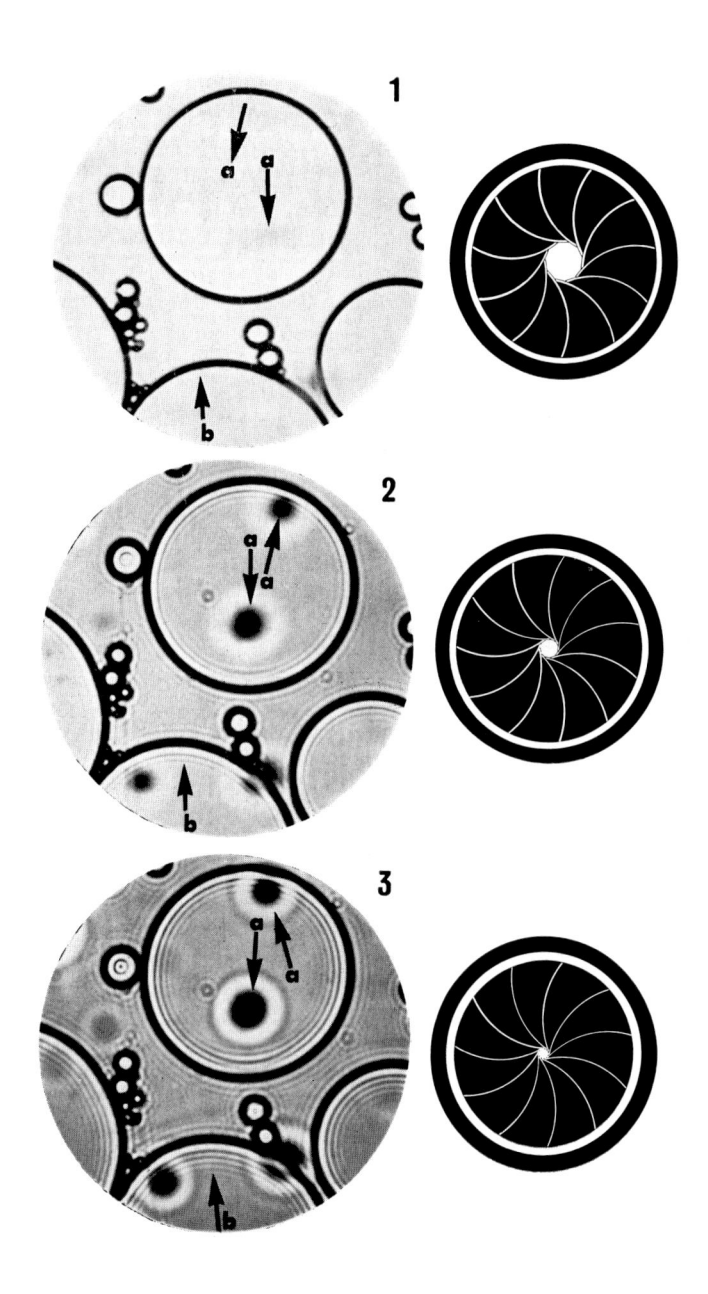

Fig. 48. Compare 1, 2 and 3.

Depth of focus

Depth of focus is the focal range over which an image is visible as a line or point and not a disc. It is not generally realised that the location of the focal plane along the optical axis is quite considerable. After carefully focusing on a fine specimen, it will be found that the focal plane (area of the ground glass screen) can be moved in two directions away from the focused point without causing any noticeable change in the focus of the image. Having moved the focused plane (ground glass screen), it is impossible to return to the position held when originally focusing the specimen. The reason for this is that the "depth of focus" in the image space is considerable. The depth of focus in the space of a lens system is given by A. E. Conrady, as

$$\frac{\lambda}{n' \sin {}^2 U'_m}$$

where λ = wavelength of the light, and n' = the refractive index of the medium in the image space, and U'_m = the angle that the marginal rays make with the axis as set out in Fig. 49. If this is applied to the image in the area of the focal plane of the camera it will be seen that the angle U'_m is a very small angle (Fig. 49). This in turn means that the angle of rays leaving the eye lens of the eyepiece

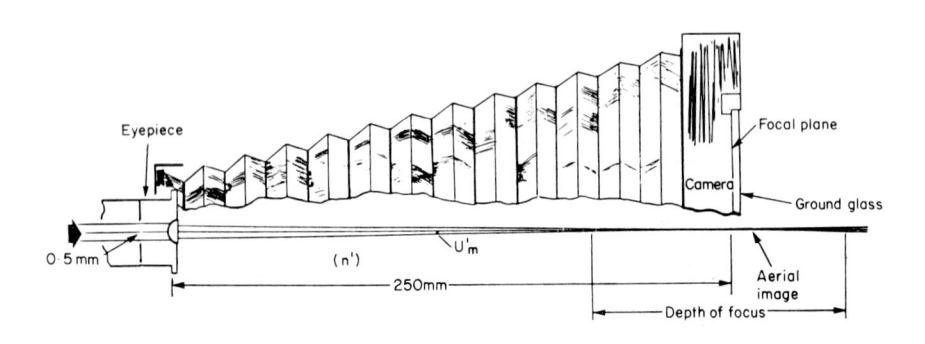

Fig. 49.

DEPTH OF FIELD

is also very narrow. Assuming (from Martin and Johnson) that the focal plane is 250 mm from the eye lens, then

$$\tan U'_m = \frac{0.5}{250} = 0.002, \text{ and } U'_m = 45 \text{ sec of arc.}$$

So that, the focal range for the plate

$$= \frac{\lambda}{n' \sin^2 U'_m} = \frac{0.005}{1 \times (0.002)^2} = 125 \text{ mm} = 12.5 \text{ cm.}$$

From the foregoing it can be seen that movement either side of the focused point in the focal plane is given by just over 6 cm, demonstrating considerable latitude without serious loss of sharpness. Depth of focus also changes with a change in N.A., the greater the angle of the rays of light transmitted from the objective the more critical is the depth of focus (see Fig. 50). The introduction of a small aperture will of course extend the depth of focus.

Depth of field

Depth of field is the range of distances in front of the objective lens in which all parts of the object must be sharp (Plate 12). It is related to "depth of focus".

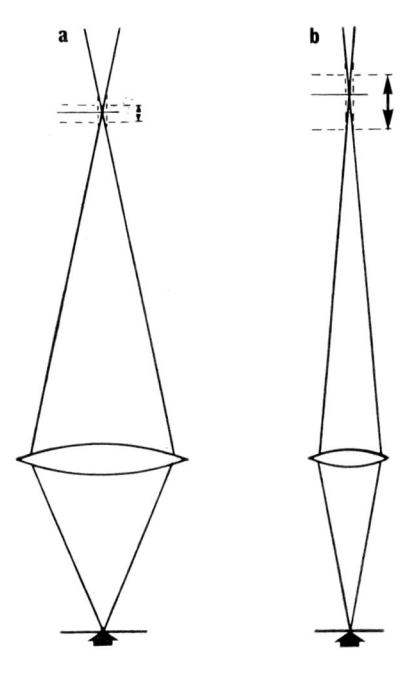

Fig. 50. Demonstrating depth of focus corresponding to N.A. (a) High N.A. (b) Low N.A.

It is a matter of lens geometry and extends over a distance and can be increased almost indefinitely by reducing the lens aperture. The depth of field expressed by an objective is very important in the practical use of the microscope and in particular in the production of a photomicrograph, revealing the depth of the specimen. By using the fine focusing adjustment, a true measurement of the thickness of the specimen is gained. To do this focusing is carried out at a point of the specimen, a of Fig. 51, and then, by shifting the point of focus, to point a^1. A change will have taken place in the lengths of the central and marginal optical paths

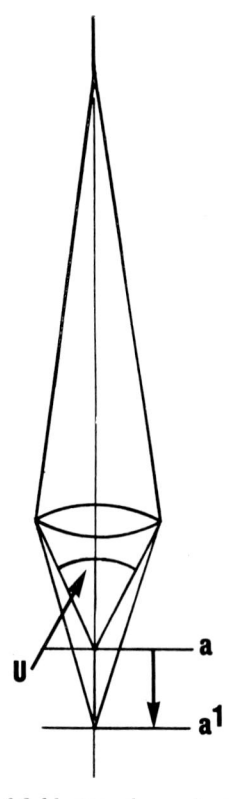

Fig. 51. Demonstrating depth of field. U is the angle that the marginal rays make with the axis.

registered in the focal plane. According to Rayleigh, if these lengths exceeded a quarter of a wavelength ($\lambda/4$), a breakdown of image quality is evident.

Calculation by Johnson on this basis shows that the total depth df is given by

$$df = \frac{\text{Allowed difference of path,}}{n \sin {}^2U/2}$$

DEPTH OF FIELD

when U is the angle made by the marginal rays with the axis in the object space, and n the refractive index of the medium on the object side of the objective. Taking the allowed difference of path equal to $\lambda/4$ and the wavelength of the light, 0·00050 mm, the following table will provide a guide to a depth of fairly sharp focus for a number of objectives.

Table 4.

N.A.	Depth of field in air (millimetres)	Depth of field in medium $n = 1\cdot5$ (millimetres)
0·25	0·0079	0·0122
0·50	0·0019	0·0030
0·75	0·0008	0·0013
1·00	—	0·0007
1·25	—	0.0004

In a camera lens there is a certain distance between a near object and infinity, where all objects are in equally good focus: "focal depth". This corresponds to a given apertal ratio (that is, the ratio of the diameter of the stop to the focal length). The focal depth is from the nearest object which will be in focus when the image on the ground glass screen is adjusted for a distant object.

The photomicrographer is only too familiar with the limitations imposed by the objective, which restricts depth of field and produces a shallow range of sharpness. (Fig. 52 and Plate 13). In visual work a shallow focal depth is not a handicap. In fact the viewer looks through the depth of the specimen and by so doing builds up a mental picture of the structure before him. But drawings (micrographs) can be made to demonstrate the depth of the subject. This does not help the photomicrographer.

The difference between, say, commercial and industrial photography and low power photomicrography is that in the former the image is at a reduced scale and the film is at a shorter conjugate focus of the lens, whereas in the application of the latter the image is magnified and therefore the film is at the longer conjugate.

This expedient can be used almost without limit in ordinary photographic work, but where magnification is required from specimen to image, then some compromise is essential as continued stopping down will adversely affect the resolution and lack of sharpness will result.

A major difficulty arises from a failure to realise that the term "depth of field" in a print has a meaning also in terms of the resolving power of the human eye. As already mentioned, the normal eye positioned 250 mm from a photographic print can just separate two points or lines 0.1 mm apart. True sharpness is only

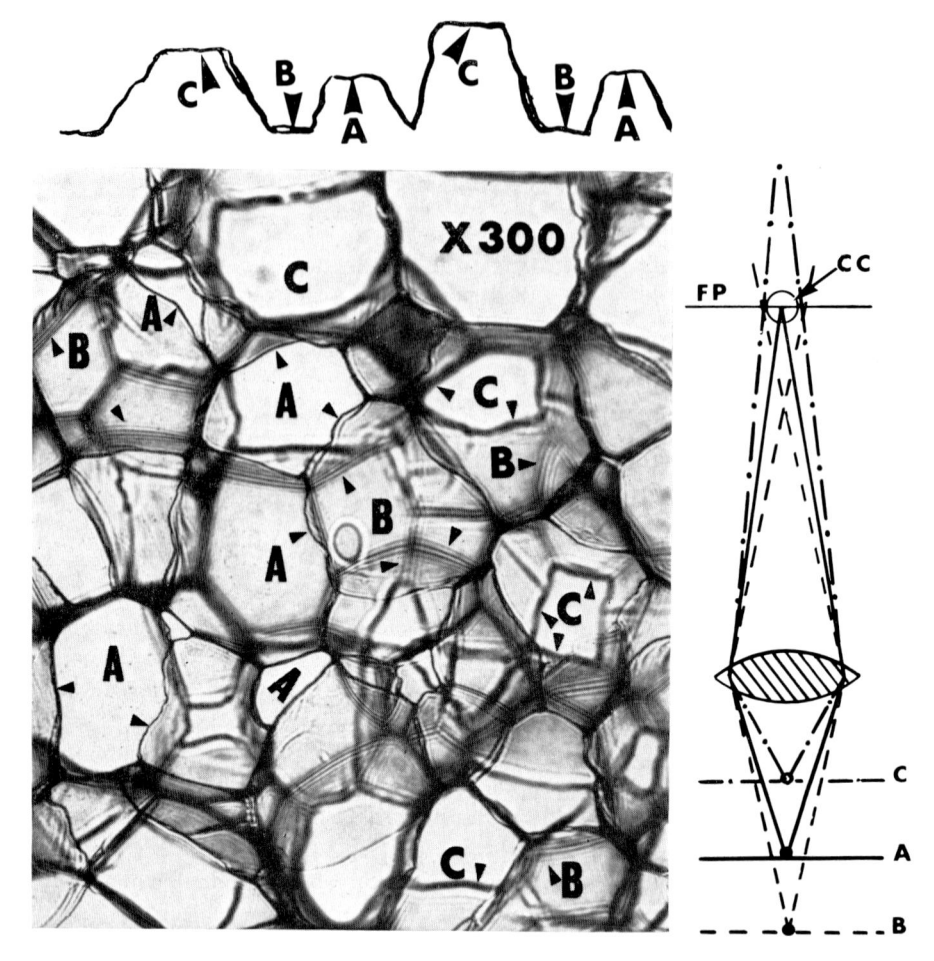

Fig. 52. Planes above and below the point of focus A. B is beneath the point of focus while C is above as indicated.

PHOTOMICROGRAPHIC DATA

Subject: polythene. Mounted: dry. Magnification: × 200 total × 300. Date: January 1967. Objective: 8 mm N.A. ·65 Apo. Eyepiece: ×6 45 mm. Tube length: 160 mm. Substage condenser: complete with top lens. Illumination: transmitted. Iris diaphragm aperture: adjusted to equal that of the objective. Camera bellows extension: 40 cm. Filter: blue. Stain: nil. Emulsion: N40. Remarks: focusing took place on the approximate middle plane.

obtained in the focal plane of the camera, FP of Fig. 52, but the eye accepts a quantity of "out-of-focus" area down to roughly the equivalent of eight lines per millimetre. The fuzz of anything less than this would be objectionable in a scientific print and would be rejected by the eye. This gives a "depth of field" both in front and behind the point of focus. This is shown in the figure where it can be seen that a point in front of a convex lens produces a point image on the screen *FP* placed at the point of focus *A*, but any point in front of *B* or behind this point *C*, gives a "fuzzy" image.

The image in Plate 9 (*C* and *D*), came from reducing the substage iris diaphragm beyond its practical limit; as a result of which all planes of focus are affected. Figure 52 was made with all optical components functioning at their maximum. Here we see images formed on three planes, *A*, *B* and *C*. *A* is clearly delineated, while *C* is fuzzy due to the point of focus being above the *FP*, and *B* is formed beneath the focused point of *A*, producing a multiplication of lines as mentioned earlier. If a print is made by enlargement the circle of confusion (Fig. 52, *CC*) is increased accordingly, and depth of field further reduced. This fact must always be kept in mind when making a negative which has to be enlarged.

In an opaque subject another problem arises associated with depth of field. The magnification varies with the distance the specimen is from the objective lens, so that the magnification at one extreme of the field can be substantially different from that at the other extreme, i.e. through the depth of the subject, so that considerable distortion can arise. This is commonly seen when photographing approximately hemispherical objects where the summit is more highly magnified than the base and the subject takes on a top heavy appearance (Plate 13).

We must accept shallow depth of field but, to get the best of this restriction, it is advisable to use the lowest magnification. Use the smallest substage aperture consistent with adequate resolution, a fine grain film exposed to light of a most suitable colour and, when possible, immerse the specimen in a medium or fluid having a high refractive index. Before making a photomicrograph it is suggested that the final magnification and size of print first be considered and then the sequence of adaptations to suit the points mentioned above can be planned. In this light the degree of limit of secondary magnification can be planned.

There are limits to the variety of problems which can be solved with the microscope and by the exceedingly shallow focus (depth of sharpness) produced by the lenses. These often cause severe restrictions in the production of the depth of sharpness in the photomicrograph, compared with the projected images seen by the eye.

A photomicrograph taken with a medium or high-power objective records an image of one shallow stratum and this is often quite inadequate, as in Plate 14. Here there is not one exposure which truly portrays the complete cell structure of these rabbit eggs. Such a series, nevertheless, is invaluable as individual exposures can be studied and comparisons made with previous batches and possible malformations detected. Any one of these photomicrographs can be enlarged and further examined.

The globe-shaped rabbit eggs were in an early stage of development and were mounted as in Fig. 53. They were photographed within a few moments of their being removed. Focusing was carried out at a point indicated by the line beneath the arrow and in each case the point of focus was lowered 0·02 mm, Fig. 54. This fine shift in focus was controlled by the focusing scale attached to the microscope so an accurate measurement through the axis of the specimen

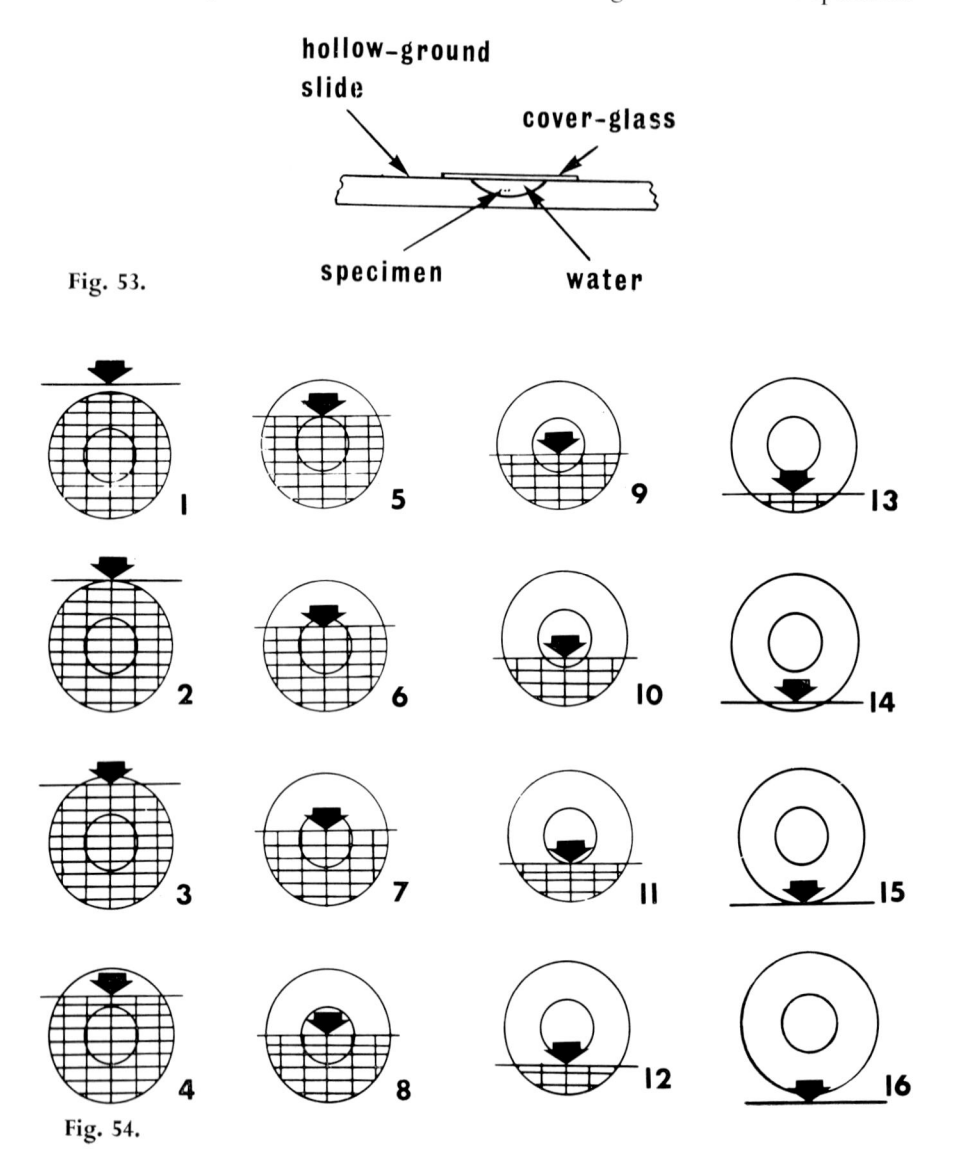

Fig. 53.

Fig. 54.

DEPTH OF FIELD

could be made. Exposures 5–8 are perhaps the most valuable in the series: exposures beyond 10 add no information but are included as an exercise. Examine recordings beyond 10 and compare with those early in the series. The toxicologist can then see at a glance with the help of a scale (Fig. 55) and drawing exactly where the point of focus lies.

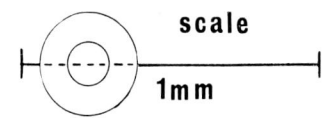

Fig. 55.

Some scientists, however, do not favour this method of individual recordings through the specimen but prefer "increased depth of field" by integrated focus, (Fig. 56), a multiplication of exposures on one negative. Such a method is not practical, since up to 20 exposures on one negative can cause confusion. Just imagine Plate 14 exposed on one negative! Where two or three exposures have been made on one negative information is just legible. When a negative has been subjected to 20 exposures it can only result in gross over-exposure, 19 times over-exposed. Further, it is impossible to see any depth, since a previously

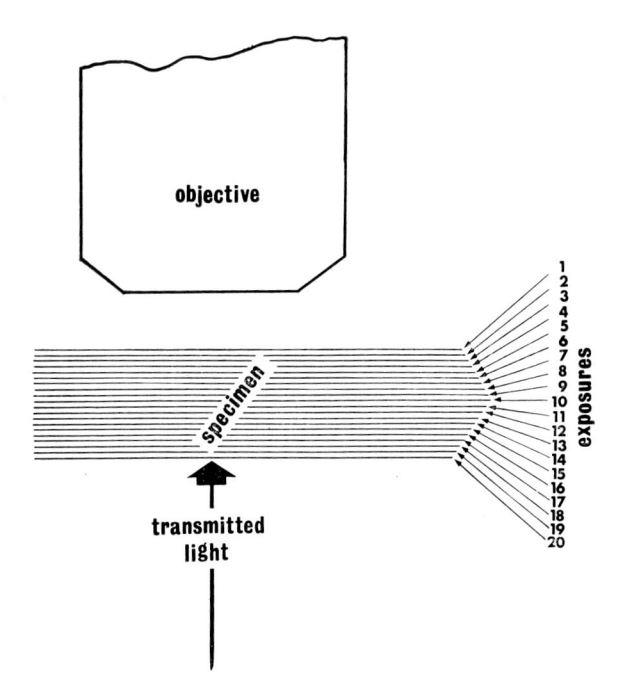

Fig. 56.

exposed sharp image is later overlaid by an out-of-focus image; i.e. an area nearer the lens (exposure 1 of the Fig.) will also be covered by an overall fuzz while recording the farthest (exposure 20 of the Fig.); exposing near and distant at the same time. The higher the magnification the more confused the images.

Refractive index

Refractive index involves the property of solids and liquids, transparent bodies and opaque, transparent and near transparent, and it is up to us to separate one body from another (when placed together). Colourless transparent substances are visible only when they differ in optical density (refractive index) from the fluid in which they are immersed. Now the greater that difference is, the more conspicuous will be the limiting surface.

A thin section of a piece of rock or a piece of bladder wort may be sandwiched between layers of mounting medium, known as Canada balsam; or in the case of the bladder wort, water.

Such subjects demand some form of illuminating medium to reveal structural detail against a dark background. The mounting medium in turn rests on a glass microscope slide and is covered by a thin cover glass. Our concern is the medium in contact with the subject matter. When it is transparent, as with small glass pellets immersed in water, it is difficult to see; if, too, a powder of mineral cryolite is placed in water it cannot be observed, appearing to have dissolved and become part of the immersion liquid. The powder becomes invisible because cryolite and water have almost the same refractive index. (We will come to a similar case later on when dealing with immunodiffusion plates). A transparent solid becomes more visible when immersed in a fluid with a different refractive index. If this is not possible the subject must be separated by means of special illuminating techniques: i.e. polarization, which will be discussed.

Minerals and the like are immersed or sandwiched in a liquid, generally Canada balsam. When both sides of the thin, prepared subject are in direct contact with the mounting medium, its texture is difficult to observe. It is often more clearly seen when observing an edge zone of the subject. The two fields differ in the information gained from the surface structure, the one in direct contact gives the indication of a smooth surface while the other indicates a rough. What has happened is that the so-called smooth area approximates to the refractive index of the mounting medium and the rough surface differs in refractive index from that of the mounting medium 1·54.

Often it is necessary to determine whether the refractive index of a mineral is higher or lower than that of the Canada balsam. This can be determined by making the Becke test. A high-power objective of the order of 8 mm or 4 mm is used and carefully focused on the edge of the subject matter, fully immersed in the mounting medium. The objective is raised and lowered on each side of the point of focus and, during the operation, it will be seen that when the subject is slightly out of focus it is surrounded by, or encloses, a ring of light. This ring expands and contracts as the objective is raised and lowered.

There are two cases to consider. The refractive index of the material is either higher or lower than the Canada balsam 1·54. (a) When the material is denser than the mounting medium, the ring of light will contract; passing from Canada balsam to material. (b) When the material is less dense than the mounting medium, the light passes from the material to the Canada balsam. In both cases, as the objective is raised, the light passes from the material with the lower refractive index, to that of the higher refractive index.

All oil immersion objectives are used with a thin layer of oil between the lower lens and the cover glass, and water is often used with a water immersion objective. Adequately refined cedarwood oil has been found suitable with the former lens, because its refractive index is approximately the same as that of the cover glass and it does not dry out quickly. A ray of white light passing through a substage condenser, oil, slide, specimen, cover glass and oil, to the objective, suffers little refraction at each glass–oil interface, as all these media have approximately the same refractive index.

Table 5. *Refractive indices of various substances*

Subject	Refractive index	Subject	Refractive index
Air	1·000	Non drying immersion oil ALP.1	1·524
Alcohol (methyl)	1·323		
Water	1·333	Oil of cloves	1·530
Distilled water	1·334	Canada balsam, "solid"	1·535
Sea water	1·343	Canada balsam	1·540
Egg-albumin solution	1·350	Clarite	1·540
Alcohol (absolute)	1·367	Permount	1·540
Chloroform	1·450	Monobromonaphthaline	1·549
Glycerine jelly	1·450	Styrax	1·582
Turpentine	1·470	Iceland spar	1·654
Olive oil	1·470	Hydrax	1·710
Caster oil	1·480	Naphrax	1·7+
Euparal	1·483	Methylene iodide	1·743
Diaphane	1·483	Pleurax	1·770
Xylene	1·497	Crown glass	1·51–1·53
Toluene	1·498	Plate glass	1·516
Benzene	1·504	Flint glass	1·53–1·65
Detel	1·510	Calcite	1·57
Cedarwood oil	1·515	Dense flint glass	1·65–1·92
Gum dammar	1·520	Fluorite, approx. average	1·434
Canada balsam in xylene	1·524	Quartz, approx.	1·544–1·553

C*

The refractive index is inversely proportional to the velocity of light within a crystal. Isotropic crystals have, therefore, only one refractive index for light of a given wavelength. Uniaxial crystals have two refractive indices and biaxial crystals have three. These indices may be determined in a manner similar to that described above. The technique, however, is very specialised and beyond this work's scope. Details of the technique, which requires polarizing and analysing equipment, are given by G. R. Rigby (1953).

It is interesting to note that when light vibrates parallel to the crystallographic axis of quartz, the refractive index is 1·553, and when the vibrations take place at right angles to this direction, the refractive index is 1·544. The difference, 0·099, is clearly a measure of the strength of the double refraction for this particular material.

Immersion Objectives. We have already seen that there are dry and immersion objectives (Fig. 57), the latter being used submerged in water, in cedarwood oil,

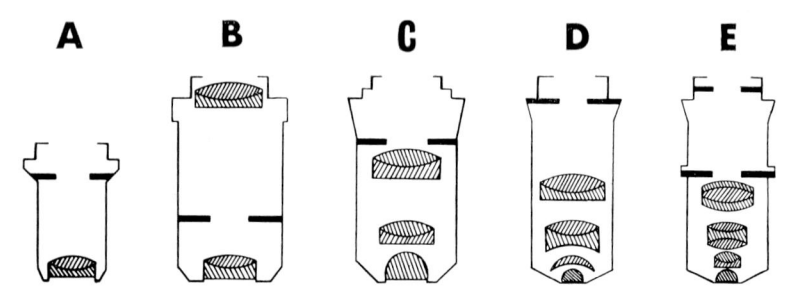

Fig. 57. Comparison in the optical construction in low, medium and high power objectives. A. Low power. B. 25 mm Holos N.A. 0·30. C. 8 mm Apochromatic N.A. 0·55. D. 2 mm oil immersion N.A. 1·30, 0·2 mm diameter of field. E. 2 mm oil immersion N.A. 1·37 with stops 1·0, 0·5 and 0·2 N.A. These stops are for use in darkground illumination. 0·1 mm diameter of field.

in suitable non-resining synthetic polymers, or in glycerine. It is claimed for some of these synthetic fluids that they do not harden like old fashioned cedarwood oil. This claim is sometimes false. Cedarwood oil makes an excellent mounting medium because of its hardening properties. It does not bleach stained specimens which can be kept for very long periods. Furthermore, cedarwood oil is an excellent property to use when examining and photographing bacteriological stained smears without a cover glass, incorporating oil immersion objectives. When immersion oil is used between the lower lens of the objective and cover glass or specimen, between underside of microscope slide and top lens of substage condenser, it is said to be a homogeneous–immersion system.

Water immersion objectives are extremely useful in research and in marine biology for photographing living and other material in water (Lawson, 1966). Water has a refractive index of 1·33 and the highest N.A. of a water immersion objective is 1·25 as against those used with oil which could have an N.A. of 1·40. Carl Zeiss, Jena, have produced a well balanced water immersion objective with an N.A. of 1·2 and long focal length. Carl Zeiss, Oberkochen, market an achromat water immersion objective, ×40, N.A. 0·75, focal length 4·6 mm and working distance 1·6 mm. A clip-on protection cap can be used. Such objectives must be used in a glass tank of sufficient depth to allow free up-and-down movement of the microscope tube without the lens touching the tank (Fig. 58). The tank should be of heavy glass, and if possible it should have an optically flat base. Water immersion lenses cannot produce the same high resolving power as those of oil immersion lenses, but this is not really a handicap when a homogeneous immersion system is in use.

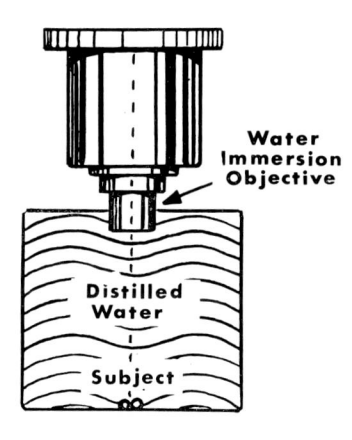

Fig. 58. (a) Focal length: 13·0 mm. Magnification: ×15. N.A.: 0·28. Working distance: 24·0 mm. (b) Focal length: 2·6 mm. Magnification: ×74. N.A.: 0·65. Working distance: 2·5 mm. Two water immersion objectives can be used.

Immersion oil. A change in the angle of light occurs whenever light passes through an interface between media of differing refractive indices (i.e. mounting medium—specimen—glass and air). Immersion oil, possessing a similar refractive index to that of glass (1·51 to 1·53 crown glass) minimizes the refraction of the light rays which emerge from the top of the cover glass, and increase the amount of light gathered by the lower lens of an objective (Fig. 59). Ray E, when passing through a dry medium, is refracted away from the lens on leaving the cover glass, but that the equivalent ray A, when passing through an immersion oil, continues at the same angle. Rays near the optical axis, such as E, are less subject to refraction. Even so, these are bent or refracted. Air has a refractive index of 1·000 and in the case of an oil immersion objective, where oil replaces the air, the refractive index is 1·524 non-drying oil and 1·515 cedarwood oil. A considerable gain is achieved in the numerical aperture, and an increase in resolution is thus brought about.

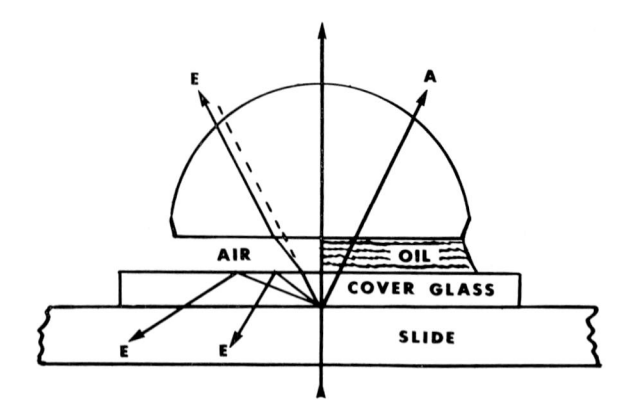

Fig. 59.

An oil immersion objective produces a depth of sharpness of a little less than one micron. The in-focus layer of the specimen material plays the main role in influencing the rays of light which form the image (Plate 15 e, f). Inhomogeneous material above or below this layer, tends to degrade resolution and contrast by diverting the illuminating and imaging bundles of light rays from the paths which they should follow.

While it is true to say that a high N.A. provides greater resolution (Plate 15 f), it does have drawbacks. It is not always possible to work with such high power and, on occasions when a wide field is necessary, one must accept a lens of lower N.A., or take a series of exposures and join them together, as illustrated in Fig. 144. It is as well to remember that the effective use of inclined rays involves increasingly greater corrections for spherical aberration. The greater the inclined rays, such as those used in high power work, the more exacting is the location of the plane of focus defined, and so objects above and below the focused point are not clearly recorded; depth of field is extremely limited; working distance is limited; field of view is limited; flatness of field is limited; and spherical correction becomes critical.

References

Lawson, D. L. (1966). The use of water immersion objectives in photographing microgel precipitation bands. *The Photographer*. July 200–203.

Rigby, G. R. (1953). "The Thin-section Mineralogy of Ceramic Materials," pp. 53–59. British Ceramic Research Association.

Cover glass

This small, seemingly unimportant, piece of glass contributes an important part in microscopy and is a "lens", in the sense that it is an addition to the already image-forming optical system, Fig. 60. There are six properties relating to the cover glass which can influence the final image definition of a photomicrograph. They are:

(a) The thickness
(b) The refractive index
(c) The dispersive power
(d) The homogeneity
(e) The surface quality of the cover glass
(f) The area between specimen and cover glass.

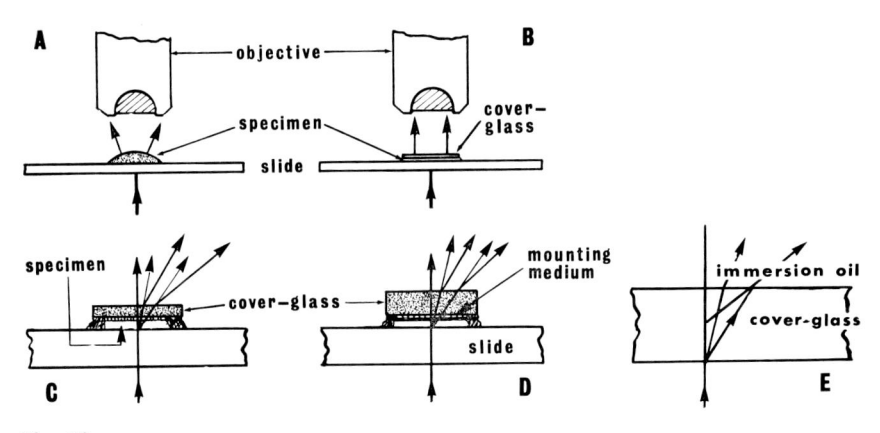

Fig. 60.

The photomicrographer should be conscious of the setbacks which might occur if consideration is not given to the exclusion of these points. Although without control of the manufacture of the glass he must always see that the correct cover glass is used and that it is mounted correctly. An unmounted specimen is likely to lead to trouble and a mediocre result (*A* of the Fig.).

(a) Manufacturers of objectives are also conscious that image defects could appear if any old cover glass were used and so they have computed their lenses to operate with cover glasses within certain limits. The standard thickness of a cover glass has generally been accepted as 0·17 mm, but seeing that some manufacturers of objectives compute their objectives for 0·18 mm thickness, it is obvious that two accepted thicknesses exist. However, whichever thickness it happens to be will be engraved on the outer case of the objective, as set out in Fig. 61. Leitz objectives are corrected for thickness 0·17 mm $\pm$ 0·03 mm. The variation in computed cover glass thickness is set out for objectives manufactured by Carl Zeiss, Oberkochen.

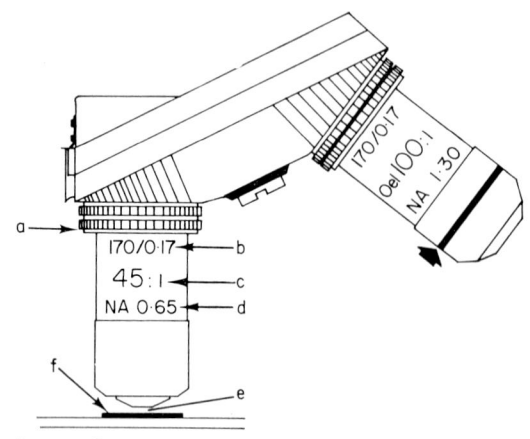

Fig. 61. (a) The distance between the objective screw flange and upper edge of the tube must be 170 mm. (b) Tube length and cover glass thickness. (c) Magnification. (d) N.A. (e) Free working distance. (f) Cover glass.

Table 6.

N.A. of objective	Admissible deviation from normal thickness of 0·17 mm	Approximate range of admissible cover glass thickness in mm
0·08–0·30	—	0 –0·3
0·30–0·45	±0·07	0·1 –0·24
0·45–0·55	±0·05	0·12–0·22
0·55–0·65	±0·03	0·14–0·20
0·65–0·75	±0·02	0·15–0·19
0·75–0·85	±0·01	0·16–0·18
0·84 (achromat)	—	1·5
0·88–0·95	±0·005	0·165–0·175

Zeiss Ultrafluars are corrected for cover glass thickness 0·35 mm. The main concern is the acceptability by the objective of marginal rays; a "thick" or "thin" cover glass can cause spherical aberration which would worsen the image, Fig. 60. The thickness tolerance is recognised by the optical manufacturers who provide an adjustable graduated correction collar built into objectives of a high aperture in the order of N.A. 0·95 (Plate 11 b), and cover glass thickness from 0·15 to 0·25 mm can be used. Zeiss Neofluar objective N.A. 0·90 × 63 has a built in correction collar for cover glass thickness 0·11 to 0·23 mm. An objective fitted with a graduated rotating sleeve is seen in Plate 11 (a). The sleeve causes a separation between the lens components (a^1) operated as indicated in Fig. 62, B and C.

COVER GLASS 57

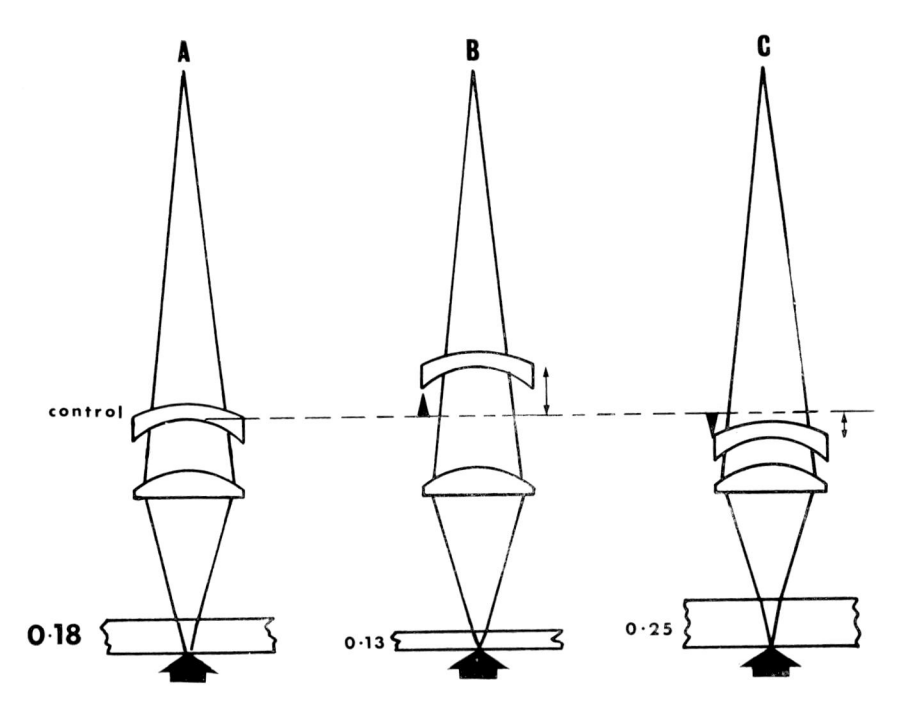

Fig. 62. Function of a correction collar, adjustment of the lens for cover glass thickness. (a). Control. (b) Thin cover glass. (c) Thick cover glass.

If the objective does not correct for cover glass thickness, then the mechanical tube length can be adjusted until a clear image is projected. As we have already seen, the objective has been designed to operate at a certain specimen to image distance, A of Fig. 63. This distance is limited by the need to change the distance at which the primary image is formed. This is brought about by changing the distance of the objective from the specimen; if the length of the body is, for instance, 170 mm instead of 160 mm. Thus the specimen to objective distance is decreased, giving extra image projection distance. The opposite effect can be brought about with the tube in the lowered position of 140 mm, here the specimen to objective distance will be greater. Of course the magnification will change, if only slightly, with the changing tube length. When making a change in the tube length due to an incorrect cover glass thickness one must be conscious of the possibility of introducing spherical errors. The necessary change (C) is brought about when using a thick cover glass and that of a thin is corrected (as in B).

The object of Fig. 64 is to demonstrate the effect of spherical aberration introduced when a cover glass thickness is too great, this effect is shown by "overcorrection" in the image. Either the tube length must be reduced or the objective correction collar adjusted. If the cover glass is too thin, the process is reversed, the tube has to be lengthened. To test cover glass thickness a test slide, comprising several cover glasses of which the underside has a coating of silver with microscopic perforations, is placed on the stage in the light train and we focus on one of the starlike perforations. The fine adjustment is moved slowly,

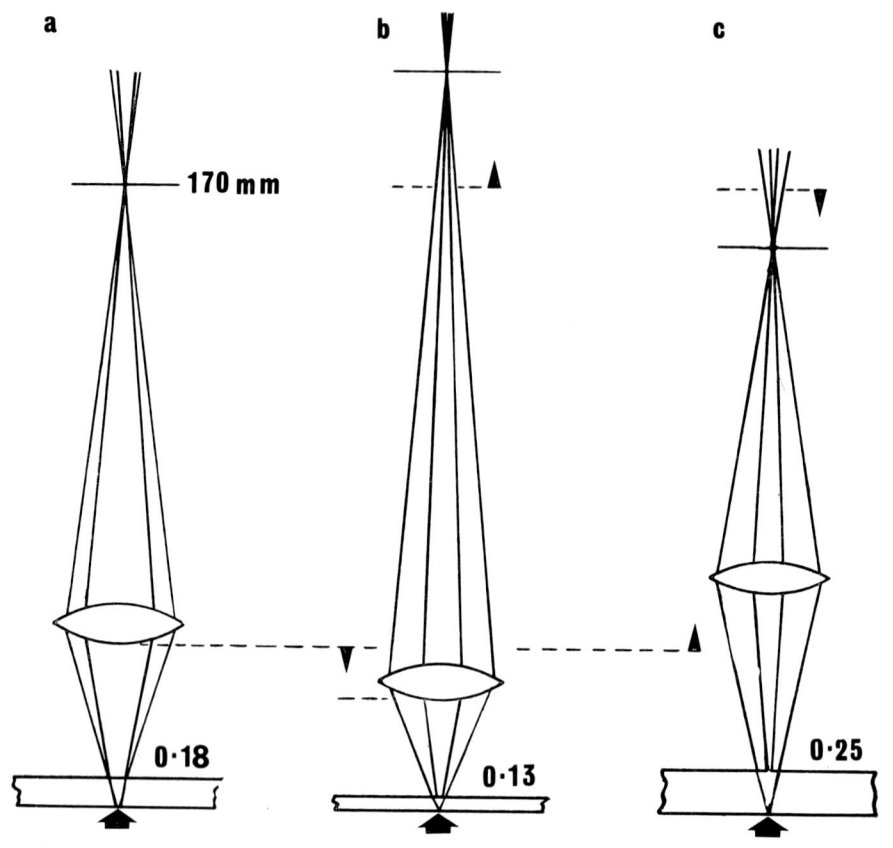

Fig. 63. Cover glass correction by tube length adjustment. (a) Control. (b) Thin cover glass and correction. (c) Thick cover glass and correction.

passing through the focal position and if the cover is too thick or too thin, a characteristic series of changes will be seen to follow those illustrated in the figures.

If the high, dry ×40 or ×60 objective has no correction collar for correcting cover glass thickness, oil can be used between the lower lens and the cover glass. Under such conditions the objective will not be so sensitive to errors due to a thick cover glass.

COVER GLASS

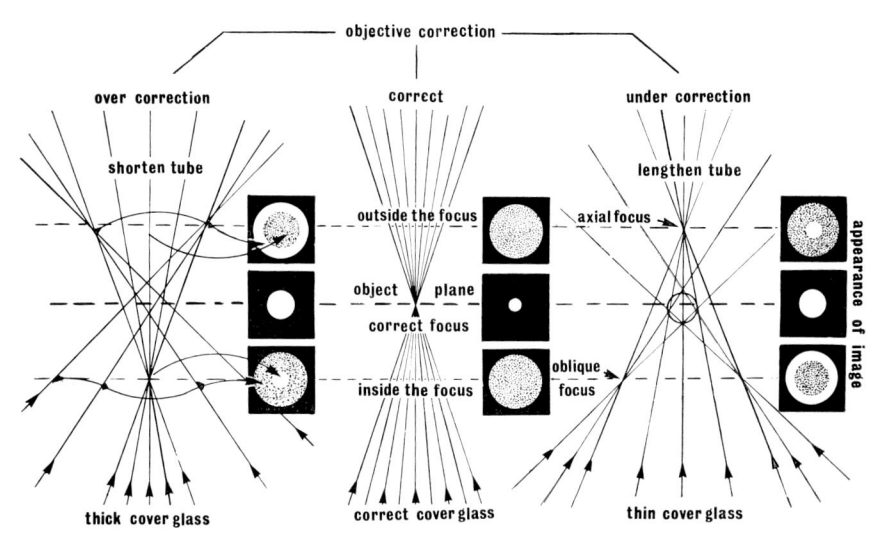

Fig. 64.

When the cover glass thickness is known and an adjustment is necessary in the draw-tube length, the following by Wild Heerbrugg U.K. Ltd. will assist in making the correction:

$$\frac{1/00 \text{ MM} \times (\text{Objective magnification}^2)}{250 \text{ mm (conventional viewing distance)}}$$

Example: Mech. tube length = 160 mm, objective ×40, cover glass thickness 0·19 mm. Cover glass thickness = 0·17 mm; difference = 0·02 mm.

$$\frac{\times = 2 \cdot 40^2}{250} = 12 \cdot 8 \text{ tube length difference}$$

Since the cover glass is thicker the tube length must be reduced to 160 mm less 12·8 = 147·2 mm. If the cover glass is too thin the tube length would have to be increased to 160 mm + 12·8 = 172·8.

Some low-power objectives are less sensitive to a variation in cover glass thickness and can operate with a "thick" cover glass when a number 2 can be used without fear of worsening the image. Chance resistance cover glasses are obtainable in squares, rectangles or circles, in the following thicknesses:

No. 0	0·085–0·130 mm
No. 1	0·130–0·160 mm
No. 1 1/2	0·160–0·190 mm
No. 2	0·190–0·250 mm
No. 3	0·250–0·350 mm

For easy measurement of cover glass thickness Leitz market a cover glass thickness gauge. It must be pointed out that when using some apochromatic objectives it is not possible to focus on the specimen through a thick cover glass. Leitz cover glasses are obtainable in one thickness and in the following various sizes:

Thickness 0·17 mm
- 15 mm × 15 mm
- 18 mm × 18 mm
- 20 mm × 20 mm
- 27 mm × 21 mm
- 15 mm diameter
- 18 mm diameter
- 20 mm diameter

(b) The standard usually adopted for the refractive index of a cover glass has been decided by the optical manufacturers who have come to the conclusion that a value of 1·524 is best. This figure falls very near to the refractive index of mounting medium and oil. High-power objectives are most sensitive to variations in refractive index; the acceptable tolerance has been fixed at ±0·0007.

A dry objective of, say, 0·95 N.A. is restricted in the angle of inclined rays it can receive from a specimen through a cover glass. This is approximately 41° and less with the axis of the glass. Beyond this figure rays strike the top of the cover glass and are reflected outwards. Some are reflected back as in Fig. 40 but when an oil immersion system incorporating an objective of, say, 1·35 N.A. is in operation, the acceptability of inclined transmitted rays is increased to approximately 65°.

(c) The dispersion values of a cover glass have a greater effect on immersion lenses, and as the value of immersion liquids in use does not exceed 48·0 the limit of the value of, say, V in a cover glass is in the order of $V = 57·0$. This, therefore, adds to a homogeneous immersion system with cover glass and oil.

(d) A value for homogeneity has been set in the refractive index, expressing the need for the same refractive index for all cover glasses within close limits. The cover glass should be homogeneous to the same fine limits, i.e. ±0·0007.

(e) It will be appreciated that all cover glass surfaces are highly polished in order to provide the necessary surface quality. Were it not for the fact that the surfaces are parallel and flat, odd effects could well occur when using polarized light, phase-contrast and darkground techniques: not to mention powers in the medium and high range in bright field transmitted light.

(f) The cover glass should be in direct contact with the top of the specimen (Fig. 60 B). Failing this, the thin layer of mounting medium interposed between cover glass and slide, C and D, is as bad as a thick cover glass. The layer of the specimen must not be over 0·003 mm. If care is not taken in this application the lower surface might lie beyond the tolerance of the recommended cover glass thickness, causing image odd effects. If air is interposed between specimen and cover glass, double refraction could occur and this is as bad as a thick cover glass.

The optical tube length of the microscope must be constant as variation will affect the spherical correction of the system. The Universal Corrector (if still available), of the type evolved by Bracey (1931) is a useful instrument.

Reference

Bracey, R. J. (1931). A universal tube length corrector. *J. Roy. Microsc. Soc.* **51**, 20.

Microscope slide

Slides used to be supplied with rounded polished edges, and they were a pleasure to handle, but most present-day slides are sold in the rough, just as they were cut, and measure 77 mm × 26 mm (3 in. × 1 in.). I was very fortunate one day to purchase some very old wooded slides, cut from teak and similar hardwoods (Plate 16). Belgian manufactured slides are sold in the following thicknesses:

No. 1	0·8–1·0 mm
No. 2	1·0–1·2 mm
No. 3	1·0–1·5 mm
No. 4	1·2–1·5 mm

Hollow ground slides have been specially made to enable marine and pond life to be photographed in their natural state (Plate 1 a & b). This slide (Fig. 53) consists of a depression ground in the face of the slide sufficiently deep to receive the subject matter, plus a few drops of water; so evaporation is not an immediate concern. A cover glass can be placed over the water. Quartz material should be used with the ultra-violet microscope.

Working distance

This is the distance between the lower lens of an objective and the cover glass and is extremely limited with most objectives. The following will give some idea of this restricted lens movement.

Planachromatic objectives	Neofluar objectives	
1·66 mm = 0·09 mm	1·92 mm = 0·12 mm	
4·4 mm = 0·18 mm	4·5 mm = 0·33 mm	
7·0 mm = 0·4 mm	7·1 mm = 0·54 mm	Working
10·4 mm = 2·7 mm	10·8 mm = 0·9 mm	distance
15·8 mm = 4·8 mm	16·4 mm = 4·0 mm	
27·1 mm = 4·9 mm	23·6 mm = 10·8 mm	
56·0 mm = 9·0 mm		

Having placed the specimen slide into position, the next thing to do is to view the specimen through the microscope: presuming, of course, that the objective has been carefully adjusted until it is almost touching the cover glass. This will mean that focusing takes place *away* from the cover glass and avoids colliding with the cover glass.

Objectives

Microscope objectives are generally classified; according to their colour correction, as achromats, apochromats and fluorite systems, etc. Such information is usually engraved on the case of the objective with tube length and cover glass thickness; as in Fig. 61. The distance between objective flange (*a*) and the upper edge of the tube length must be in accordance with the noted information (*b*) as must be the thickness of the cover glass (*b*). The magnification of an objective is indicated (*c*) and the N.A. indicated at (*d*). The free working distance is indicated here by (*e*), and the top of the cover glass is at (*f*). Leitz oil immersion objectives are marked with a black narrow ring indicated by the arrow.

Objectives for use with opaque specimens. These are commonly used with normal built-in incident illumination, such as that of the Ultropak of Leitz, and a free standing light source which projects a beam at an oblique angle to that of the optical axis of the system of the microscope. Plates 13 and 18 demonstrate that opaque subjects need care and skill in handling and directing the lights, in these cases, obliquely. As a rule no cover glass correction is necessary but if a source is projecting a beam, at a low angle of incidence, refraction takes place and the cover glass is best removed. Subjects in the rough, such as sand or eggs, are recorded without a cover glass. There is great flexibility in the mechanical tube length and some special microscopes are inverted for this particular application. Zeiss have manufactured a range of objectives for use in the metallurgical field which are corrected for infinite tube length, but they must be used with a telescopic eyepiece system.

Achromatic objectives. Chromatic and spherical aberration can be reduced to a minimum by the introduction of a negative flint glass and a positive crown glass, suitably shaped and cemented together. Such is the case in this particular objective, which is computed to remove spherical aberration and corrected for the medium colour—yellow-green, Fig. 66—and over corrected for blue rays and under corrected for red rays. Chromatic aberration is reduced to the extent of merging blue and red rays (Fig. 32). The secondary colours of the spectrum have been directed to a point near to the blue–red, 486 nm–656 nm (Fig. 65). This difference in focus is not noticeable, and some excellent results can be obtained in black and white by using a yellow-green filter in the light train, though perhaps a tricolour green is more suitable in photomicrography. Figure 66 demonstrates the peak transmission of this lens and others (compare 1 with 3 of this figure).

OBJECTIVES

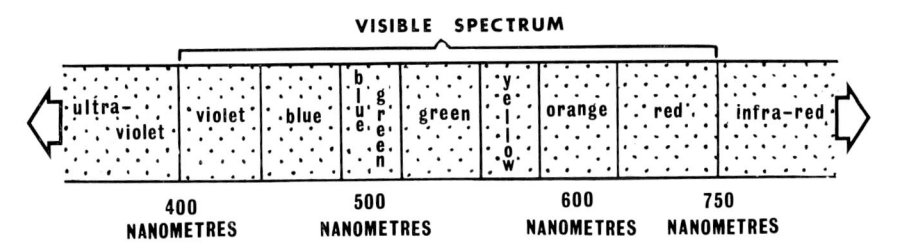

VISIBLE SPECTRUM

| ultra-violet | violet | blue | blue-green | green | yellow | orange | red | infra-red |

400 NANOMETRES   500 NANOMETRES   600 NANOMETRES   750 NANOMETRES

Fig. 65.

Apochromatic objective. This is the best for photomicrography, since it reproduces quality images of colours of several wavelengths (including blues, red and yellow) at a common focus. Violet, indigo and long red rays are also directed to a useful focus—as illustrated in colour range 1. Usually this lens is made up of several kinds of glass, including such as mineral fluorite, or fluospar which permit the attainment of a high N.A. Multi-stained specimens recorded on colour transparencies are best made through an apochromatic lens. Semi-apochromatic objectives (colour range 2) consist of fluorite lenses with apochromatic correction and are, therefore, suitable for photomicrography.

Plano-objective. This is a well designed multi-element lens system, made by Leitz, Wetzlar, designed to eliminate curvature of field and made with achromatic and apochromatic correction. Thus, when functioning in conjunction with a Periplan plano-compensating eyepiece, especially manufactured for use in photomicrography, a near perfect image of a large flat field is obtained.

Ultra-violet objective. We have already seen that the resolving power of an objective is inversely proportional to the wavelength of light used, and a comparison was made between blue light (475 nm) and red light (650 nm), (Fig. 65). The effective change in wavelength is also proportional to the N.A. of an objective and, when the maximum resolving power has been achieved with any one objective using a tungsten source, further improvement can only follow through a change in wavelength. Theoretically the resolving power at, say, 250 nm, will be double that attained with the same objective subjected to 500 nm. From this it can be seen that additional information can be recorded with UV radiations, whereas with visual light this is not possible, Fig. 66 (Plate 22, e, f).

Glass is unsuitable for these lenses due to the absorption of all radiations of the short wavelengths. UV objectives are therefore made of quartz which is transparent down to 200 nm. The commercially available UV lenses, e.g. those made by Cook, Troughton and Simms (now Vickers), are monochromatic for wavelength 275 nm and 253 nm and have a very narrow "band width" (Fig. 66). The performance of these lenses deteriorates considerably if the wavelength is more than 2 nm from the figure for which they are corrected. Some newer glycerol immersion objectives have been made with a slightly wider band width, totalling about 100 nm, as in the figure.

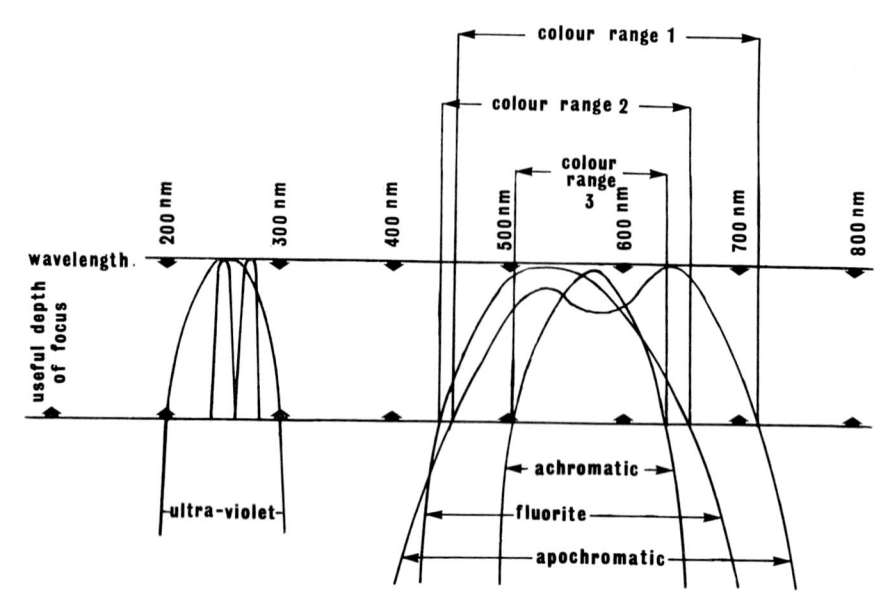

Fig. 66. Illustrating various colour ranges typical in objectives. The figure demonstrates vertically the range of wavelengths most suitable in the image formation of apochromatic (colour range 1), semi-apochromatic, fluorite (colour range 2) and achromatic (colour range 3). The approximate colour peak can be followed where the various colour rays focus. By way of comparison other ultra-violet bands and peaks can be seen in Fig. 195.

Parfocal objectives. Those who do not have parfocal objectives know only too well that, when changing from one objective to another to increase or decrease the magnification, the tube must be raised before turning the nosepiece. But this and similar operations are not necessary with parfocal objectives which are optically and mechanically designed, to ensure a regulated distance between the specimen and the front lens, and provide constant aerial image. Before making an exposure, however, it is always advisable to check the focus of the image by using the fine adjustment. Note: parfocal series of eyepieces must be used in conjunction with this type of objective, and the manufacturers' recommended tube length should not be overlooked.

Mirror objectives. Reflecting mirror objectives consist of a convex and concave surface-coated mirror; some being complex such as the Beck (Fig. 67). These objectives are surface aluminized for use in the ultra-violet and infra-red region of the spectrum. One advantage of this type of objective is that it is almost free from chromatic aberration. Zeiss, Jena produce a less complicated mirror objective, but with a high N.A. This should be very useful in electro-magnetic radiation. The number of reflecting and refracting surfaces have been kept to a minimum to reduce glare and light scatter (see reflecting microscope).

OBJECTIVES 65

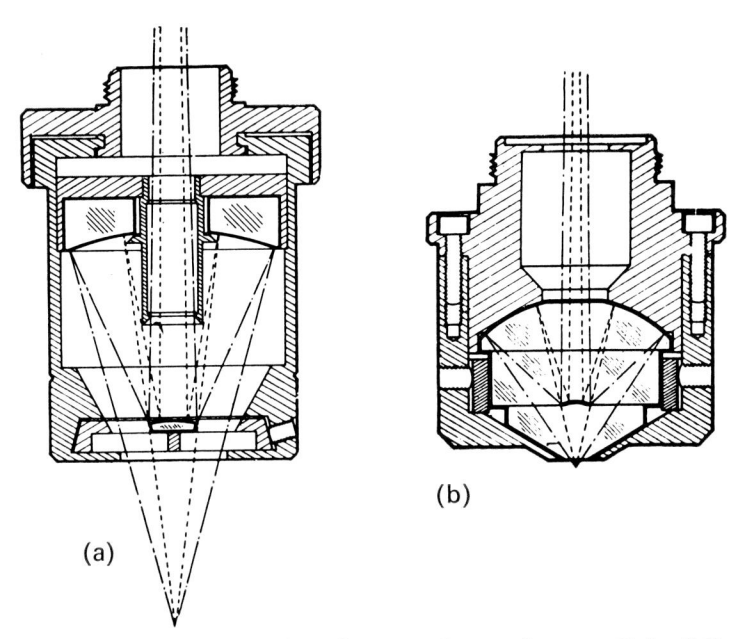

Fig. 67. Left: Ealing Beck long working-distance reflecting objective. Right: Ealing Beck oil immersion reflecting objective.

The magni-changer. Of the very wide range of high quality microscopes and accessories manufactured by the American Optical Co. Ltd. (handled by British American Optical Co. Ltd.) the magni-changer is unique. This feature is built into the microscope immediately above the apochromatic objective and beneath the body, offering instant selection of sixteen achromatically corrected optical elements, some of which are seen at the top of Plate 19. By operating a simple "dial-in" (lower) a change in the magnification is brought about. Each optical element is in true alignment with the principal objective and eyepiece. To make the changeover easy from one magnification to another should not require further focusing as the position of the image is always constant. As a safety measure, it may be advisable to refocus. The working distance can range from 38 mm, 10·2 cm and 20·3 cm. The maximum working distance is a great asset in this field. Specimens measuring as little as 0·00254 mm can be recorded with extreme clarity.

Zoom optical system. Bausch and Lomb, Rochester, N.Y. (with a London office) have given much thought and time to the production of a zoom lens system for some of their microscopes (Plate 20). The actual range of magnifications with a ×10 objective and ×10 eyepiece extends freely from ×100 up to ×200. The ×20 objective with a ×10 eyepiece zooms smoothly from ×200 to ×400. Figure 68 illustrates how such magnifications are produced. On the left of the

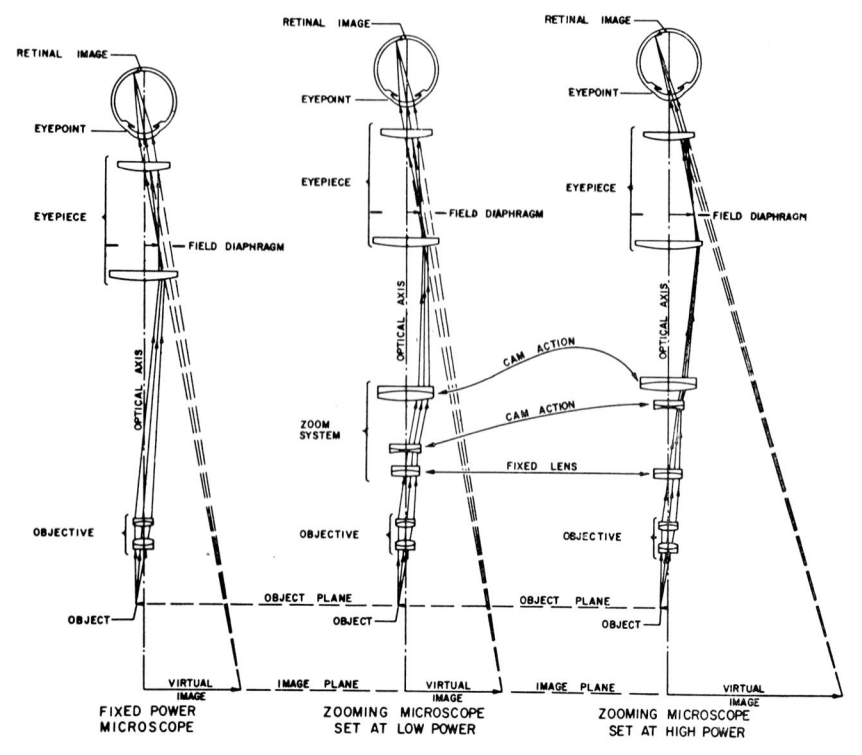

Fig. 68. Bausch and Lomb zooming microscope.

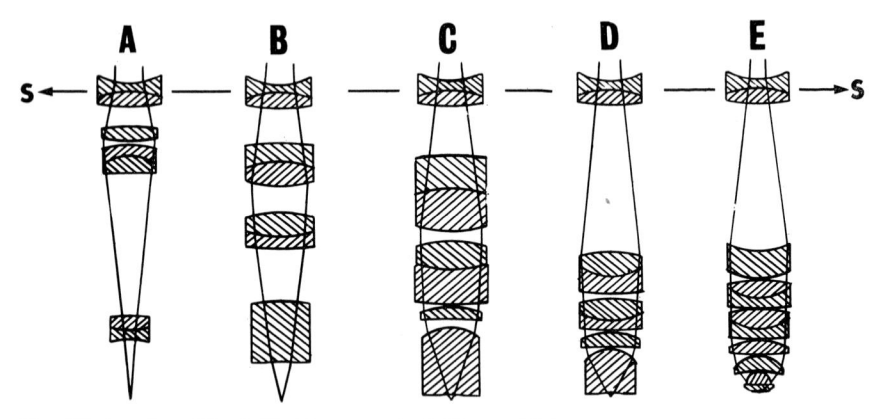

Fig. 69. A series of flat field objectives. A. 4.50 mm field ×4 N.A. 0.9. B. 1.80 mm field ×10 N.A. 0.25. C. 0.90 mm field ×20 N.A. 0.50. D. 0.45 mm field ×40 N.A. 0.65. E. 0.18 mm field ×100 N.A. 1.25. (See Plate 68).

figure is the conventional type of compound microscope already discussed; the centre and right illustrates the layout of the zoom microscope for low-power and high-power. The zoom lens system does not form an intermediate real image but simply moves the image outwards which, when zooming, is held in constant focus by cams controlling every movement. Maintaining this sharp image is a great benefit, especially for those who have not known the freedom of varying the magnification, without focusing after each turn of the body adjustment.

Flat-field objectives. As already discussed, curvature of field is an unwanted quantity in photomicrography. Extended research by Bausch and Lomb has now produced flat-field objectives, replacing the conventional achromatic lens. The newly designed lens system includes an extra—a $\times 5$ negative doublet lens system situated in the plane of the nosepiece (Plate 68 and S of Fig. 69). This lens is an accurately computed extra component for the objectives which assists in the production of the magnified image. Wide flat-fields follow: Fig. 69 illustrates the construction of the elaborate and efficient lens systems so necessary to this new application. It also illustrates the function of the top doublet which remains in position in the upper half of the nosepiece. Unlike other objectives the high-power flat-field objectives have a thick meniscus element facing the specimen, whereas the front lens of the old types have a hemisphere. According to Bausch and Lomb, this was one of the main causes of curvature of field. Another well known flat-field series of objectives is the Planachromats made by Zeiss, Oberkochen, which includes seven dry and three oil immersion lenses of $1 \cdot 0$, $1 \cdot 3$ and $1 \cdot 25$ N.A. The $\times 40$ N.A. $1 \cdot 0$ is a bright-field objective (see darkground objectives). Under this heading it is well worth mentioning the anastigmats made specially for photomicrography.

Object marker. This resembles an objective but has a diamond set into it, replacing the near lens. A circle can be scribed upon the firm cover glass around any portion of the specimen which it is desired to mark. The diameter of the circle is adjustable in steps of $0 \cdot 1$ in. to $0 \cdot 5$ in.

Ink-marker. "Miami" equipment combines microscopic examination at high magnification with facility for semi-automatic ink marking. The makers claim that it is the only equipment of its kind available as a standard manufactured product. One application for this equipment is in microcircuit work. Here slides containing hundreds of circuits are examined under the microscope for flaws and defects, faulty circuits being ink marked for subsequent rejection.

The equipment consists of a high quality microscope using long working-distance objectives, a micrometer stage and a micromanipulator, mounted electromagnetic inker, the whole assembly being mounted on an antivibration base. By a special mounting the microscope is converted from stage focusing to limb focusing which is essential for micro-manipulation work. In this way focusing is achieved without disturbing the relative positions of the specimen and the inker. The objectives are suitable for use with normal incident illumination and, in addition to providing a working distance which is an order of magnitude longer than conventional objectives, they give a high optical quality.

Condensers

Bright field. The function of the substage condenser is to collect, project and focus a uniform cone of light (through the specimen) for the objective. And to produce an image of the light source, via the plane side of the mirror, or direct from the source, into the plane of the specimen (Fig. 70). If the beam of light from the top lens of the condenser is of a similar angle to that required to fill the

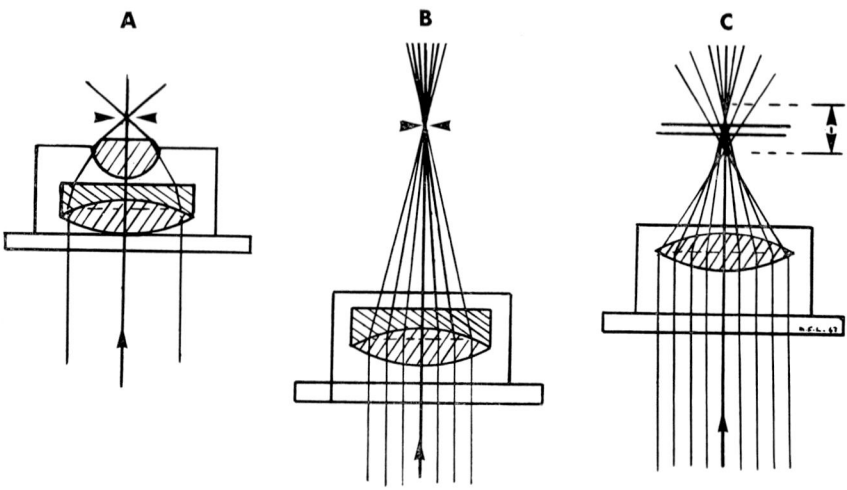

Fig. 70.

objective, then the maximum numerical aperture of that objective is used (Fig. 71 and Plate 9). On the other hand, if the beam of projected light is too narrow, due to the misuse of the substage condenser iris diaphragm, the resolving power will fall off, resulting in an inferior negative image (Plate 9 C, D and E).

Sometimes a "flare spot" appears on the emulsion or is seen in the ground glass screen. This may be because of rays reflected from one of the surfaces of the objective lenses. If the small bright central spot moves when turning the objective, the fault can be quickly located; but if the spot is stationary when turning the optical systems, then the spot may originate from within the tube.

It is equally detrimental to the final image if the cone of projected light leaving the condenser floods the objective as this causes glare (Fig. 72) (Plate 9a), which

CONDENSERS

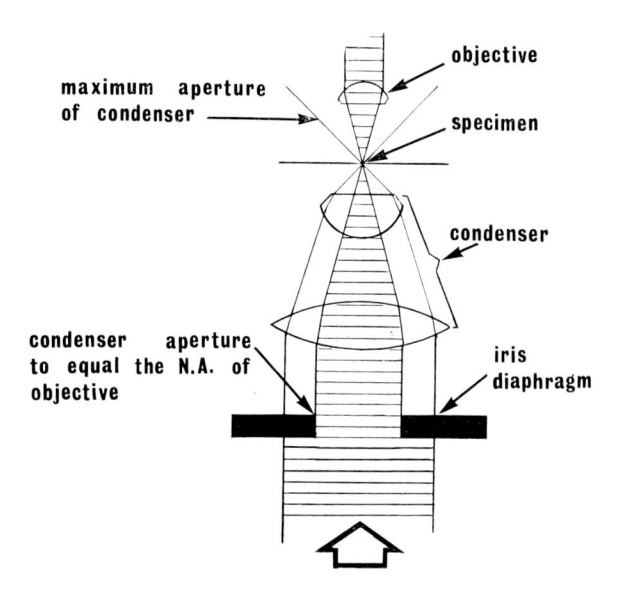

Fig. 71.

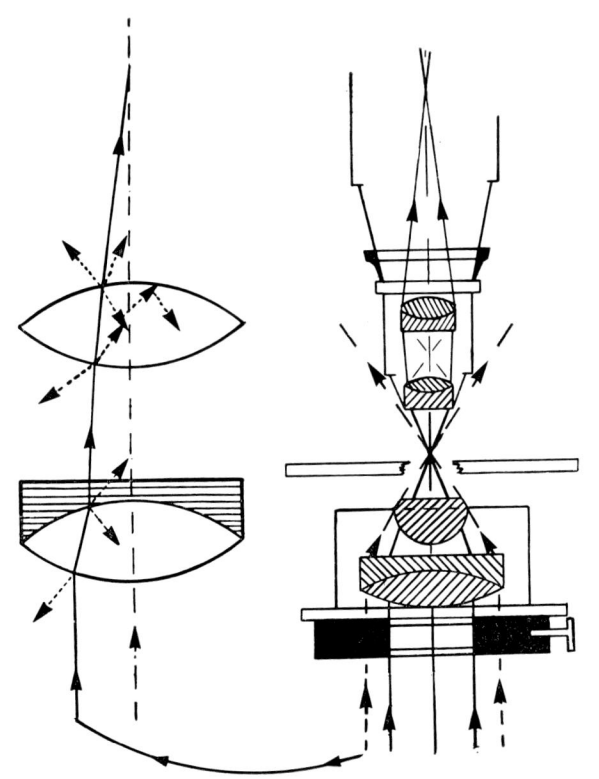

Fig. 72.

is as bad as the lack of resolving power. From these simple illustrations it is clear that the condenser contributes to the quality of the end product, whether through lack of resolution or the introduction of diffraction rings or bands. With bright field illumination the correct functioning of the objective depends on the use of the condenser (Fig. 48).

For ideal conditions the aperture diaphragm is usually positioned in the focal plane of the condenser. In this position the condenser aperture is a conjugate of the image focal plane of the objective and is a controllable "opening" or exit pupil for the whole optical system. As illustrated, misuse here will ruin the end product: a fact that cannot be over emphasised. Sometimes it is impossible to position the aperture in the focal plane of the condenser and the deviation will have little importance when medium and high power objectives are in use, since the image of the iris diaphragm will lie close to the front lens of the objective.

To meet the demands made upon the condenser the top lens is nearly always removable: some unscrew, while the top lens of the Ortholux condenser is removed by a "flip-top" action. Low power work necessitates a much wider cone of light consisting of rays at a lesser angle (Plate 21). Removing the top lens of the condenser causes the focal length to change and so it is necessary to refocus. At times the whole condenser can be dispensed with and the light source condenser used instead.

The position of the condenser aperture will, however, have an effect on low power transmitted bright field photomicrography, because the focal length of the condenser is increased when the top lens is removed (Fig. 70 A and B). This phase in the production of a photomicrograph is not often given the attention it demands. In such cases the exit pupil of the substage condenser will be positioned at a finite distance away, nearer the light source. If refocusing of the condenser does not take place after removal of the top lens, the projected rays will be converging rays and will illuminate the specimen with oblique marginal rays and, of course, produce an inferior image (C of Fig. 70).

The size of the specimen field should be adequate for the objective in use, and likewise the illuminating cone of light from the condenser. Objectives with a numerical aperture of 0·40, and larger, should really be illuminated with the complete condenser unit, that is with the top lens in use. The full numerical aperture of an oil immersion objective can only be retained when oil is sandwiched in front of the top lens and beneath the microscope slide (Fig. 59).

Abbe condenser. Generally, the Abbe type of condenser in common use today gives rise to some problems for it is expected to do more than the design permits. Aberrations of the common type provide difficulties and give poor picture quality. It is strange that the man associated with the troubled condenser originally designed (1872) the achromatic system, in itself suitable for the needs of any objective, for it was corrected for spherical aberrations (Fig. 70).

The present day condenser consists of two simple lenses of approximately

11 mm focal length and N.A. of 1·2. While this is acceptable for visual purposes, it presents problems to the photomicrographer. The image will be unevenly illuminated and the edges of the picture unsharp because, at times, the lens is incapable of covering the whole field. Sometimes chromatic aberration is evident around the image edge particularly noticeable in colour work.

Darkground condensers. Darkground condensers have a "patch-stop" placed in the centre which rays have to traverse, forming a hollow cone. In many cases the stop is placed into the condenser in the plane of the iris diaphragm and each objective and condenser requires its own stop. For incident light, darkground illumination, the system is designed with both annular mirrors and prisms. Both transmitted and incident methods produce a darkground with a highly illuminated specimen (Figs 73, 75 and Plate 17f).

Aplanatic condenser. This simple condenser is more suited to photomicrography, as it is corrected for spherical and chromatic errors (Fig. 75). Whilst it is true to say that some of the projected rays do not meet at a common point, a large proportion do form a cone of light, referred to as the "aplanatic cone", B of the figure. The greater the correction of the lens the more are errors eliminated: the greater the aplanatic cone, the more efficient it becomes until it equals its numerical aperture. For high-power work, where extremely critical illumination is required and when objectives of N.A. greater than 0·65 are used, an aplanatic or holoscopic oil immersion condenser is advisable.

Aplanatic–achromatic condenser. It must now be clear that optimum results can follow only when a fully corrected optical system is in use. It is particularly important in colour expression and this condenser is about the best for the purpose. In theory the condenser must have the greatest possible aplanatic aperture, but this is not so. In practice, its aperture floods the back lens of the objective and produces glare. The aplanatic cone should therefore be reduced very carefully.

Achromatic condenser. This can be used both dry and with oil and is fully corrected for objectives of the highest power. As a rule these condensers have an N.A. of 0·95–1·0 when used dry: this is increased to 1·4 when oil is used. A typical condenser of three lens unit may have a focal length of 1·2 mm but this will be in the region of 5·5 mm when the top lens is removed. The Zeiss achromatic condenser of N.A. 1·0 has a focal length of 12 mm, and the Leitz achromatic condenser N.A. 1·3.

Oil immersion. As already indicated, these condensers are for high-power oil immersion objectives. When oil is used between objective lower lens and the specimen (underside of slide and above top lens of condenser, Fig. 59), it permits a cone of light to pass through the specimen without deflection.

The correct use of an oil immersion objective and oil immersion condenser sets the highest limit on the resolving power of the optical microscope (Fig. 40). What was said earlier of the eye is now true of the optical microscope, for a point is reached when, owing to the magnification, it becomes difficult to record

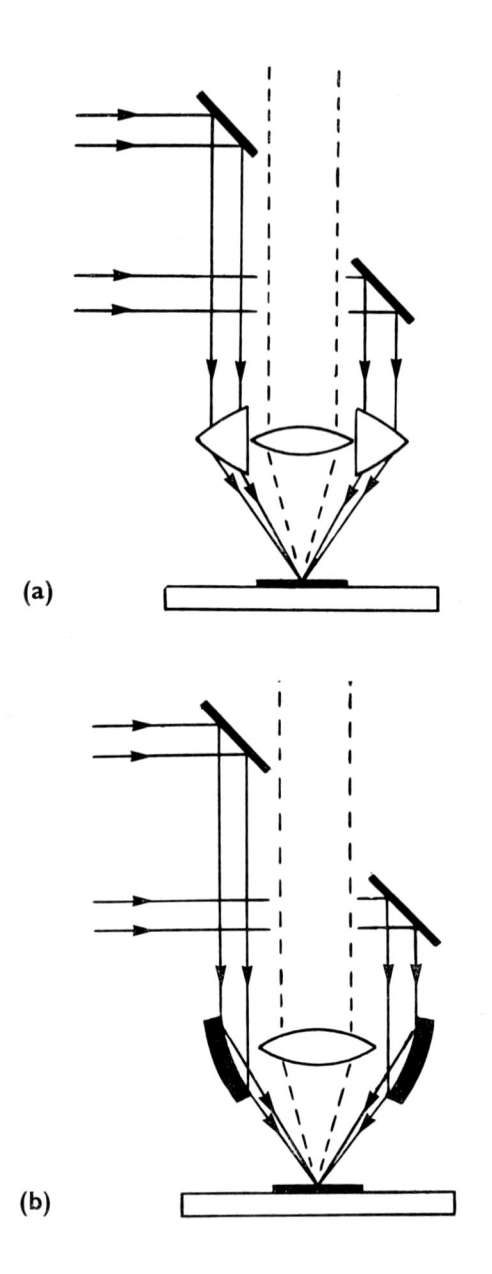

Fig. 73. Darkground incident light systems. (a) Leitz. (b) Reflecting condensers by Beck and Chapman.

CONDENSERS 73

single lens   Abbe

Abbe   aplanatic

achromatic   aplanatic-achromatic

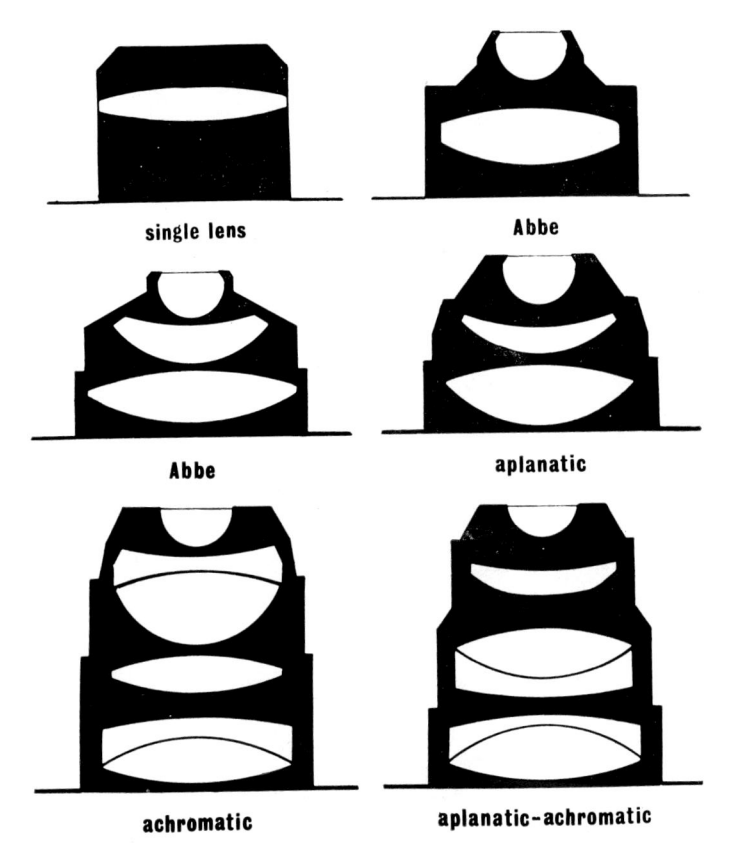

Fig. 74.

extremely small objects: in fact, the higher the magnification the less we see of the object. Further magnification does not provide any more details, and so your attention is drawn to Plate 10. Where the electron microscope takes over and provides the necessary information. This limit is in the region of 250 nm and can be further reduced to 75 nm by the introduction of ultra-violet source. Such images, not seen by the eye, are rendered visible only by photographic means. Rays of short wavelengths are strangely absorbed by glass. This means that the condenser optical systems for use in the ultra-violet are made of quartz. Always clean the lenses after using oil and before putting the microscope away.

Less condenser. By way of contrast, it must be stated that some subjects are best photographed without a condenser—in particular wide fields and subjects such as seen in Plate 32.

74　　　　　　　　　　　　　　　　　　　　PHOTOMICROGRAPHY

ABBE　　　　　　　　　APLANATIC

ACHROMATIC　　　　　OIL IMMERSION

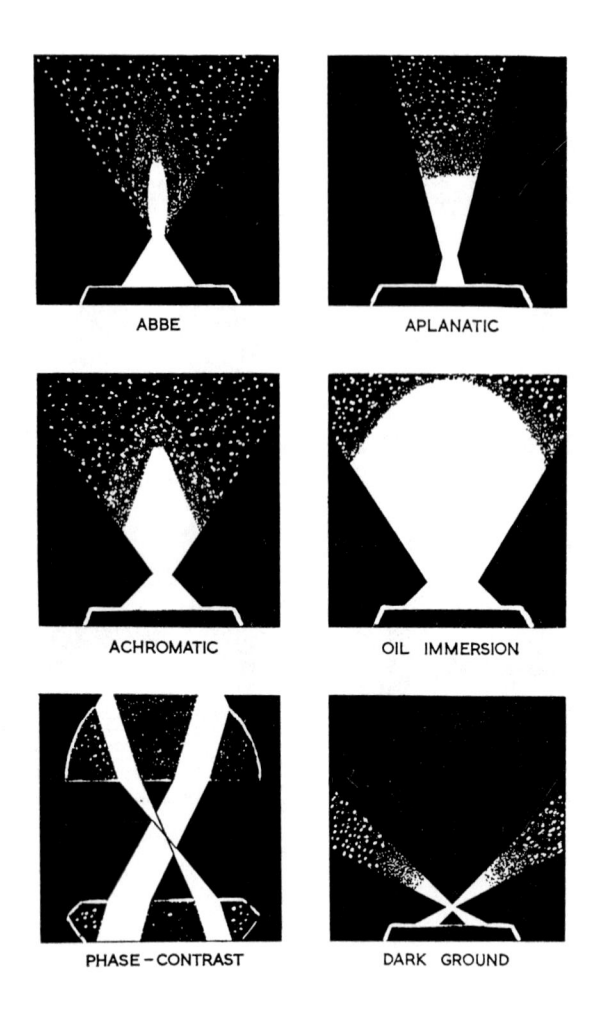

PHASE-CONTRAST　　　DARK GROUND

Fig. 75.

Eyepiece

The microscope eyepiece (Fig. 76) is mounted in a cylindrical tube and fits loosely in the microscope draw-tube (Fig. 12, 17). An eyepiece projecting flange prevents the eyepiece from falling further down the tube. This method of seating in the draw-tube is universal in vertical and inclined microscopes and this ensures that the objective works at the correct tube-length (Fig. 77).

In general, the function of an eyepiece is to magnify the real intermediate image, pass on and enlarge. The eyepiece produces what is known as an aerial image. In turn the image projected by the eyepiece must be suitably placed to enable it to be recorded, and to form an image on film or plate. In addition, some degree of correction is often introduced into the eyepiece to remedy slight aberration and curvature of field, which might be present in the image projected by the objective. The eyepiece cannot improve the effects of poor resolution but will only make defects more apparent. It is not advisable to use eyepieces with high × numbers with low-power objectives though it is recommended that a low × number eyepiece be used with such objectives. I use low × number eyepieces with medium-power objectives with very good results.

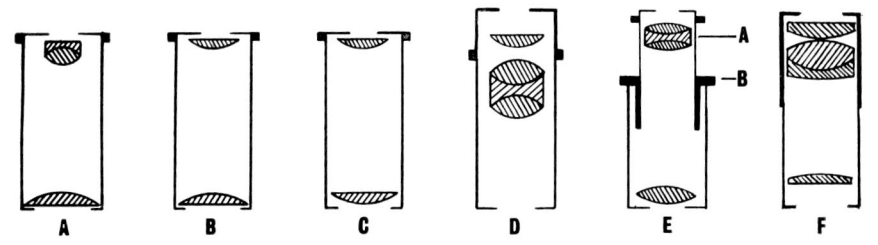

Fig. 76. Eyepieces. A. Chromatized Ramsden eyepiece. B. Ramsden type eyepiece. C. Huygen type eyepiece. D. Compensating eyepiece. E. Projection eyepiece. F. American Optical Co. widefield eyepiece.

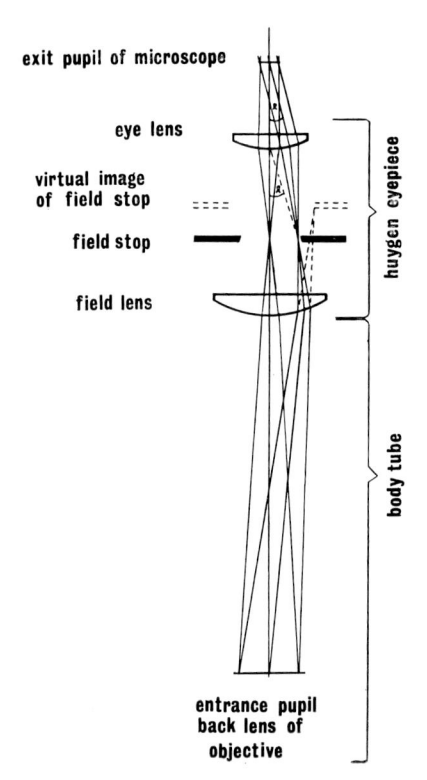

Fig. 77.

We have already seen that a change in the focal length of an objective brings about a change in the scale of the real intermediate image and finally the image provided by the eyepiece (Figs. 24, 25, 27). This change is brought about in the focal length of the objective, the distance between the specimen and objective, without changing the distance between the objective and eyepiece. An increase or decrease in the final projected image size may be shown. Another method of varying the projected image size is by a change in the eyepiece, while the distance between the specimen and objective remains the same, Fig. 85. Photomicrographic equipment, fitted with a positive projection eyepiece, projects an upright image; while the negative projection eyepiece projects an inverted image.

The eyepiece has been less subject to modification since the seventeenth century than has the objective. The type most generally used for photomicrography is the Huygen (designed by Sir G. B. Airy), (Fig. 76 c) and the radii of both plano-convex lenses were positioned in accordance with Christian Huygen as early as 1690 (Fig. 78 a).

These lenses are arranged one above the other with a circular diaphragm between them (Fig. 78, indicated by b in each figure). The lower lens is known as the field lens. Its function is to collect rays of light from the objective at as wide an angle as possible and to focus them near the plane of the diaphragm

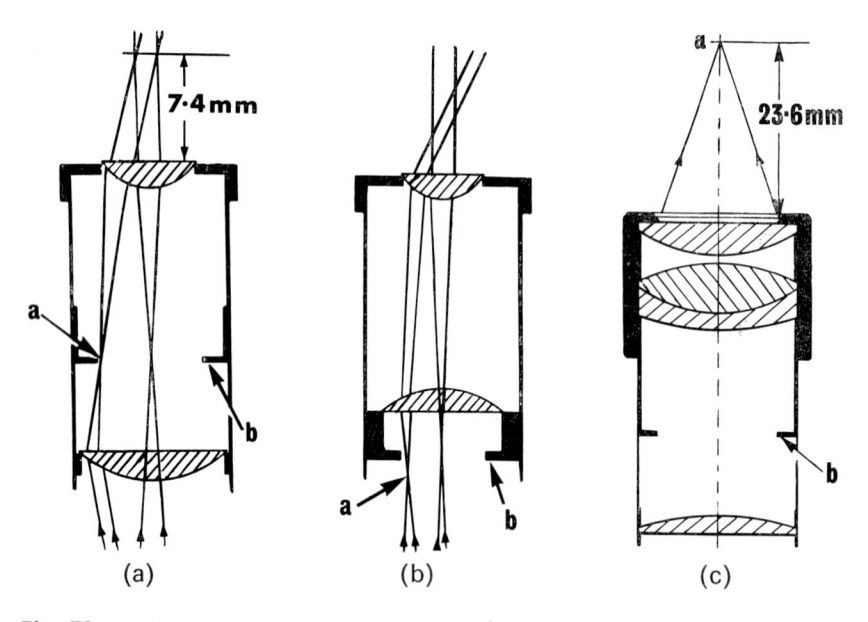

Fig. 78. (a) Huygen eyepiece (negative type). (b). Ramsden eyepiece (positive type). (c) Wide flat field eyepiece, *a* is the point of intersection of light rays, *b* indicates the circular diaphragm in all eyepieces. In (a) and (b) *a* indicates the plane of focus in the plane of the diaphragm.

(*a* of Fig. 78, a and b). The upper eye-lens is then used to magnify the image within a cone of rays, either accepted by the eye or focused in the focal plane of the camera (Fig. 27). Huygen eyepieces are "negative", in the sense that the principal focal plane is positioned between the lenses. Usually the distance between the two lenses is twice the focus of the eye-lens and the focus of the field lens is three times that of the eye-lens. With such an eyepiece it must be remembered that the linear magnification in the plane of the field stop is not that of the objective alone, but is also due to the field-lens which modifies or changes the image.

Other eyepiece systems have been introduced over the years. Figure 78 (b), designed by Jesse Ramsden, 1783, is for micro metric purposes. The Ramsden eyepiece also contains two plano-convex lenses but, unlike the Huygen eyepiece, the convex surface faces are opposed and the field diaphragm, *b* of the figure, is placed below the field lens. This eyepiece system is positive in the sense that the focal plane is situated below the lower component: i.e. between the eyepiece and the objective. Such a lens system is mainly for measuring and counting particle size distribution, when a suitable graticule with cross wires is placed in the field diaphragm. The intermediate image is formed here and a squared pattern of lines, circles and scales of a set size is projected simultaneously with the object image (Plate 22). For very many years the Huygen and Ramsden eyepieces held their own, but today more complicated lens systems have been introduced (Fig. 78 c).

Field of view and viewing angle. The overall image produced by the objective of the compound microscope is sharply limited by the diaphragm in the eyepiece (Fig. 77) which is known as the eyepiece field stop. Its diameter depends on the focal length and type of eyepiece and is limited by the inside diameter of the tube housing the eyepiece lenses. This diaphragm allows only a certain portion of the real intermediate image to be viewed. The diameter of this field is known as the "field of view number", and each particular eyepiece has its own "field of view number". This number makes it possible to determine the specimen area which can be recorded with the eyepiece field of view. The "specimen view" is determined by dividing the field of view number by the initial magnification of the objective in use, if necessary making allowance of a factor due to body magnification or an auxiliary optical system. The field of view number and the eyepiece focal length also serve to determine the angle in which the eye sees the complete projected field. If S is the field of view number, then the viewing angle W results from the expression

$$\text{tg}\frac{W}{-2} = \frac{S}{2} \cdot f_{\text{oel}}$$

Increasing the angle of view. We have already seen that the aperture of the eyepiece controls the angle of view, whether this is seen by the eye or projected on to a ground glass screen, approximate values of 25° at ×5 and 35° at ×15. An increase in the overall area (angle of view) is often necessary without changing the magnification of the eyepiece or objective. It can be made without decreasing the resolving power though we have already seen that low power objectives produce problems, in that astigmatism, curvature of field and a decrease in resolving power, are sometimes evident in the real intermediate image. This disturbing factor makes the photomicrographer cautious when he wants to increase the angle of view and so he uses a lower power objective.

Eyepiece magnification and type	Field of view number	Angle of view
×5 compensating eyepiece	20	23°
×6·3 compensating eyepiece	18	26°
×8 compensating eyepiece for micrometer discs	16	30°
×10 compensating eyepiece	16	36°
×12·5 compensating eyepiece for micrometer discs	12·5	36°
×16 Kpl eyepiece	10	36°
×12·5 wide-angle eyepiece	18	50°
×20 Kpl eyepiece	8	36°
×10 wide-angle eyepiece	18	41°
×25 Kpl eyepiece	6·3	36°

The eyepiece diaphragm in the Huygen eyepiece controls the overall projected field and, as seen, it intercepts a certain amount of the outer area of the real intermediate image. However, it is possible to introduce into the eyepiece a larger diaphragm, say from 20 mm to 21·5 mm. This is best suited to a low-power eyepiece operating in conjunction with a wide flat field objective. Thus the two optical elements produce an increase in the angle of view, or overall projected image on the emulsion, without any loss of image quality. The Leitz Periplan ×10 eyepiece projects an extra wide field of view.

The point on the axis above the eyepiece (Fig. 78) where all the principal rays of light intersect, can be found by holding a piece of finely ground glass at about 7·4 mm above the eyelens of the Huygen eyepiece, and 23·6 mm above the eyelens of the wide flat field eyepiece.

Compensating eyepiece. The compensating eyepiece (Fig. 76 D), or orthoscopic eyepiece as it is sometimes called, is rather more complicated in construction and is intended mainly for use with apochromatic objectives. On the other hand, this eyepiece can be used with certain achromatic lenses because of its ability to compensate the chromatic magnification differences peculiar to high power achromatic objectives. Abbe designed the true orthoscopic eyepiece. This type provides superior colour correction for most objectives.

Projection eyepiece. The Zeiss Homal projection eyepiece system (Fig. 76 E) consists of a negative focal length. This eyepiece forms the intermediate image produced by the objective, as well as that of the top lens: both are virtual images. For this reason there is no field diaphragm in the eyepiece. As a rule the field diaphragm controls the format size of the negative and the projected image is of a wide field. The Homal eyepiece is most suitable for correcting curvature of field and in this connection each eyepiece of a given magnification has been computed to operate with a set objective, as set out:

Zeiss Objective	13 mm	9 mm	6 mm	4 mm	2 mm
Zeiss Eyepiece	×6·3	×8	×10	×12·5	×16

The projection distance of this particular eyepiece is 160 mm beyond the real intermediate image produced by the objective, and again it should operate at the manufacturers' set tube-length distance, not overlooking cover glass correction.

Wide flat-field eyepieces. Four element, wide flat-field eyepieces (Fig. 78 C) are now made by the American Optical Company and Leitz, and are manufactured in sizes ×10, ×15 and ×20. The eyepieces are parfocalized, ensuring their interchangeability and reliable high quality, to produce a wide flat-field. They are fully chromatic and correct an amount of curvature of field and distortion, as well. In addition, these eyepieces can be used for particle size measuring, as a specially calibrated graticule can be inserted and used in conjunction with a specially calibrated micrometer disc. To calibrate these two, focus on the stage micrometer and move it until one of the graduations corresponds exactly with one of the divisions of the eyepiece micrometer. The true distance, X, seen on the stage micrometer, which corresponds to the number division Y of the eyepiece micro disc (Fig. 79) is then read, and if this is divided by the number of divisions of the eyepiece micrometer, we find the distance each division subtends

$(C = X/Y)$.

The number of divisions covered by the specimen multiplied by the calibration constant C gives the length of the specimen.

Plano-eyepiece. Specially designed to operate with plano-objectives, these objectives are corrected for curvature of field, have achromatic and apochromatic correction and project a wide flat field.

Quartz eyepiece. Quartz eyepieces are used in conjunction with monochromatic objectives in the ultra-violet. Perhaps this particular eyepiece is better known as the quartz projection eyepiece, since direct visual observations with UV radiations are not generally permissible. This eyepiece has been computed to project an image at a distance of 30 cm.

Parfocal eyepieces. There will be a change in the tube-length when changing an eyepiece unless, of course, a parfocal system is employed. Such a series of eyepieces have been constructed so that the focal plane comes at a set distance

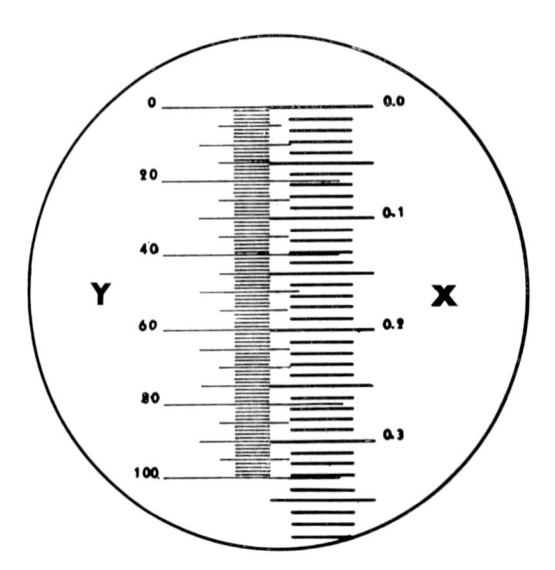

Fig. 79.

beneath the end of the draw-tube. A series of eyepieces such as these have a common focal plane (Fig. 80). However, the optical tube-length is not constant with a change in objective because of a change in the position of the back focal plane (Fig. 15).

Direct focusing with ultra-violet. Some eight years ago the Bausch and Lomb Optical Company in the United States developed the first ultra-violet microscope, permitting visual focusing with the new "ultra-scope tube", made by the Radio Corporation of America. Photomicrographers can now focus the image visually in the ultra-violet and need no longer use a substitute visual wavelength to estimate the region of focus. This microscope is a great advance and should be of tremendous value in research.

Visual eyepiece for photomicrography. An eyepiece of the type illustrated in Fig. 81. A is used when photographing and focusing an object right up to, and during, the exposure of film or plate (Plate 23). A small percentage of the image forming rays from eyepiece B is reflected when coming in contact with a right angle beam-splitting prism, C, and entering a well planned optical system which could be termed a unit-power telescope. This is made up of two positive eyepieces which have a common focal plane where a graticule D is positioned. The focal plane is a considerable distance from C, as compared with the distance of A from C, and yet the eyepiece is so computed that the image, sharply focused on the graticule, is at the same time a sharp image in the focal plane of the camera. The graticule is designed to indicate the limit of the image field covered by film or plate.

Comparison eyepiece. This is the outcome of yet another demand placed upon the

EYEPIECE

microscope as a recording instrument (Fig. 82). It is true that one's normal observations are seen by both eyes as a single unit, each eye providing a slightly different picture of the same subject—stereo vision. This is not so with the

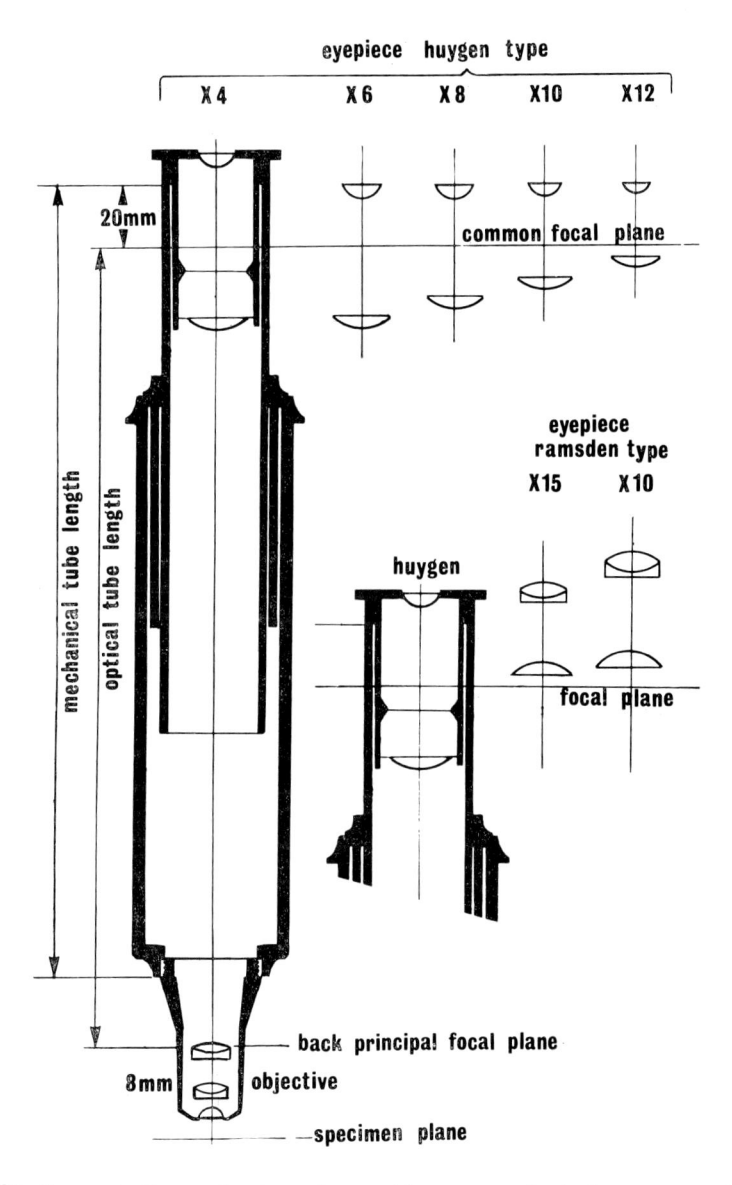

Fig. 80. Demonstrating parfocal eyepieces with common focal plane compared with Ramsden type eyepiece.

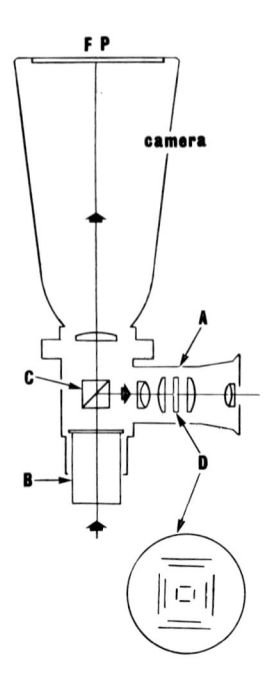

Fig. 81.

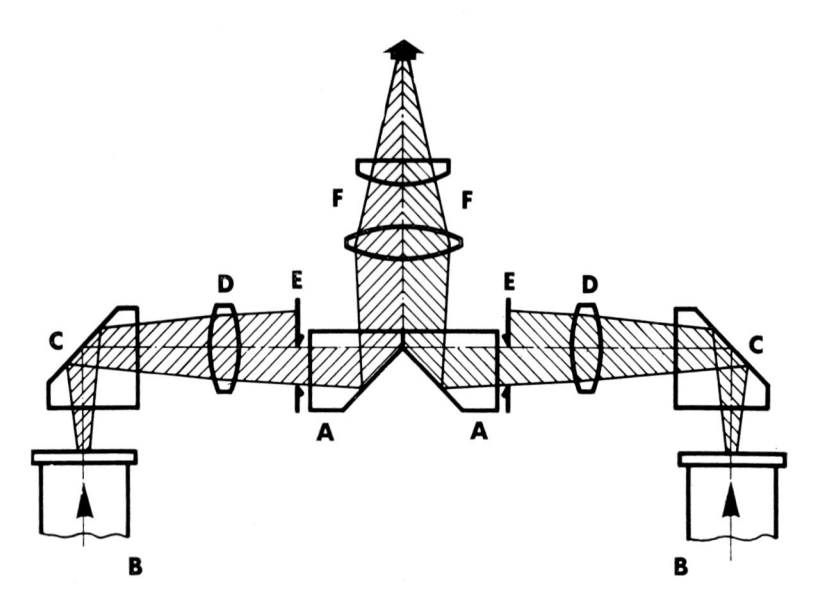

Fig. 82. Comparison eyepiece by Zeiss, Jena. A. Deflecting prism. B. Eyepiece. C. Deflecting prism. D. Correcting optical systems. E. Iris diaphragm. F. Observation and projection eyepiece.

comparison eyepiece. These instruments permit observation of images from each of two microscopes, placed side by side, seen in one field of view. The optical systems are manufactured by Zeiss, Jena, and also the British American Optical Company.

The instrument fits into each of the microscopes as illustrated in Plate 24. The two image planes from the two microscopes are formed adjacent to each other by means of deflecting elements, enabling direct comparison (Plate 25). The optical instrument is bridged to both microscopes, either inclined or upright tubes. In each case the exit pupils of the eyepieces are in the reflecting surfaces of the two deflecting prisms. From the illustration it can be seen that the central double prism reflects upwards the right half from the left microscope: and the left half reflects from the right. Thus, both halves of the projected image are side by side and seen as one. In compensating the tube extensions, the manufacturers have placed cemented lenses and half lenses between the deflecting prisms.

The comparison eyepiece is easily attached to microscopes with eyepieces of the usual diameter, and are fixed by two felt-padded clamping rings which are pushed on to the eyepiece sockets or tubes, finally resting in the two collars fitted onto the eyepiece covers. The location of attachment and the eyepiece covers must not be changed, so as to preserve the position of the exit pupils at the first reflection surface. To both makes of instruments a camera of any size can be attached, and phase contrast and polarised light used, as well as a wide range of objectives. The exposure is made in the normal way and it controls the exposure of both images.

Pointer eyepiece. This is a ×8 Huygen type eyepiece with a built-in pointer at its focal plane, controlled from a light external indicator, operated while viewing the specimen through the eyepiece or on the ground glass screen. Such a pointer is extremely useful when sharing one's observations. A forerunner to this eyepiece was in existence at the turn of the century.

Micrometer eyepiece. A Kellner positive type eyepiece is made to accommodate a graticule in the focal plane so that an image of the graticule appears superimposed on the film with that of the specimen (Plate 26).

Centering telescope. This is a small telescope inserted into the tube instead of the eyepiece, and serves for observation of the magnified exit pupil of the microscope objective, as well as aperture stop image appearing in it. This provides accurate checking of the objective aperture. Further, the centering telescope is indispensable for adjusting the annular phase-contrast diaphragm in relation to the phase annulus in the objective.

Eyepiece—camera light trap. It is most necessary to have some means whereby stray light is prevented from entering into the camera via the eyepiece aperture. The tube light-trap (Fig. 83) is interposed between microscope eyepiece tube and the support arm of the pillar stand of the camera. Such light traps are manufactured to accommodate the various tube sizes (Plate 27).

D*

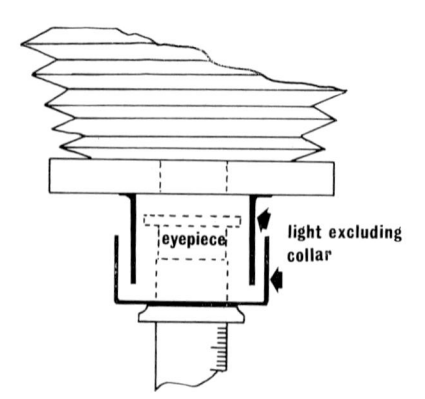

Fig. 83.

Magnification

This is the ratio of the apparent linear size of a specimen seen through the microscope, the virtual image, to the size of the specimen as it appears to the unaided eye at a distance of 250 mm (10 in.). It is not the case in photomicrography when having a ground glass screen attached to a camera for here an image of the specimen is projected upon the screen. When using a bellows camera (Fig. 84), the magnification will vary directly as the distance does of the projected image from the eyepiece: a zooming effect. Figure 85 illustrates a fixed focus microscope camera. The projected image is at a constant distance from the exit pupil of the eyepiece but magnification has been changed by changing the eyepiece as indicated. On the other hand, Fig. 86 illustrates magnification changes introduced by changing the objective magnification. The magnification relationship is therefore a straight line: magnification is measured in linear. Magnifications can fluctuate more when working with a bellows camera (Fig. 84) but fixed-focus cameras need the magnification in the objective and eyepiece to arrive at the initial image magnification.

There is a growing tendency to identify objectives by their primary magnifications, rather than by their focal lengths (Fig. 61). The former is now engraved upon the mount. This is a logical method of identification since, owing to the variation applied in optical tube length, the primary magnification is ensured merely by dividing the mechanical tube length by the equivalent focal length of the objective.

From the photomicrographs of human hair (Plate 28) it is not difficult to deduce that the numerical aperture of an objective determines the resolution and this in turn influences the magnification. Such can be traced by examining a test preparation of diatom, pleurosigma angulatum, with objectives of various focal length, numerical aperture and magnification. This test will prove that such a specimen can only be resolved when using an objective with numerical aperture over 0·75, indicating a magnification of approximately ×600. How the

MAGNIFICATION

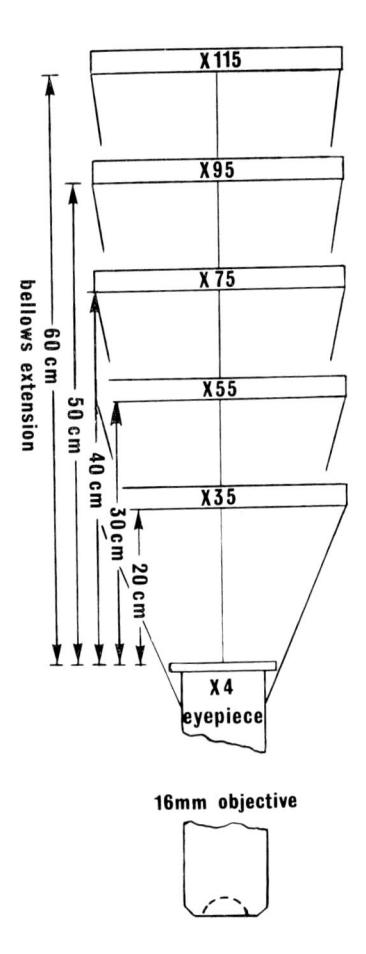

Fig. 84.

final magnification is reached is entirely under the control of the photomicrographer. Such a magnification can be obtained by using route one, a ×25 objective with a ×25 eyepiece. Note: this is a medium-power objective and a very high-power magnifying eyepiece.

A magnification of ×600 can also be reached by using a ×60 objective together with a ×10 eyepiece. The total is approximately the same as the above but the route as different as its end. The print made from route one will lack detail while route two photomicrograph should resolve the structure of pleurosigma. Route one objective has numerical aperture 0·30, and route two an objective of 0·85 numerical aperture. We therefore return to the angle of the cone of light which, proceeding from a sharply focused specimen point, is received by an objective. As already seen, such an angle is measured by $n \sin u =$ Numerical Aperture.

In testing the optical system of a microscope, the first question should be *not* what is its maximum magnifying power? *but* what is the lowest magnifying power under which *it will* record lines of a given degree of difficulty? It is always advisable for the purpose of quality, to employ the lowest power objective with

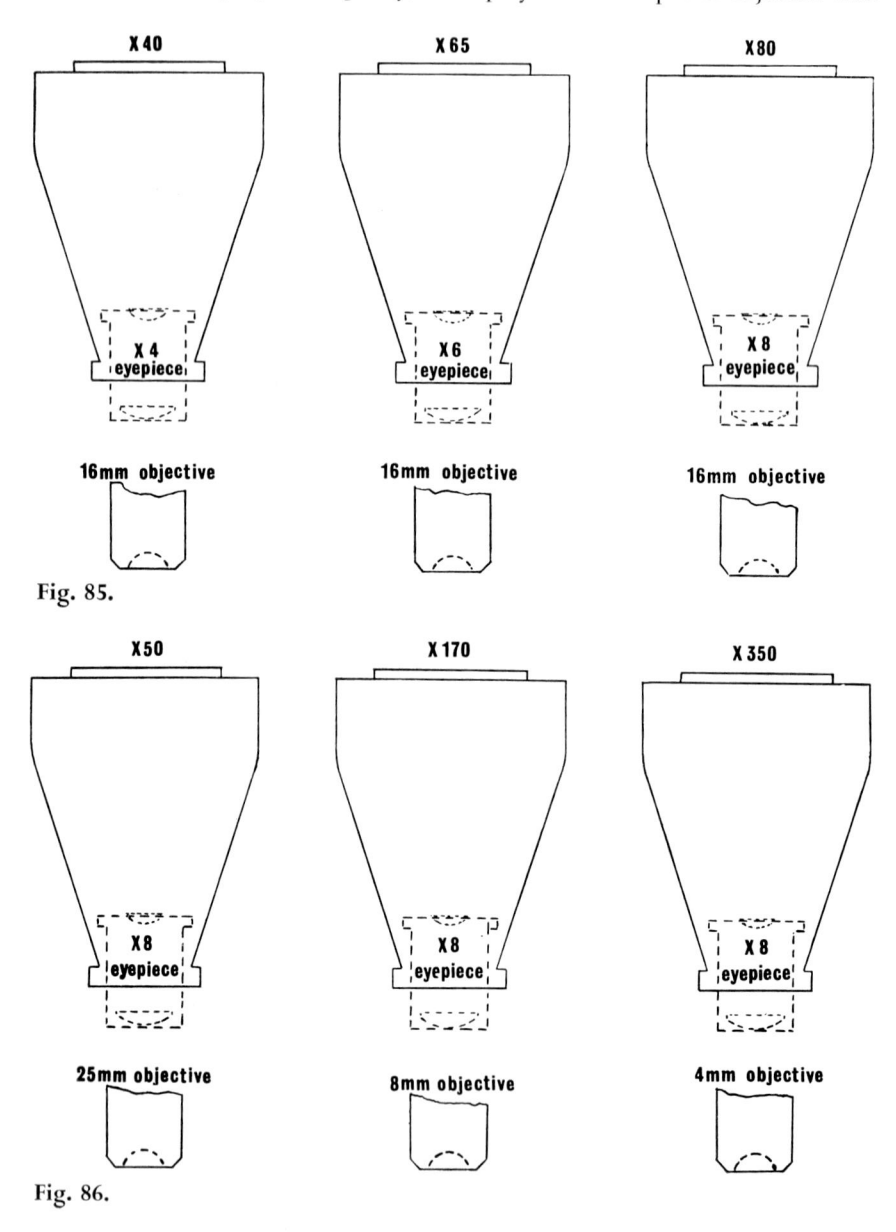

Fig. 85.

Fig. 86.

MAGNIFICATION

which the details of structure can be clearly recorded: considering its state after enlargement. The lower the power, the less is the liability to error from false images and the better can the mutual relations of the different parts of the specimen be appreciated.

For the accurate measurement of magnification a stage micrometer is necessary (see graticules). This is in fact a scale, engraved on a microscope slide, and offers the quickest and most accurate method of measurement when replacing specimen slide for micrometer slide. Further, such a scale can be photographed and reproduced by enlargement or contact. A magnification table can be made for easy reference as illustrated (Table 7).

Table 7. *Approximate magnifications*

Objective focal length	Eyepiece	\multicolumn{6}{c}{Camera extension (cm)}
		10	20	30	40	50	60
		\multicolumn{6}{c}{Diameters}
75 mm	×4	—	—	12	16	20	24
75 mm	×6	—	—	15	20	25	30
75 mm	×8	—	14	21	28	35	42
75 mm	×10	—	16	24	32	40	48
50 mm	×4	6	12	18	24	30	36
50 mm	×6	8	16	24	32	40	48
50 mm	×8	10	20	30	40	50	60
50 mm	×10	12	24	36	48	60	72
25 mm	×4	10	20	30	40	50	60
25 mm	×6	15	30	45	60	75	90
25 mm	×8	20	40	60	80	100	120
25 mm	×10	23	46	69	92	115	138
16 mm	×4	19	35	55	75	95	115
16 mm	×6	22	44	66	88	110	132
16 mm	×8	28	58	88	118	148	178
16 mm	×10	34	68	102	136	170	204
8 mm	×4	35	70	105	140	175	210
8 mm	×6	50	100	150	200	250	300
8 mm	×8	66	132	198	264	330	396
8 mm	×10	77	154	231	308	385	462
4 mm	×4	80	160	230	300	370	440
4 mm	×6	97	194	291	388	485	582
4 mm	×8	132	264	396	528	660	790
4 mm	×10	175	350	525	700	875	1,050
2 mm	×4	150	300	450	600	750	900
2 mm	×6	208	416	624	832	1,040	1,248
2 mm	×8	270	540	810	1,080	1,350	1,620
2 mm	×10	334	668	1,002	1,336	1,670	2,004

Measurements were taken from the top of the eyepiece mount to the ground glass screen.

Table 8. *Objectives and compensating eyepieces* †

Apochromatic objectives			Magnification with eyepieces				
Focal Length (mm)	Numerical Aperture	Initial Magnification	42 mm ×6	30 mm ×8	22 mm ×11	15 mm ×17	10 mm ×25
40	0·11	3	17	27	32	55	85
16	0·35	10	55	80	100	160	255
8	0·65	20	115	170	205	345	545
4	0·95	40	215	320	390	650	1,020
3	1·2	55	300	440	525	900	1,400
2	1·3 Fluorite	90	500	720	875	1,450	2,250
‡3·75	0·95	45	240	350	425	700	1,110
§3·75	0·95	45	240	350	425	700	1,110

† R. and J. Beck Ltd.
‡ Oil immersion
§ Water immersion

All factors determining the magnification and its relationship to the resolving power given by the aperture of an objective (range of useful magnification) can be represented diagrammatically, as is shown in the following example of a typical series of objectives and eyepieces. The horizontal lines in the diagram (Fig. 87b) represent the steps by which the total magnification increases in accordance with the standard series. The solid lines from the lower right-hand corner to the upper left-hand corner are guide lines for the objectives, the magnification and numerical aperture of which are indicated beside the lower end of the guide line. The guide lines for eyepieces are dash-dotted, running from the lower left-hand corner to the upper right-hand corner. The eyepiece magnification and the field-of-view number are indicated beside their lower end. The guide lines for the eyepieces intersect those of the objectives at the steps, which indicate the total magnification of the corresponding combination. The frame limits the range of useful magnification as defined by Abbe (Fig. 87b).

When operating with a free standing camera, as illustrated in Fig. 87a, to record, say, the exact magnification of a specimen suspended in water, perhaps in a hollow slide or a large watch glass, it is advisable to use an eyepiece graticule in place of a micrometer. If the specimen is replaced by a micrometer and focusing takes place, the objective will move downwards towards the scale. At the same time the eyepiece will move away from the focal plane of the camera which is at a fixed height, producing a higher magnification than that recorded of the specimen. Conversely, a specimen lying in a small watch glass (B of the figure) will lie well beneath the plane of a micrometer and, when focusing on the scale, the objective and eyepiece will move upwards towards the focal plane of the camera

MAGNIFICATION 89

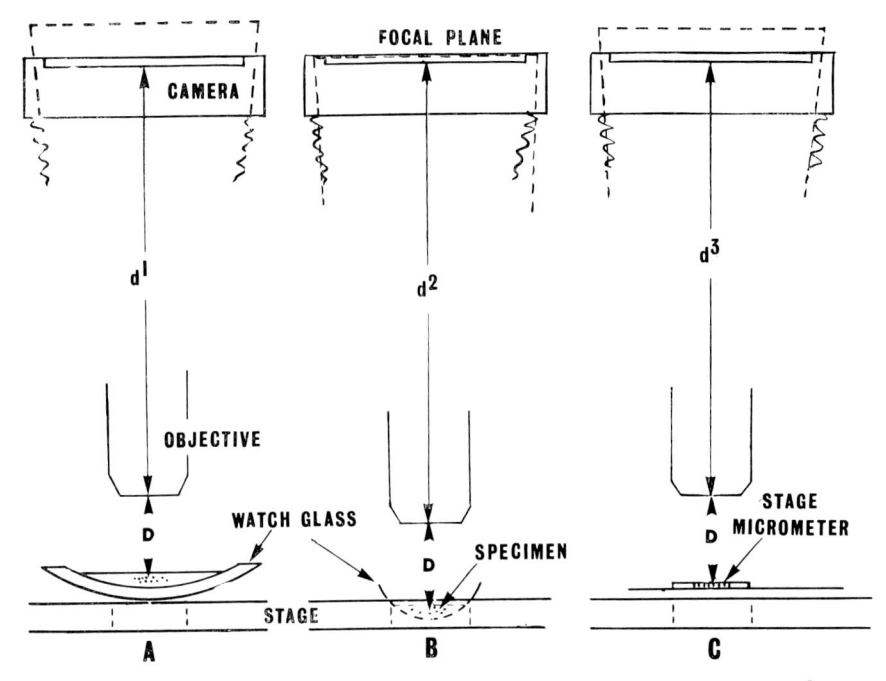

Fig. 87a. Variation in objective to focal plane distance, when recording magnification with a micrometer scale. Distance D remains the same but distance d^2 varies as the objective is moved upwards or downwards, producing a difference in the recorded magnification of specimen and scale. Dotted line represents camera attached to the microscope. Here the camera moves with the microscope tube when d^1, d^3 is constant, but the unattached camera registers a change in magnification.

as the distance D remains the same; d^2 changes. Again, a false reading of the magnification of the specimen will be made. Such changes in magnification from specimen to micrometer do not occur when a camera is attached to the microscope for the camera then moves with the objective. Compare A, B and C of the figure, the dotted lines indicating that the camera is attached to a microscope. Here the distance between objective and focal plane D remains the same. The change in distance from the plane of focus to the focal plane does not take place when the stage has a height adjustment, as fitted to the Leitz Ortholux and the specimen is moved up or down in the line of the optical axis.

When the power of the objective is multiplied by the eyepiece, the magnification is indicated, this being correct for a 10 in. bellows extension. The bellows extension can be termed as one unit (1) and any change in the distance above or below the 10 in. can be written in terms of 1: e.g. a 12 in. extension would be 12/10 and a 6 in. extension 6/10 and so on. Small cameras are not 10 in. from the eyepiece: therefore the magnification of the image on the

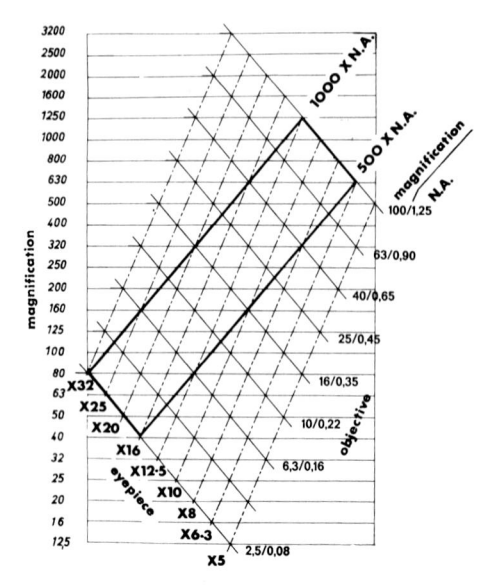

Fig. 87b.
Eyepiece magnification: ×32, ×25, ×20, ×16, ×12·5, ×10, ×10, ×8, ×6·3, ×5.
Eyepiece field-of-view number: 5, 6·3, 8, 10, 14, 18, 18, 18, 18, 20.
(wide angle)

negative must be worked out correctly, and subsequent enlargement to 7 × 5 is used to bring the image to a workable size. The exact number of times enlargement is noted (linear). This enlargement is then multiplied by the magnification of the negative image which will give the correct magnification. The equation to be used is as follows

$$\frac{\text{Tube length (mm)} \times \text{eyepiece}}{\text{Aperture of objective}} \times \frac{\text{extension (in.)}}{10}$$

$$\frac{160 \times 10}{4} \times \frac{10}{10} = 400$$

If a 35 mm camera is used it would be as follows

$$\frac{160 \times 10}{4} \times \frac{2}{10} = 80 \times 5 = 400$$

A five times linear enlargement of the 35 mm negative to a 7 × 5 print would bring the magnification to ×400, equal to that of a 7 × 5 print image obtained in 10 in. extension.

Secondary magnification. When making a secondary magnification (i.e. making an enlargement from the original negative) it is important to see that no noticeable loss in image quality occurs. Some seem to think that making a small overall negative image by low-power objective is the best route to a good enlargement but this is not always a success, as in Plate 28 illustrating human hair. Here is a damaged piece of hair, the total magnification of which is ×250. (Take a piece of your own, compare the size and see how fine a piece of human hair really is.) Plate 28 (a) was taken by using transmitted light, with a 25 mm objective and a ×6 eyepiece and bellows extended to 32 cm, which produced an initial magnification of ×50. This has been enlarged to give a magnification of ×250 as can be seen in the illustration. Plate 28 (b) was recorded with an 8 mm objective, ×8 eyepiece and bellows extended to 37 cm. No further enlargement was made. The two show two methods of arriving at the same destination. But what of the quality of (a)?

An apochromatic objective used with correct illumination will resolve lines

$$\frac{1}{90{,}000 \times \text{N.A.}}$$

in. apart. The image projected by the objective must undergo further enlargement by the eyepiece and, as already mentioned, it is sometimes yet further enlarged by extending the camera bellows extension; and then by the enlarger. It follows in many instances that the lack of resolving power in a print becomes apparent to the eye. If a 20 in. × 16 in. exhibition print is made the initial negative image *must* be correct. Near enough is not good enough. It must be remembered that as the numerical aperture increases it becomes increasingly difficult to obtain a critical focus over a large area. It does not always follow that an enlargement made from a negative image of low magnification is inferior and, by way of contrast, your attention is drawn to Plate 15, a much heavier subject with less lines to resolve. It shows the cell formation in a stem and the data for obtaining each of the four photomicrographs is

Table 9.

	A	B	C	D
Objective	16 mm	16 mm	Enlarged	8 mm
Eyepiece	×4	×4	from A	×4
Camera bellows extension	23 cm	48 cm	to	23 mm
Magnification	×48	×100	×100	×100

From these illustrations it is difficult to choose which is the best, (b), (c) or (d). Here there is not a vast difference in the resolving power and lines per mm resolved through the two objectives. Also, the degree of secondary magnification is not as high as in the plate illustrating the damaged human hair. The optical system used for the hair and botanical subjects is shown below.

Table 10.

Objective	Focal length	Primary image	N.A.	Approx. resolving power, μ	Lines/mm resolved	Approx. useful magnification
Low-power	25 mm	×6	0·21	0·25	588	150
Medium-power	16 mm	×10	0·28	0·75	820	200
Medium-power	8 mm	×20	0·45	1·23	1,333	350

Empty magnification. This refers to high magnifications which increase the overall size of the image, usually after enlargement, without enhancing the resolution of the subject detail. On the other hand, it is sometimes desirable to employ empty magnifications, since much depends upon the purpose for which the photomicrograph is required. A picture made for viewing in the hand is useless for viewing at a distance of 7 ft—as at an exhibition.

Illuminating the specimen

Subjects can be divided into two categories: those that are opaque, and those that are partially transparent, becoming transparent when placed in the light train of the microscope. As a rule, near transparent specimens are viewed and photographed by means of a light passing through from beneath the stage via the condenser (Fig. 88 F, E), (Fig. 89 A, B). If the specimen area is large, the substage condenser should be removed to allow a much wider cone of transmitted rays direct from the lamphouse. When the structure of a near transparent specimen is made visible, i.e. seen by visual observation or by projection onto a screen, the term transmitted bright-field illumination is used. If the refractive index of a near transparent specimen differs from the mounting medium surrounding it and virtually no detail can be seen, we can render structure visible by using darkground illumination. Further, if the specimen is of the same refractive index as the surrounding medium, and is barely visible by transmitted bright-field illumination, then perhaps the introduction of phase contrast (Plate 29) or interference techniques (Plates 30 and 79) may make visible the structure of the specimen, even if no differences in colour or density exist. X-ray is used when wishing to examine the structure of a substance (Plate 30). In certain cases specimens are stained causing them to fluoresce when viewed by blue light, while ultra-violet is restricted in its application. In the examination of certain crystals, minerals and others, polarized (converging bundles of sets of parallel rays) light is used, often displaying beautiful colours (Plates 17 and 73 upper). Opaque specimens can be illuminated by a free standing source (Fig. 89 C, Plate 31 and 74) which projects a beam

ILLUMINATING THE SPECIMEN

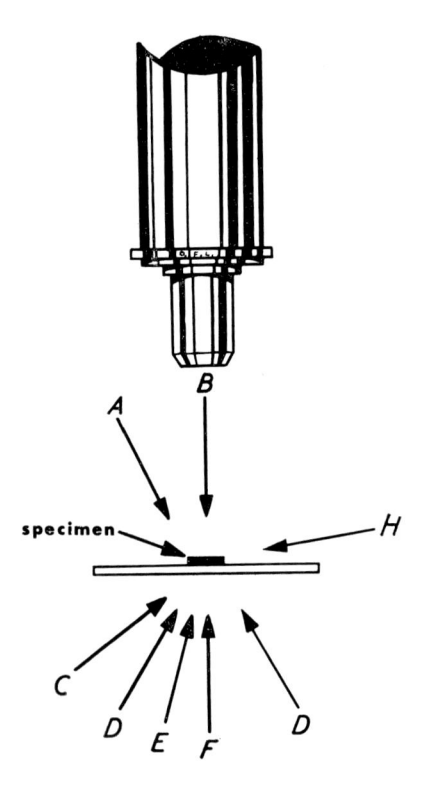

Fig. 88.

of light at an off centre angle, or by a built-in incident source directed onto, and reflected from, the plane of the specimen D of Fig. 89.

Illumination by transmitted light. In considering the illumination of the specimen by transmitted light, the characteristics of the optical system must also be taken into account. There are two sets of planes which are of importance in the formation of an image. One conjugates at the exit pupil of the eyepiece, and one at the actual plane of the specimen. They are brought about by what is known as the Köhler method of illumination.

In the case of a compound microscope, the exit pupil of the eyepiece reveals an image of the aperture of one of the objective lenses. The field of view is normally controlled by the aperture stop in the focal plane of the eyepiece and the primary function of the substage condenser is to ensure that the whole projected field be illuminated evenly. Between the top lens of the condenser and lower lens of the objective lies the specimen. A photomicrograph demands complete even illumination of the specimen. The cone of rays from the condenser should be of sufficient angle to fill the back aperture of the objective with

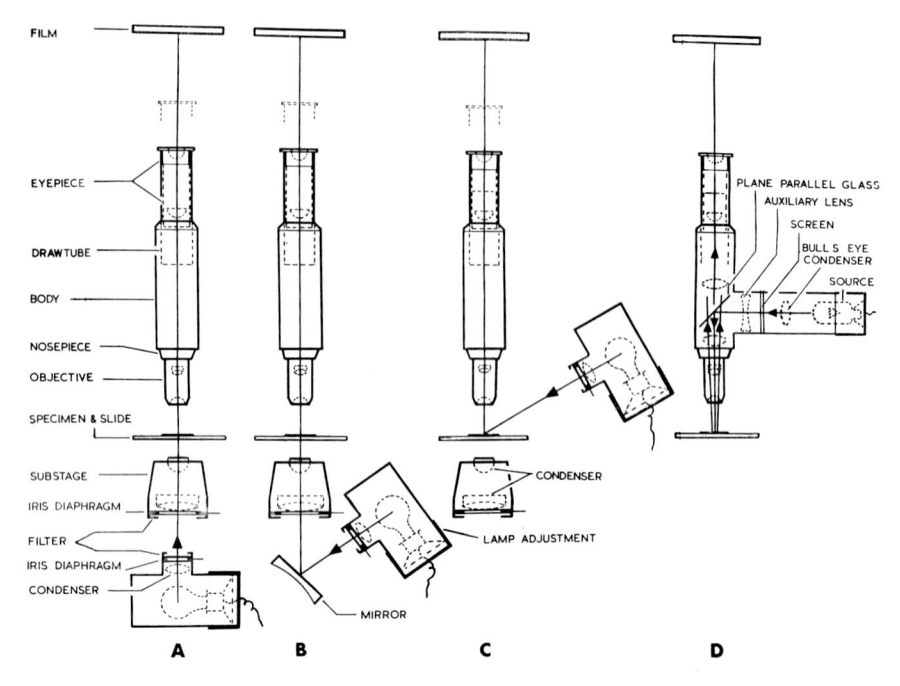

Fig. 89.

light but avoiding glare. In order to accomplish this the cone of rays should be capable of forming an image of the source, or of the illuminated iris diaphragm, and should be free from aberrations in the plane of the specimen. The cone of rays should be controlled by an iris diaphragm positioned immediately beneath the substage condenser and another in a plane near to the specimen, usually in front of the source in the lamphouse.

It is convenient for the substage condenser to be interchangeable. The two-element Abbe substage condenser has an N.A. 1·2. Nevertheless, care must be exercised when using it to illuminate a "large" subject. Extra care must be taken when recording in colour as there is no chromatic correction. If the top element is removed to provide a wider cone of light further adjustment will be necessary. The lower spherical surface of the lens is calculated to be of such radius that the object point must be positioned to lie in the aplanatic surface. Flare must be avoided if good results are to follow. Here, the substage condenser is the key. The three-element Abbe substage condenser N.A. 1·4 (Fig. 74) is of a higher standard but, odd as it may seem, it has no chromatic correction. I suggest that a blue or green filter always be used with these lens systems.

Top illumination. When investigating the appearance of opaque or semi-opaque

ILLUMINATING THE SPECIMEN

specimen, some form of illuminating apparatus is required to direct light downwards and sometimes around the specimen from vertical and oblique angles. Making a recording by reflection is not commonly exploited in the field of photomicrography and yet there is much opaque subject matter. Perhaps the lack of interest in this particular application is due to the difficulty of evenly illuminating a small opaque area, or overcoming the shallow depth of field offered by the optical system. Reflection of light occurs at surfaces where there is a considerable difference of refractive index between materials as, say, crystals and metal: seen more clearly when illuminated by lighting directed at an oblique angle, Fig. 90.

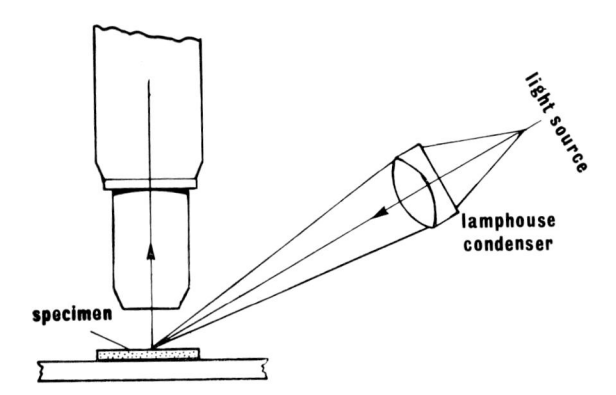

Fig. 90.

In the case of high-power photomicrography, where only a small area of the specimen needs illumination, light reflection is comparatively simple, since there are several fairly satisfactory types of vertical incident light illuminators available (Fig. 92).

Built-in incident-light illuminators are restricted in that, as a rule, they do not illuminate a wide field. Furthermore, the method can suffer from a defect in dealing with irregular surfaces (Fig. 91) in that there is very little relief of "plasticity" revealed. In other words, there is a false appearance of flatness (Plate 21b). A much better rendering of relief in irregular surfaces is obtained

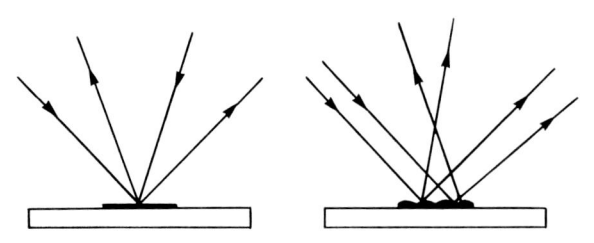

Fig. 91. Reflection from a smooth surface.   Reflection from a rough surface.

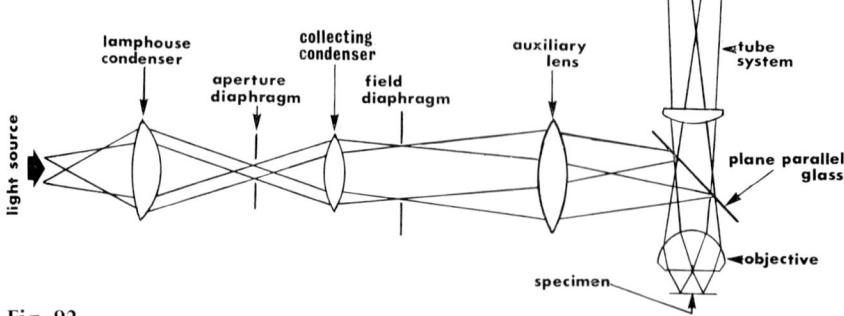

Fig. 92.

by the tungsten light ring illuminator (Preston, 1931) and the 1925 darkground condenser of Professor F. Hauser. Even so, the former leaves much to be desired as it is restricted in its ability to create full contours—light and shade. Darkground illuminators are still to the fore and capable of illuminating a variety of subjects. Ring illumination can only be used with small specimens of about 1 cm in diameter, which cannot be moved far from the central position without obstructing the light. The light from a flash ring illuminator is too even and fails to implant necessary highlights and shadows—modelling. Instead, the entire irregular area of the specimen is evenly illuminated and a roundness is lost. Of course there are many subjects suitable for ring illumination: it is a matter of using the right tool for the job.

The parabolic mirror, marketed by Gillett and Sibert (Fig. 93), is an achromatic mirror objective, ×4·7 0·12 N.A. and very easy to manipulate. This piece

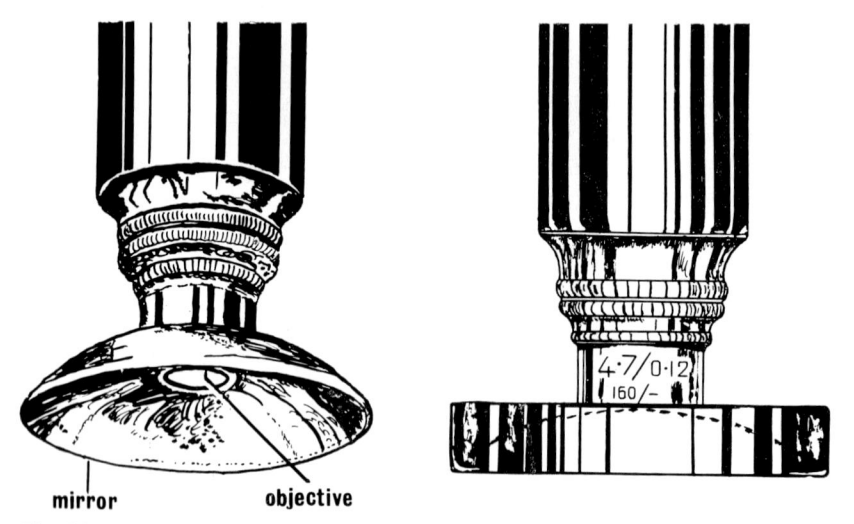

Fig. 93.

ILLUMINATING THE SPECIMEN

of equipment operates with only one focal length objective which limits its usefulness. Figure 94 illustrates a similar application. (Nikon also market a similar mirror with a 65 mm and a 35 mm objective.) The principle (as shown in Fig. 94) is that the specimen is positioned above an opaque disc and incident parallel light passes through the glass microscope slide. On striking the underside of the reflector the light is redirected to provide bright, shadowless lighting over the entire surface of the specimen. The Hauser instrument (Fig. 95) differs in that it causes the incident light to be redirected twice, before illuminating the specimen by means of scattered light at a more grazed angle. No actual light rays enter the objective. Nikon also market the Half-Reflecting objective, in focal lengths 65 mm and 35 mm, used for on-axis shadowless illumination.

Low-power objectives have restricted angular aperture and, if consideration is not given to this and to the area covered, images of the specimen are likely to be recorded from different points of view. The two opposite extremes of the marginal rays will result in a photomicrograph that has a number of dissimilar

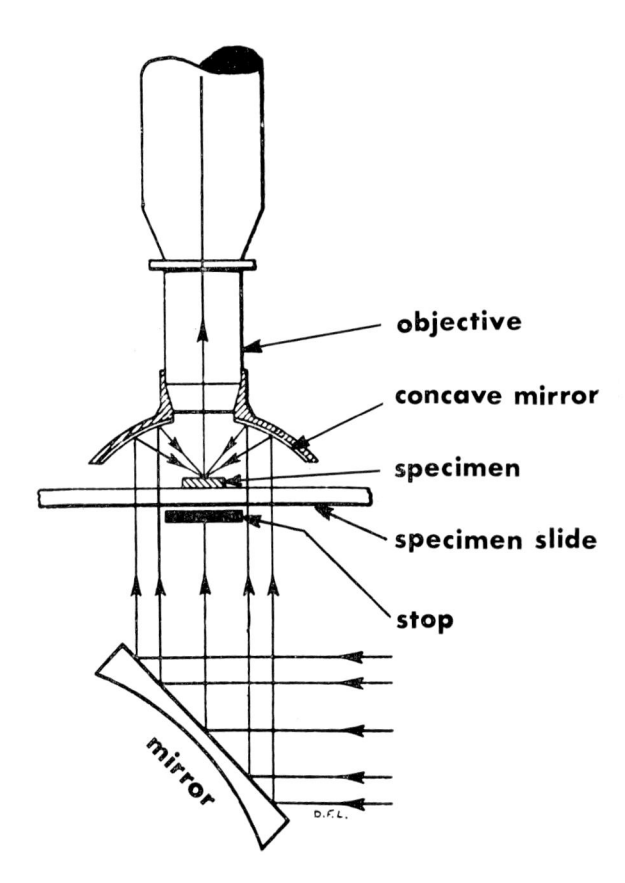

Fig. 94.

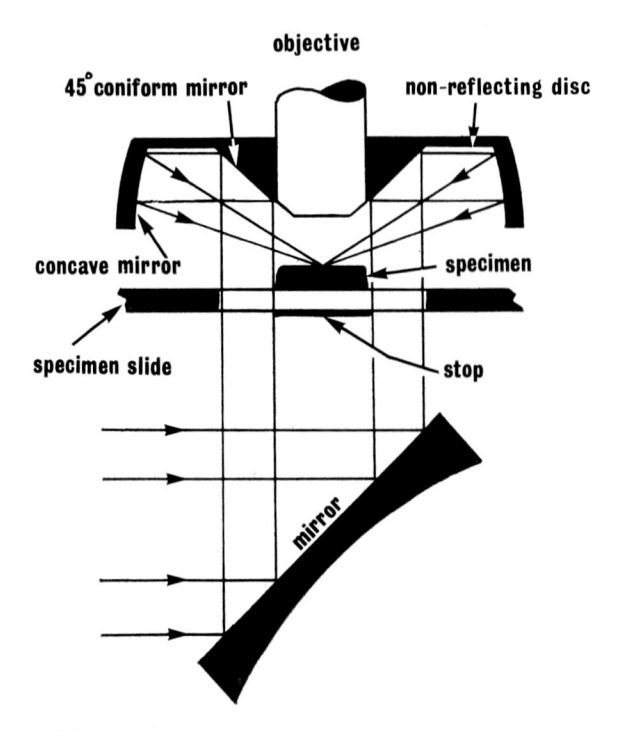

Fig. 95.

perspectives of the specimen recorded together. Some recorded large transverse sections exhibit a two dimensional or stereo effect at the extreme ends, revealing the subject's depth as if it were tilted when photographed, Plate 32. Perspective and angle of view are particularly noticeable when recording a large specimen by means of a series of photomicrographs, forming a line overlap (Fig. 144) or mosaic, for here it will be found that only the centre area of the picture can be used, ensuring a true overlap of angle of view and magnification. In the fields of low-power photomicrography of opaque objects the illumination must be much more adaptable than in the case of high-powers, because of the much greater variety of surfaces which require investigation. It is in this respect that most illuminators fail.

The suitability of illumination should be judged by the following considerations:

(a) The area of illumination must be uniform over a comparatively large field, Plates 33 and 34, Fig. 88 and the Frontispiece.

(b) The angle of incidence of the source directed upon the specimen must be variable between wide limits, so that it can be used to advantage for both regular and irregular surfaces, Fig. 88, Plates 31, 35 and 60.

(c) Independent lights should be capable of height adjustment, making provision for near vertical specular incident illumination, Fig. 96 and Plate 39.

ILLUMINATING THE SPECIMEN

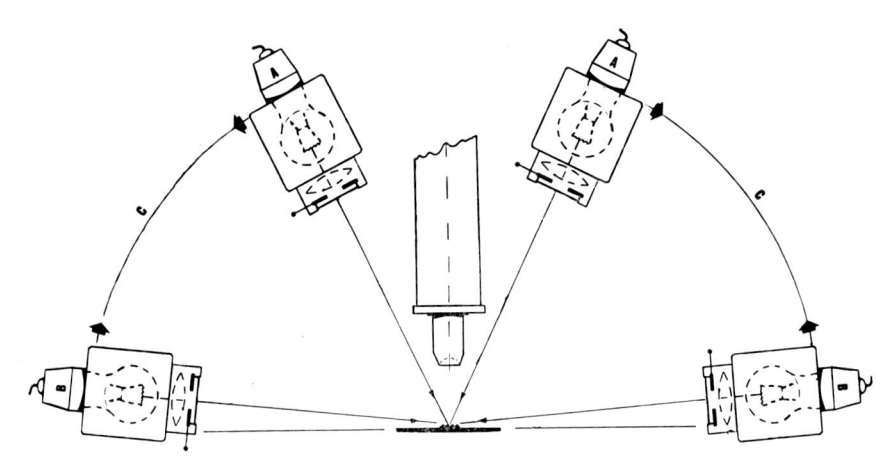

Fig. 96.

(d) A *good* condenser and an adjustable iris diaphragm are necessities, Fig. 12 and Plate 17.

(e) The lamp should provide a means of focusing the source.

(f) There should be provision for a filter.

(g) The lamp must be capable of being operated from all sides of the specimen and on occasions more than one lamp is necessary, as is a reflector.

(h) A rheostat is a useful means of controlling the light output.

(i) The specimen should not be limited in size and, further, it must be possible to move it in a controlled way so as to bring different areas into the field of view.

(j) The source should be of low voltage and wattage (see table on page 147), to free it from heat which could damage the objective and specimen. The type of lamp most suitable for illuminating an opaque subject is illustrated in Figs 90 and 93. (The photomicrographs in Plates 2–31 demonstrate the usefulness of these lamps, as do the colour Plates 73 lower, 74 upper and 70 upper.)

(k) A of Fig. 96 is the minimum angle of incidence of the light related to the optical axis of the microscope; and C is the range between maximum and minimum angles of incidence. The maximum angle B can be referred to as grazed illumination. In general, the greater the angle the more slight inequalities of the surface will be shown, Plates 21(c) and 35(c), and the greater will be the relief seen in the specimen. A case in point is Plate 21(b), where vertical incident illumination does not convey the depth on the subject matter.

Of course the shorter the focal length of the objective in use, the more inclined will be the angle of the light rays with or without a direct source. Care must be taken to prevent incident rays from being redirected into the objective.

The effect of a single source where the illumination is low and from one side will be to present a false appearance, or over emphasise shallow stratum. However, it can be used to exaggerate certain features for identification purposes. Then it is advisable to demonstrate the effect by the use of two methods of illumination, which brings us to illumination azimuth.

Illumination azimuth may indicate the angle subtended by the projection of the aperture angle of an illumination cone to the focusing plane. Thus, illumination azimuth is the angle measured in the focusing plane, between a preferred direction determined by the objective and the principal direction of the incident light. If lines or scratches on a specimen are illuminated in the direction of the lines, as in "optical staining" page 132, the illumination azimuth will be 0°, Fig. 97. An illumination normal to the lines, corresponds to an illumination azimuth of 90°. In the case of direct illumination, azimuth is 360°. When the diaphragm obstructs the annular cone with central angle, the angle will have the value of the azimuth of illumination. Azimuth effects expresses strong unilaterally increased contrast contours: depending on the direction of the contours in relation to the incident light, usually at right angles.

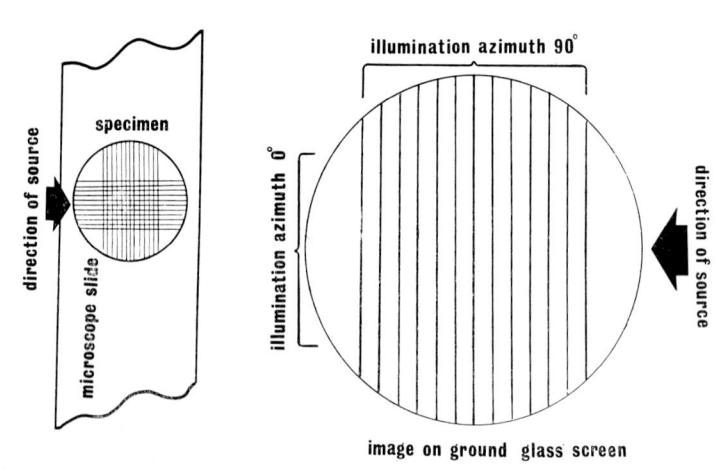

Fig. 97.

The disadvantages of shallow depth-of-field expressed in an objective is strongly evident when an image is produced at a relatively low magnification and in particular by reflected light. Subjects with considerable depth-of-field are seen in Plates 21, 31 and 33. It is not generally accepted that such subjects are more difficult to record than thin histological sections photographed by medium and high power objectives incorporating transmitted illumination. It is desirable that such recordings for technical purposes should exhibit sharp depth of, say, 3–4 mm which will produce an image up to approximately ×15 linear. Here definition refers to resolving power, grain, sharpness, tones and detail.

Depth-of-field is determined by the aperture. In the application of photomicrography, the size of the aperture affects both depth-of-field and resolution: in particular the latter because a change is brought about in the N.A. Improvement in one is at the expense of the other and therefore an aperture should be used which is compatible with accepted resolution.

One must also have a correct balance of highlights and shadows created by the use of an independent light source. Before exposure, contours and highlights can be studied in the viewing eyepiece or ground glass screen of the camera. So can the angle and height of the light beam as it is changed, or suitable material is interposed to reflect a portion of the light back to fill in especially if the shadow areas are too dark. Control over depth of field and contrasts is necessary as light and shadows are apt to be accentuated in this type of photography. B of the figure has a depth of 6 mm when the iris diaphragm and the optical system used were reduced to one-third the full aperture. Any further reduction would mean a loss of definition. The out-of-focus area is accepted by the eye because it is not part of but only allied to the subject.

Oblique transmitted light. As already indicated, great care should be exercised in the alignment of the optical system. Even so when finally adjusted a surprisingly disappointing photomicrograph, such as (a) of Plate 37, is sometimes recorded. Although everything required by the textbook had been carried out, (a) lacks image quality and is flat and uninteresting. This is brought about because the refractive index of the crystal is approximately the same as the mounting medium in which it is embedded. In its most simple form, the compound microscope consists of only two compound lenses, the objective and the eyepiece: the former receiving the rays of light direct from the specimen via the source, Fig. 89A. By custom a substage condenser is placed beneath the specimen but this is often unnecessary. In fact, on occasions, it does not assist in producing an overall clearly delineated area. Microscopes may be turned to good account in more than a standardised mode; but there are limits to the use which can be made with advantage. A modification must be used with caution.

Unusual conditions often exist when a specimen is suspended in a fluid, or fixed in a mounting medium as in Fig. 98. In order to bring the subject to life, and to produce an image in relief from its background the source was directed at an off-centre angle (Plate 37b). Figure 98 demonstrates this. A substage condenser can be employed, but with caution. The source no longer transmitted a pencil beam of light in the line of the optical axis, but instead a series of rays reflected from the edge of the system. This has illuminated each individual crystal from an oblique angle by means of scattered light. Comparisons can be made between Fig. 96 *A* and *B*. Whenever such a technique is introduced in the production of an image it is always advisable to include both methods of producing an image, as set out here.

Owing to the fineness and transparent nature of some specimens (Plate 4d),

102   PHOTOMICROGRAPHY

it may be necessary to illuminate from an oblique angle. Oblique angle could mean anything between 0° to 90° and one's results can vary as do the number of degrees of tilt. If the angle is as much as 70° (Fig. 99) it may result in a misrepresentation of the subject in which a double image, especially with a thin specimen, may be recorded, resulting in lack of definition. An angle of 30° or

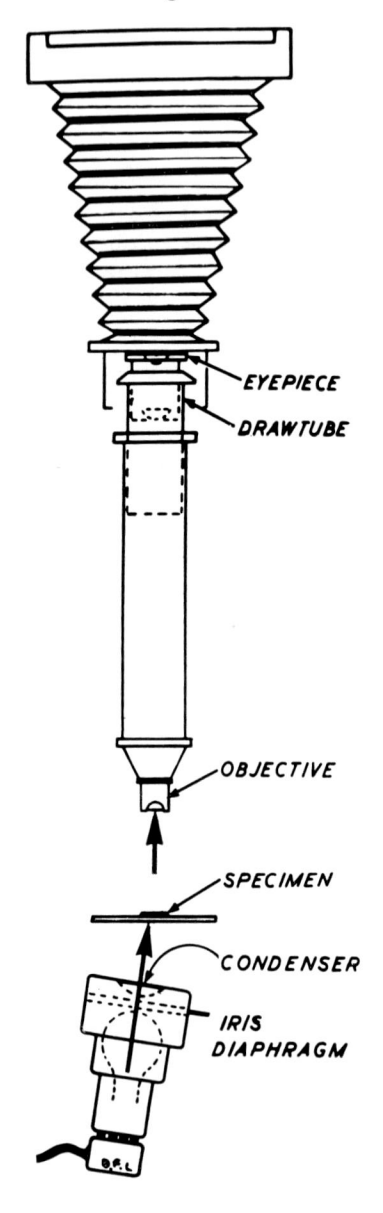

Fig. 98.

ILLUMINATING THE SPECIMEN

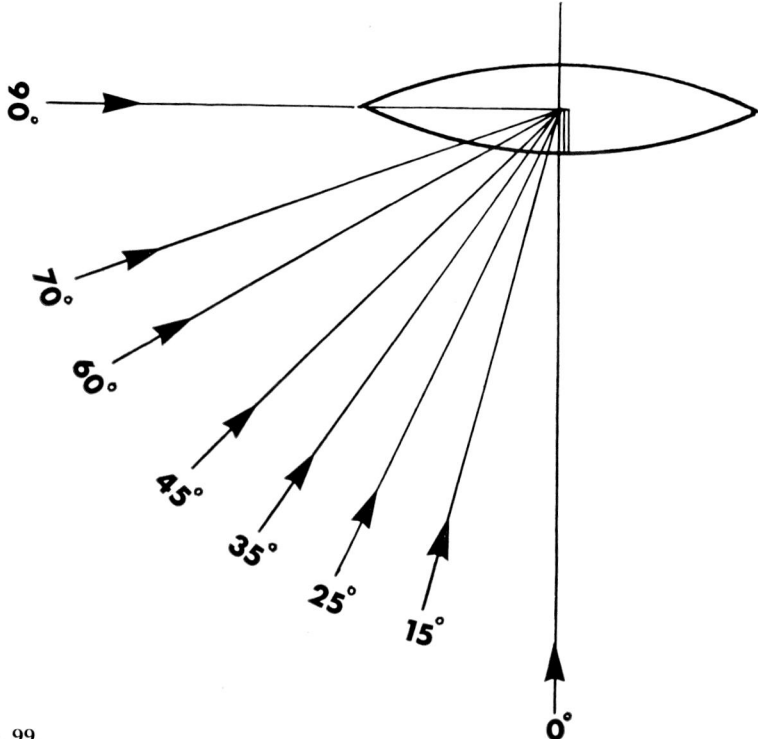

Fig. 99.

25° will enable fine transparent detail to be more clearly resolved than at other angles. A further improvement in the image quality can be achieved when incorporating a blue filter (365 nm).

Oblique illumination with a yellow mercury line would resolve, according to Abbey's formula, a distance of

$$\frac{\lambda}{2 \text{ N.A.}} = \frac{577}{2 \times 1\cdot 3} = 0\cdot 22 \; \mu$$

but ultra-violet will show a great improvement over this figure. R. Neuhauss (1907) and Köhler (1909) obtained excellent results with oblique ultra-violet radiations at wavelength 275 nm. The greater the angle of the transmitted light, the higher must be the transmission (nm) of the filter used. Even illumination should be achieved, revealing all anatomical details.

A behind-the-lens iris diaphragm (the Davis shutter is not now available) can be employed with a low-power objective, but if a small aperture is introduced it can lead to the misinterpretation of appearances peculiar to the specimen, even when the very best optical system is in use. Thus, the sharpness of the outline of the transparent specimen could be impaired by a change in the course of the rays that pass by its edges, producing diffraction (Fig. 46). It is obvious that some slight diffraction must exist in the rays transmitted through the specimen, as well as along the shadow edges. If one is not cautious, such rays will interfere with the perfect distinctness, not merely of their outlines but of their images and the various markings. On the other hand, the presence of vertical transmitted light may fail to reveal an actual shadow substance. In addition to the possibility of diffraction bands, there is a peculiar phenomenon when applying oblique illumination. At certain angles and in one direction double images are produced, or a kind of overlaying shadow of the specimen or a part of it. This image is not unlike a secondary image formed by reflection from the outer surface. A specimen is thus illuminated by two different sets of rays, transmitted light which fan through it obliquely from the source of the illumination to the opposite side of the objective, and those of the radiated light, which being intercepted by the specimen, are given off from it again in all directions.

This method is of great importance in the demonstration of certain structures and, on occasions, a greater obliquity is required than can be applied by directing the source at a slight angle. One of the very earliest methods devised for obtaining off-centre "transmitted" light, was the eccentric prism of Natchet which was interchangeable with the microscope condenser. By rotating the eccentric system oblique rays were directed upon the subject from all sides.

Incident-light illuminators. The incident-light method of illuminating the surface of a specimen is of great value and is widely adopted today in the examination of flat surfaces at both low- and high-power magnifications. It is used in the study of metals, ceramics, fibres and minerals. In the pharmaceutical field, one of its uses is to examine the surface of tablets.

These illuminators produce an axial beam of light. When this is reflected from a plain polished surface, any pits or scratches upon the surface of the specimen will appear dark. Vertical illuminators are readily interchangeable and fit between the body tube and the objective. The light enters the illuminator at right angles to the optical axis of the instrument. Special objectives with a condenser built annularly around them are available from several manufacturers and will fit any microscope. A built-in source is usually supplied with these objective–condenser lenses, known as Leitz Ultrapak. The condenser can be bought on its own and is known as Agnee condenser. A small built-in source is available for visual purposes and, by the manipulation of a flip-top device, a much stronger source is brought into operation for photography. This optical system can illuminate any opaque subject and its use in the research laboratory is unlimited. The dry wide-angle model gives a grazing incident beam so the illuminated surface of the specimen is recorded in relief (Plate 39).

The mirror type of incident-light illuminator shown in Fig. 100 contains a small metal mirror at an angle of approximately 45° to the optical axis. This

ILLUMINATING THE SPECIMEN

Vertical illuminators
(a) mirror; (b) prism; (c) glass slip

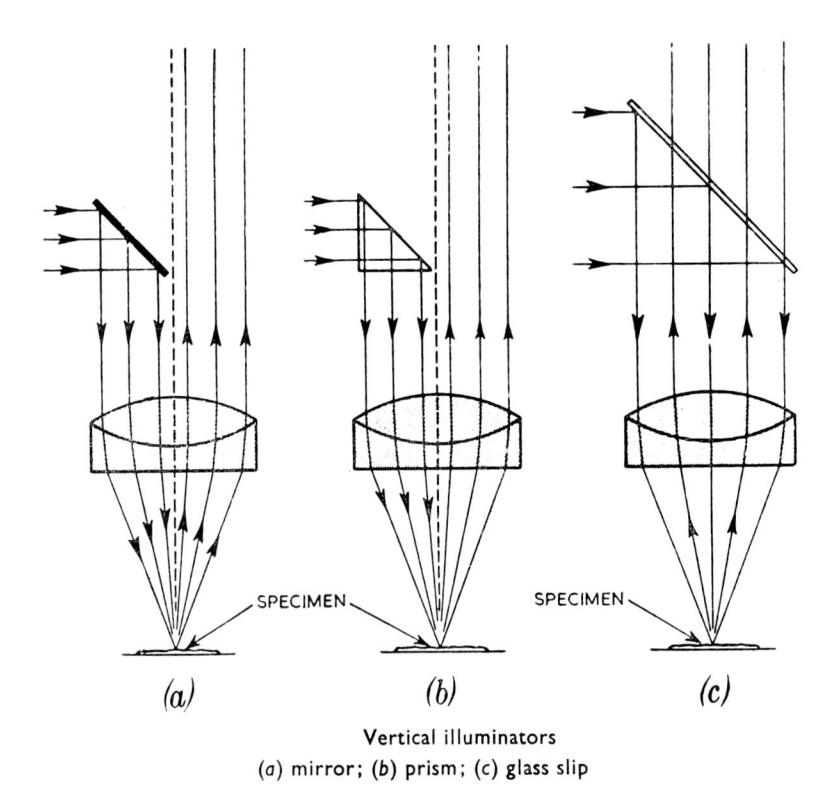

Fig. 100.

directs the light beam through the objective and onto the surface of the specimen. A total internally reflecting prism may be used in place of the mirror as shown in the figure. Both these types of illuminator suffer the disadvantage that the light incident upon the specimen is not perfectly parallel to the optical axis of the microscope. A more serious objection is that the obstruction which they produce reduces the numerical aperture and the resolving power of the objective by 50 per cent. These limitations are avoided when a thin cover glass is used at 45° to the optical axis, just behind the objective (Fig. 100). This acts as a reflector to illuminate the specimen via the objective and then transmits a proportion of the light reflected back from the specimen. The Beck illuminator is of this type and enables the full numerical aperture of the objective to be achieved, together with truly axial illumination. To enable a greater depth of field to be obtained when operating with low-power incident–light objectives, a detachable behind-the-lens iris diaphragm is now available.

The Beck–Chapman opaque illuminator (Fig. 101) is entirely portable and

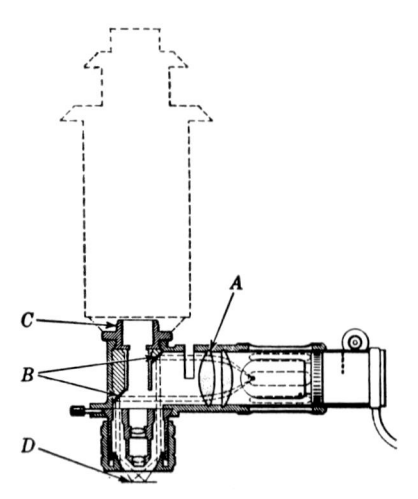

Fig. 101.

easily attached to the base of the body tube after first removing the objective. The image is viewed and photographed in the normal manner. There are many subjects for which this apparatus can be usefully employed. The construction of the illuminator is shown in the figure, in which it can be seen that the condensers at A transmit the light from a 6 V 15 W bulb. Only the marginal rays B are used, and these are focused as an annulus of light on to the surface of the specimen D. A central circular stop removes any light which would suffer direct reflection back into the objective from a flat surface. In this way a darkground effect is produced. The general result is of contrast as compared with normal vertical illumination. C replaces the objective.

The Wrighton–Beck Metallurgical illuminator shown in Fig. 102 can be attached to any microscope for work by vertical illumination. The light from A is controlled by a diaphragm C set behind a condenser with one ground surface B. The illuminator screws into the microscope between body tube and condenser. At a lower point within the tube is a further condenser D and iris diaphragm E. The image of the lens is formed at the same distance from the reflector H as the image formed by the eyepiece, to ensure critical illumination. Lens G assists in forming an image of the light being redirected by mirror F close to the back lens of the objective at J. Most work in reflected light is conducted without a cover glass over the specimen and it is important that high-power objectives are suitably compensated.

Reference

Preston, J. M. (1931). A New Top Light Illuminator. *J. Roy. Micros. Soc.* **51**, 115–118.

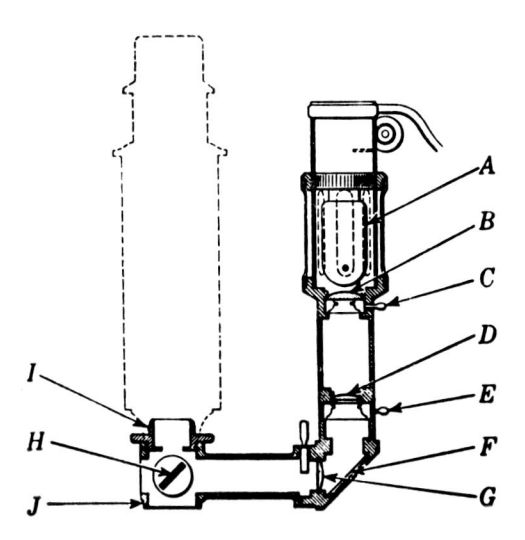

Fig. 102.

Balphot metallurgic microscope

Perhaps one of the revolutionary changes in design in microscopes is the inverted microscope (Fig. 238). Such a microscope enables the surface of the specimen to be prepared for specular reflection, as the specimen is lying face downwards, and the objective pointing upwards, as if it were replacing the substage condenser in the vertical position. Such objectives are corrected for infinity and for use *without* a cover glass. In addition, they are also corrected for a much longer tube length, to that of infinity, whereas standard bench microscopes are corrected for 160 mm and 170 mm. This design introduces added optical systems for the collimated image beam. However, the final image is not degraded and is on a par with that formed by a converging beam. The optical system can be followed from Fig. 103.

Polarizing vertical illuminators are obtainable from a wide market. The Bausch and Lomb model is small but effective, the main beam traverses a special calcite prism before entering the objective (Fig. 104). The beam is divided: one, a now image forming ray, is polarized perpendicularly as for calcite (Fig. 173); the other is reflected at the cemented surface, the extraordinary beam passing through the calcite prism, a quarter-wave plate and objective, to the specimen. On the return, the image forming beam again passes through the quarter-wave plate, thus the beam is totally reflected by the calcite prism and is directed at the silvered surface and directed to the eyepiece. Exposures will be fairly long due to the large amount of absorbed light within the system.

E

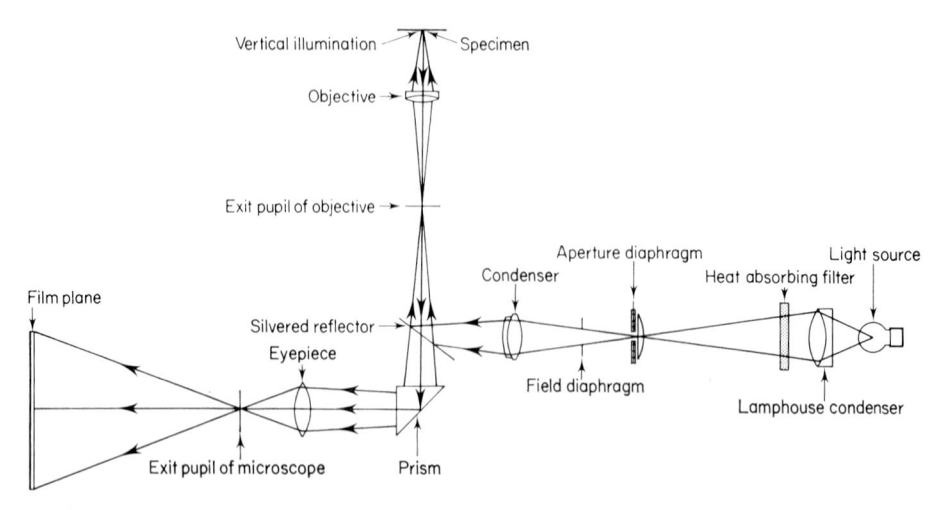

Fig. 103. Optical system of Bausch and Lomb. Balphot vertical illumination for metallography.

Specular highlights

When illuminating opaque subjects by reflected light there is a danger of creating specular highlights and, even by moving the source, little improvement can be made. On the other hand, specular highlights are sometimes necessary to illustrate a point, as in Plates 31 and 39. However, in some instances unwanted highlights can be eliminated by using a polarizer such as a Pola-screen, or Polaroid filter. This can be placed in the filter carrier in front of the lamphouse condenser or in a similar convenient place in the light train. If there is no provision for housing it can be positioned behind the microscope objective lens and rotated to the azimuth excluding the highlights. If a large amount of work is to be done, a makeshift method becomes trying and it is, therefore, suggested that the microscope be fitted with a good quality polarizer in the *illumination beam* and an analyser in the *image forming beam*.

When inspecting certain metals, the vertical illuminating beam is placed in the axis of the viewing beam, thus creating specular highlights. The viewing axis cannot be changed but the illuminating beam can be moved to an oblique angle by incorporating a reflector at an angle of approximately 45°.

Sometimes specular highlights must be introduced when photographing small faint zones of inhibition on the surface of agar. To the uninitiated in the biological field this result is unobjectionable though, unfortunately, often met with scorn.

Kohler illumination: incident light

Incident light Köhler illumination necessitates a different arrangement to that

KÖHLER ILLUMINATION: INCIDENT LIGHT

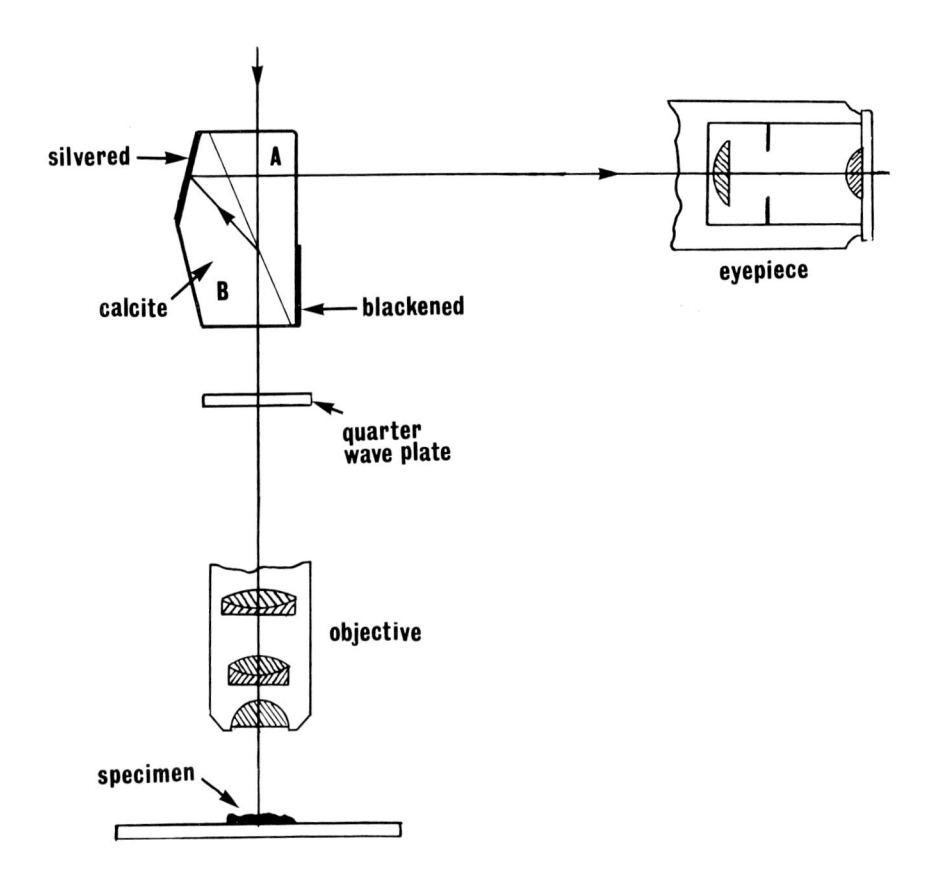

Fig. 104. Polarizing vertical illuminator, Bausch and Lomb.

of Köhler transmitted light and both require an optical system of a high standard.

A standard illumination arrangement has been devised (see Fig. 105) enabling a regular high quality recording. The built-in system is adjusted as follows: 1. The light source is imaged on 3, the aperture diaphragm through a condenser seen at 2. The whole of the condenser is used transmitting a maximum cone of light. The optical system, 4 and 6, image the aperture diaphragm to 8, a plane–parallel glass placed at 45° angle: this directs the image into the back lens of the objective 9. The field diaphragm is imaged by the optical system 6 at infinity, used by the objective in the specimen plane 10. The infinite image distance of the objective images the specimen in the focal plane of the eyepiece. Focusing is carried out with a near closed field diaphragm 3, when the diaphragm outline will be seen in the area of the specimen. Having focused, the diaphragm 3 is opened until the specimen is completely illuminated.

Fig. 105.

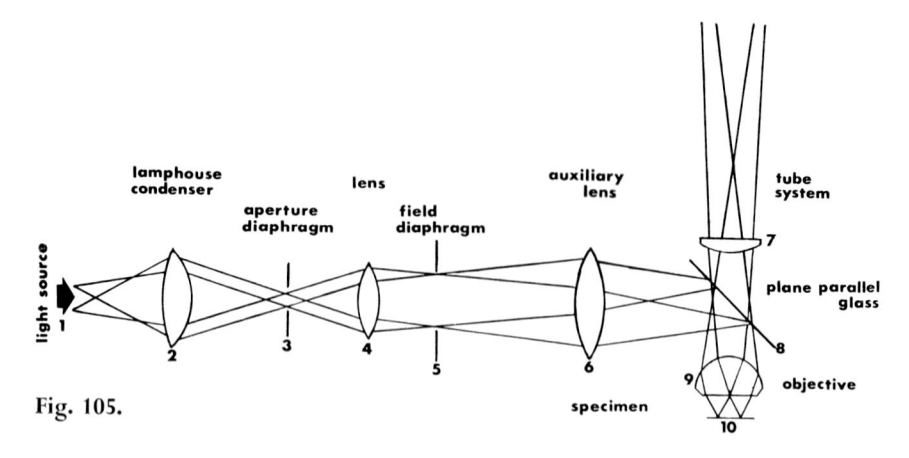

Some independent lamps

The range of free standing lamps and associated equipment available is now almost unlimited, and from these the photomicrographer must select the particular system best suited to his needs. Further, he must select a source which will provide several methods of illumination, such as darkground, phase-contrast, bright-field, fluorescence and polarized light.

An efficient and popular independent type of illumination is the Versalite Lamp (Plate 36a). The lamp is a 6 V 15 W and is carried in a precentred lamp holder, which can be focused to produce an image of the filament, enabling Köhler condition of illumination to be applied. There is a condenser system and iris diaphragm, and also provision for 5 cm × 5 cm (2 in. × 2 in.) filters. Plates 13 and 31 demonstrate the success one can have with similar lamps. Figure 106 demonstrates another. A higher wattage, 6 V 48 W lamp, such as the Tenslite (Beck), is also free standing and the intrinsic brilliancy, when at full voltage, is adequate for most techniques. Plate 36(b) demonstrates a small, unique and efficient source, 6 V 15 W lamp. The lamp fits beneath the substage condenser and is run off a transformer as are most lamps. Figure 107 illustrates a tungsten lamp.

One main feature in the success of the picture is that all lenses are carefully aligned. With nearly all up-to-date microscopes the centering of the source is made possible by the very construction of the apparatus, which involves built-in illuminating systems. Centering the cone of rays is a necessary function and not something to be quickly glossed over.

Lamp adjustment

When using a large "pearl" bulb of the domestic type, no special care is necessary in orienting the source to the mirror, or substage condenser. The

LAMP ADJUSTMENT

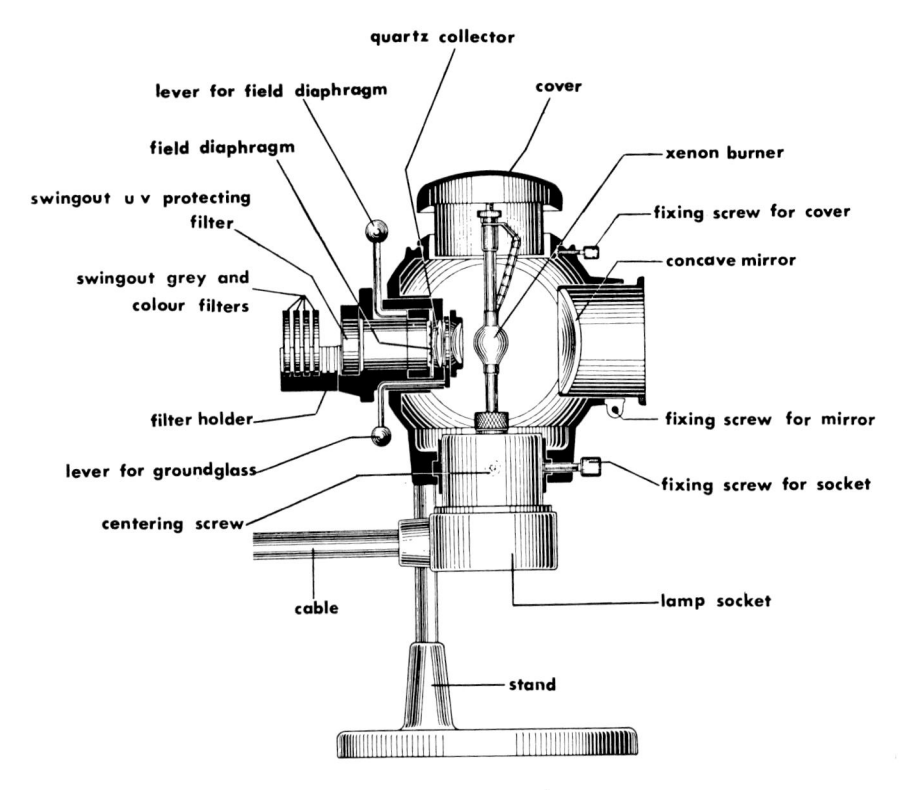

Fig. 106. Wild Universal Microscope Lamp, xenon burner.

source must not be too close, because of the intense heat dissipated. It is generally accepted that the source is not really suitable for serious work such as photomicrography.

The source placed behind an optical system should produce a controlled, evenly illuminated surface of high intensity, and should be large enough, when imaged by the lamphouse condenser, to fill the field diaphragm. Low voltage bulbs, such as the 6 V 15 W and 6 V 30 W are used because of their usefulness to do just this. In addition, there is no excess heat to cause alarm. The filament area is small and requires a lamp condenser to enlarge and pass on the intensity. The source is centred (Fig. 108) by projecting a beam on to a white card and focusing the image of the small ribbon filament which, if correctly adjusted, should image the filament indicated by a faint circular image of the lamp iris, which is now wide open. The filament image should be in the centre of the iris diaphragm image.

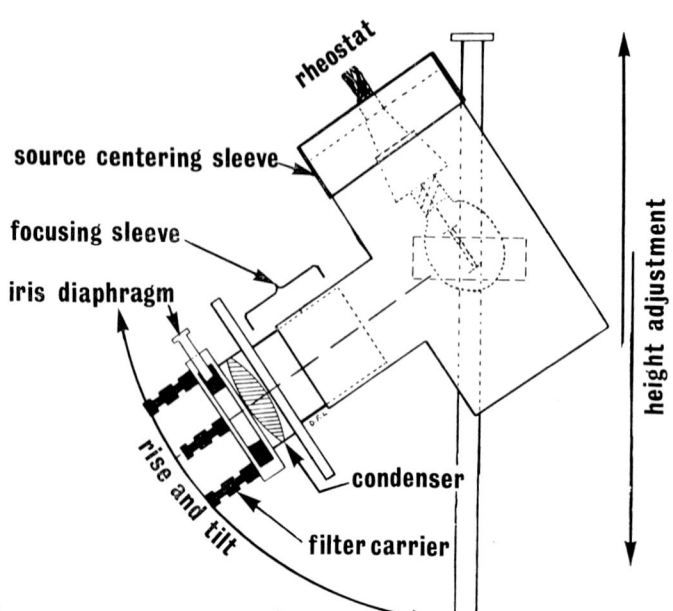

Fig. 107.

Centering the condenser for Köhler illumination in transmitted bright-field

The rotating stage may need to be adjusted so that it is centred to the body tube. This is done by placing a centering slide (Fig. 109) on the stage, focus on the centre, and by rotating the stage, the centre point should remain in the same position during rotation of the stage. If the source is too bright a piece of ground glass can be placed in the light train to reduce glare.

The substage condenser and iris diaphragm are one unit and, as a rule, are centred by the manufacturer. The supporting collar ensures that the complete unit can be returned centrally, in the line of the optical axis of the microscope. Condensers used for darkground illumination and phase contrast techniques usually have centering adjustments. The substage can be raised to a short distance from the underside of a slide. By opening and closing the field diaphragm we can see if the aperture is concentric with the particular projected field. A simple control is easily carried out by looking at the diaphragm image in the exit pupil of the objective after removing the eyepiece. When the diaphragm is correctly positioned it will occupy the area of the exit pupil of the objective, as seen in Figs 110 A and 111 (a).

When removing the top lens of the condenser to allow a wider cone of rays to illuminate the specimen, the condenser *must* be refocused or uneven illumination will result. It must be remembered that by removing the top lens the focal

CENTERING CONDENSER FOR KÖHLER ILLUMINATION

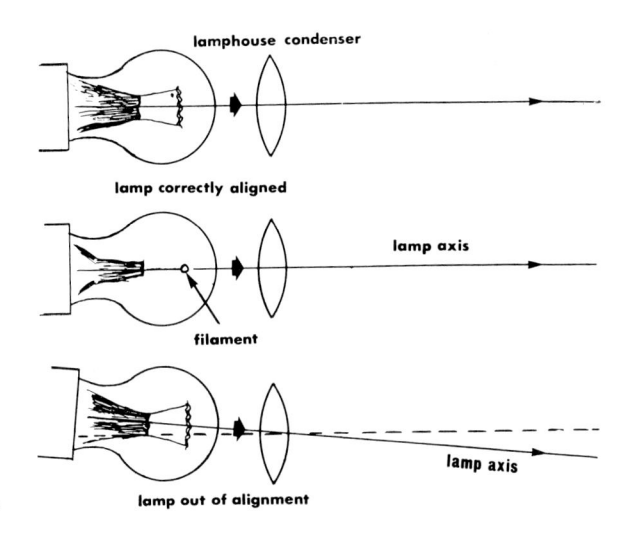

Fig. 108.

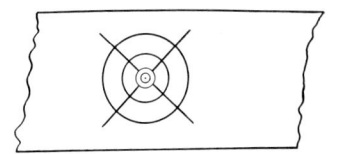

Fig. 109.

length has changed from short to long. Further, it is a bad practice to lower the condenser because the light from the top lens does not fill the specimen area and an unevenly illuminated field will follow, as in Fig. 110 C. A worsening of the image is created if a small aperture is used and, with the eyepiece removed, the back of the objective is seen, as in B. When the condenser is too close to the specimen the objective is not filled with an even cone of light but instead a ring is formed in the back lens, illustrated in D. The conditions of B, C and D create uneven illumination and poor image quality.

The Köhler condition of illumination should be applied whenever possible (named after Professor August Köhler who first introduced it while at Zeiss, about 1900). Köhler illumination requires minimum size illuminant so optimum light performance can be reached with a given optical system. The size of the light source aperture is controlled by the particular objective in use—low- or high-power. This in turn will control the aperture of the substage condenser which must equal that of the objective, Fig. 71. It must be borne in mind that the plane of the source should be imaged in the entry pupil of the microscope before maximum illumination quality is achieved in the exit pupil.

Fig. 110.

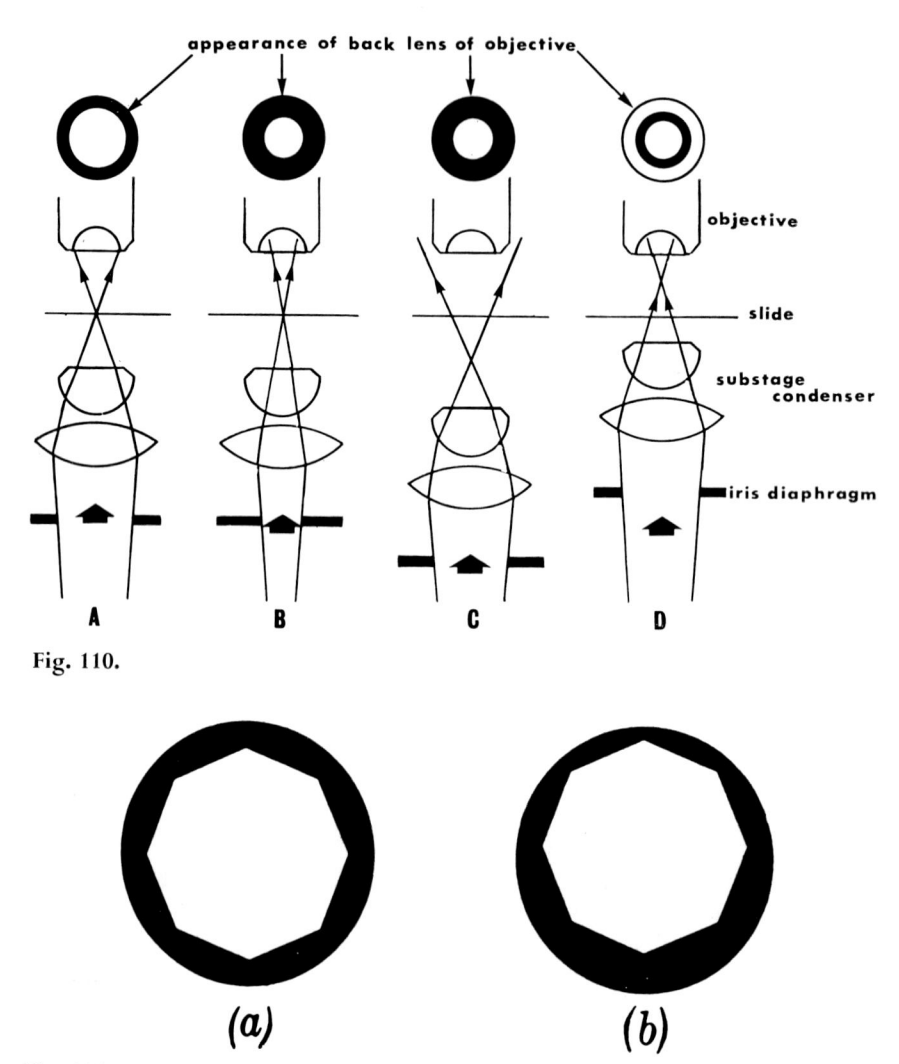

Fig. 111. Image of Iris diaphragm formed in back of objective lens. (a) In critical alignment. (b) Out of alignment.

Köhler illumination in transmitted bright-field

The series of adjustments which lead to the Köhler method of illumination is illustrated in Fig. 112. Here a mirror is demonstrated, as well as a direct light source.

(1) Set the microscope in front of the lamphouse with some 20 cm (8 in.) between lamp and mirror or 25·5 cm (10 in.) between source and substage condenser,

KÖHLER ILLUMINATION IN TRANSMITTED BRIGHT-FIELD

Focal plane image of field diaphragm

Exit pupil of microscope image of filament and aperture of diaphragm

Eyepiece

Intermediate image eyepiece field diaphragm

Back focal plane of objective image of filament and diaphragm

Objective

Specimen plane image of field diaphragm

Substage condenser

Lower focal plane of condenser diaphragm image of filament

Entrance pupil of microscope

Mirror

Field-diaphragm

Lamp-house condenser

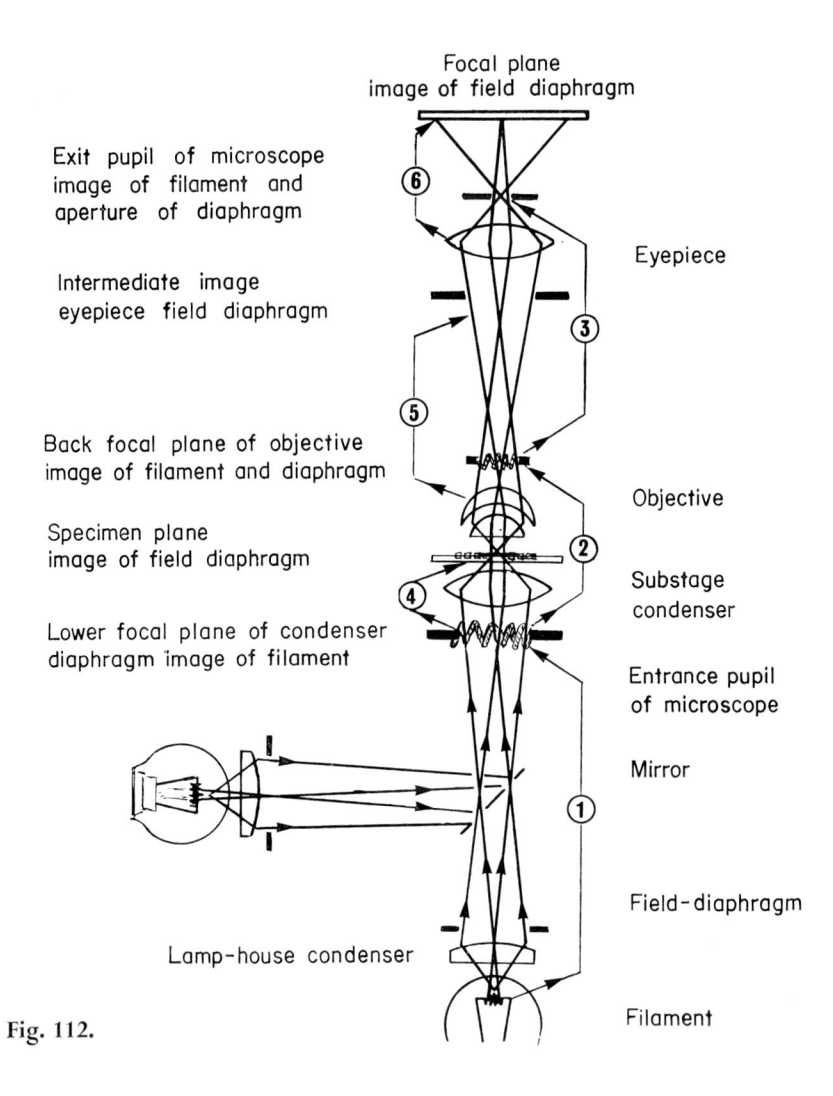

Filament

Fig. 112.

if a direct source is used. Use the plane side of the mirror and a 25 mm objective. The filament of the source is centrally imaged by the lamphouse condenser (Fig. 108) in the aperture diaphragm plane of the substage condenser. This can be termed the entrance pupil of the compound microscope. With the Köhler method of illumination it is important that the image the light source formed on the back of the condenser should be large enough to give a cone of light at the required numerical aperture; and hence for high power work a short focal length condenser is preferable. The focal length of the condenser increases as the specimen area increases. It must be possible to centre the condenser laterally so that it is in

E*

line with the optical axis of the eyepiece and objective. Adjustment of the iris diaphragm in the line of the optical axis is also desirable (Fig. 113).

(2) After passing through the condenser and objective a second image of the filament of the source is formed in the centre of the back focal plane of the objective. The diaphragm image (Fig. 111) is best observed after removing the eyepiece and, when opening and closing the iris diaphragm, the bright circular field seen in the objective changes in diameter (Fig. 112, 3).

(3) A third image of the filament is formed, this time by the eyepiece, in the exit pupil of the microscope. After the completion of these three important steps (Fig. 112, 4), the field diaphragm of the lamp is imaged by the condenser, in the specimen plane. At this stage it is advisable to look down the draw tube while adjusting the size of the substage iris diaphragm, until the back lens of the objective appears to be three-quarters filled with light. The circle of light should be complete and evenly illuminated (as in Fig. 110, A) and shown in practice (Plate 40). If it is not a uniform circle the source must be slightly adjusted to correct the axiality of the light (Fig. 108).

(4) Following this an image is now formed by the objective in the real intermediate image plane.

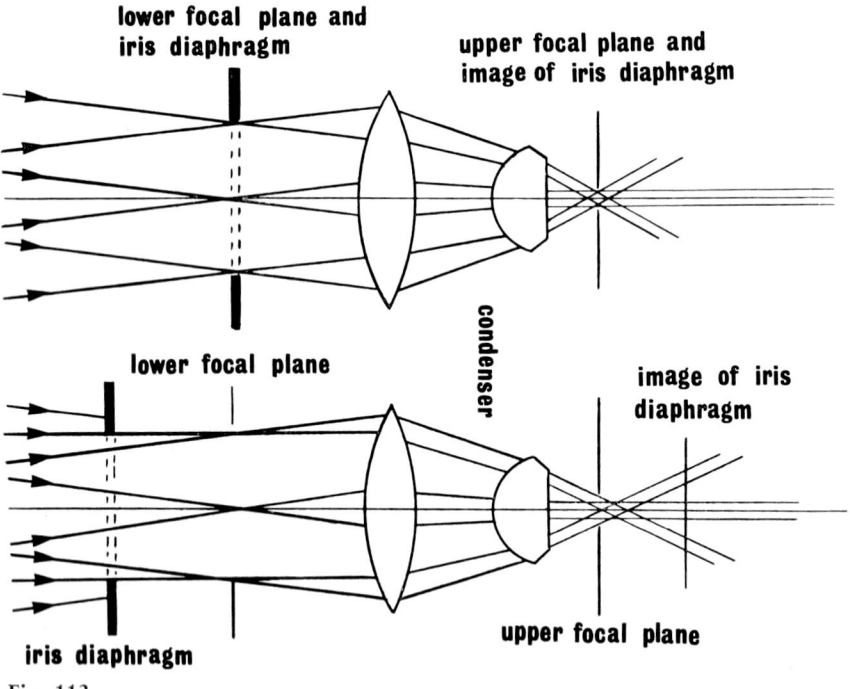

Fig. 113.

(5) A further image is now formed by the eyepiece in the focal plane of the camera.

The advantages achieved by the Köhler method of illumination is that the entire area of the lamphouse diaphragm has the same luminance power as the small filament: thus only a low-power lamp is required. The area illuminated in the specimen plane corresponds to the one which is imaged in accordance with the field stop of the eyepiece. Maximum use of the cone of rays is ensured and light scatter is eliminated; evenly illuminated specimens with maximum resolution are a particular feature of this method of illumination.

Tilting stage

The tilting stage (Plate 11) method of producing a stereo pair of photomicrographs is very satisfactory for both low- and high-power magnifications. Single photomicrographs can also be made with two exposures on one negative: one exposure with the specimen tilted to the right and one with specimen tilted to the left (Plate 7 b and d). It may seem strange that this method should be so satisfactory since, in axial transmitted bright-field photomicrography, endeavours are made to keep the slide truly at right angles to the optical axis of the system in use. The only piece of equipment required to produce stereo photomicrographs by this method is a tilting stage which is quite easy to make, Fig. 114. The specimen is photographed in two positions, with the stage 7° to the right for one negative and with the same degree of tilt to the left exposing another negative without moving objective or eyepiece. This method can be used with oblique reflected and transmitted lighting. This tilting stage has been made in the USA to the author's design (as in the figure), in this instance the stage being used in the production of electronmicrographs. It has been reported from the United States that tilting the specimen for exposure produces far better electronmicrographs than when the subject is horizontal.

Once the illumination has been adjusted, nothing further is necessary. The illumination should be centralised when the subject is aligned at right angles to the optical axis of the microscope. This operation may appear to be strange, especially as the subject will be in a tilted position when photographed, but at this stage even illumination is ensured and unevenness is prevented, as would be the case if adjusted when tilted to one particular side. Great care must be taken when centering the specimen to ensure that it coincides with the point of pivot, and the pivot is in line with the axis of the objective, D and A (Fig. 115). Tilting the specimen presents two perspectives to the optical system, and the two angles taken on planes B to B and B^1 to B^1. The axis of tilt must occupy the vertical bisector of the mounted image A. The actual taking of the photomicrograph does not differ from axial vertical transmitted light photomicrography. The picture obtained by the left tilt must correspond with that obtained by the right tilt,

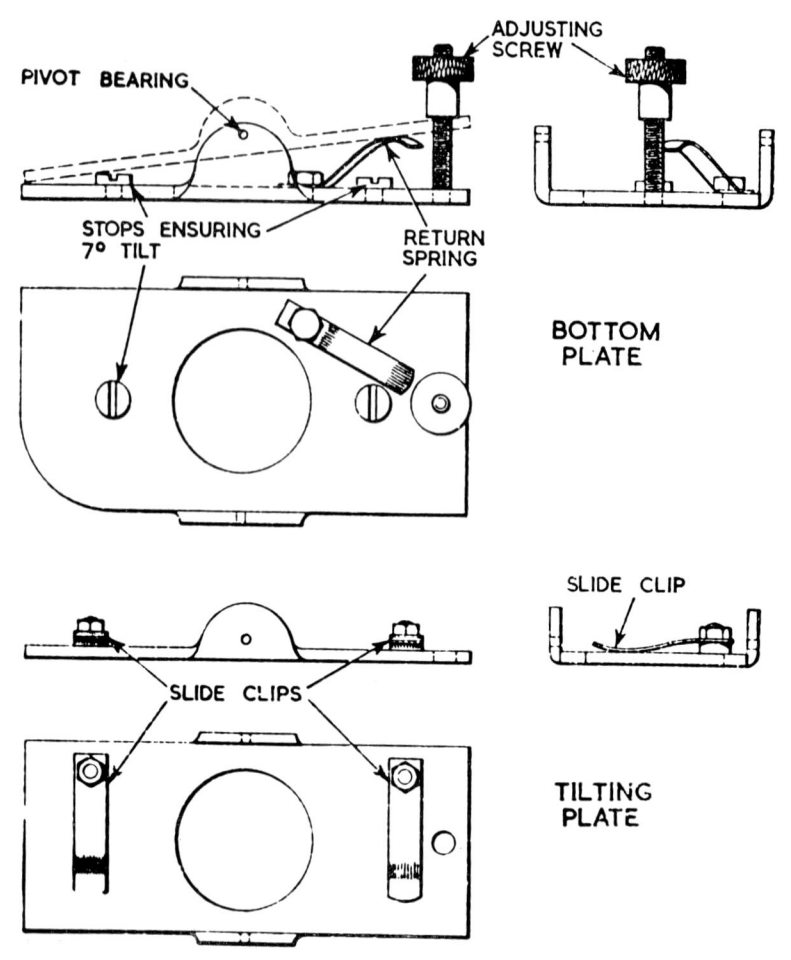

Fig. 114. Diagram of tilting stage.

the field being duplicated apart from its angle of tilt. The exposure of each of the two negatives must be the same, as must be the development time. On certain exceptional occasions it may be necessary to refocus on the subject when about to take the second shot. In such cases great care should be taken to see that this is carried out on exactly the same spot as the first exposure. If focusing takes place on two different points, some considerable confusion will be caused when viewing these two views as one.

In this type of photomicrography extra attention should be paid when sticking the left and right-hand prints down so that they are mounted on the correct side. If wrongly mounted, the subject is viewed under complete reversal conditions,

LIGHT SOURCE   119

known as pseudo-stereoscopic: shadows become "embankments" and vice versa, as when viewing aerial stereo pairs which have been wrongly presented.

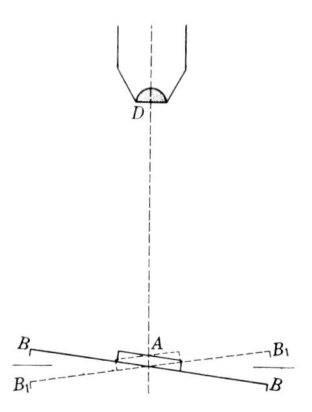

Fig. 115.

Light source

High pressure mercury lamp, 200W. Over 50 years ago Kuch and Ratschinsky wrote "Annalen der Physik", in which they described experiments they had made with high pressure mercury vapour (HPMV) lamps in which a pressure of several hundred millibars was used. Thus an effective light source could be made with the light produced by the flow of an electric current through vapour mercury. During the past 25 years the HPMV lamp has become a practical proposition for everyday applications and, in particular, for the radiation of microscope specimens. Fluorescence microscopy has been referred to as ultra-violet radiations but, mostly, this is incorrect. The application is now known as fluorescence.

The technique invariably used is to insert a deep blue glass filter into the radiations or light train. This blue filter, generally of the Chance 1·5 mm or 3 mm BG 12 type, effectively eliminates the majority of ultra-violet light; the final spectral energy available being controlled by the filter (Figs 116 and 117),

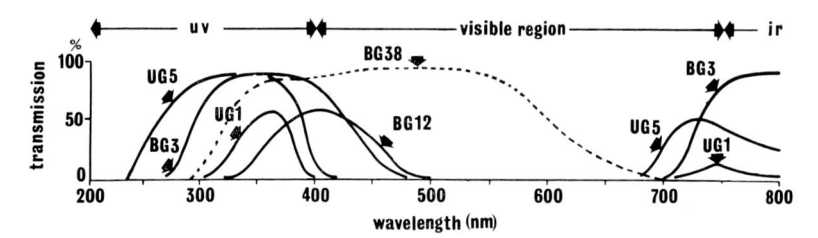

Fig. 116. Transmission curves for Schott and Gen. exciter filters. Emission spectrum Mercury lamp HBO 200 W.

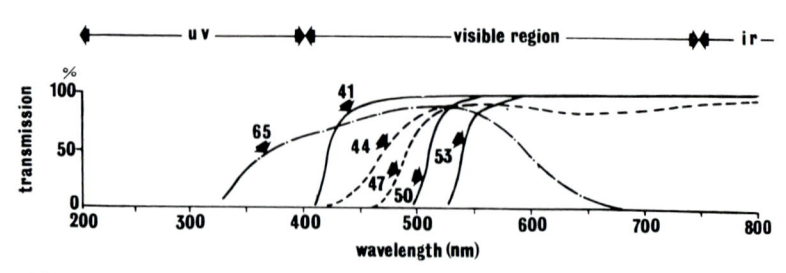

Fig. 117. Transmission curves for Schott and Gen. barrier filters. Emission spectrum Mercury lamp HBO 200 W.

which is therefore a common factor. In fluorescence the specimen under investigation is either luminescent, emitting characteristic rays, or it has been treated with fluorochromes which produce a secondary radiation where certain areas fluoresce, usually in an attractive colour against a dark background.

A strong light source is necessary. The high pressure mercury lamp, with radiation of high intensity, is fitted to the Leitz Wetzlar Fluorescence Microscope. In addition, a 6 V 30 W low voltage lamp transmitting "White" light is also fitted. This lamp emits light as a result of an electrical discharge through a rarified inert gaseous atmosphere and, once the lamp is warmed up, a stable discharge results from ionised mercury vapour. It is this vapour that is primarily responsible for the spectral output of the lamp. In consequence, because of double, triple and other ionised states of the mercury vapour, the output appears in the form of peaks of narrow spectral width, Fig. 118. Beyond wavelength

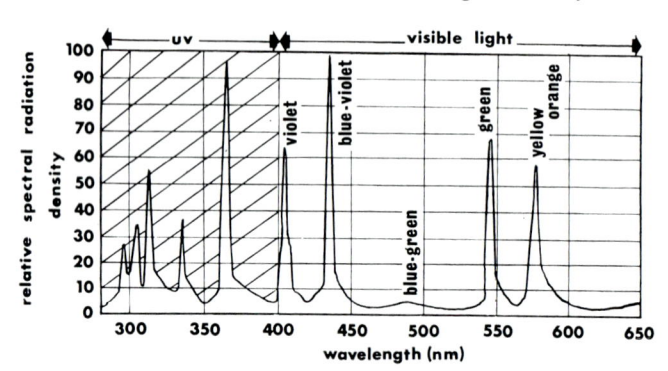

Fig. 118. Emission spectrum for Mercury lamp HBO 200 W.

650 nm the relative spectral radiation density remains at zero.* There is too, a small, but nevertheless irregular continuum, which covers most of the usable spectrum for fluorescence photomicrography. This part of the lamp's output is

* Wavelengths originate in units of nm. 1 nm = 1 Nano metre = 10^{-9} metres = 1 mu (millimicron) = 10 Å.

LIGHT SOURCE

probably that part of the available energy used for useful excitation of a fluorescent specimen. Owing to the high proportion of ultra-violet radiations, it is necessary to use an interference line filter or green VG9 filter for black-and-white photomicrography.

The lamp, in common with all electrical discharge devices, requires an elaborate electrical power supply which must be capable of supplying the "striking" voltage to initiate the discharge in the lamp. This power source must have controlled "regulation", achieved by the use of an ion cord choke. The lamp takes a period to warm up after initially striking a discharge, during which time it is unsuited for photomicrography.

Leitz supply the following set of filters to be used in this application.

Blue filter	1·5 mm BG 12	UV filter	1 mm UG 5
Blue filter	3 mm BG 12	UV filter	3 mm UG 5
Blue filter	5 mm BG 12	Red suppression	4 mm BG 38
UV filter	1 mm UG 1	Diffusing disc N	
UV filter	2 mm UG 1	Grey 0·2% transmission	
	Heat absorbing 2 mm KG1		

The following table sets out the filter combinations.

Table 11.

Exciting radiation	Exciting filter (Schott)	Suppression filter**	Type of illumination
Ultra-violet	1 mm UG 5°*	Edge K 460 Edge K 490	Darkground
	3 mm UG 5°	Edge K 430 Edge K 460 Edge K 470 Edge K 490	Darkground
	1 mm UG 1°	Edge K 430 Edge K 460	
	2 mm UG 1°	Edge K 430 Edge K 460	Darkground
	2 + 2 mm UG 1°	Edge K 430 Edge K 460	Darkground or bright-field
Ultra-violet + blue	3 mm BG 3	Edge K 470 Edge K 490	Darkground
Blue light	1·5 mm BG 12 3 mm BG 12 5 mm BG 12 3 + 5 mm BG 12°	Edge K 530	Darkground or bright-field Bright-field

* The exciting filters marked ° must be used in conjunction with a red suppression filter BG 38 4 mm. This filter is used in place of the earlier copper sulphate cell for the suppression of the residual red transmission of the exciting filters.

** The figure following the letter K indicates the wavelength in nm at which the pure transmission is 50%.

The Ortholux microscope is equipped to enable the photomicrographer to quickly change over to fluorescent radiation by inserting the HBP 200 W or CS 150 high pressure mercury burner. Great care must be exercised and goggles should be worn when handling these and similar burners.

The specimen is immersed in a fluorescent impregnant for periods ranging from seconds to 20 min or more. Rays from a deep blue filter induce fluorescence in many unpigmented plant and animal tissues and cells, which can be photographed in colour when good colour differentiation is obtained. Care must be exercised when viewing the specimen and the time kept to a minimum to avoid bleaching of the specimen. Serious bleaching is due to denaturation, through ionisation from irradiation by unnecessary ultra-violet energy, which penetrates the barrier filter (Fig. 210) by "shock transmission".

Since the effective light emitted by a "stained" fluorescing specimen is deceptively low in brightness, high speed colour film is necessary, preferably daylight film. Low speed films need long exposures, during which time bleaching of fluorescent "stain" colours can take place. Further, when exposing for long periods colour changes can take place—for instance, reds reproduce as orange or brown. In addition to a short exposure the film must enable morphologic details to be rendered. However, depending on the staining of the specimen Edge filters K 380 or K 420 will prevent rapid fading.

Tungsten halide lamp. This lamp is of recent development and operates at a high filament temperature far in excess of that produced by a projection bulb. Like the carbon-arc source, one of the most important features is the intense brilliance which provides maximum image brightness. Unfortunately, because of the heat, a heat absorbing filter (Fig. 210) should be used which will slightly lessen the light output. Tungstic iodine is dissociated at the hottest part of the filament, situated in the centre. Tungsten tends to build up preferentially at those points which hitherto were the points of failure. The operating temperature of the lamp filament approaches melting point of tungsten. Because of the reactivity of the iodide with alkaline components in ordinary soda or soda lime glass in lamp manufacture, the tungsten halide lamp uses a fused silica envelope. The envelope not only facilitates continual operation of the chemical reaction but also as increased optical transmission in the shorter wavelength. This lamp is a static source and remains optically aligned with the microscope throughout its life. There is no warm up period. The spectral output from the lamp is a continuum and, by incorporating a blue light filter, fluorescent radiation is introduced. For black-and-white photomicrography a VG9 green filter should be employed. There is no explosion hazard as may exist with the mercury lamp. Light output as a function of time remains substantially constant, there is no blackening of the bulb and no degeneration in stability of discharge through mercury vapour, as occurs when the residual trace gas in the mercury lamp is "cleaned up" by absorption and "gettering" (disintegration of

LIGHT SOURCE

electrodes under bombardment) processes normally occurring in the discharge lamp.

The energy from the lamp is conveniently available in those parts of the spectrum for which the conventional blue glass filters are employed and, as the energy is from a spectral continuum, the advantages for routine and research will be obvious. This lamp offers emission characteristics (Fig. 119) that cover fore-

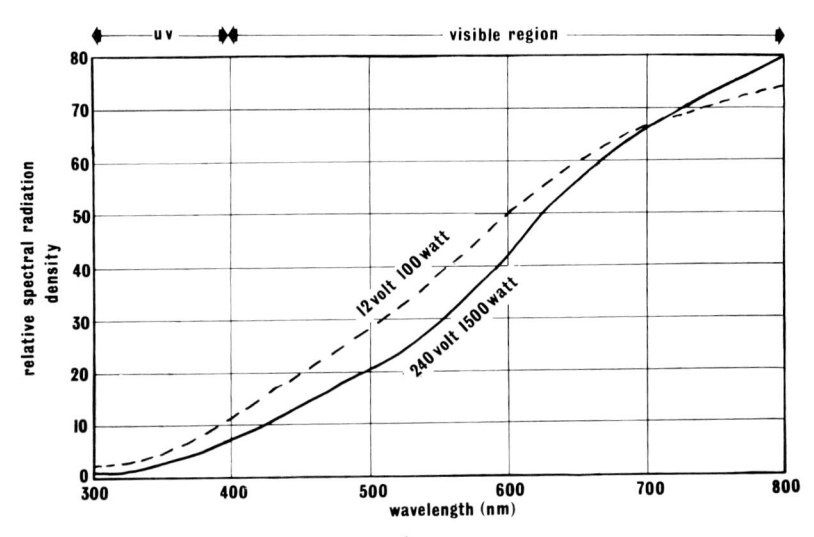

Fig. 119. Emission curves for iodine quartz lamps.

seeable future developments in fluorochromes and conjugation techniques. Although for any particular spot in the spectrum, in the deep blue and near ultraviolet region, the gross energy may be rather less than in the mercury lamp, this discharge disadvantage is more than offset by the ease with which the energy output of the iodine quartz lamp may be controlled. In practice this manifests itself as a slightly lower overall brightness of image but control of lamp brilliance facilitates control of contrast between fluorescing and non-fluorescing components in a specimen. Here non-fluorescence refers to non-specific staining. Therefore as non-specifically stained material is gradually reduced in brightness, by reducing the lamp intensity, the ratio between fluorescing and these components tends to infinity.

Fluorescent tubes. These are most suited for low-power photomicrography, giving a better photographic rendering than the long lived high pressure lamps. The condition of these tubes is the result of a coating of certain fluorescent materials, making the ring illuminator extremely useful in the visible spectrum. Daylight colour film is best suited to the colour correction of the tube, which has a luminosity of 0·3 to 0·7 cd/cm^2.

Mercury compact source. This is a very high pressure discharge lamp (Fig. 120), made by Philips Electrical Ltd. and is known as the ME/D 150, and is a 250 W lamp. The lamp was specially designed for Leitz fluorescence–microscope equipment. The discharge tube is of quartz and the small envelope is of special hard glass, which is transparent to ultra-violet. The tube uses mercury vaporised in argon

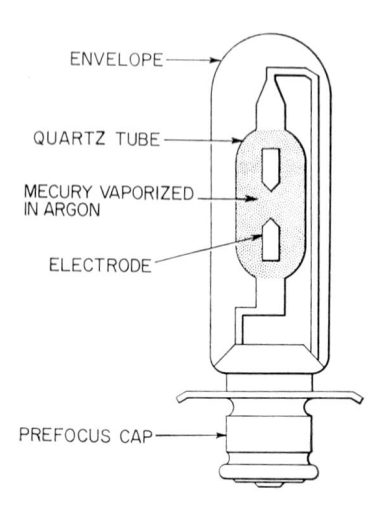

Fig. 120.

and is not equipped with any means of cooling. The 150 W has a continuous spectrum through the visible and near ultra-violet region. A special feature of the lamp is the annular disc which has been added to minimize movement of the arc. The lamp is burnt in the vertical position and its life is estimated to be 200 hr during which time there is no change in light output. In the case of a mercury discharge lamp, the peaky spectral output results in serious bleaching of the specimen due to the denaturation from the ionisation from irradiation of unnecessary ultra-violet energy, which penetrates the blue glass barrier filter (Fig. 210) by "shock transmission".

Zirconium arc lamps. Zirconium is a metallic element, very suitable for electrodes. When an electric arc is struck between such electrodes, in hydrogen or ammonia, the metal fuses and melts into grey drops on to the negative electrode. The Sylvania enclosed concentrated arc lamp (Fig. 121) has a permanent metallic anode and a specially prepared refractory zirconium–oxide cathode. These two elements are sealed within a glass bulb filled with argon, an inert gas.

When an arc is established between the two electrodes, the oxide surface of the cathode is raised to its melting temperature and molten zirconium is liberated. As a result of the high temperature to which the cathode surface is raised, a brilliant white light, emitted by the molten zirconium and a cloud of vaporized zirconium and zirconium oxide, extends for a few thousandths of an

LIGHT SOURCE

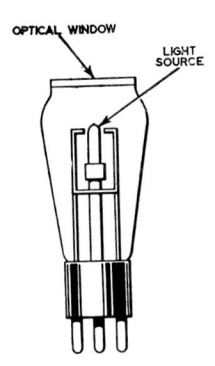

Fig. 121.

inch from the cathode. Slightly larger than the cathode is a metal ring which forms the positive electrode. The light is emitted through the aperture in the anode, forming a narrow intense beam of light with a continuous spectrum and a colour temperature in the region of 3,200°K (Fig. 122). There is no particular

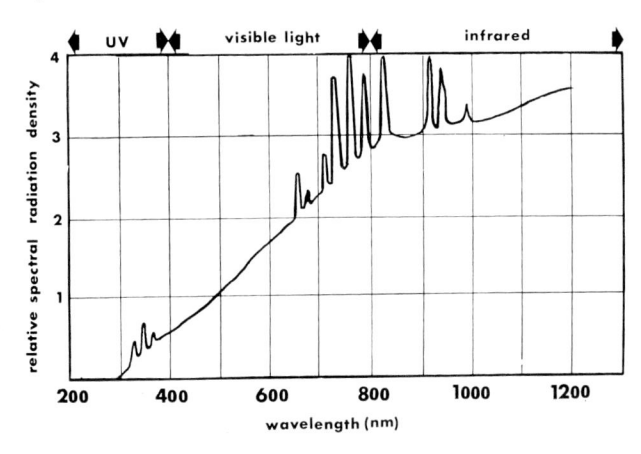

Fig. 122.

position in which this lamp must be burnt, as with others; the molten metal is held in position by surface tension. Its radiations extend from one end of the spectrum to the other; from 250 nm through the visible region of the spectrum, reaching a maximum at 1,000 nm, and then further into the infra-red.

The concentrated arc lamp is ideally suited for photomicrography, as the light-source sends out a very narrow beam of light. Perhaps, at times, it might be thought too critical because it shows up any imperfections, such as fine dust, finger marks, slight air bubbles and stresses in the glass—scratches and striae in glass show up as luminous bodies.

Halogen arc lamp. The 250 W halogen arc lamp produces white light of a true colour rendering in visual observation. For black-and-white photomicrography a 546 nm interference line filter or VG9 green filter should be used.

However, this lamp is unsuitable for colour photomicrography due to its spectral energy distribution. The basic continuum is superimposed by the narrow emission bands as set out in Fig. 123.

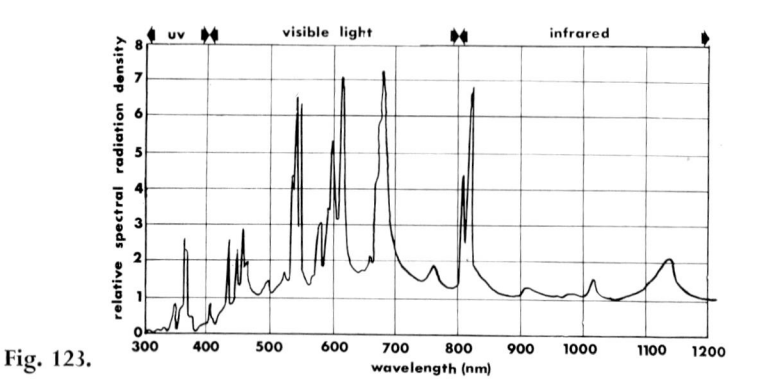

Fig. 123.

Xenon-arc flash tube. This British Thomson–Houston product (Fig. 124) is a dual purpose source capable of giving a continuous output and an instantaneous flash of a very short duration. The high amount of infra-red radiation is much less

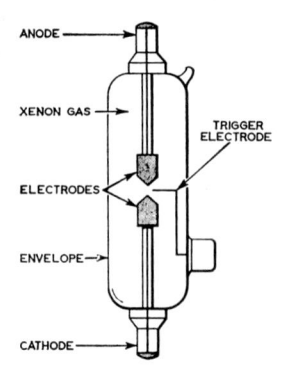

Fig. 124.

than in the xenon high pressure lamps, due to the lack of glowing tungsten electrodes. The low infra-red is perhaps an advantage when operating with live fast moving pond life. Exposures are in the region of 1/100 to 1/500 sec and capable of arresting anything to be viewed through the microscope. The FX–33 (USA) Xenon flashtube is small and straight with a 3 mm gap, mounted in a suitable illuminator fitted with a 6 V headlight bulb.

Xenon high pressure lamp. A leading advantage of this gas discharge lamp, Fig. 125, is that it can run off the mains, AC or DC, and produce a very high luminosity. The colour temperature is 5,200°K to 6,000°K and the light intensity per unit

LIGHT SOURCE

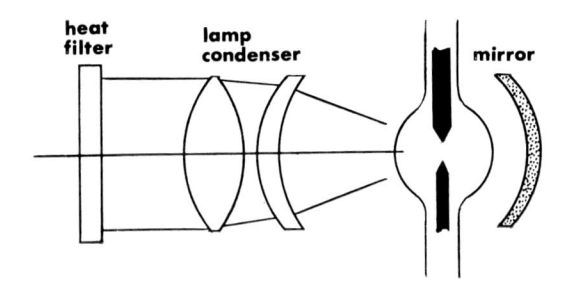

Fig. 125. Optical arrangement and high pressure Xenon lamp fitted to Leitz Lamphouse 250.

area near to the cathode exceeds 100,000 stilb, but in the centre of the arc approximately 25,000 stilb is registered. Immediately the lamp is switched on it yields full output and, further, the on and off position operates the lamp without causing damage to the burner. Its life is estimated to be in the region of 1,200 hr. The lamp has a continuous spectrum in the visible range and strong lines in the short and long-wave region. The spectral range used is demonstrated

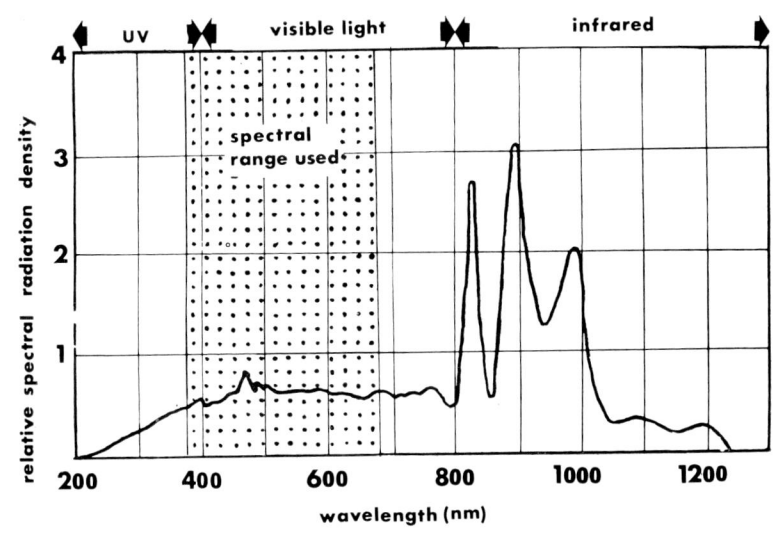

Fig. 126. Emission spectrum for H.P. Xenon lamp 150 W. Colour temp. 6000°K.

in Fig. 126. The Kodak 30R filter can be used with great success without affecting the film speed. Should a compensating filter be necessary see Table 12.

The Xenon Combination Lamp 150 W is an alternative to the standard low wattage lamp and must be used with a rectifier connected to 220 V a.c. The ignition requires 20 kV and this is arranged automatically by means of a sliding

Table 12.

Diffusing disc R	For homogenizing the light source at low magnifications
Diffusing disc N	For homogenizing the light source at high magnifications
Grey filter 0·2% transmission	Suppression filter for visual observation
Grey filter 5·0% transmission	Suppression filter for visual observation
Green filter 1·5 mm VG9	For black-and-white photomicrography
Heat absorbing filter Calflex B1/K2	Built into Leitz Lamp Housing 250

contact. After it has been switched off, while still hot, the lamp may be switched on again, immediately becoming operational. The optical arrangement of the lamphouse gives an arc of the xenon high pressure discharge lamp which is reflected into its own plane, compensating for a slight decrease in brightness. The special three lens collector system permits a beam of light to be directed into the substage condenser and, when the lamp is used for long periods, the Schott K.G.1 heat absorbing filter should be inserted. A special feature of this lamp is the inclusion of four filters which can be brought into operation independently, by an easily operated switch. The fitted filters are as follows:

Grey filter, 1% transmission (2·3 mm. NG3).
Grey filter, 0·5% transmission (2·4 mm. NG9).
Panchromatic green for monochrome photomicrography.
A diffusing disc.

This is to blur the structure of the source, if operating without a substage condenser, thus ensuring a far greater degree of even illumination (Fig. 136). The BG 12 filter (Fig. 116) can also be inserted. It is required for fluorescent work when the SC 150 ultra-violet high pressure mercury-vapour lamp is in use.

Low-voltage filament lamp. Coiled filament bulbs are manufactured with one, two or three filaments and their ability to provide an even spread of light is very necessary in photomicrography. A range of these lamps can be compared in the Table 17, page 147, and comparisons made in colour balance (Figs. 127 and 137). A conservative estimation of the life of one of these bulbs is about 100 hr. The heat dissipated by these lamps is very low and very rarely requires a heat absorbing filter. A wide range of automobile-type bulb is available and a pre-focus cap commonly facilitates its use. Most microscope lamp holders have an adjustable sleeve to ensure the source is aligned with the lamp condenser, Fig. 107.

Carbon arc lamp. Although this lamp is no longer made, the form of illumination is very popular in certain fields of research. The light is intense and nears that of daylight quality. Either AC or DC arcs may be used. The lamp gives out an amount of heat, produces unwanted nitrogen gas and needs constant adjusting and maintenance. The arc lamp is usually adjusted to fine limits and the circular

LIGHT SOURCE

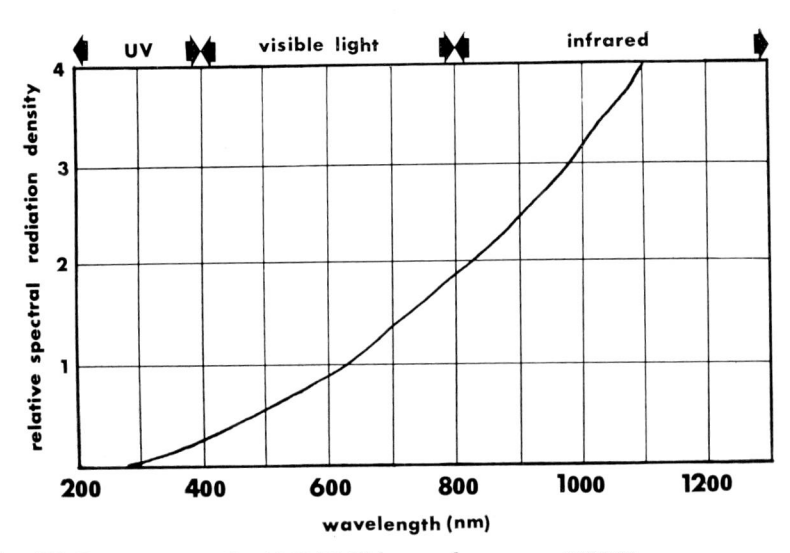

Fig. 127. Emission curve for 12 V 30 W lamp colour temp. 2900°K.

crater of carbon rod, in line with the axis, is used when the light source ensures a perfectly uniform illuminant. A clockwork mechanism is commonly provided to feed the carbons together as they are consumed, and the positions of the carbons must be periodically checked to make sure that the second electrode has not advanced to block the light from the first. A heat absorbing filter is necessary.
Electronic flash. The equipment needed for this specialized branch of photomicrography consists of a microscope fitted with an observation or viewing eyepiece to enable one to watch the specimen right up to the time of exposure. The microscope with a viewing eyepiece (Fig. 81) produces two images of the specimen, one projecting an image on to the photographic material and the other image being viewed from the side: the adjustment of both images is synchronized.

A rotating stage assists in following a moving specimen, but up and down movement makes focusing difficult. The moving specimen can be illuminated by a secondary light or pilot light, sufficient to do the necessary focusing etc. Since both hands are fully occupied when arranging the picture it is suggested that some form of foot switch should be used to fire the flash and camera shutter. The source tube fitted to the combined lamp house should have a flash duration of approximately 1/500th sec and the output at least 150 joules. The light is directed through a 6 mm aperture, close to the tube, and this directs a beam of light into a lens system some 50 mm (2 in.) from the source. A directional tube can also be fitted between the lamphouse and the substage to control the whole output. Unlike the camera, the iris diaphragm of the microscope is wide open which means that the specimen receives the full duration of the flash (Plate 57): unless of course a shutter unit is incorporated in the area of the lamphouse condenser.

If such a shutter is not used the depth-of-field is affected through the dimension of the flash tube in relation to the focal length of the condenser. Further, if the output is too great the only way of controlling it is to reduce the substage iris diaphragm though this causes ill effects on the final image. The exposure determination is more difficult to estimate than with most forms of illumination. Trial and experiment and a strict log of all work carried out are the best tools here.
Micro-flash unit. The flash unit by Leitz consists of a flash source with collector, and the mirror housing for the various microscope stands. A special adapter is not required. All electronic components including flash tube, collector, and reflector are housed in the small unit (Fig. 128). The mirror housing contains a

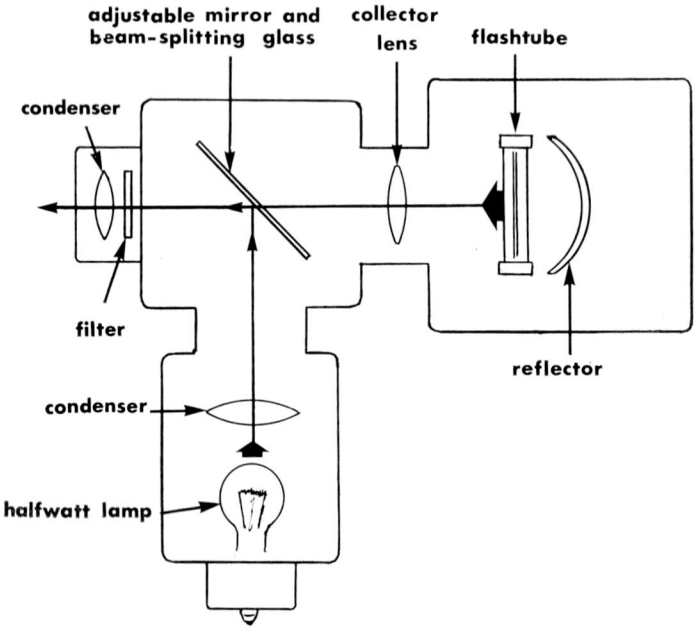

Fig. 128.

beam-splitting glass inclined to the beam at 45°, and a hinged mirror. It has a special aperture with clamping device for the low-voltage lamp required for aligning and focusing the object.

The photomicrographer will work generally with the mirror turned out. In this position 10% of the available halfwatt light and 90% of the flash light will be directed to the specimen by a beam-splitter. As a result of this even distribution, almost the entire intensity of the flash is used, while the adjustable proportion of halfwatt light is sufficient for focusing etc. in bright field. In very weak lighting conditions (darkground or phase contrast) observation is possible at full intensity of the halfwatt light by turning in the mirror. This facility of adjustment has the additional advantage that the flash device can remain on the microscope. The flash output is from a small capillary flash tube, behind which is a cylindrical reflector. The flash duration is 1/1000 sec. The light colour temperature of 6,000°K (Fig. 126) permits the use of daylight films without compensating filters.

The Zeiss, Oberkochen, flash unit is very small and contains an auxiliary tungsten lamp used for prefocusing the specimen. The unit has a three lens collector and there are two levels of power, 30 or 60 W-sec. The former gives a flash duration of 1/3000. The second full strength or 1/2000th second at half-watt.

The American Speedlight Corp. produce an efficient flashtube which stands erect when in use and the output is directed through a small aperture in an upright panel. The F.T.221 works in conjunction with a 300 W control bulb. Both can be incorporated by the makers into a lamphouse of one's own design.

The Leitz Ortholux has provision for flash illumination. The flash tube is housed in a special metal case from which a light beam is emitted just under 6 mm in diameter, with an energy rating of either 150 joules at 1/1000th of a sec, or 300 joules at 1/500th of a sec. The flash tube is so arranged that the regular microscope illumination can pass it as a secondary illuminator, so alignment does not pose any difficulties.

The EG and G flash illuminator, Model 516, is a concentrated strobe lamp produced by Edgerton, Germeshausen and Grier. Inc. of Boston and is designed to produce short duration, high intensity flashes of light for both photomicrography and the simple microscope. Its suitability for subjects covers a wide range from metallurgy, to delicate diagnostic close-ups of blood vessel structure in the human eye to moving insects. The relative coolness of the micro-flash illuminator is of particular importance in cases where the subject is susceptible to damage caused by heat.

The flash illuminator consists of a power supply and flash lamp assembly, incorporating EG and G FX–33 xenon flash tube with a vycor envelope. The flash duration is 150 microsec at 100 W-sec, having a recycle time of 5 sec. Fitted behind this very small flash tube is a 6 V tungsten lamp of variable intensity for focusing and alignment, suitably housed in a modified microscope lamphouse.

The EG and G flash illuminator, Model 517, a model of the FX-33, is simply the flash tube housed within a special pyrex holder, together with a built-in reflector. The assembly is protected by a heavy rubber sleeve which permits the flash lamp to be held in the hand or clamped in position or even in an aquarium. The light source size is 4 mm I.D. $\times$ 38 mm (1·5 in.) in length, neatly fitted into the lamp assembly which measures 33 mm $\times$ 213 mm ($1\frac{1}{4}$ in. $\times$ $8\frac{3}{8}$ in.) in length. The spectral output of the FX–33 extends over the entire visual range and is suitable for both black-and-white and colour processes.

Special light sources for Reichert camera/microscope. Modern photomicrography with its wide range of special techniques demands light sources capable of meeting all requirements in respect of brightness, spectral light distribution and operating convenience. High intensity light sources have been provided for different application. Since they are normally used alternatively with low-voltage lamps, all the light sources are mounted so that they can be readily interchanged and used to their best advantage. The lamp housings are fully adjustable and connected to the microscope either directly or through a sturdy optical bench. They are provided with a suitable condenser, iris diaphragm, and set of colour filters. The actual burners are fitted with a metal base for location and after initial adjustment will retain their performance without servicing.

The "Lux U" Lamp has three interchangeable lamp holders to enable a quick changeover from low-voltage lamp to zirconium source or sodium vapour source. The "Mercurius" Lamp can house the HBO Mercury Vapour, Xenon or Low-voltage lamp.

Colour expression

First and foremost, photomicrography should portray all details in a clear and precise manner and, whenever possible, the area of greatest importance should be orientated in a way which can be readily understood. Photomicrographs should be aesthetically sound, of first class quality and with a good background which, when it enhances the subject, usually passes unnoticed because it is unobtrusive. A background that is not entirely satisfactory calls attention to itself and, as it were, clamours for the limelight.

Optical staining

Colour photomicrography presents a greater range of both problems and possibilities for background and subject matter than does black-and-white photography. Instead of the intensity of tones, between black-and-white expression, colour stain, or optical methods, can introduce colour and emphasize one area against another. (Colour plates 70, 71 and 72).

Low power optical staining was first introduced by Rheinberg in 1896, and not until 1945 did others (Royer *et al.*, 1945) report their methods of optical staining. Rheinberg used what is known as a "Mikropolychromar" system, Fig. 129, which incorporated annular colour filters mounted in a special substage unit, manufactured by Zeiss. The method devised and described by Royer (1945) is a form of darkground or phase contrast illumination, in which the substage condenser stop is replaced by a transparent colour disc, and the surrounding annulus is filled by a second filter of a paler, but contrasting, colour. Specimens viewed through the microscope were illuminated by the colour of the annulus,

OPTICAL STAINING 133

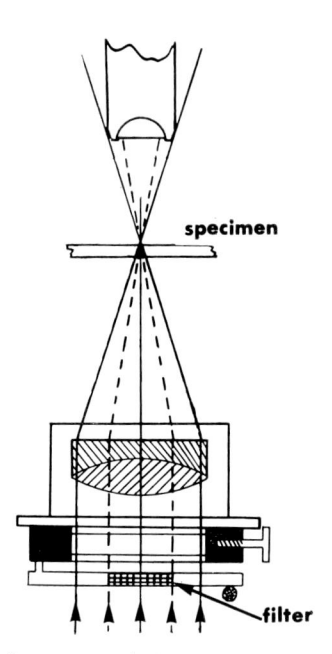

Fig. 129. Rheinberg optical staining technique.

while the background recorded the colour of the "stop". With increasing powers of objectives, so the system became increasingly difficult, and the highest magnification published is ×25 (Royer et al., 1945).

Another application applied to optical staining is reported by Payne (1949). Sir T. R. Merton (1949) suggested a method in which a blue darkground image is viewed against a background illuminated in its complementary colour (deep yellow). The yellow filter, in the form of a circular disc of just sufficient diameter to illuminate the full numerical aperture of the objective, is surrounded by a clear annulus. The intensity of the bright-field is carried by the use of an adjustable wedge-shaped blue filter, mounted in front of the lamp. By varying the density of the lamp filter a variation from pure darkground to almost pure bright-field effects may be obtained.

From this technique, which I made practical (Plate 41), it will be seen that colour filters can be used to good effect without a special instrument to record colourless subjects which cannot be chemically stained (Lawson, 1964). Here added detail and an aesthetic appeal can be achieved too.

Method of illumination. My method of optical staining is brought about by the use of two illumination techniques, (a) direct transmitted bright field and (b) indirect refracted light at off centre, illuminating the specimen from beneath. Both (a) and (b) are used together as one means of illumination (Fig. 130). The two 6 V 30 W lamps are operated with light filters in the light train, but at the

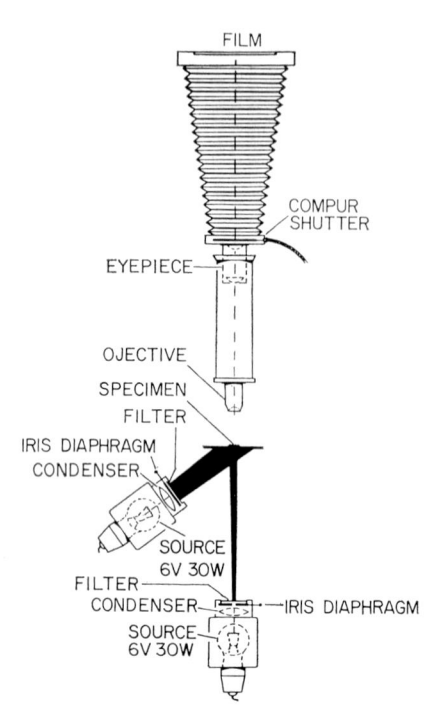

Fig. 130. Lawson optical staining technique (see Plate 71).

same time they are controlled independently in both the annular size of the cone of light and its colour. In this method of optical staining the direct light produces transmitted bright-field, while the indirect filtered light produces the specimen image of refraction. The two together give a controllable colour image.

The two light sources are fitted with an adjustable condenser, iris diaphragm, filter carrier and height and tilt adjustment. Colour filters fit over the condenser in a carrier, as indicated in the figure. The indirect light "stains" the specimen with a pre-selected colour and the direct light provides the background colour of one's choice.

It was not possible to use a substage condenser with the microscope. Nevertheless, low and medium power magnifications can be made successfully, but it has been impossible to produce a satisfactory high power magnification with this method because the angle of the indirect light has to be very shallow and close to the base of the slide, which the lamphouse does not permit. Furthermore, high-power photomicrography demands the use of a substage condenser.

Operation. It was discovered that the direct pencil beam of light must be much smaller than would normally be applied with the particular objective in use, transmitting nearly parallel rays. As the substage condenser is not in use the

OPTICAL STAINING

source is focused in the plane of the specimen. It is absolutely essential to see that the vertical source, whatever the bulb may be, is in correct alignment with the lamphouse condenser. If this is out of alignment in any degree the condenser will not project a beam of light in the line of the optical axis of objective and eyepiece, and therefore uniform illumination will not result. Instead of uniformity, therefore, there will be colour bands around, or along, the sides of the specimen and this can be especially noticeable in this particular method involving colour film.

Stage 1. A choice of field is made (Fig. 131). Direct lamphouse iris diaphragm is opened to its full aperture, the specimen illuminated and focusing carried out. At this stage the direct light is switched off and the lamphouse iris diaphragm remains fully open.

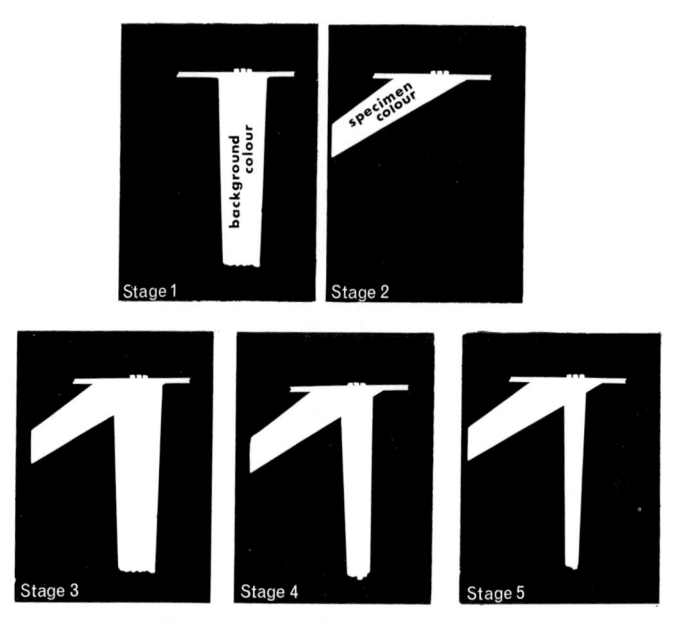

Fig. 131. Demonstrating "optical staining" by Lawson.

Stage 2. The indirect light is positioned beneath the stage at an angle of approximately 40 degrees and at a distance of approximately 15 cm (6 in.) from the specimen. The iris diaphragm aperture is wide open and a cone of light is adjusted to fill the specimen field with unfiltered light and, at the same time, focused just beyond the specimen. The marginal field was also evenly illuminated. At this stage the specimen image is bright against a black background. The source is switched off and filters inserted in both lamps. On this particular occasion a red is attached to the indirect lamp and a green to the transmitted bright-field lamp.

Stage 3. The direct and indirect lights are next switched on, and focusing on the specimen repeated. This is necessary because a fundamental difference in the formation of the image of the specimen and the background becomes apparent. At this stage both iris diaphragms are wide open and only a little colour is visible in the specimen, due to excess light from the bright-field source.

Stage 4. The iris diaphragm aperture of the direct light is slowly reduced and at the same time the image is observed in the ground glass screen or viewing eyepiece.

Stage 5. It will be noticed that, as the aperture is closed, the specimen becomes more and more colourful, until the particles exhibit the colour of the indirect light, revealing this method of optical staining at its best.

To obtain colour prints or transparencies of an unstained specimen which, in its natural tones, is against a "stained" background, the combination of indirect white light and direct filtered light will produce what is required (Plate 71),

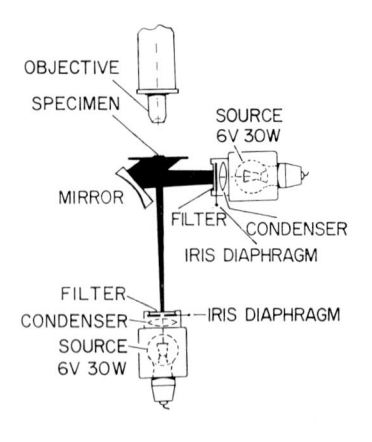

Fig. 132.

Figs 131 and 132. In terms of black and white images it will be similar to the specimen image of Plate 71 upper.

Use of mirror. If space beneath the microscope stage does not permit the use of the indirect source, as indicated in Fig. 130, it can be used in conjunction with the microscope mirror to reflect and direct the beam (Fig. 132). However, when photographing subjects which demand medium-power the mirror is necessary. I have made successful "optically stained" colour photomicrographs of antibiotic material during trials which show impure material when other methods have failed.

Refraction and transmission. Having seen results obtained by this method of optical staining we must now examine the action of the light to see what happens when the colourless specimen and surrounding area appear as one colour against

OPTICAL STAINING

another. If a semi-transparent or transparent specimen is illuminated at an angle from beneath the light rays, which are too oblique to pass directly into the objective, leave the background field dark. Thus features of the specimen which show abrupt changes of refractive index appear as self-luminous bodies on a background colour of one's own choice. When the indirect light rays strike the object, the rays are bent, or refracted, in various directions (Fig. 133) and

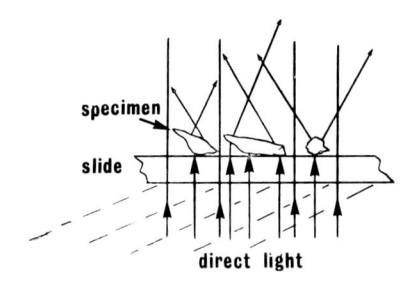

Fig. 133.

some of the image-forming light rays do not traverse the objective optical system as do the transmitted bright-field rays. Most of the uninterrupted indirect rays, those that are not refracted by the particles, continue on a straight line to a point beyond the microscope or are totally reflected from the underside of the slide and do not affect the background colour. The direct rays of light illuminate the background and are transmitted vertically through the glass slide and cover glass (if used) into the objective, and form an image at the exit pupil of the microscope. Some of these rays become extinct on reaching the specimen, whilst others pass through the flat areas of the specimen (see Plate 41, 71 and compare individual crystals).

Double images have been observed when photographing fine particles at higher magnifications. These multiple images are probably formed by refraction, brought about by the shallow angle of the indirect light, or reflection within the glass slide, sometimes causing a displacement of the image. There are, however, some flat crystals and similar materials which, in a way, act as a homogeneous medium and permit a small percentage of the direct light to travel through, while the rough outer edges exhibit the full colour of the indirect light. This can be seen in colour photomicrographs (Lawson, 1964, 1965, 1966). Needle crystals and scratches lying parallel with the indirect light do not redirect its rays when working at a magnification of say, $\times 300$ (see Fig. 97).

A wide range of subjects, such as crystals, transparent sections, moisture droplets, emulsions, textiles and textile chemicals, fibres, hair striations on the surface of transparent materials, can be photographed in this way.

References

Lawson, D. F. (1964). *J. Soc. Cosmet. Chem.* 641–656.
Lawson, D. F. (1965). *Visual*, Vol. 3. 1. 25–28.
Lawson, D. F. (1966). *Management Today*, July/Aug. 62.
Payne, B. O. (1949). *J. Roy. Microsc. Soc.* **69**, 209.
Rheinberg, J. (1896). *J. Roy. Microsc. Soc.* Ser. 11, **16**, 373–388.
Rheinberg, J. (1899). *J. Roy. Microsc. Soc.* Ser. 11, **19**, 142–146.
Rheinberg, J. (1899). *J. Roy. Microsc. Soc.* Ser. 11, **19**, 243–245.
Royer, G. L., Maresh, C. and Harding, A. M. (1945). *J. Biol. Photogr. Ass.* **13**, 123.

Monochromatic light (filters)

There are five main reasons for using a good light filter: to enhance colour rendering in the specimen image, i.e. reduce or increase contrast; to correct the image colour to suit that of the particular emulsion in use, suppressing or enhancing a particular stained area; to improve image quality by shortening the wavelength of light used, i.e. blue light provides approximately twice the resolving power to that of a red light; to balance the film with a particular light source with a light correction filter for accurate colour reproduction; to reduce the intensity of the transmitted light by the use of a neutral density filter (pages 140 and 150).

This form of illumination is perhaps the simplest and most effective kind of light to use. It contains only one colour, that introduced by the photomicrographer, and all its light waves are the same length. In contrast, white light emitted from a tungsten, or coiled filament bulb, produces a mixture of colours in no set order, each colour having a different wavelength. Some doubt exists whether there is a standard "white" light, for experience has shown that a wide variation occurs from cool white light, daylight, Philips 55 and white fluorescent lamps. This difference is no doubt due to the red emission. It adds to the difficulties encountered in colour photomicrography, particularly when recording subject matter which is almost transparent and when surrounded by a clear mounting medium. Heavily stained preparations covering a complete field can disguise any breakdown which may be visible in "white" light. Monochromatic light is more controllable and easier to handle, as its waves move together more or less in phase. In this way the pencil beam of light is strengthened and the effective power, usefulness and controllability of the beam is increased. It is said to be easier to control an army marching in step than one out of step and this is a most apt analogy. So, whenever subject matter permits, it is advisable to use a good filter for black and white photomicrography (Plates 42 and 43) and, when possible, an extremely faint one for colour. The colours to choose are mentioned, not because of their attractiveness, but because the final result with high resolution is most satisfactory. Blue or green improves quality and assists in recording fine detail. Stained specimens are best recorded with a filter of a complementary,

MONOCHROMATIC LIGHT (FILTERS)

or near complementary colour, and this will produce the desired contrast so often required in black and white expression (Plate 42 and Fig. 134). By "contrast" I mean not "soot and whitewash" prints, but a difference in the rendering of colours recorded in various tones.

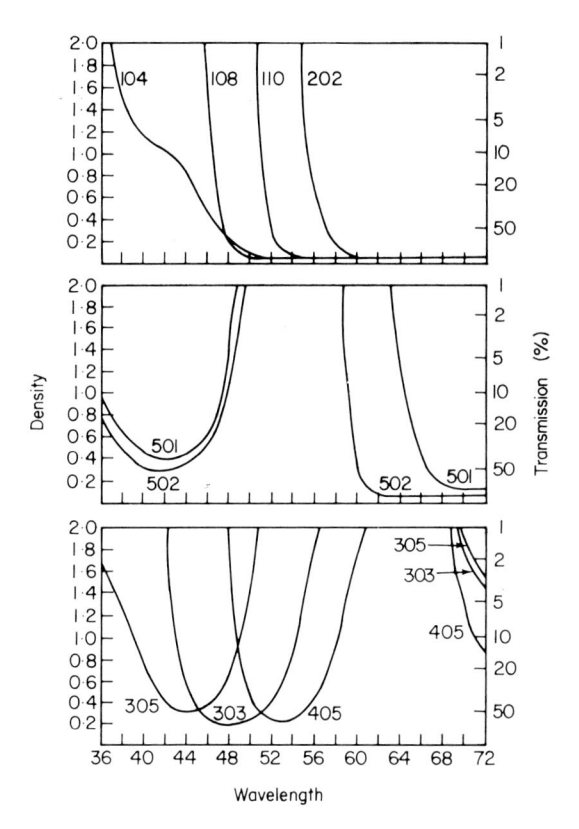

Fig. 134. Curves relating to Ilford filters. Upper: 104, 108, 110, 202 (Micro 8, 9, 4 and 5). Centre: 501 and 502 (Micro 6 and 7). Lower: 303, 305 and 405 (Micro 2, 1 and 3).

It is sometimes necessary to reduce contrast in one particular colour in order to emphasise or enhance it in another. For example, blue stained and red stained specimens produce a similar density through the medium of monochrome photography and a filter with a complementary colour must be chosen to create a difference in tones. Sometimes it will be necessary to use a yellow filter to get detail but, unfortunately, the yellow band of light does not resolve as many lines per mm.

F

Table 13. *Maximum contrast filters for stained specimens*

Stain	Colour	Filters recommended
Acid Fuchsin	Magenta	Deep Green + Yellow-orange
Aniline Blue	Blue	Deep Green + Orange
Azure I	Violet-blue	Deep Green + Orange
Basic Fuchsin	Magenta	Deep Green + Yellow-orange
Carmine	Light Magenta	Deep Green + Yellow-orange
Crystal Violet	Lavender	Deep Green + Yellow-orange
Eosin	Orange-red	Yellow-orange + Blue-green
Gentian Violet	Deep Violet	Deep Green + Orange
Haematoxylin	Deep Blue	Deep Green + Yellow-orange
Light Green S.F.	Light Green	Deep Red
Methylene Blue	Light Blue	Purple + Yellow-orange
Methyl Green	Blue-green	Deep Red
Methyl Violet	Deep Violet	Deep Green + Orange
Orange II	Light Orange	Medium Blue
Saffranin O	Orange	Yellow-orange + Blue-green
Sudan III	Reddish-orange	Yellow-orange + Blue-green

Filter	Function
Blue	Increases contrast in yellow and orange areas.
Light blue	Used as a background colour for red stained specimens: does not affect the image but enhances it.
Blue-green	A very useful filter for increasing resolving power: wavelength is 470 nm.
Light green	Will improve contrast in pale pink.
Green	Used with Ilford Chromatic emulsion. Will also increase resolving power and contrast in reds.
Yellow/Orange	Both of these filters will improve contrast in Eosine, Carmine, Haematoxylin. Will also increase faint images against a yellow background.
Red	Extreme contrast filter for blue and green specimens. Will lighten red, orange and yellow, and will darken blue, blue-green and green.
Neutral density	For varying the intensity of illumination. Available in a number of different densities, to be used with transmitted light.

The use of the correct filter in photomicrography increases the contrast of the photographic image and the definition, owing to reduction in chromatic aberration. This is particularly true of an achromatic lens system combined with a yellow-green 530 nm–580 nm filter.

An objective having an N.A. of 1·37 resolves 66,041 lines per cm with white light, but when a green filter (500 nm) is used with this objective, it resolves 71,585 lines per cm. An objective having an N.A. of 1·30 increases from 62,666 to 67,927 lines per cm when a green filter is introduced into the light train. These figures show that the correct filter can improve the resolving power of an objective.

Liquid filters. These are in some respects better than glass since it is possible to control their characteristics over a wide range but they are not so convenient to handle. Liquid filters also serve as heat absorbing filters. A good liquid green, known as Zettnow's filter, comprises saturated aqueous solutions of copper nitrate and chromic acid. The strength of these two solutions can be varied to control the density of the green. There are, however, many chemicals which will give a green solution and many which will produce a blue-green. Iodine-green, acid-green and methyl-green are the most popular and can provide very dense filters. Acetate of copper (green) and bichromate of potash (yellow) with a strong solution of glacial acetic acid, will produce a yellow-green. A very useful liquid blue, suitable for fluorescence, is prepared by the addition of concentrated ammonia solution to a saturated solution of cupric sulphate until the precipitate of hydroxide is re-dissolved. This method produces a deep blue solution known as cupric tetraminosulphate or cuprammonium sulphate.

Gelatine filters. These are more suited for use with low-power objectives than for work in the higher power range. Also, gelatine cemented between optically worked glass is suited for work in the lower range of magnifications. A further class of gelatine filter is that cemented between optical flats. These are subject to the most rigorous tests and is adequate to the highest quality work and where the utmost precision is required.

Position of filter and effect on focusing. A colour filter in the form of gelatine film has no appreciable effect upon focusing changes but with cemented glass filters there is a definite though small effect on the image forming rays. In general the alteration of focus is negligible when the filter is placed in front of the lens as that of a simple microscope. If, however, a filter is placed in the light train above the substage condenser it is as well to focus with the filter when in position. It is therefore necessary to pay due regard to this to obtain the sharpest focus.

The subject which shows multiple colours can be improved by using an Ilford Gamma 402 filter to give correction to all colours on a par with visual luminosities. The colour-correction filter (Addacolor C.C. filter) manufactured by Harrison and Harrison can be used with similar effect. If the choice of negative and processing is not well used, the results may not be satisfactory in black and white.

When photographing a colourless subject, such as diatoms, it is advisable to use a blue filter coupled with a process emulsion or similar negative material; the details therefore appearing dark. The use of a green filter with a chromatic emulsion gives approximately the same results. It is as well to remember that the shorter the wavelength of light the greater the resolving power. Should chromatic aberration appear through the optical system a deep yellow filter with orthochromatic emulsion, or a green with panchromatic film, will be necessary to counteract it. The exposures will of course be long.

Gelatine filters, such as the Wratten "M" series (Table 14) sandwiched between glass, are indispensable for this type of work.

Table 14. *Uses of Wratten "M" series filters*

Name	Code	Colour	Transmission per cent	Uses
A	25	Red	15·0	For infra-red photography and for objects stained red and brown.
B	58	Green	26·0	For maximum contrast in transparent specimens and red-stained specimens—particularly useful in medical work.
C	47	Blue-violet	3·2	Enables high resolution to be achieved with orthochromatic plates, also enhances contrast of colourless specimens (Plate 50).
D	35	Purple	—	Rarely used alone (requires long exposures).
E	22	Orange	34·0	General purposes, isolates the yellow band from various light sources.
F	29	Red	7·0	Enhances contrast in brown subjects.
G	15	Deep yellow	67·0	Enhances detail and contrast of blue stained specimens and is often used with a green filter.
H	45	Blue-green	5·07	Increases resolution and contrast of colourless and orange-yellow specimens.
XI	11	Light blue	40·4	Gives high resolution with panchromatic material; used with high-intensity tungsten light.

Corresponding Ilford filters are listed below:

Table 15.

No. 305	Micro 1 (blue-violet)
No. 303	Micro 2 (blue)
No. 405	Micro 3 (green)
No. 110	Micro 4 (deep yellow)
No. 202	Micro 5 (deep orange)
No. 501	Micro 6 (purple)
No. 502	Micro 7 (magenta)
No. 104	Micro 8 (yellow)
No. 108	Micro 9 (pale yellow)

Figure 134 illustrates the curves relating to these filters. All filters should, of course, be placed between the subject matter and the light-source, not between subject and objective.

Continuous running filter monochromator. The main feature of the Zeiss, Oberkochen, continuous running filter is that it is a special type of interference monochromator (Plate 44). This continuous running filter is a longitudinally banded filter made to traverse across the front of a light condenser and transmit light in spectral selectivity from violet (approximately 400 nm) to infra-red (approximately 750 nm). In so doing there is practically a linear relation between the spectral colours and their location in the filter so that a slight shift of the filter is a controlled area of 0·63 mm and corresponds to a shift of λ max. of 1 nm. The half width (HW), the breadth in tenths (TW) and the maximum transmittance (T max/%) in relation to the slit width having the following values:

Table 16.

Slit width mm	HW/nm	TW/nm	T max/%
0	12	30	30
2	12·6	31	29
4	14·5	33	27

This filter can be used in a special diaphragm assembly on a microscope with built-in illumination or it may be fitted to a high performance free-standing microscope lamp.

Infra-red. The objective, being fully corrected, brings the blue-violet, yellow and blue-green into focus on the same plane. It is advisable to use a compensating eyepiece to ensure the colour correction of the particular objective in use. This is not possible when using infra-red material as there is a vast difference between the visual focus and the infra-red focus. The sensitivity of the eye falls off at approximately 750 nm but the sensitivity of photographic material goes beyond this point, having a peak at approximately 900 nm. In view of this it is recommended that an Ilford Tricolour Red filter or a Wratten 25 "M" series filter, be used when focusing and replaced with an infra-red filter (Fig. 135) before the exposure. It must be borne in mind that focusing has taken place in light of a wavelength less than the infra-red and that some further focusing must be done by careful adjustment. It is recommended that the fine adjustment be used, the degree of movement being determined by trial and experiment. It is as well to remind ourselves here that the objective should be racked *away* from the

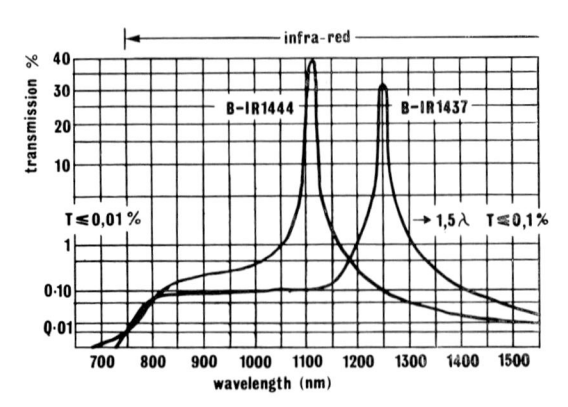

Fig. 135. Spectral curves for Filtraflex B–IR 1444 and B–IR 1437 filters.

specimen. There are a number of filters which transmit beyond 700 nm, as follows:

Wratten 88A transmits from 720 nm;
Wratten 87 transmits from 740 nm;
Ilford 207 + 813 also transmits from 740 nm.

The exposure factor is governed by the colour sensitivity of the particular emulsion in use, the type of light used to illuminate the subject and the filter used. Since a colour filter functions by absorbing a part of the light which falls upon it, it follows that the exposure required will be greater than if it were made without the filter, and the actual ratio of the filtered exposure to the corresponding unfiltered exposure is known as the "exposure factor". The numerical value of the exposure factor for a given filter is not fixed, but depends upon the colour sensitivity of the photographic material used, and upon the quality of the light illuminating the object. The exposure factor is only an approximation, a starting point (see Plate 43 for further information).

Ultra-violet. These filters are visually opaque but transparent to short- or long-wave radiations. Their function is to make a film effectively insensitive to visible light (see Fig. 65). The eye is not sensitive to light of wavelength less than 400 nm but the special ultra-violet lamp, already referred to, can be used in the visible region of the spectrum (see Fig. 184). The lamphouse is fitted with a filter which transmits ultra-violet and absorbs all visible rays. The filters capable of doing this are Corning 586 and Wratten 18A. The Wratten 17A filter is slightly inaccurate with panchromatic film because it allows a small proportion of visible rays to pass but it is effective with non-colour-sensitive emulsions as the transmission is in the region of 200 nm. The Wratten 32 filter, which is minus-green, absorbs heavily the two most prominent bands from the mercury lamp, 546 nm and 550 nm. A liquid filter, housed in an optically flat cell, is a concentrated solution of nickel sulphate. The Corning Glass filter 9863 is just as effective, as is Filtraflex R–UV series.

Ultra-violet absorbing filters are the yellow, yellow-green and green. The extreme end of the mercury arc, that is the red transmission, may be absorbed by a solution of cobalt salt (see Ultra-violet photomicrography, page 231).

Interference filter. The Fabry–Perot interference filter consists of two highly reflecting silvered surfaces separated by a very thin and uniform spacing layer of air or cryolite. With this type of filter it is possible to select a very narrow band of radiation in any part of the spectrum. Dye filters cannot approach the selectivity of a Fabry–Perot filter and the effect obtained with a gelatine filter is therefore not so pronounced as with an interference optical filter (see pages 236 and 238).

The multiple reflection filter. This consists of two semi-transparent films of silver separated by a dielectric spacer, which is transparent and of a low refractive index. The three layers are fixed on glass by evaporation in vacuum. The thickness of the dielectric layer controls the spectral position of the transmitted light. The spacing of the reflecting layers is comparable with the wavelengths of light and interference phenomena enable the filter to be highly selective.

Heat-absorbing filters. The special Pyrex glass heat absorbing filter, blue-green in colour, absorbs in the infra-red heat rays as well as part of the visible red. Care should be taken when obtaining one of these filters to ensure that it is free from air bubbles and defects in the glass. These Aklo filters can be purchased in light, medium and dark shades of 2–3 mm in thickness. Some loss of light is experienced, the light colour heat absorbing filter transmits 25%, the medium 60% and the dark heat absorbing filter 42% of the incident light. They do not absorb all the heat, as can be seen from the following: 75% is absorbed by the light heat absorbing filter, 93% by the medium and 96% by the dark heat absorbing filter. The Chance heat absorbing filter is very efficient and is used extensively in research (see Fig. 190).

Neutral density filters are listed on page 151, and in Fig. 139.

Diffusion disc in the field plane

When some form of diffusion is necessary, especially when photographing a "large" field by transmitted light the specimen is often illuminated direct by the source, dispensing with the substage condenser system. The specimen is mounted on an opal slide 77 mm × 25 mm (3 in. × 1 in.), or a piece of fine ground glass surrounded by a large dark disc, such as a washer. Though unwanted light is excluded from round the specimen area there is still considerable flare and, if the correct negative material is not used, a flat result follows. Of course the lamphouse iris diaphragm can be reduced to lessen the cone of projected light. A diffuser (Table 12) can be positioned away from the specimen as illustrated in Fig. 136.

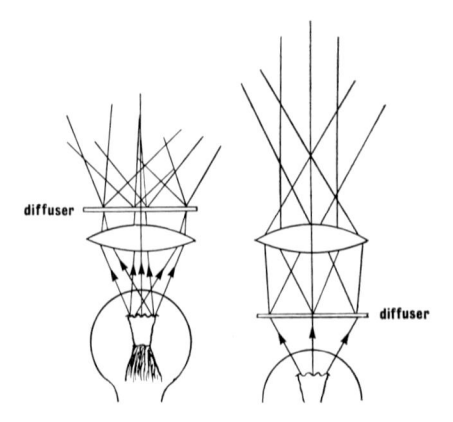

Fig. 136. Left: Diffuser positioned in front of lamp condenser. Right: Diffuser positioned between lamp and condenser.

A diffuser in conjunction with flashtubes serves a very useful purpose. Some lamphouses are fitted with a diffuser which becomes an acting source, positioned against a small lamphouse iris diaphragm. Opal diffusers are made of a Plexiglas material of approximately 2 mm thickness. The use of ground glass is not so efficient, some non-diffused light is transmitted while much is scattered as seen by the figure. Further, it is difficult to make an evenly ground surface area suitable for close-up projection. The iris diaphragm cannot regulate the light intensity but merely controls the amount of light passing through to the diffuser.

Colour of light

Sensitivity of "light colour" is characteristic of reversal film, forcing in us a consciousness of colour casts, which are seen far too often in the final transparency. The results of the filament lamp when viewed by a light source of daylight colour, or colour balance, are often poor and if the transparency is viewed with an ordinary tungsten filament lamp the colours appear to change towards the yellow bands. What has happened is that the colour temperature of the viewing source has changed. The temperature of a source is expressed in degrees Kelvin (°K) and is the measure of light colour. The approximate colour temperature of various light sources is:

COLOUR OF LIGHT

Table 17.

Source			°K
Candle flame (yellow light)			1,900
Incandescent bulb	230 V	60 W	2,500
Microscope bulb	6 V	12 W	2,700
Microscope bulb	6 V	15 W	2,850
Microscope bulb	6 V	30 W	3,000
Microscope bulb	12 V	60 W	3,000–3,400
Microscope bulb	12 V	100 W	3,200
Projection bulb			2,900–3,200
Iodine-Quartz lamp	12 V	100 W	3,300
Clear flash bulb			3,400
Halogen arc lamp		250 W	3,400
Photoflood bulb		500 W	3,450
Warm fluorescent tube			3,500
Winter sunlight			5,000
Flash tube			5,500
Xenon high-pressure burner		150 W	5,200–6,000
Blue flash bulb			6,000
Xenon lamp			6,000
Daylight, sun obscured			9,000
Daylight, blue sky (blue light)			15,000

Low colour temperature expresses yellowish light while high colour temperature is a bluish tint. Both will be encountered during the course of making colour photomicrographs and so, before venturing into this field, it is advisable to determine the colour temperature of the light source in use. Figure 137 shows the spectral energy distribution curves for some tungsten filaments at different

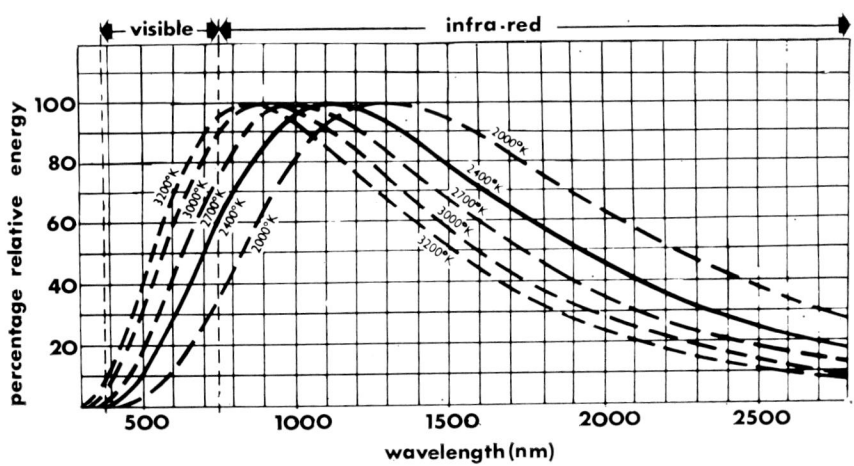

Fig. 137.

F*

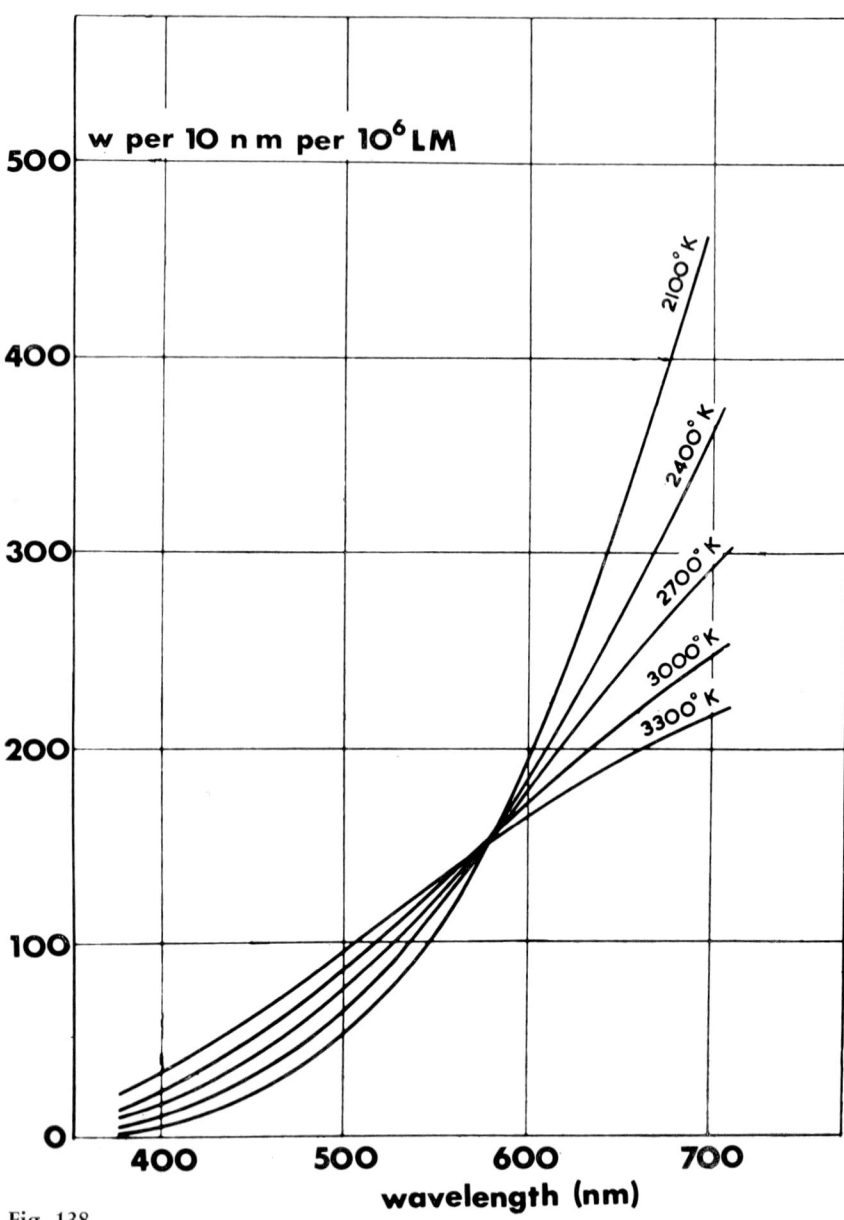

Fig. 138.

temperatures. The curves illustrate the relative spectral energy and shows how the curve moves further into the infra-red as the filament temperature increases. Figure 138 curve is for accurate work in which energy figures are required for such purposes as calculation. Tungsten halogen lamps have similar spectral distribution curves, but in the case of iodine additive lamps there is approximately 6% absorption in the 550 ± 10 nm waveband. It can be seen from the conversion table for light balancing filters that if the colour balance of the source does not correspond with that of the colour film, it is necessary to use a correction filter and, without a filter, it will be impossible to produce a high quality transparency. Over the years I have discovered that 6 V 12 W and 30 W lamps give good results, provided of course that all other considerations have been satisfactorily worked out. However, during winter, voltage changes can affect the colour balance of the photomicrograph, particularly in the colour background areas. Nevertheless, a constant standard can be obtained if the source is modified by the correct filter. As in Table 17, one source will produce a blue effect while another a yellow. The photomicrographer must adjust this balance. The use of filters in colour work needs much more care than in black and white and more experiments are usually called for here than when filtering light for black and white work.

A change in colour temperature can also be brought about in the apparatus itself. In some instances this is most evident when switching from low power, without a substage condenser, to high power, with the use of the substage. Such changes, perhaps unnoticed by the eye, are due to the many optical elements in high power photomicrography. Theoretically, the filter required for the particular source may not correct the light falling on the film so some trial exposures

Table 18.

Wratten filter No.		Exposure increase in stops	Colour temperature of source °K	Colour temperature of source °K
82C	82C	1 1/2	2,490	2,610
82C	82B	1 1/2	2,570	2,700
82C	82A	1	2,650	2,780
82C	82	1	2,720	2,870
82C		2/3	2,800	2,950
82B		2/3	2,900	3,060
82A		1/3	3,000	3,180
82		1/3	3,100	3,290
81		1/3	3,300	3,510
81A		1/3	3,400	3,630
81B		1/3	3,500	3,740
81C		1/3	3,600	3,850
81EF		2/3	3,850	4,140

(Converted to 3,200 / Converted to 3,400)

may be necessary. Of course if there is a yellow cast a blue filter is needed, and vice versa. A strict log ought to be kept of all details of the optical system used: bellows length of camera, filter, etc. Then, when another subject demands the same set-up, it will save considerable time if the previous details can be referred to.

Filters for colour photomicrography can be divided into four classes:

(1) The light balancing filters, commonly known as correction filters.

(2) The use of material exposed to a source contrary to its colour balance: i.e. daylight film with artificial source. Here either a Wratten 80B (light blue) filter should be used, except by daylight source on a film colour balanced for artificial light, or a conversion filter 85B (orange). In practice, the daylight material exposed to artificial light through the use of the filter is less successful than the artificial light film exposed to daylight.

(3) Neutral density filters provide a valuable means of controlling light intensity (Fig. 139) but do not affect the colour rendering of the subject. These filters should be used when the light intensity, seen on the focusing screen or transmitted through the viewing eyepiece, is in excess of normal vision. In addition, an intense or uncontrolled light causes a poor negative image quality (Plate 9(a)). Neutral density filters can be supplied in which instead of colloidal carbon the absorbing material is silver grain, in a state of division similar to that in a photographic image. The scatter is considerable and their use lies mainly in special applications likely to be encountered in research, where it is desirable to simulate effects.

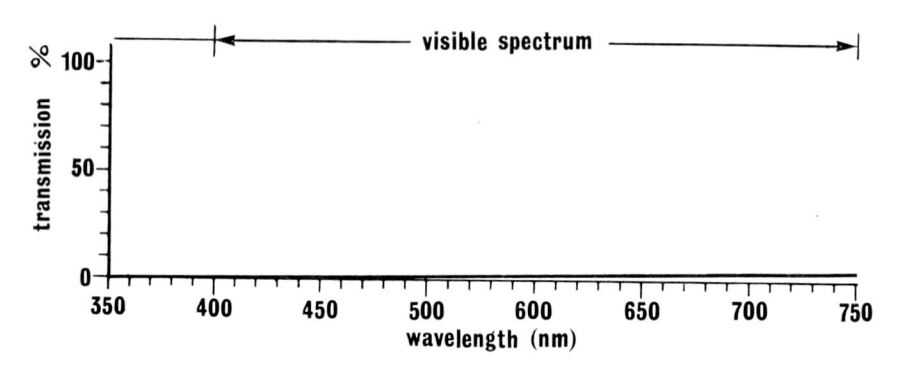

Fig. 139. Transmission curve. Neutral density filter NG/9 1 mm thick.

Neutral Wedge Screens are supplied in two forms, (a) with uniformly increasing density and (b) in step form; that is, a series of uniform steps, each a definite amount denser than the previous one. Neutral density filters can be used singly or collectively and may be purchased from Kodak in several densities as set out:

Table 19.

Density	Percentage transmission	Factor
0·1	80	1 1/4
0·2	63	1 1/2
0·3	50	2
0·4	40	2 1/2
0·5	32	3
0·6	25	4
0·7	20	5
0·8	16	6
0·9	13	8
1·0	10	10
2·0	1	100

(4) Background colour correction. Sometimes a mounting medium produces a yellow/amber overall hue which can be obtrusive viewed on the screen. If so it should be counteracted by the use of a pale blue filter which will produce a very slight, but pleasant, blue-tinted background.

Vario variable neutral density filter incorporates in one unit the ability to vary the amount of light passing through by ×2 to ×10. Two polarising units snap into each other and the first unit is calibrated in filter factors and has only to be rotated to the marks indicated to obtain the desired density. The base unit may also be used as a polarising unit.

Stained specimen

A stain in a bottle, a blot on a piece of filter paper, or a stained specimen, may appear as something quite different when reproduced in colour through the microscope. One reason is that some of the colour saturation is lost in the specimen, this being much more evident in some subjects and with certain particular stains. The fact that the specimen is subject to intense transmitted light has a bearing on the loss of colour, as does the magnification. The latter is mentioned particularly, because the higher the magnification the less saturated becomes the colour. One may think of a fairly dark blue balloon, which when blown up, becomes a much lighter blue. If the balloon is enlarged further it is, of course, no longer seen. It is always advisable to heavily stain minute subjects such as bacteria. If this is not done the image will not be seen at ×2,000. A short while ago the writer received two prepared slides of bacteria and was asked to reproduce them in colour but these, when held towards the sky and viewed through a strong eyeglass were almost colourless and so it was impossible to comply with the request. However, there is a danger that an overall colour cast may appear in heavily stained subjects, and this can extend to adjacent details of a different colour. Any colour can give a cast to an intrinsically neutral, white background suggesting that the specimen was not washed after staining. Unevenly stained, unevenly cut and mounted specimens, are not the best subjects to photograph, either in colour or in black and white. Flare within a lens system will cause colour light scatter, producing a bias in the whole area, including the clear area around the specimen.

In some cases pigmentary colours are the general rule. Many butterflies and moths owe their colours to the physical structure of their wing scales but when they are viewed by high power transmitted light they are often entirely colourless. Unfortunately, some subjects exhibit their colours at low magnification only. Shimmering colours seen in the shells of lobsters and other similar surfaces are further examples of beautiful colours, physical in origin, though difficult to arrest through the microscope.

Colour. Some kind of "cockpit drill" should be applied when making colour transparencies, every point being carefully checked. The following are suggested:

(a) All optical components be aligned and adjusted.

(b) Light source adjusted and aperture made operational—check for odd effects and presence of dust on optical systems.

(c) Apply Köhler illumination.

(d) If necessary insert correction filter.

(e) Focus the specimen and finally check for even illumination.

(f) Make an exposure recording with light meter.

(g) Take a series of exposures: i.e. one each side of the recommended meter reading, plus that recommended.

(h) Record exposure times and relevant data.

(i) Select the best colour transparency and record the exposure and correlate the exposure with the meter. Form a table of exposures for future use.

(j) Project transparency and check for colour casts or possible defects.

(k) If colour cast is detected, place the necessary correction filter in the light train of the microscope. Remove dust, etc. if necessary.

(l) Make a further exposure with the correction filter in place, and if the final result is correct, record all details for future use.

Camera

The main function of the camera is to house the emulsion in the plane of focus, which is known as the focal plane, or the film plane. Some cameras operate at a selected distance from the eyepiece, and this may be extended to twenty inches, or thereabouts. Focusing is either carried out by viewing the image in the viewing eyepiece (Fig. 140) or the image on the ground glass screen (Fig. 141). There is no need for the camera to have a lens attached as the microscope objective is the taking lens.

CAMERA

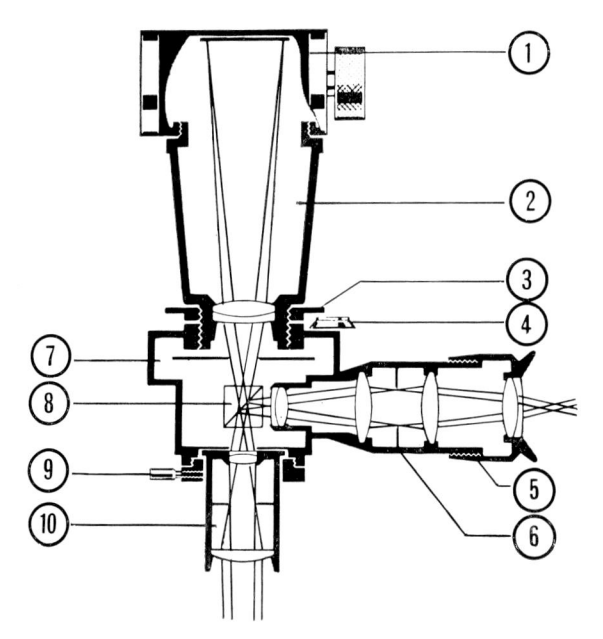

Fig. 140. Leitz 35 mm camera and attachment. 1. Leica miniature camera. 2. Intermediate adapter ×1/3. 3. Setting ring for orienting the attachment on the camera. 4. Dial of central wheel shutter with flash synchronization. 5. Correction mount of the telescope eyepiece. 6. Diaphragm with eyepiece graticule. 7. Central wheel shutter. 8. Swing-out deflecting prism. 9. Fixing screw for clamping micro attachment to the microscope tube. 10. Eyepiece.

The camera must be rigid which is particularly necessary when it is extended to, say, twenty inches. Exposures for such cameras are usually long so a shutter is not often used. The exposure can be controlled by the electric light switch which also controls the microscope lamp. Of course controlled shutter speeds are essential when exposing colour film. Some cameras provide for a $\frac{1}{4}$ plate, or 12·8 × 10·2 cm (5 × 4 in.) negative (Plate 46), while a popular size is the 35 mm (Plate 47a).

A 35 mm negative must not be expected to produce the same print standard as that made from a $\frac{1}{4}$ plate or 12·8 × 10·2 cm (5 × 4 in.) negative. A 35 mm negative must be enlarged five times to arrive at a 15·2 × 12·8 cm (7 × 5 in.) print if the whole of the negative format has been filled, whereas a $\frac{1}{4}$ plate negative is only slightly enlarged to achieve the same dimensions. A $\frac{1}{4}$ plate negative enlarged five times will give an overall area measuring 40·7 × 50·8 cm (16 × 20 in.), as opposed to an enlargement of fifteen times linear of a 35 mm negative. From this it is obvious that there will be a difference in the image quality of the photomicrograph if both negatives are processed aright.

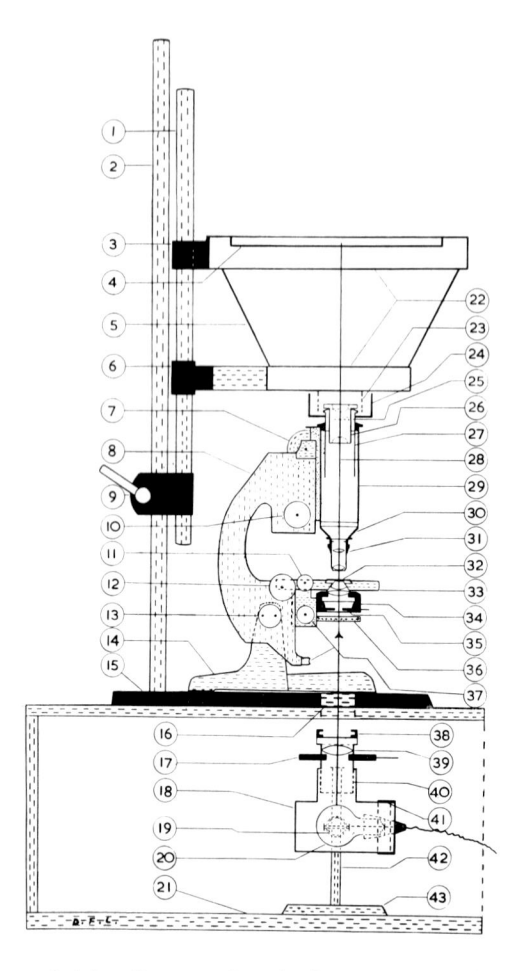

Fig. 141. 1. Camera height adjustment bar. 2. Camera support. 3. Height adjustment. 4. Plate holder and ground glass screen. 5. Bellows (treble length). 6. Height adjustment. 7. Coarse adjustment. 8. Limb. 9. Main height adjustment clamp. 10. Fine adjustment. 11. Stage adjustment (lateral). 12. Stage adjustment (forward movement). 13. Joint. 14. Foot. 15. Metal base board. 16. Aperture in base board and cabinet. 17. Lamphouse iris diaphragm. 18. Lamphouse body. 19. Height and tilt adjustment screw. 20. Coiled filament bulb. 21. Cabinet. 22. Camera. 23. Lens mount. 24. Light protective ring. 25. Primary image formed. 26. Eyepiece. 27. Draw tube. 28. Optical axis. 29. Body tube. 30. Nosepiece. 31. Objective. 32. Specimen. 33. Stage. 34. Substage condenser. 35. Substage iris diaphragm. 36. Filter carrier. 37. Substage focusing, adjustment and mirror attachment. 38. Filter carrier. 39. Lamphouse condenser. 40. Focusing sleeve. 41. Bulb adjustment sleeve. 42. Height bar. 43. Foot.

CAMERA

With regard to the larger special photomicrographic units, which provide a wide range of possibilities, the vertical arrangements, such as the Zeiss Ultraphot (Fig. 142a), the Leitz Wetzlar Orthoplan and Ortholux, have proved to be the most popular. A 10·2 × 12·8 cm (4 × 5 in.) camera can be used as well as a Polaroid Land system, or a 35 mm camera. In addition, the Leitz model can be used for ciné-micrography and a highly sensitive light meter with a large

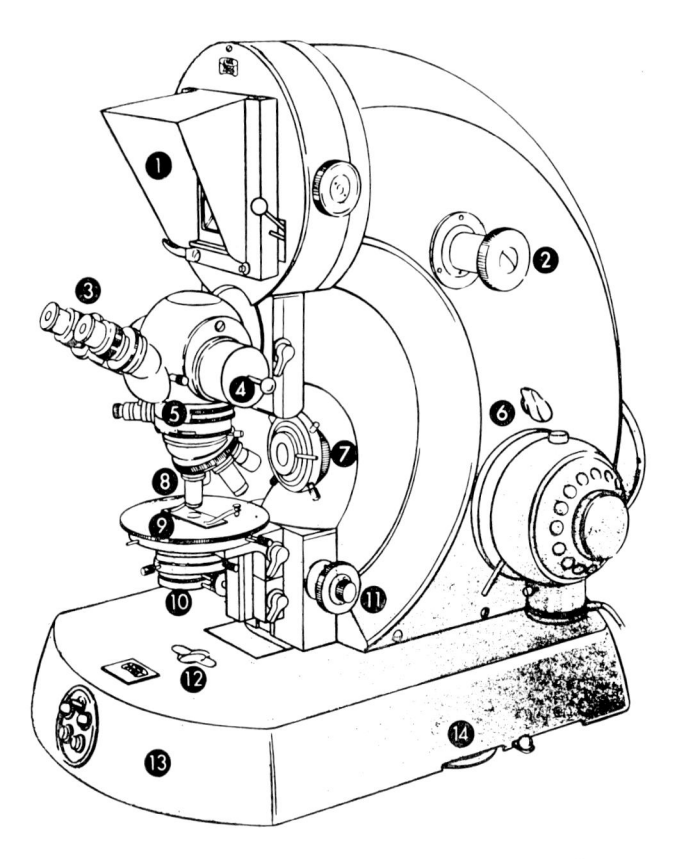

Fig. 142a. Zeiss Ultrophot. 1. Photo-head 9 cm × 12 cm with built in setting lever for shutter. 2. Adjustment varying length of camera 30 cm. 3. Normal oblique viewing, rapid changing mechanism for quick interchange of tubes. 4. Adjustable ray director. 5. Optovar with 3 magnification stages and focusing microscope. 6. Lamp switch for various forms of illumination. 7. Constant centering lamp. 8. Objective fitted with lower lens protection. 9. Rotating stage for fine, rapid adjustment in all directions; interchangeable stage-carrier and carrier for illumination apparatus. 10. Adjustment for illumination apparatus. 11. Coarse and fine adjustment. 12. Constant centering of light source for microscope and filter carrier. 13. Push button for shutter and automatic camera. 14. Illumination field diaphragm for trans-illumination.

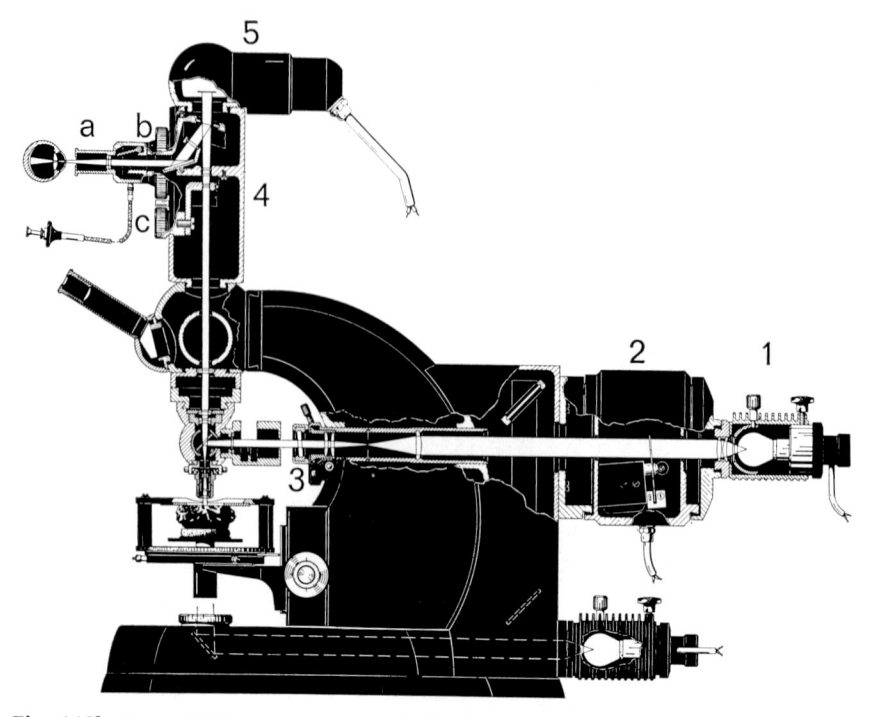

Fig. 142b. Zeiss MPM microscope. 1. 60 W illuminator with collecting lens. 2. Light modulator. 3. Continuous interference-filter monochromator. 4. Photometer head, (a) viewing magnifier with cover, (b) revolving disk with photometer stops, (c) revolving disk with projection lenses. 5. Detector housing with detector.

measuring range easily covers all exposure times likely to occur in photomicrography (Plate 45).

By way of contrast, Fig. 140 illustrates the Leica camera attachment. The camera lens is replaced by a micro adaptor which can be purchased from the manufacturers. This adaptor can be used with any make of microscope and can also be used with any eyepiece. The length of the camera, from film to eyepiece, should be 125 mm. If shorter, it may lead to spherical errors. The distance can be increased by adding extension tubes. High-power eyepieces can be used for distances exceeding 147 mm without impairing the image quality.

This specially designed eyepiece camera attachment is the connecting "rod" between microscope and the Leica camera. It is designed specially for use with a Leitz microscope, though not limited to this make. The camera attachment seen here fits firmly into the camera lens mount and holds the eyepiece firmly at the other end. The camera lens is not used. The specimen is viewed through the viewing telescope (Fig. 140, 5). At the same time the optical system renders the camera parfocal with the microscope. Exposure is controlled by a built-in

CAMERA   157

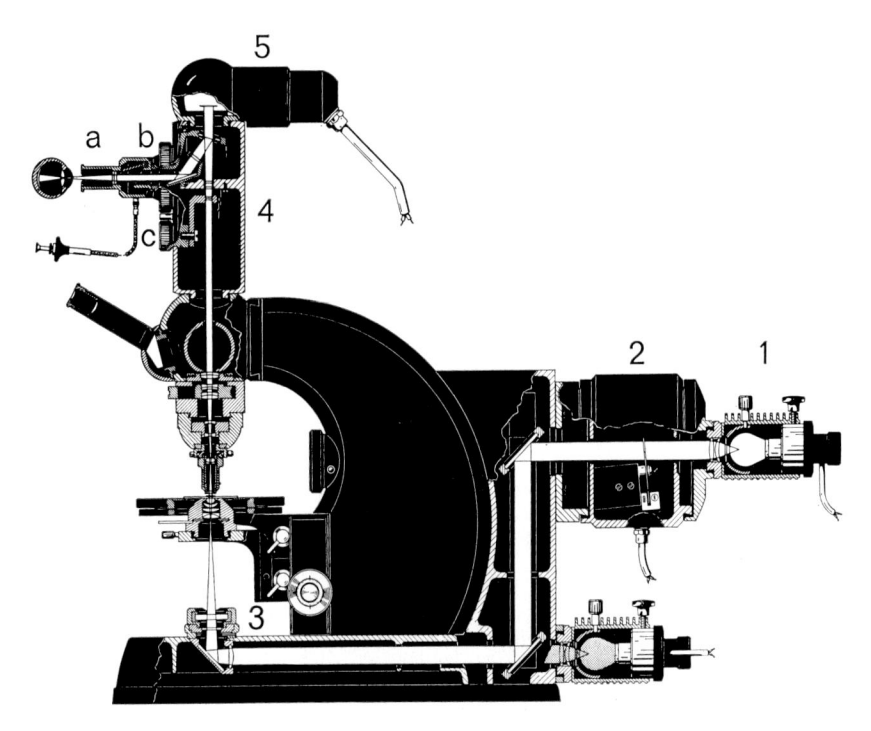

Fig. 142c. Zeiss MPM Microscope with Photometer, path of rays with reflected light (see caption for Fig. 142b for explanation of numbered parts).

shutter on the housing. This, since it permits continuous viewing, is especially suitable for holding in focus objects which otherwise may swim out of focus, or sink slowly out of the plane of focus. This combination of camera and attachment is used extensively in medicine, in research, in industry, in schools and colleges. It is an advantage to have two cameras—one loaded with colour and one with black and white film.
Miniature camera with bellows extension. Some single lens reflex cameras can be used with near focusing bellows device to take photomicrographs. The apparatus is simple and works according to the principles of the plate bellows camera. The bellows fit over and engulf the eyepiece, excluding all other light. The bellows can extend to 125 mm or 250 mm. When extending the bellows, a change in magnification takes place without changing the microscope optical system, Fig. 84. In this way attention can be directed to a specific area or to a part of the specimen.
Exakta camera. There are three main methods applying this camera to photomicrography. The first consists of a collar which is clamped around the microscope draw-tube, linking camera to microscope. The top part of the collar is

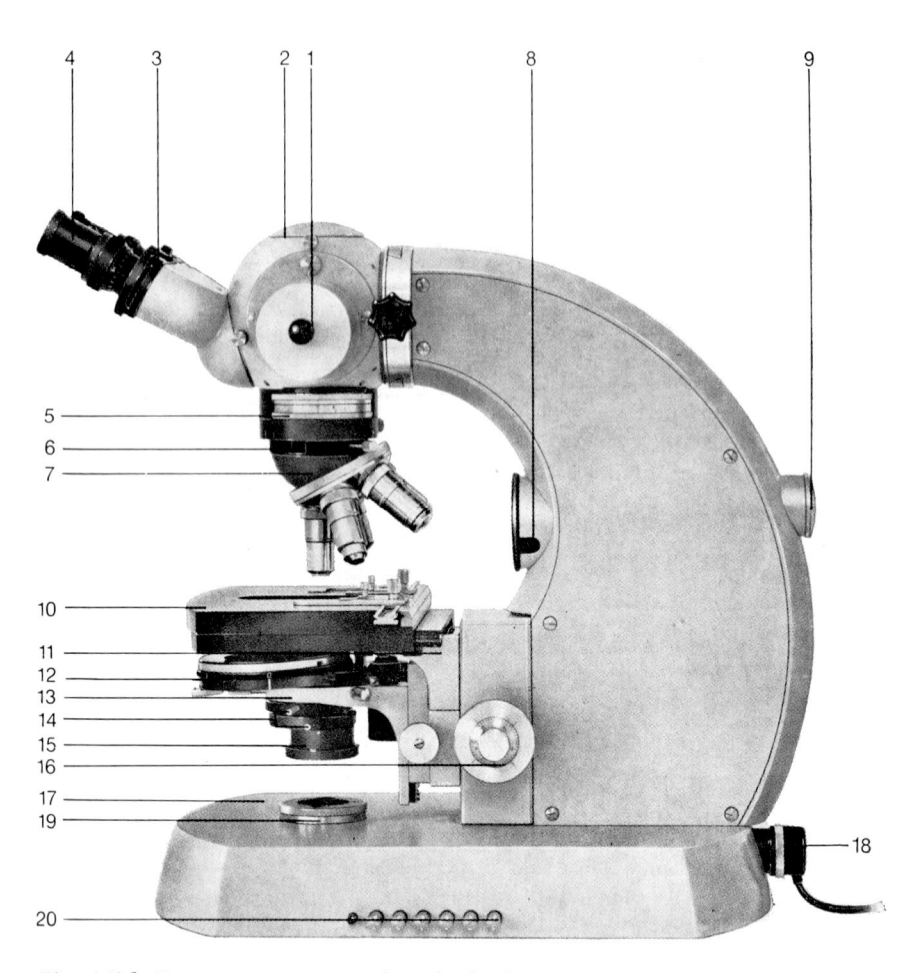

Fig. 142d. Zeiss microscope. 1. The tube head contains an adjustable prism system permitting the beam to be directed as follows: All light transmitted to observer's eyes; All light reflected upwards as in photomicrography; 1/3rd of light reflected to eyes and 2/3rds upwards. The tube head shown is the Optovar, which can be changed for a tube head without it or the Luminar head. 2. Dovetail ring (concealed) can serve for attaching a vertical tube. 3. The binocular body is interchangeable with special purpose bodies. 4. Eyepieces. 5. The Optovar is an optical system allowing variation of the total magnification by the factors 1·25, 1·6 and 2. It also contains a focusing Bertrand lens for viewing the exit pupil of the objective. 6. Slot for compensators or retardation plates. Slot for analyser, interference-contrast slide, etc. visible in Fig. 13. Both slots may be closed with dust plugs. 7. The quintuple revolving nosepiece can be exchanged for single nosepieces or vertical illuminators. 8. Opening for reflected-light aperture-stop insert. 9. Connection for 60 W surface illuminator. 10. Large, graduated, square mechanical stage. This can be exchanged for rotating and centering mechanical stages, glide stage, auto-levelling stage for polished specimens, universal stage. 11. Attachable stage carrier. 12. Achromatic–aplanatic bright-field/dark-field phase-contrast condenser V/Z, 1·4 N.A. It can be

CAMERA

hinged and a catch releases or locks the hinged section in position. The top of the hinged section is threaded to take the standard extension tubes and bayonet camera adaptor. In this way the actual projection distance from the eyepiece to film plane can be adjusted to the image viewed.

The hinged section, which is attached via the tube or adaptor to the camera bellows, allows visual inspection of the specimen, as does an interchange of eyepiece powers, without having to remove the whole set-up. The utility of this feature will depend on the construction of the microscope which should be rigid. The camera is clamped firmly over the eyepiece but does not touch it. The bayonet ring adapter has a clamp collar permitting the alignment of the camera in relation to the microscope, vertical or horizontal.

The second is an adaptor attached to the microscope draw-tube by spring pressure, exerted by turning a capstan collar containing three contact surfaces evenly around it. There should be no slackness in the draw-tube. The upper part of the link has a fitting taking an intermediate ring with a curved dovetail fitting which fits to the draw-tube link and is locked into position. The intermediate ring is screw-threaded at its upper end and takes the standard Exakta extension tubes or camera bayonet adaptor ring, according to the magnification: i.e. film distance from the eyepiece. With direct attachment of the camera to the microscope eyepiece a reduction of about $\frac{1}{3}$rd of the overall magnification, projected by the viewing eyepiece, is registered in the film plane. Like most cameras, the projection distance is about 250 mm.

The third type of adaptor is the "Shell" support. This system works in conjunction with the Ihagee Sliding-Bellows Unit. The front panel of the bellows unit is threaded to take a metal hood which acts as a light trap around the eyepiece and camera. The eyepiece is removed and placed in the "Shell" and in the draw-tube. The bellows unit is then lowered over the "Shell" when all outside light is excluded. Centering must be carried out to ensure a sharp overall image on the negative.

To increase the magnification without changing the microscope optical system the bellows unit can be extended. Here the camera will be fitted to a strong vertical metal column which stands away from the microscope.

---

exchanged for other condensers. 13. Attachable, vertically adjustable substage with centering screws and dovetail ring. 14. Filter holder. 15. Auxiliary condenser lens. 16. Focusing is achieved by racking the specimen stage up or down. The precision reduction gearing is actuated by coaxial controls. One graduation interval on the control is equivalent to a vertical displacement of 2 μm. The focusing lever serves for rapid lowering of the specimen stage and comfortable relocation of the specimen in the focal plane defined by the surface of the base acting as a stop. 17. The standard base contains: 18. A 6 V, 15 W illuminator, a lamp condenser and an adjusted reflecting mirror; 19. A diaphragm insert with iris field stop; 20. A filter control with holders for six filters of 32 mm diameter.

Kopil extension tubes achieve the same object as the Kopil bellows extension units. Both tubes and bellows units fit the Pentax, Leica, Exakta and Praktica. The Kopil microscope adapter enables the camera to be fitted to the microscope tube. The hinged joint allows the camera to be swung into or out of the operating position (Fig. 143). Of course there must be no slackness in the microscope tube.

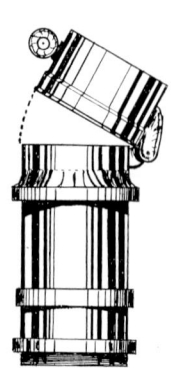

Fig. 143.

Exposure determination

It is essential to determine the optimum exposure time and the choice of method depends on experience, economic consideration and seriousness of the exercise. (The latter is included because some require merely any image as a basis for a design.)

Test exposure. When using cut film or plate it is possible to make a series of test exposures, consisting of strips across the negative, from long to short exposure, developing for a set time and then selecting the exposure which conforms to one's printing techniques. When a negative is used for a test, a strict log of the various exposure times should be made as the strip is exposed: each receiving double exposure of the preceding one. 35 mm film can be exposed in the same way, or, if one is extravagant, several whole frames. After development, the choice of the "correct" exposure can be made. A simple multiple exposure attachment is available for 9 × 12 cm and 13 × 18 cm bellows type cameras. If, after the "correct" exposure has been found, a change is made in the lamphouse iris diaphragm or substage iris diaphragm, then a further test exposure must be made. With filters, the filter factor (in relation to negative material) must be added to the "correct" exposure.

A successful manual method of exposure determination is obtained by using a multiple exposure attachment which displaces the 35 mm camera in the film plane. A series of test exposures can be made without the previous exposure being affected by the second and so on. When the test exposure is made a blind moves across the film, covers the exposed area and uncovers an area for a fresh exposure. Some six exposures can be made on one piece of negative material.

For large format negative material, such as 9 × 12 cm, it is possible to have a series of thin masks, which, when used one after the other, exposes the whole film area. This means that after each test exposure a different mask must be loaded over the film face in the film holder. Of course this operation must be carried out in the dark.

A successful method, particularly useful when it is necessary to calibrate non-automatic equipment, is to photograph a standard test specimen with a known exposure time, and increase or decrease exposures of the unknown according to its density.

To find the "correct" exposure, it is advisable to record lighting conditions, film speed, development time and further points contributing to the exposure, to save making tests another time. A change in the optical system will alter the exposure time as will increasing or decreasing the camera bellows. Of course to some, the image density on the ground glass screen is an indication as to the "correct" exposure. Light measurement of exposure time is always superior providing one gets to know what the exposure meter is "saying".

A photometer with good optical and electronic systems will accurately measure transmission. The relation between absorbance and transmission is an exponential function. Some meters, however, are unable to reproduce all the information transmitted from the projected image. Slight mistakes seem to be inevitable and measuring, including reading, errors as large as 1% transmission, are common. Middle range scale readings are the most accurate. The photocell is placed over the top of the eyepiece in use, or it can be placed against the ground glass screen, measuring the projected image. Perhaps this latter method can be more selective, as the photocell can be placed against a "dark and light" area and a mean reading taken. The usual size of the measuring probe (spot measurement) is approximately 1/100 of the projected field of view. Each individual component of the specimen can be measured separately and thus not only the exposure time but also the contrast may be determined.

Leitz Wetzlar have produced a model of the Microsix–L exposure meter, in which a light-sensitive probe replaces the eyepiece and is joined electrically to the body of a light meter. This meter covers the whole range of lighting met with in photomicrography; including low and high power and readings from low-level fluorescence. Such meters are absolutely necessary since colour film reacts to slight changes in exposure time by considerable shifting of colour values.

Some microscopes are fitted with a rheostat. With this the illumination intensity on the film plane can be adjusted to a constant light value, or brightness, making exposure times more constant.

In darkground and fluorescence photomicrography the total light available depends far less on the brightness of the individual particles (which alone determines the exposure time) than on their numbers. Only structure measurement, which allows individual particles, or their bright fringes, to be measured separately from the background, can give accurate results in such cases. This requires an extremely small measuring surface. However, under certain conditions, spot measurement will give a better relationship between specimen and background and can even, with sufficiently large bright particles, produce a more accurate measurement. But particularly in darkground it is often just as makeshift a method as integral measurement.

Attempts have been made, particularly with automatic cameras, to improve integral and spot measuring techniques by estimating the area of the field actually covered by the specimen, then correcting the exposure time by an appropriate setting of a regulating device. In bright-field photomicrography, however, this small part of the specimen is of little consequence as far as measurement is concerned, for measurement, with and without a thin specimen gives results which do not differ significantly.

More than eleven years ago Zeiss, Oberkochen, brought out a research microscope with a built-in miniature camera capable of turning out, completely automatically, correctly exposed colour transparencies and black and white negative material. In this instrument the exposure problem was solved by means of a built-in CdS cell long before the appearance of the automatic cameras in use today. Frame-by-frame film transport, too, was already automatic.

Now Zeiss have developed a new model, Fig. 142d, whose design shows notable improvements through the use of experience. The automatic control is operated by a built-in photomultiplier which now permits exposures down to 1/1000 sec instead of 1/100 sec as before. This means that the electronic control, which automatically measures the available light, determines the appropriate exposure time and opens and closes the shutter accordingly. The film in the cassette is stopped, when used up, automatically.

In addition to the integral exposure determination over the whole picture area, selective measurement is now possible on specific details of the subject. For this, the exposure control switches over automatically to a sensitivity corresponding to the smaller area now under measurement. The automatic control can be set to any film speed within a range of 5 to 40 DIN (2·5 to approximately 800 ASA) in steps of 1 DIN. Cassettes are interchangeable so that different types of film can be used as desired. A synchronising contact is built in for flash exposures with the micro flash unit which can be triggered when the shutter is operated by hand: i.e. without the automatic exposure control. Built into the microscope are four neutral density filters, one green filter and one conversion filter (Fig. 142(d) 20).

By replacing the normal tube of the photomicroscope by a Luminar head, using the Contarex single lens reflex camera, low magnification photomicrographs can be taken with the instrument used as a "simple microscope": that is to say, *without an eyepiece*.

Ciné can be incorporated with the microscope, and a television camera whereby the transmitted microscope image can be relayed to any number of screens simultaneously.

The Autolynx automatic photo head by Gillett and Sibert has been designed to fit their Conference Microscope and, with the latest electronic techniques, coupled with special electromechanisms, a 35 mm film can be correctly exposed and operated.

Primarily, the Autolynx has been devised to meet the demands of routine photomicrography: 35 mm in format. The specially designed head measures the light intensity and determines the exposure to allow one frame to be exposed every 6 sec. The film is transported automatically throughout, and on the conclusion of each exposure the film movement is registered on a frame counter.

Watson semi-automatic exposure control unit. The unit provides a means of estimating the required exposure time and of controlling the duration of the exposure, using any of the automatic cameras of the Microsystem 70 range. The two functions are mechanically linked, so that the action of determining the exposure time automatically selects the appropriate speed.

The exposure estimating device uses a CdS cell as a light-sensing element, incorporated in a Wheatstone bridge type circuit. The other three resistances of the bridge are used for (a) calibration, (b) film speed control and (c) balancing. The electrical balance of the bridge will be affected by a variation in any of the four resistances. In practice, the calibration resistance is fixed, the film speed resistance depending on the light level. Electrical balance is achieved (and indicated by a microammeter) by varying the balancing resistance.

The exposure timing device uses a delay circuit in which a condenser is charged through a variable resistance. The rate of charge, and consequently the exposure duration, is determined by the value at which this resistance is set. The bridge balancing resistance and the timing resistance are both varied in a fixed proportion to each other by the simple expedient of being mounted on the same control shaft. The resistance values have been selected so that when the bridge circuit is in balance the delay circuit will cause the correct exposure time to be given for the particular combination of light level and film speed.

The unit will accommodate a range of film speeds of 3–800 ASA for black and white, and 25–640 ASA for colour reversal film in steps of one third of a stop. It has a range of automatically controlled times of 1/30 sec to 60 sec in 50% increments and has provision for externally timed exposure of any longer duration.

Leitz auto head. The Orthomat microscope camera (Plate 45) is a fully automatic 35 mm microscope camera and offers the advantages of the well-tried unit-component: microscope and Orthomat can, whenever required, be combined to form a fully automatic photo-microscope or used as independent units. The beauty of this complete unit is its readiness for action. It is fully automatic in the exposure control, even in ill-lit applications such as darkground, phase contrast and fluorescent light (see Plates 77 and 78). It has a fully transistorized electronic system with photomultiplier as detector. Practically unlimited range from 1/200 sec. to long time exposures for weakly fluorescent specimens. Exposure control is adjustable individually to the characteristics of each specimen; even nearly point-sized specimens can be exposed exactly.

Automatic compensation of brightness changes in the specimen (as in Plate 6), are compensated, and adjustments made. In addition, adjustments in exposure times are made in both low and high power techniques. Interchangeable film magazines for black-and-white and colour photomicrography, can be made with ease. The specimen can be studied throughout the exposure.

Photomicrography with phase contrast equipment differs in no way from ordinary brightfield. The "Object Dark/Light Ratio" control takes care of the varying size of the specimen. Specimens illuminated in polarized light must be exposed either as in brightfield or as in darkground illumination depending on their size and nature. The operator decides after brief inspection whether to use the integrating or detail measurement method.

Optical bench

In contrast to using the microscope vertically for photomicrography, as illustrated throughout this series, the horizontal optical bench is still preferred by some. But this is limited in its application, failing both to record a wide range of subjects and in illumination techniques. This means that only permanently prepared specimens can be photographed as the microscope slide is on edge when positioned for photography.

Pond creatures in water in a hollow slide (Fig. 53), or wet mounted specimens such as algae, gelatine, emulsions, marine life, pickles, soft drinks and a host of other subjects, cannot be photographed. Also, unmounted subjects such as sand, sugar, aspirin (Plates 31 and 42), penicillin crystals, powder, contents of various tablets and so on, would simply roll off the microscope slide.

Endeavouring to discover new methods of investigation is one of the most important duties of every photomicrographer, and to communicate these must be the anxious desire of every earnest teacher in the science. In an attempt to become a Fellow of the Institute of British Photographers (now the Institute of Incorporated Photographers) I submitted many photomicrographs of pond creatures, in a free swimming state, and dry mounted crystals, particles in liquids, ingredients in soft drinks, sand and so on: all illustrated with special techniques. But these were not accepted and on meeting one of the Adjudicators, who was proud to show his horizontal optical bench in a basement, I asked: "How do you photograph pond creatures, dry crystals, sugar, liquids and so on?" The reply: "I've never done any in my life, and I shouldn't know how to go about it," needlessly to say, ended any consideration of that practical aspect.

This example shows that a Judge, a medical photographer, had been confronted with work in which he was completely inexperienced, the scope of the subject being so vast. The rejection is a judgment on the critics, but those entering this field need all the encouragement that can be mustered and practical experience passed on to them is a job well done. Everyone who has experienced the success of original researches naturally desires to encourage others in the same course, which cannot be better done than by showing as clearly as is possible the final photomicrograph. It is incumbent upon all to facilitate the communication of knowledge in every possible way.

Line overlaps in photomicrography

Through movement the eye has a wide field of vision. By the aid of the fore-and-aft and lateral movement of the specimen under observation the eye can cover an "enormous" area in a very short time. If it is desired to examine any part of a specimen in detail, it is necessary to concentrate the eye on the selected objects in turn and thus risk missing something which is not included in the concentrated field of vision. Also the eye is adversely affected by fatigue, glare and discomfort, and on occasions by imagination. The body has no means of accurately reproducing the mental picture of what has been seen and transferring it to paper and no human observer could record or even remember more than a small fraction of the details seen in a line overlap such as illustrated here (Fig. 144). The original mounted series of photomicrographs measured 137·3 cm (4 ft. 6 in.). (The same series measuring 457·2 cm (15 ft.) was exhibited at Europhot.) In the former each used area of the print measured 15·24 cm (6 in.) across and in the latter 53·34 cm (21 in.). The original specimen area represented in each print is approximately 1 mm. The overall length of the specimen measured 1 cm and instructions were given that the reproduction must be $\times 100$. This meant the finished photomicrograph or series of photomicrographs would be over 121·92 cm (4 ft.) in length. There are many routes to arrive at $\times 100$: let us look at some set out in Fig. 145. These can be compared to the simple microscope since only one lens was used. But what of the end product?

a. and b. methods are out of the question because the initial magnification was far too low.

c. has possibilities, the ends of the specimen-image are sharp but the projected image lacked detail in the dark area.

d. and e. have more to offer. It is possible to produce a negative $\times 9$ and $\times 12$. These would need considerable enlargement to arrive at $\times 100$ but the image quality would not permit such an enlargement.

f. produced a magnification of $\times 15$, fitting into a half plate negative. Figure 146 (a) After enlargement up to $\times 100$ proved that the image quality at the ends was not on a par with the method illustrated in Fig. 146 (d).

Fig. 144. Line overlap.

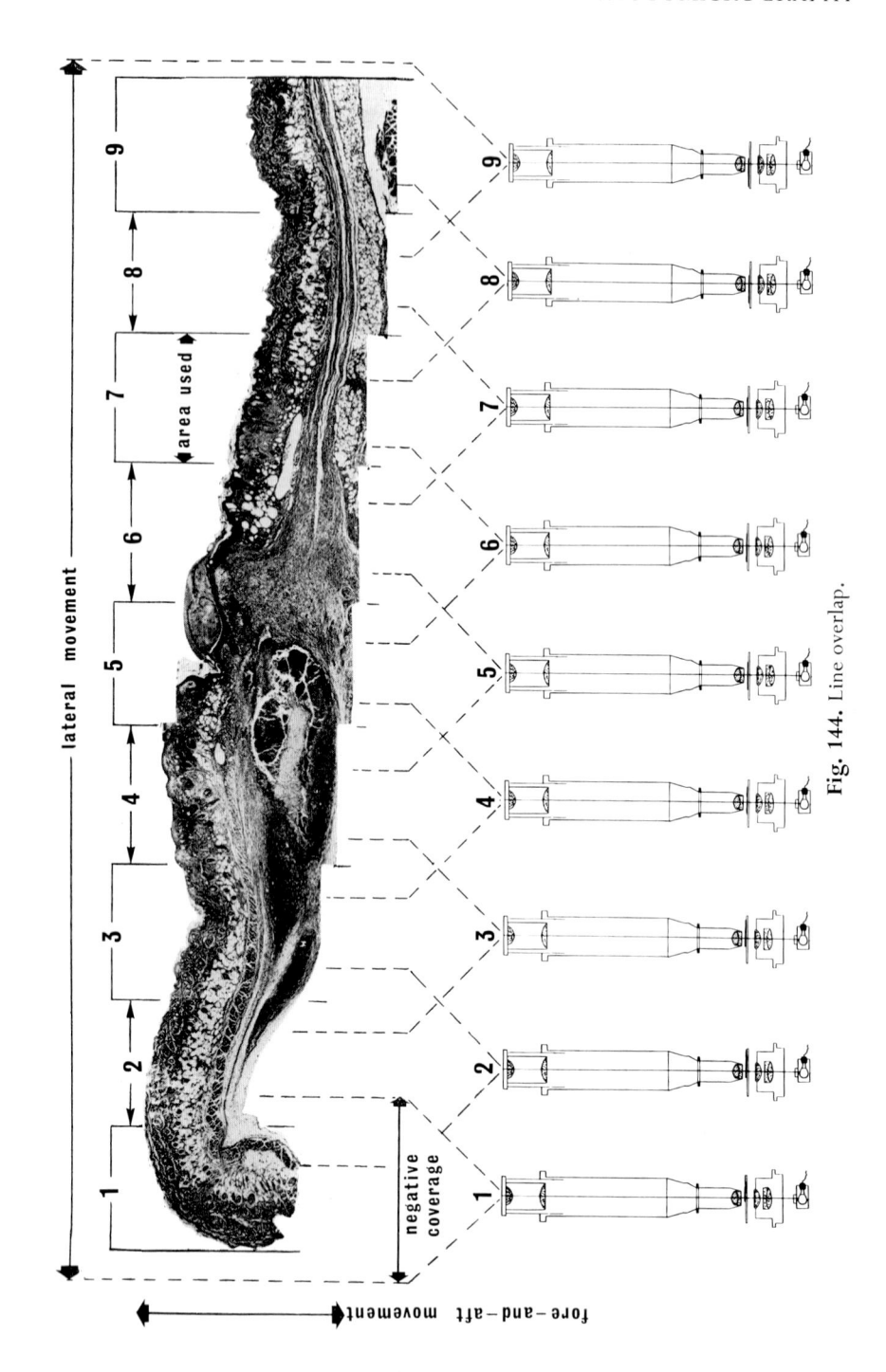

LINE OVERLAPS IN PHOTOMICROGRAPHY

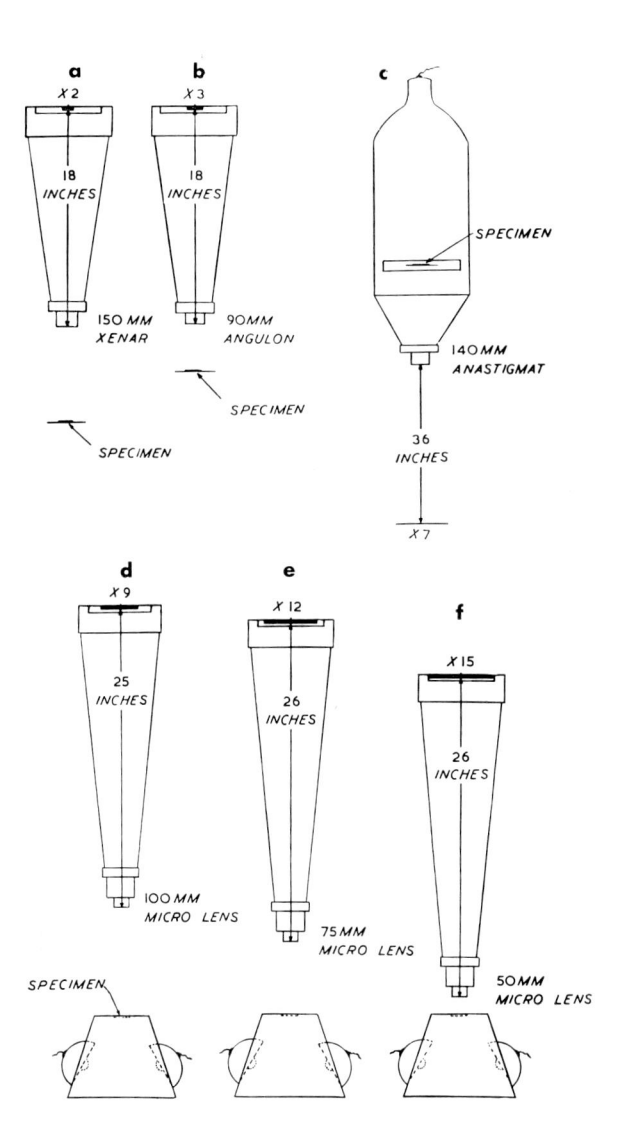

Fig. 145.

We have already seen that in relation to magnification, low-power objectives do not resolve fine detail as do higher power objectives. Another point is the quality of the extreme ends of the photomicrograph. Experiments have been carried out with a wide range of microlenses, enlarging the extreme ends to ×100, the magnification of the line overlap. All methods fail to come up to the results of the method illustrated in Fig. 144. A microscope fitted with an eyepiece and

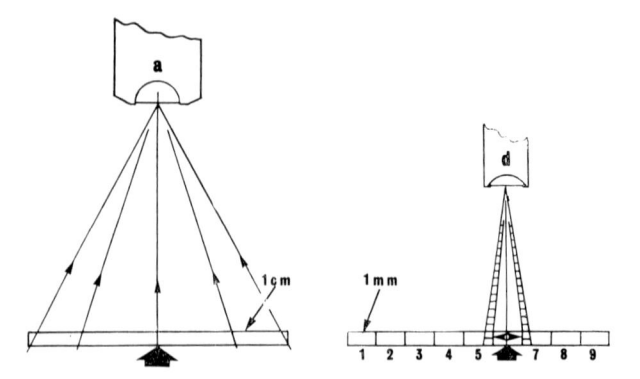

Fig. 146. (a) The whole area of 1 cm covered by a microscope fitted with a low-power objective 0·08 N.A., the eyepiece does not make an accurate recording of the ends of the specimen. (d) Complete area covered by the objective N.A. 0·30 used to record the L.O. The arrow indicates the area used in each print.

low-power objective (Fig. 146 (a)) cannot produce quality on a par with that produced by a medium-power objective illustrated in (d) of Fig. 146.

The detail in the resultant photomicrograph can be studied at leisure by more than one person (Plate 41) and information extracted from it under pleasant conditions. Although the actual exposure of the film is one of the first essentials in the production of a photomicrograph, it is by no means the most difficult. The various operations which precede and follow it call for a considerable degree of technical knowledge and skill and are fraught with opportunities for failure. Practice and experience are as essential for thorough proficiency as is theoretical knowledge. As regards the two main objectives: the technical quality of the photomicrograph depends chiefly on the care with which the microscope and allied equipment are adjusted and the negative and paper treatment.

Standards to be attained are:
(a) a high degree of resolving power,
(b) even illumination,
(c) critical definition,
(d) high quality negatives and prints,
(e) an understanding of the specimen,
(f) good presentation.

These must always be the objectives in photomicrography: so much so, indeed, that they become second nature. The value of a photomicrograph entirely depends on the amount of useful information which can be obtained from it which in turn depends upon the quality of the finished product.

It is most essential that even illumination be applied (Fig. 147). If this is not done the consecutive photomicrographs will be unevenly exposed, resulting in a series of light and dark prints: i.e. the right hand edge on print one is the left hand edge on print two, and so on, down the line. For the benefit of those who have not the latest advanced microscope/camera, let us consider the point of even illumination in some order. For the sake of illustration, we will commence at the bottom of the equipment and work our way to the top. Having examined the specimen, a decision is made regarding the type of illumination to be used. In the case of the L/O it was vertical transmitted light (Fig. 144). Having placed the camera and baseboard into position the source of light is centred on the ground glass, the microscope is placed in position, correctly centred by the light projected on to the ground glass. This, without the optical system being attached to the instrument. Having ensured absolute vertical illumination, the substage condenser and objective is inserted in position. At this point the light projected by the condenser may not be as true as without and, if so, adjustment is necessary. The lamp condenser, with iris diaphragm wide open, is now focused on to the plane of the substage iris so that an image of the lamp filament is clearly seen. When the specimen slide is placed in position, the condenser is focused on the underside of the subject until the lamp iris (now almost closed) is seen sharply in focus. The eyepiece is added then the substage iris is opened and the specimen brought into focus on the ground glass, when even illumination should follow. The basis of this series of adjustments can be applied to the modern built in illuminators.

When focusing the specimen, it is necessary to make sure that there is no displacement of the slide, fore-and-aft or laterally. If there is a sharp focus will only be obtainable on one side, making it impossible to match the prints when laid down. Having found the exposure by a test exposure, the film is processed, and if correct, this trial run can also be used as a final check for evenness of illumination and sharpness. Finally, before starting the run of exposures, move the specimen from one end to the other (laterally) at the same time watching the specimen to see if any fore-and-aft adjustment will be necessary. This proved to be the case with this illustration. The lateral overlap is determined by passing the image seen on the right side of the ground glass, to the left (between exposures) allowing for ample overlap. In this way certain sections of the specimen are photographed twice, on one negative on the right and on the following negative on the left. After a final check has been made all over the apparatus, the series of exposures is made. It is well to remember that the two opposite extremes of the marginal rays result in a photomicrograph that has a dissimilar perspective of the specimen in relation to the centre. It is possible that the result may be a stereo effect, evident at the extreme ends, and will not scale when laying the prints down. Before printing, a check on the enlarger would be worthwhile: even illumination and dust-free condensers are essential.

It is advisable to make a test L/O by using the test prints which can be attached to one another by sellotape. From the series of photomicrographs now portable and in one piece, an idea of the area required for mounting and just where the

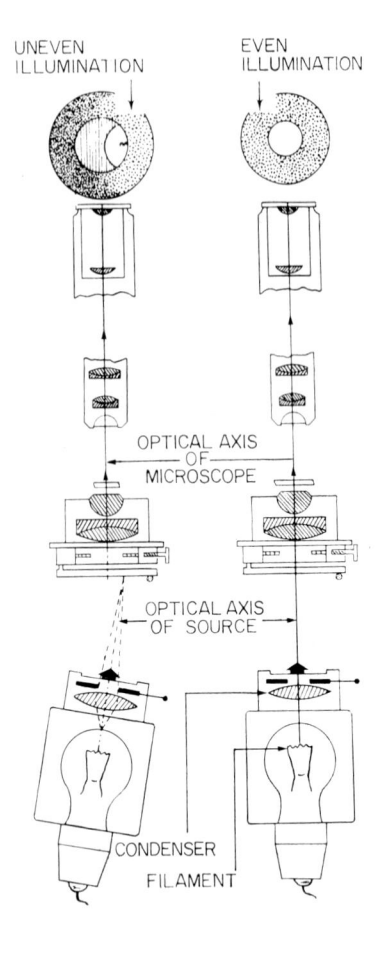

Fig. 147.

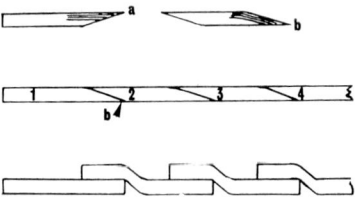

Fig. 148.

first print is to be mounted on the board can be obtained. From the figure, it will be seen that the extreme edges, right and left, of the L/O are not used; the right hand edge of print one is overlapped by the left of print two (Figs 146 (d) and 144).

Before mounting the prints it is suggested that these be chamfered at either end (Fig. 148). This operation is carried out by first lightly cutting the glazed surface of the print with a sharp razor blade. The print is then held and the unwanted edge is torn back (back slid), leaving a shallow bevelled edge. This makes for a continued flat line overlap, free from steps caused by the edge of the prints, as in Fig. 148, lower. Gum is the best mounting medium and a roller adds to the flatness of the prints.

Carefully annotate the mount and give information which is fitting to the particular L/O. The magnification must accompany the prints and a neat scale line assists in a quick and reliable measurement of any particular area.

"Wedge" method for recording particles in a liquid

Although this particular method involved photographing fine particles in broth at various stages of fermentation, the technique could be used for many such subjects and will, perhaps, give new ideas to those in similar research. The photographic optical system used is like a simple microscope.

A special glass container, Fig. 149 and Plate 48, is made out of four 26 mm × 77 mm microscope slides, the front slide being 0·79 mm (0·35 in.) in thickness

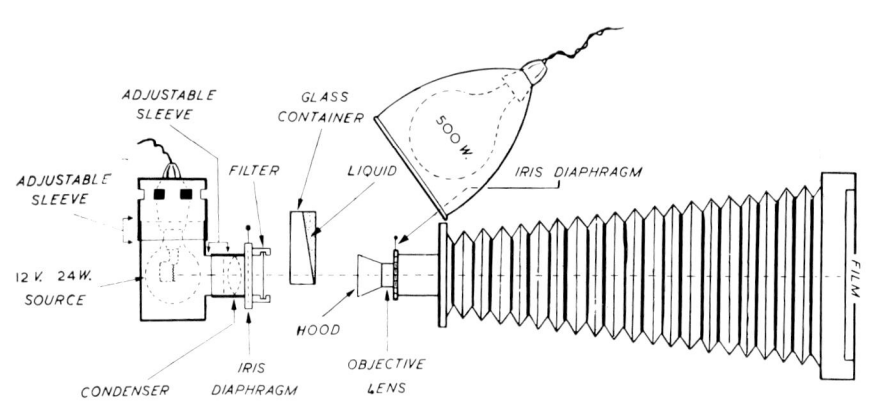

Fig. 149.

while the others are 1·5 mm (0·06 in.) thick. Two slides formed a wedge, Fig. 150. These were sandwiched between two vertical slides. The idea of a wedge shape watertight container was to allow the minute particles to separate from others of variable size and permit recording at the same time by reflected and transmitted light (Plate 49). A wooden wedge was made, around which the four slides were stuck, the glass container was then placed in an oven for twenty-four hours to enable the adhesive (Araldite) to set. This was because the container would have to be washed after each sample had been photographed. The wooden

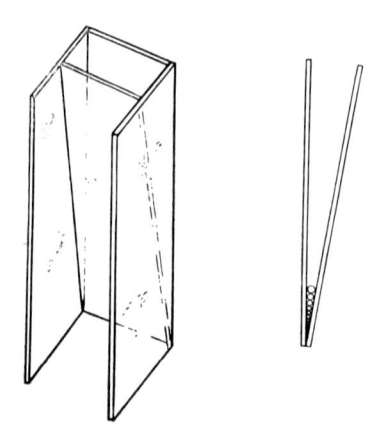

Fig. 150.

wedge support was then removed. A rigid housing was provided for the glass container, so transmitted and reflected illumination could be used without the subject moving, and at the same time bringing the subject into alignment with the optical axis of the lamphouse and lenses. Also, provision was made to prevent stray light from entering the taking lens as well as for housing a filter.

The lens used was an objective of 100 mm N.A. 0·08, ×0·7, set in a mount (Plate 3) with iris diaphragm and lens hood. A magnification of ×8 was attained by incorporating an extension of 63 cm in bellows length. Of course a microscope, less eyepiece and draw-tube, could have been used in a horizontal position and camera positioned at the eyepiece in the normal way. Transmitted light was from a small microscope intensity lamp, Fig. 149, incorporating a 12 V 24 W single-coiled filament bulb. Advantage was taken of the adjustable sleeve, indicated in the figure, enabling the source to be centred behind the condenser and the source focused on the subject: made possible by the behind-the-condenser adjustable sleeve. Use was made of the iris diaphragm and a light blue filter was used throughout.

Reflected light. A 500 W Photoflood was positioned as in the figure. Exposures were of a few seconds, whereas with transmitted illumination the exposures were greater than for reflected light.

Standard orthochromatic material was used for exposures made by transmitted illumination. Fine detail was obtained and excess in contrast were avoided. Extra contrasty material was used with reflected illumination in order to increase the contrast in both broth and pellets.

Photographing precipitation bands

The original method of photographing precipitation bands and the implications involved in the experimental procedure has been reported (Lawson, 1957). I

PHOTOGRAPHING PRECIPITATION BANDS

include a little of the technique here. This method provides a study in electrophoresis plates (Plate 50), precipitation characteristic of antigens and antibodies in agar gel plates, prepared on 7·7 cm × 2·6 cm (3 in. × 1 in.) microscope slides and cover-glasses, in capillary tubes and in petri dishes. All this has the advantage of simplicity and convenience. The original work in the technique of immunodiffusion plates was carried out by Ouchterlony (1949), and the method named after him. Characteristic of this particular technique is the formation of lines or bands varying in width, intensity and shape, within a flat, thin layer of agar gel. The bands are composed of antigen–antibody precipitates (Plate 50c and b) laid down, where two components diffuse toward each other, from small wells or shallow cups, cut out in the gel, meet in optimum proportion.

The arrangement (Fig. 151) was designed to illuminate the fine precipitation bands or lines, as luminous bodies against a black background. When a white base is used the high intensity lights cause the agar and bands to appear completely transparent. Agar is translucent (a turbid media). With a black base, the scattered light rays strike the bands from all angles, illuminating them within the thin gel layer, as illustrated in Plate 50 (e) and (f). The thicker the gel the greater

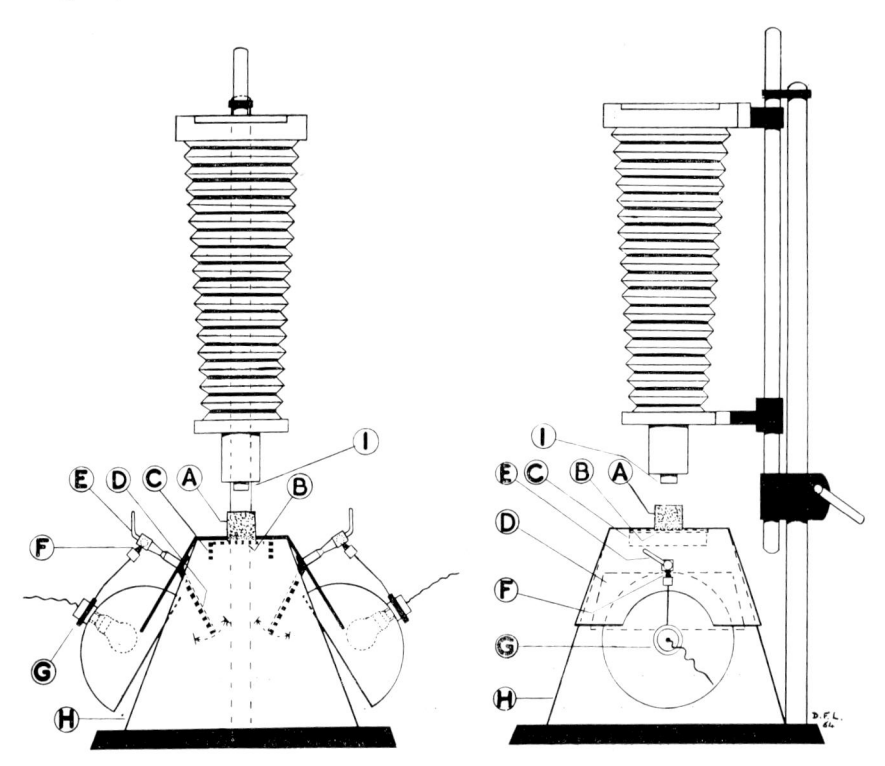

Fig. 151.

the light scatter will be. Of course, the glass container should be free from defects, if not, wavy lines appear in the final print.

The apparatus depicted in Fig. 151 illustrates the arrangement. A vessel containing the specimen is placed on a glass plate B located at the top of the opaque metal housing. The latter has an aperture in its centre just large enough to allow a petri dish, watch glass etc. to fit into it, so that stray light is thus prevented from escaping round its periphery, and entering the lens of the simple microscope. A baffle plate C prevents the curved glass edge of the container from being flooded by uneven light. D indicates an adjustable baffle plate and a reflector which directs an even scattering of light through the underside of the glass container. It is also a means of excluding direct transmitted light from the source and, at the same time, controls the background area. (The wider the coverage by the objective, the larger is the area necessary for the base). A ball joint lock E secures the lamp adjustment arm. A 500 W photoflood lamp fitted with an adjustable 250 mm (10 in.) reflector G. For use in transmitted darkground illumination the lights are directed onto a black velvet base. The light rays are then reflected from this, and from the white surface D and from the reflector H. A white base replaces the black when bright field transmitted effects are required (see Plate 50f). Surfaces D and H cause a scattering of light which passes through the suspension in the glass vessel at off axial angles, thus clearly illuminating any contained minute objects. This simple microscope arrangement allows the use of low power microscope objectives which are interchangeable together with a behind-the-lens iris diaphragm attached by the standard microscope thread. The objective screws into a specially made mount and, with the use of the extending camera bellows, initial magnifications can vary from $\times 4$ to $\times 45$.

The use of a water immersion objective in photographing Micro-gel precipitation bands. Since developing and introducing the method for recording precipitation bands, it has been found necessary to make certain modifications which now incorporate a water immersion objective to enable a wider range of subjects to be photographed and to produce higher magnifications according to the demand of the subject matter.

To some, water immersion objectives are a thing of the past associated with marine biologists, who would attach them to a microscope and operate them with an eyepiece. In this particular case the water immersion objective was divorced from the microscope. The objective, coupled with the camera bellows extension, produced an image size of $\times 10$ and upwards. After, enlargement followed as can be seen from Plate 51.

It may be that the introduction of the apochromatic or similar objectives, offering high resolution and a corrected image, have caused the loss of interest in water-immersion lenses which cannot produce the same high standards. On the other hand, there is no need for such high resolving power when, for

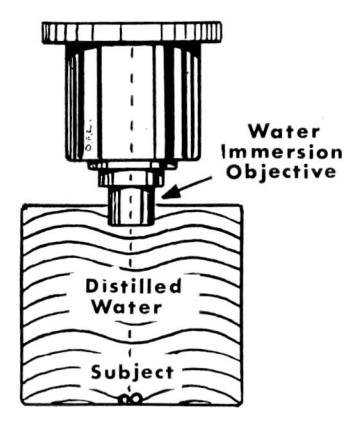

Fig. 152.

instance, a homogeneous immersion system is in operation, as here (Fig. 152).

Water immersion objectives are divided into two types: low-power and high-power. The latter gives a medium range of magnifications. The low power lens is particularly useful because it has a long working distance and, incorporating a behind-the-lens iris diaphragm, enables a reduced aperture to produce a considerable depth of field which is so necessary in this particular application. A low-power water immersion objective, as illustrated here, must be used with some kind of glass tank with sufficient depth to allow free up and down movement without the lens mount touching the tank. It is also advisable that the tank be of heavy glass and, if possible, it should have an optically flat base.

The subjects illustrated here could not be photographed on a microscope slide as would a temporarily wet mounted or living specimen, due to the shape and numerous reflections exhibited from the small glass capillaries. The drying out process and refraction took place. This also applied when photographing precipitation bands in immunodiffusion plates prepared on small circular microscope cover glasses laid on a microscope slide.

At first it was thought that the subjects could be submerged in glycerine, contained in a small glass cell as illustrated. This was not very satisfactory, focusing was difficult and, when the lens was raised, the glycerine was drawn up by the lens. In addition the viscosity of the glycerine caused a bumping effect (which took some time to settle down) and the image of the subject disappeared from the ground glass screen. However, there were no reflections from the glass to cause confusion with the precipitation patterns. Distilled water was then used in place of glycerine. Here the problem was not as much the loss of resolution as the loss of visibility, it being difficult to observe the image of the bands. The use of stale distilled water should be avoided as it contains carbon dioxide and is damaging to some specimens. Exposures were short, even when using orthochromatic material.

When moving the objective during focusing, a slight disturbance was set up in the distilled water which caused the small capillaries to roll. To prevent this, glass strips, acting as furrows, were fused to the glass base of the cell.

References

Lawson, D. F. (1957). "Photographing Precipitation Bands", *J. Photogr. Sci.* **5**, 1.
Ouchterlony, O. (1949). *Ark. Kemi. Min. Geol.* **26 B**, 1.

Transmitted darkground effects

When placing certain specimens in the light train of the microscope, the specimen, small as it is to the naked eye, is sometimes seen as a luminous body surrounded by a dark area as happened when placing rabbit blastocysts under the microscope for Plate 54. Surprisingly enough the minute spheres appeared to exhibit bioluminescence: the biological emission of light from living organisms. Although the subjects recorded here do not exhibit bioluminescence, for bioluminescence is produced by an organism which relies on vestigal mechanism, it does, however, demonstrate that the effects are not dissimilar.

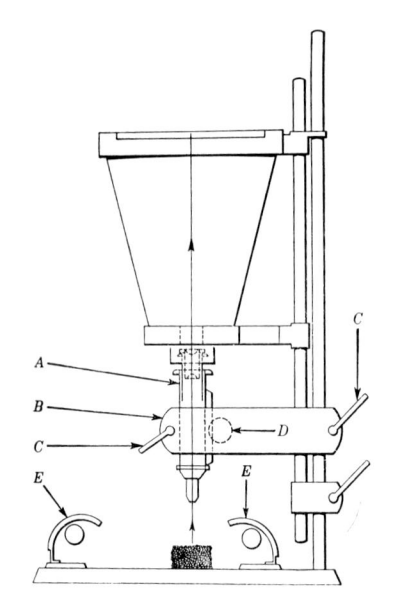

Fig. 153.

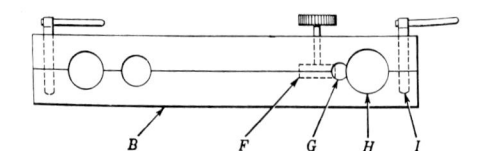

Fig. 154.

The subjects here proved to be too large to photograph as a colony under the microscope when using a low-power optical system consisting of eyepiece and objective. When viewed singly in transmitted light illumination, some cellular details are apparent but only part of the specimen remained in focus at any one time. Also, there was far too much opaque material for the particular photographic method to be of value.

Instructions in this instance were not to damage the blastocysts in the slightest degree and that the photographic operation must take the minimum of time. However, if photographed using the apparatus described in Fig. 151, the embryonic area can be immediately identified, showing up a dense whitish area quite distinct from the trophoblast, which endows the rest of the blastocyst with a typical granular appearance.

The author designed the apparatus so that no illuminating beams should shine directly into the objective. Specimens such as large crystals and gemstones, bands in an Ouchterlony plate and many other subjects, are seen illuminated against a black background. By this technique photographic representations of the specimen are obtained with far greater definition and clarity than by other means (Plate 52). However, it must be pointed out that this method of darkground illumination leads to better visualisation of fine detail but not its resolution.

The scientific data relating to Plate 52 is as follows: Virgin female New Zealand white rabbits were mated naturally with bucks of proven fertility. The time of coitus was recorded and estimations of age were made at random. No further treatment was given to the mothers prior to death. After 144 hr (6 days) the mothers were killed and the entire reproductive tracts were removed. The uterus and uterine tubes were cleared of fat and flushed with Krebs bicarbonate ringer solution containing 10% rabbit serum. The blastocysts were washed into a petri dish and were just visible to the naked eye. They were carefully transferred to a clean watch glass containing a small amount of ringer solution and photographed without delay.

In the rabbit, implantation usually takes place on the 7th day p.c. These blastocysts, therefore, were approximately 24 hr pre-implantation. The only method of examining the embryonic area of developing blastocysts previously has been by the flat-mount technique devised by Mogg and Lutwak-Mann (1958). This meant fixing and mounting of the blastocysts, and such treatment obviously prevents any further studies being carried out with these blastocysts.

Using the author's technique (Fig. 151), the shape and size of the embryonic area and the blastocysts as a whole, enable one to pick out immediately those which are grossly distorted in development. Further, this method enables examination and recording of the blastocysts prior to transfer to recipient females, and, in making this record with the simple microscope, the minimum amount of handling takes place so no foreign substances are added to the incubating medium. A change in power of the objective enables a more detailed study of any particular area, see Plate 53. The deformed area can be studied. Once established, the technique is quick and would hardly interfere with the routine of a transfer operation. (The rat embryos (Plate 52, b) were photographed by the same technique as the rabbit blastocysts.)

The mothers were sacrificed 14 days after mating had been confirmed and the embryos removed from the uterus, complete with amniotic sac. They were transferred to a petri dish containing saline, the amniotic sac punctured and the embryos floated out. The umbilical cord was severed and the individual embryos were transferred to a watch glass containing saline.

The resultant photographs show the following features:

The naso-maxillary figure is clearly visible, as are the developing forearm buds. The tail shows well developed somites, but the hind limbs are still rudimentary. The viscera are well developed at this stage, but an obvious hernia is present. In normal animals this would regress until at birth the abdominal wall completely encloses the viscera.

Perhaps the most interesting feature of this technique of photography is the way in which deformities can be detected, even though organogenesis is not yet complete. The two embryos on the left of Plate 52 (lower) are deformed, and should be compared with the two normal embryos on the right half of the photograph.

The skulls and backbones of the normal embryos are perfectly smooth, indicating complete closure of the neural plate, whereas those of the deformed are obviously misshapen. The nervous tissue is exposed, which gives a jagged appearance to the skull. The dorsal view of one of these embryos reveals the continuation of the cleft down the backbone, a condition known as craniorachischisis.

No difficulties arise in interpreting the photographic images as this apparatus does not produce other line distortions such as "doubling" of the image. It is, however, necessary to see that the apparatus is standing on a firm base and the base, in turn, is on a solid floor for vibration could cause movement of the suspended objects.

Reference

Mogg, F. and Lutwak-Mann, C. (1958). *J. Embryol. Exp. Morphol.* **6**, 57–58.

Photographing large bulk materials

It is not always possible to cut pieces off a specimen, and some manufacturers object to their samples being defaced. To overcome this, I devised the "strap" method of producing a photomicrograph. Illuminating a subject as in Plates 3 and 33 is not always easy and more than one lamp is often necessary. In such cases the microscope, stage and foot, restricted height adjustment and limited stage area often restricts movement.

Any type of microcamera can be used, the bellows extension (Fig. 153) will permit a more variable magnification (Plate 60), filling the negative format with a selected area. A heavy base is absolutely a necessity in such work to ensure rigidity. The microscope body is taken out of the limb, making the stage, condenser and foot unnecessary. (Of course a microscope tube (Fig. 155) can be purchased from almost all microscope manufacturers.) The microscope tube is held *rigid* and *vertical* by a strong light alloy "strap" B, and D provides coarse adjustment, while C held the two sections together. This "strap" can be seen in more detail in Fig. 154. A hole was drilled through the halves to house the microscope tube, coarse adjustment and camera-supporting pillars. Major adjustments are made at C and final focusing by adjustment F. Low power objectives can be used there being *no* restriction in the height movements. A much wider field can be recorded by the objective without the inclusion of an eyepiece. A small area in a specimen the size of a house brick can be photographed but the specimen must be moved manually.

Two tubular lamps, 63 mm in length, 25 mm diameter and 50 V, are housed in adjustable reflectors E, illuminating the subject from a low angle. Of course

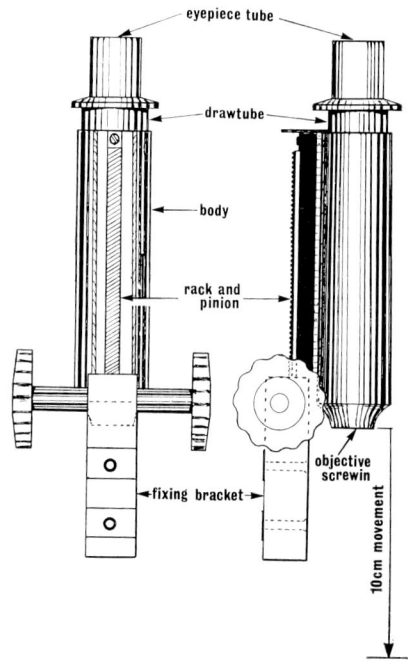

Fig. 155. Most useful body tube and supporting bracket.

any type of free-standing light source can be used. Whatever the source, care must be taken to ensure even illumination which, in this application, becomes more difficult as the magnification is increased.

Recording gas bubbles inside a bottle

As we have already seen, a subject will, frequently, not fit onto the microscope stage. In such cases ways and means must be devised as in Fig. 156. This arrangement was made to photograph gas bubbles on the inner surface of a bottle

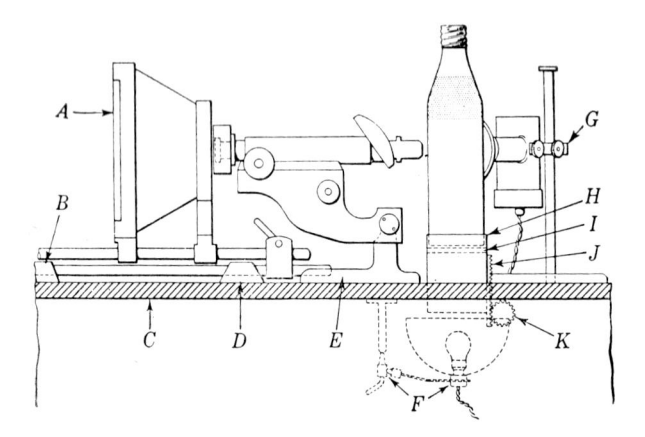

Fig. 156.

of aerated soft drink in order to detect whether the condition of the surface was clean or contaminated. The figure illustrates the microscope, less stage and condenser, with the foot in a reversed position to enable the objective to operate close to the bottle.

A quarter plate microscope-camera was used for this particular task and was modified to operate in the horizontal position. Three brackets, two of which are seen in Fig. 156 at B and D, held the camera firmly in position. Focusing on the inner side of the bottle was carried out, the image seen on the ground glass screen. A special stage housed the bottle resting on a piece of Perspex, I, held within the metal tube H. Glass would, of course, be better but is more difficult to cut. The stage could be racked up and down by J and thumb-screw K (Fig. 157), to enable a survey over a large area. The bottle was turned manually. The main source of illumination came from F, a 500 W Photoflood bulb, situated beneath the bench top. This lamp was fully adjustable and could be locked in position on a ball joint. A second source was provided at G. This top light played an important part in illuminating the area to be photographed, but care and patience were required in establishing its position, since a great deal of reflection occurred at the curved surface of the glass. When the light was correctly positioned, the bubbles were seen in relief, making observation of the hydrophilic and hydrophobic surfaces quite simple, Plate 55. Shutter speeds in work of this sort must be short, because the bubbles are in constant motion.

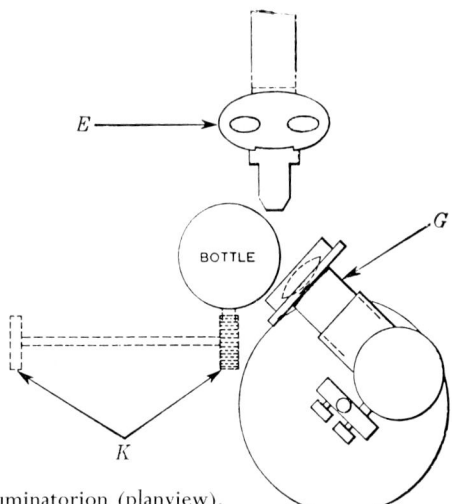

Fig. 157. Bottle illuminatorion (planview).

Back lighting with "simple microscope"

There is a place in the laboratory for the simple microscope and here it was put to good use, revealing irregular shoulders of small glass bottle tops (see Plate 56). These and similar defects prevented the cap from sealing (Fig. 158) allowing the gas content to escape. A piece of flat metal was placed on top of the bottle and illuminated from behind as indicated in Fig. 159. The metal was ground flat to approximately 0·025 mm (1/1000 in.) and used in the examination of scores of bottles. The photographic image was superimposed on the manufacturer's blueprint to see if a breakdown had occurred and where.

The source (Fig. 159), was a 500 W Photoflood, positioned approximately 12·8 cm (5 in.) from the back of the screen as illustrated. The strong illumination revealed the uneven glass tops, small areas of defective glass showed clearly above the shoulders, and defects in the bottles' casting could also be seen beneath the flat metal.

In addition, some bottle tops were not level but sloped away to the rim, and here the piece of flat metal acted as a good "spirit level". These were the defects preventing the crown tops from sealing, allowing the gas content to escape.

A 100 mm short mounted objective lens, together with a behind-the-lens iris diaphragm (Plate 3), was used at a reduced aperture with a bellows camera extended to approximately 42·5 cm (23 in.). The negative included both

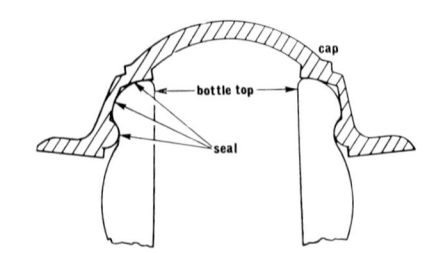

Fig. 158.

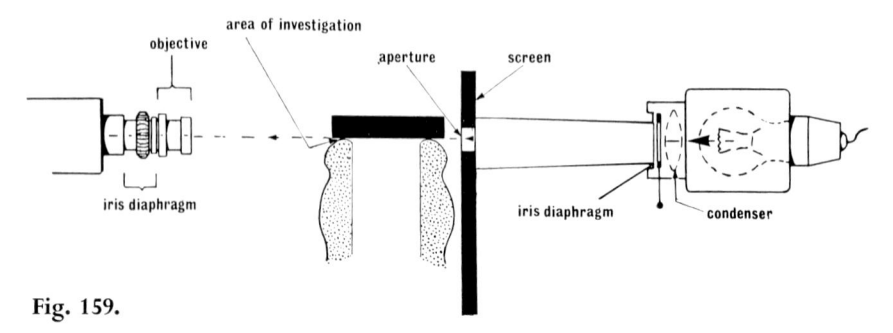

Fig. 159.

shoulders, but only one is shown as space does not allow for the reproduction of the whole. Focusing took place on the area, left to right, which meant that the point nearest the objective was out of focus.

Reproduction ratio for magnified images using a lens with bellows extension, is as follows

$$\frac{\text{Image size}}{\text{Object size}}$$

Example:

$$\frac{24 \text{ mm}}{48 \text{ mm}} \text{ or } \frac{36 \text{ mm}}{72 \text{ mm}} = 1:2, \text{ and the magnification} = \tfrac{1}{2} \text{ or } \times 0\cdot 5.$$

The required lens extension through intermediate mounts, rings, or focusing bellows, is determined by means of the reproduction ratio and the lens focal length (f). A reproduction ratio of $1\cdot 5 = \tfrac{1}{5}$ or $0\cdot 2$. The required extension is therefore f/5, or f $\times$ 0·2. For a 90 mm lens: 18 mm.

The exposure increase factor must be multiplied by the normal exposure without extension to obtain correct increased image exposures. This exposure increase factor from the formula $(M + 1)^2$, where M is the magnification. Thus

for 1:1 we find $(1 + 1)^2 = \times 4$.

Or

$EF = (1 + 0\cdot 2)^2 = \times 1\cdot 44.$

Depth-of-field expressed through the simple microscope range lies approximately half in front and half behind the point of focus. As the behind-the-lens iris diaphragm is reduced, depth-of-field will show a marked increase away from the lens. Depth-of-field (for any given f/stop) and the reproduction ratio are determined solely by the reproduction ratio or magnification. Thus, for 1:5 these figures would be the same for any focal length objective lens. Recommended apertures for this type of photography are in the region of f/11 and f/16. Further reduction may cause a deterioration in the image quality.

Table 20.

Magnification and reproduction ratio	Exposure increase factor	f/8	f/11	f/16	Approximate field stop in inches
0·1 = 1:10	×1·2	2·4	3·2	4·8	9·5 × 14
0·2 = 1:5	×1·4	0·6	0·8	1·2	4·7 × 7
0·3 = 1:3·3	×1·7	0·3	0·4	0·6	3·2 × 4·7
0·4 = 1:2·5	×2	0·2	0·25	0·4	2·4 × 3·6
0·5 = 1:2	×0·3	0·12	0·16	0·24	1·9 × 2·8
0·6 = 1:1·67	×2·6	0·10	0·12	0·20	1·6 × 2·4
0·7 = 1:1·4	×3	0·08	0·10	0·16	1·3 × 2
0·8 = 1:1·25	×3	0·06	0·08	0·12	1·2 × 1·8
0·9 = 1:1·1	×3·6	0·05	0·06	0·10	1·1 × 1·6
1 = 1:1	×4	0·04	0·06	0·08	0·95 × 1·42
1·5 = 1·5:1	×6	0·024	0·03	0·05	0·63 × 0·95
2 = 2:1	×9	0·015	0·02	0·03	0·5 × 0·7
3 = 3:1	×16	0·01	0·013	0·02	0·3 × 0·5

Phase contrast

The principles of phase contrast illumination are not new, although they have been applied to photomicrography only since the work of Zernike in the early 1930's. He was able to advance and retard the main beam of light by $\lambda/4$ after it had traversed the specimen. The diffracted rays were unaffected and no change in wavelength took place.

In brief, phase contrast can be created by the combination of

(a) An annular diaphragm, d, positioned beneath the microscope condenser, which directs a hollow cone of light through to the specimen (Fig. 160).

(b) A diffraction or phase plate (Fig. 160) is positioned at the back focal plane of the objective, the image appears as a greyish ring.

(c) An aperture viewing unit which is easily inserted into the optical system to assist in viewing the phase plate and annulus image.

(d) A centering phase telescope (eyepiece system) which can be used in place of the aperture viewing unit.

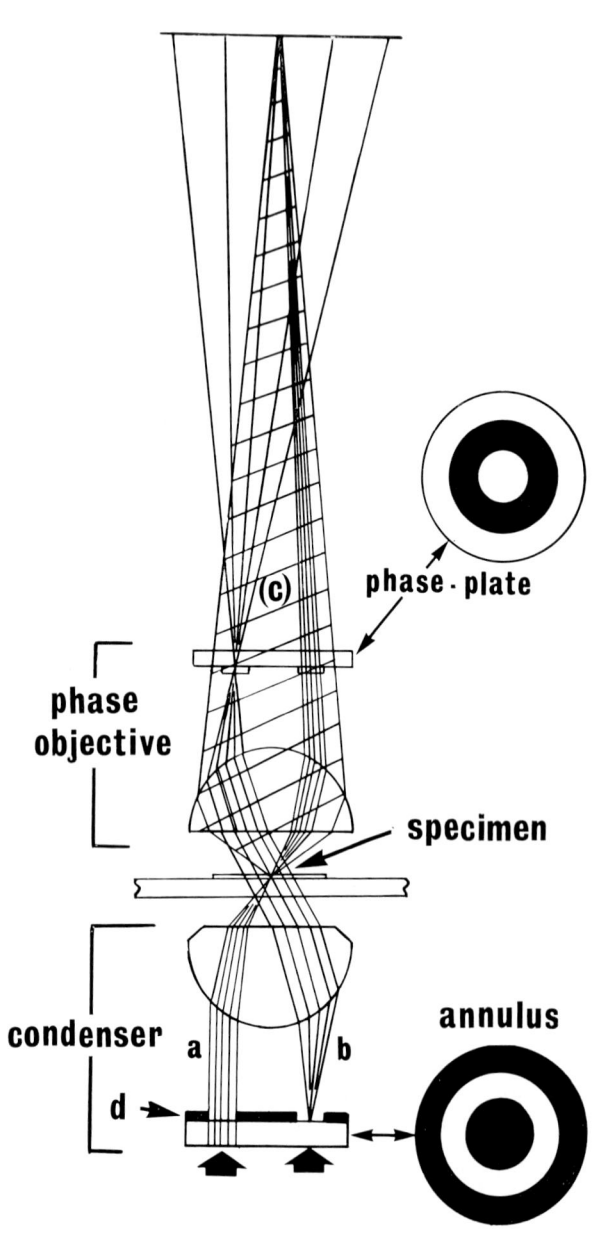

Fig. 160. Beck transmitted light phase contrast.

PHASE CONTRAST

Some of the light passing through the almost transparent specimen is diffracted by slight differences in the optical path (refractive index × thickness) and moves so as to be distributed over the whole aperture of the objective. The balance of light passes directly through the specimen as a cone of concentrated light toward coincidence with the "ring" of the phase plate. The phase plate alters the intensity, and the phase relationship, of the diffracted and direct light so that when they recombine to form an image, invisible specimen optical path differences are converted into visible light intensity differences.

The one object of phase contrast is to improve the visibility of the specimen under investigation by separating it from the surrounding medium, possibly, by a small amount of optical path only.

The American Optical Company have designed their phase optical system to give a choice of optical differences through the phase plates as in Fig. 161. In addition there is a phase objective for bright and one for dark contrast images.

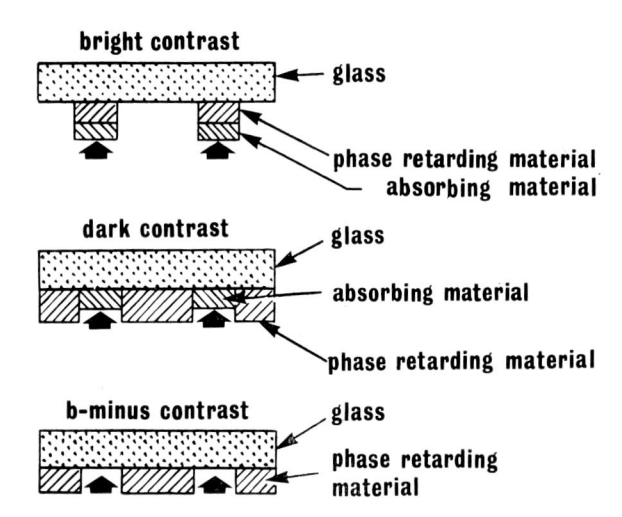

Fig. 161.

A specially designed rotating turret condenser contains four individually centred annular diaphragms and one open aperture. With one annuli correctly aligned, any setting in the turret can be used in quick succession. Another important feature of this apparatus is its substage condenser system. Three condensers (Fig. 162) provide a choice of working distance and magnification, accommodating a wide or narrow specimen according to the objective in use.

Phase contrast fluorescence combines the advantages of phase contrast with those of fluorescence. A phase contrast image is superimposed on the fluorescence image to make two, the colours of which are usually well differentiated and appear simultaneously in the microscope. This combination calls for a mixture

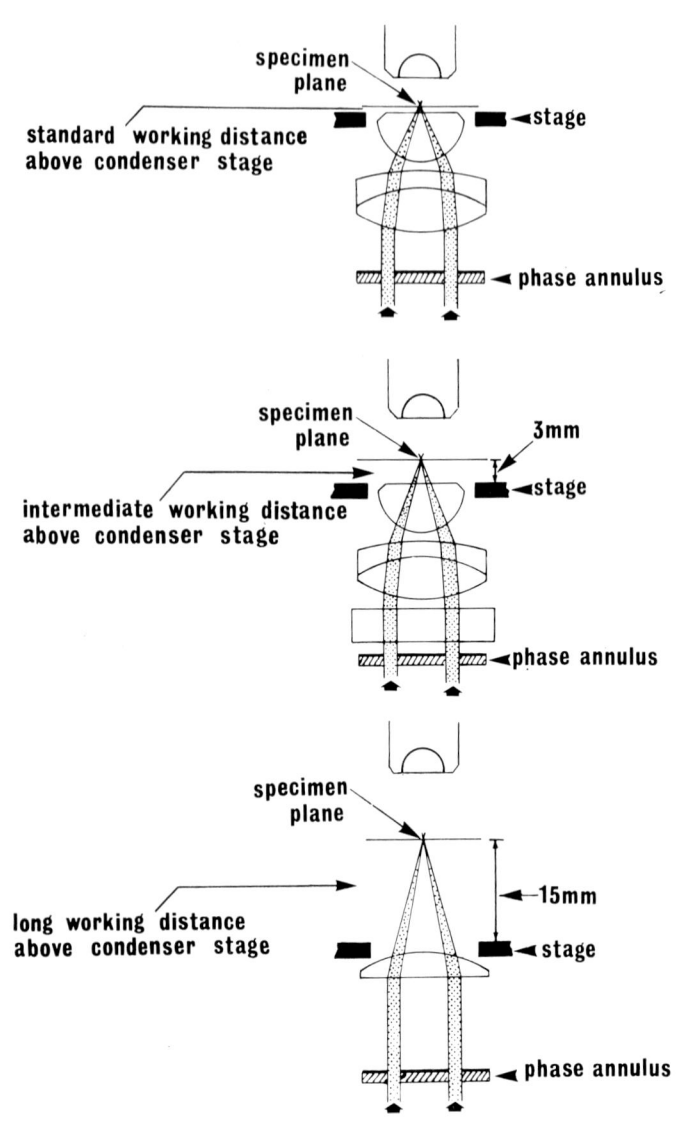

Fig. 162.

of ultra-violet or blue exciter radiation with the existing illumination system. Such a lamp-house as already set out, meets the twin requirements. By means of a built-in beam-splitter, mixed light consists of the exciting radiation and a variable proportion of the light from the source for the excitation of fluorescence and the representation of non-fluorescent portions in phase contrast respectively.

PHASE CONTRAST

It is impossible to give a full explanation in this work. However, the following may be of assistance when considering briefly the theory of phase contrast.

One way of approach is to consider firstly the happenings of light through a partially absorbed specimen and through a transparent specimen, both as viewed in the microscope. Turn to Fig. 163 where A represents a light wave falling on a

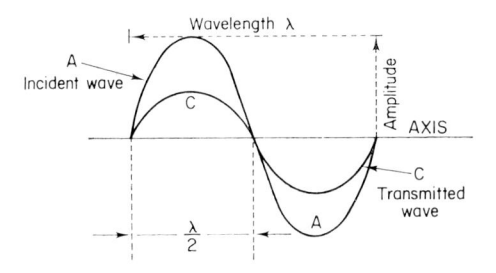

Fig. 163. Image formation in a partially absorbing specimen. A represents the incident wave, C the transmitted wave.

specimen. Here the intensity of the light will be proportional to the square of the amplitude of this wave. Having passed through a partially absorbed specimen the intensity of light is reduced and the emerging wave is now indicated by C in the same figure. It will be seen that the amplitude is lower than that of A.

The emerging wave can be illustrated in another way. It can be regarded as the sum of the original wave and of another wave C, illustrated in Fig. 163. If the two waves A and C are used together they will give C. That is taking into account upward displacements as positive and downward displacements as negative. The waves A and C of Fig. 164 have a real physical existence, for A is

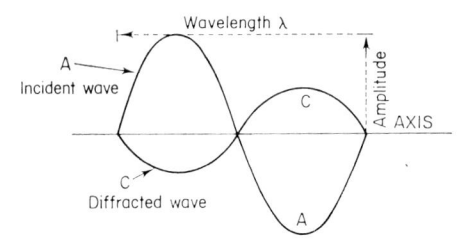

Fig. 164. The diffracted wave is shown to be 180° out of phase with the incident wave. The image formation of a transparent specimen.

the incident wave and C is produced by diffraction at the point of the specimen. The resultant image is obtained by the interaction of interference of these two waves.

Now we will consider waves passing through a transparent object. Since the wave passing through such an specimen has lost no energy, the amplitude of that wave is unchanged. However, owing to the retractile properties of the object the transmitted wave will now be either retarded or advanced in phase in relation to the incident wave.

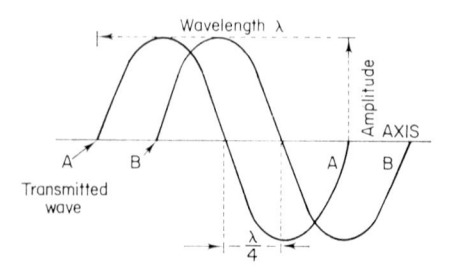

Fig. 165. Image formation of a transparent specimen. The transmitted wave *A* is unaffected in amplitude but is delayed in phase. *A* can be represented as the sum of *B* and also the diffracted wave illustrated in Fig. 166, these two waves are 90° out of phase. Thus if a further difference of 90° is added, the two waves will be 180° out of phase. This will give the wave seen in Fig. 167.

Figure 165 illustrates the transmitted wave *A* and is shown retarded in relation to wave *B*. You will see that, unlike Fig. 164, the waves have the same amplitude. In the case of an obsorbing specimen the transmitted wave *A* can be represented as the sum of the incident wave *A* and a diffracted wave, as in Fig. 164. An important difference between the diffracted wave *C*, Fig. 164, and the wave *A* of the illustration is obvious. In the case of the obsorbing specimen the diffracted wave is half a wavelength out of phase with the incident wave. It can be seen that the crest of wave *A*, Fig. 167, corresponds with the trough of *C*, Fig. 167. However, the wave diffracted by a truly transparent object is out of phase with the incident wave, Fig. 166, which depends upon the thickness of the specimen and the refraction index over and below the mounted specimen. The phase difference for most transparent objects is in the order of one quarter of a wavelength, as in Fig. 165.

Having touched upon phase differences between specimens, we will now turn

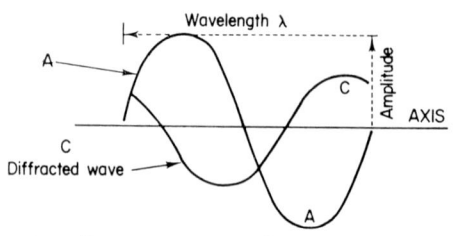

Fig. 166. Image formation of a transparent specimen.

PHASE CONTRAST

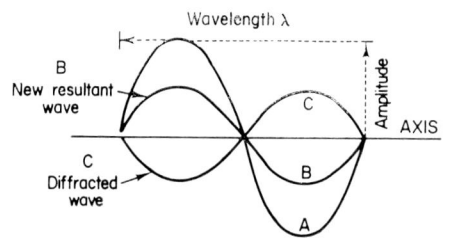

Fig. 167. Image formation of a transparent specimen.

to the optical system of a phase microscope. There the phase difference between the diffracted and incident wave is increased from one quarter of a wavelength (90°) to half a wavelength (180°), so that the two waves cancel each other out. Figure 167 illustrates that the phase of the diffracted wave C has been changed so that the crests correspond in position with the trough of the incident wave. The sum of these waves is the new resultant wave B (Fig. 167).

Any difference in refractive index and thickness of a *transparent* specimen may be clearly seen in a black and white photomicrograph. The human eye is insensitive to these phase differences, so that special methods have to be employed for image formation. There are many ways in which the devices to produce phase contrast may be introduced but the basis of the method remains unchanged. The result may be positive or negative phase contrast, depending upon whether the direct ray is accelerated or retarded by the necessary quarter wavelength ($\lambda/4$). From the practical point of view these methods differ as do positive and negative images. Figure 160 illustrates a simple phase contrast arrangement for work in transmitted light. A diaphragm, having a narrow circular iris or annulus, is placed beneath the condenser and a phase plate in the back focal plane of the objective, as indicated. Through the annulus, which gives a bright ring of light, the specimen is illuminated by a hollow cone of rays. The objective forms an image of the annulus in its back focal plane, as in the diagram, by the bundle of rays b. In this plane, a phase plate, consisting of a disc of glass d with a depressed ring matching the annulus below the condenser is fitted. The amount by which the phase ring is depressed changes the phase of the light by $\lambda/4$. By this means the light passing through the depression is out of phase with the light scattered or diffracted by the specimen, bundle c in the diagram, and by interference with the light indicated by the undeviated bundle a, is imaged on the film. In use, the image of the substage diaphragm must be focused exactly and centred upon the ring in the phase plate. All the light transmitted directly through the specimen will then pass through the ring in the phase plate and receive the necessary acceleration or retardation. All the light received by the objective will be projected upwards and only a very small proportion of this light will pass through the ring in the phase plate. The bulk of the diffracted light will pass through the remainder of the phase plate and suffer no phase change. The two beams are then received in the eyepiece where they interfere. For maximum interference the intensity of the direct beam must be reduced to match that of the diffracted beam. This is accomplished by the deposition of a thin film of metal over the ring on the phase plate to act as a neutral density filter.

The particular arrangement shown in the figure varies in detail according to the equipment of the various manufacturers, but the principle remains unaltered. Manufacturers invariably provide the necessary information for alignment when supplying any piece of equipment. A typical instance of the value of phase contrast technique is provided in Plate 29. Under transmitted light, it is very difficult to discern any structure at all. The technique can also be applied to the examination of opaque materials by reflecting light, when it is capable of differentiating between very small differences in the level of the surface. This method, first described by F. S. Cuckow (1949), is a very valuable tool in metallography, and can also be applied to the study of fibres, crystals and other materials. M. Francon (1954), describes a method by which two movable quartz plates have been cut at 45° to the optical axis and interposed between polarizing units, and could be used to form a colour-contrast technique. This method has not yet been widely adopted.

Figure 168 shows the arrangement of the microscope for phase contrast with an opaque specimen, illuminated by incident light. In this case the annulus is held in a centering mount on the vertical illumination fitting. The annulus is then imaged, after reflection by the specimen, in the back focal plane of the objective, indicated by the bundle of rays b in the diagram. The directly reflected bundle a passes through and is retarded by the phase ring and therefore interferes with the scattered bundle c, enhancing the contrast of the image. A levelling mirror is provided for the stage, to permit the annulus image to be accurately centred with regard to the phase plate in the back focal plane of the objective. A second levelling table on which the specimen can be mounted is also provided, so as to ensure that the phase condition is maintained. A different annulus for each phased objective is required and these are held in separate mounts with quick change fitments. The range of objectives are those supplied for imaging transparent specimens, but for opaque work they are corrected for uncovered objectives and the surfaces are bloomed.

A. Wilska (1954) introduced a modification of these methods by the use of annular zones which were coated with lamp black in place of the conventional phase rings, the ring apertures were much larger than normal. He claims that this method gives a much better contrast and resolving power than does conventional phase contrast equipment.

The Leitz phase contrast apparatus with the Heine condenser permits both narrow or wide annular illumination for phase contrast and darkground or intermediate stages of illumination to be achieved by suitable adjustments of the condenser.

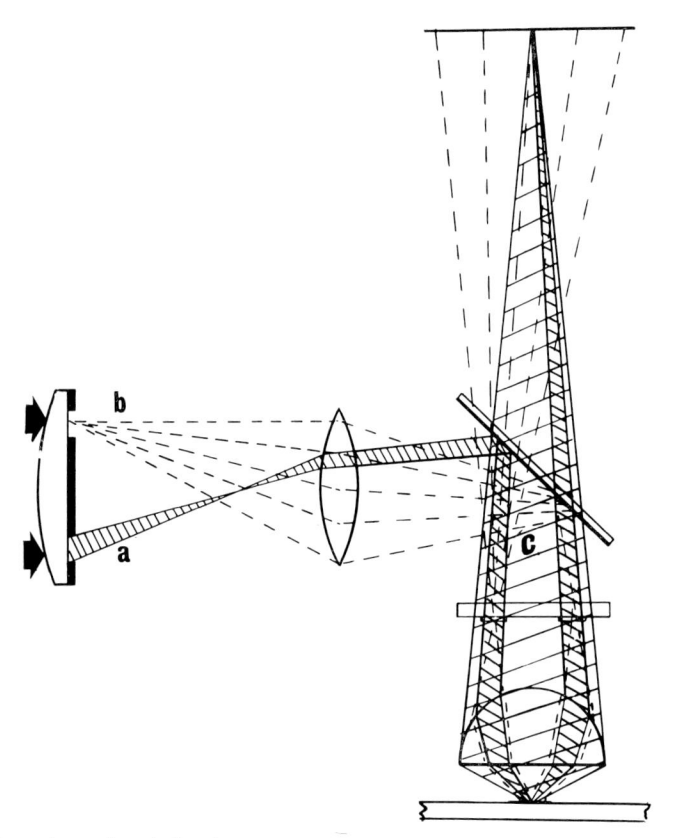

Fig. 168. Beck incident light phase contrast.

Carl Zeiss, Oberkochen, have designed a special phase contrast planachromat objective for imaging tissue cultures. The planachromat has a magnification of ×40, an N.A. of 0·60 and a working distance of 1·5 mm. This optical system has been designed for the examination of cultures in culture flasks or petri dishes placed on an inverted microscope where observations are made from below, through the bottom of the glass vessel. The objective is corrected for working through a glass thickness of 1·3 mm and provided with a correction mount for compensating deviations in a thickness of ±0·2 mm. In addition, the ultrafluar objective N.A. 0·40, ×32, and objective N.A. 0·85, ×100 have also been designed for use in phase contrast techniques.

If a phase microscope is not available, or is too expensive, a phase adaptor can be used with great success. This small piece of equipment screws into the microscope body after the nosepiece has been removed.

Setting up the microscope for phase contrast. As in Köhler illumination (page 115), the iris diaphragm, in front of the lamphouse condenser, forms the field stop and the lamp filament will be imaged in the field of the condenser diaphragm. The microscope should be ready for use: i.e. the eyepiece, phase objective and substage condenser are attached, the stage has been adjusted so that it is centred to the body tube and this, in turn, is adjusted to the recommended length in conjunction with the objective and finally the specimen slide is placed on the stage. Do not omit to check that the source is in correct alignment with its condenser. Set the microscope in front of the lamphouse with some 8 in. between lamphouse iris diaphragm and microscope mirror. This distance is approximate for an Abbe N.A. 1·2 substage condenser. Both lamp and mirror are adjusted to ensure that the projected beam fills the centre of the flat side of the mirror and is then redirected through the centre of the substage condenser. This can be checked by directing the light onto the underside of the closed substage iris diaphragm and adjusting until it illuminates its centre. The lamp condenser is now closed one third. The next step is to position the substage condenser by moving it in line with the optical axis until it is correctly focused on the underside of the specimen slide. When this is done the light will automatically be in alignment with the objective. If an immersion substage condenser is in use, oil should be placed on the top lens and the condenser gently raised until the slide and condenser are homogeneous. Phase contrast photomicrography demands that the light source fills the conjugate area of the diffraction plate at or near the focal plane of the objective. The substage iris diaphragm is opened beyond that of the lamphouse iris diaphragm. The specimen is now fully illuminated.

Bring the objective down to within a short distance of the cover-glass and then focus on the specimen. This will, of course, take place away from the specimen. If the specimen is incorrectly illuminated it may be necessary to open the condenser diaphragm in relation to the conjugate area of the diffraction plate to detect the specimen. This is carried out by removing the eyepiece and by replacing it with a centering telescope. This, in turn, is focused on the coating of the diffraction plate and the diaphragm adjusted until an image of the outer fringe of the condenser diaphragm is seen over the conjugate area of the diffraction plate. The eyepiece is now returned and focusing takes place. Should the substage condenser be positioned too far from the specimen it may cause an obstruction and prevent the full light intake from illuminating the specimen making the centre of the specimen field dark or exhibit colour zones. At this stage it is as well to examine the lamphouse iris diaphragm. If this has been inadvertently reduced it will contribute to the above. An ill-adjusted substage condenser will also introduce spherical aberrations and it will be impossible to image a sharp overall area. If the final magnification is to be high through the phase objective, it may be an advantage to locate the field with a lower power objective before centering the point of interest and then to dial in with a high power phase objective. This will save time and make high-power work less fatiguing. Now we have a phase objective and matching substage condenser.

PHASE CONTRAST

Check on the lamp filament. Make sure it is focused on the iris diaphragm of the substage condenser. This can be carried out by focusing the lamphouse condenser: i.e. rack it to and from the condenser until focus is achieved. If a diffusing disc is used in front of the lamphouse condenser it should be removed for this operation but replaced after the adjustment has been made.

The next step is to open the diaphragm of the lamp until the specimen field is fully illuminated. Now substage condenser and phase objective should be aligned in the optical axis as well as being central. The substage iris diaphragm should now be adjusted so that the image of the annular opening (beneath substage condenser) is also centred on the conjugate area of the diffraction plate. After this all that is necessary is to centre the diaphragm. If the condenser is out of alignment, the image will be expressed by *a* of Fig. 169 and when correctly centred it will take on a different shape, as in *b*. After final alignment for phase contrast the substage diaphragm should be adjusted until the entire conjugate area of the diffraction plate is seen. Alter until the image of the opening is centred upon the conjugate area. If this is absorbing light and the phase diaphragm set at off-centre it will be indicated by *c*, and if the phase diaphragm is out of centre with the diffraction plate it will be expressed by *d*. When the whole unit is off-centre it will be indicated by *e*. After centering the diaphragm with the

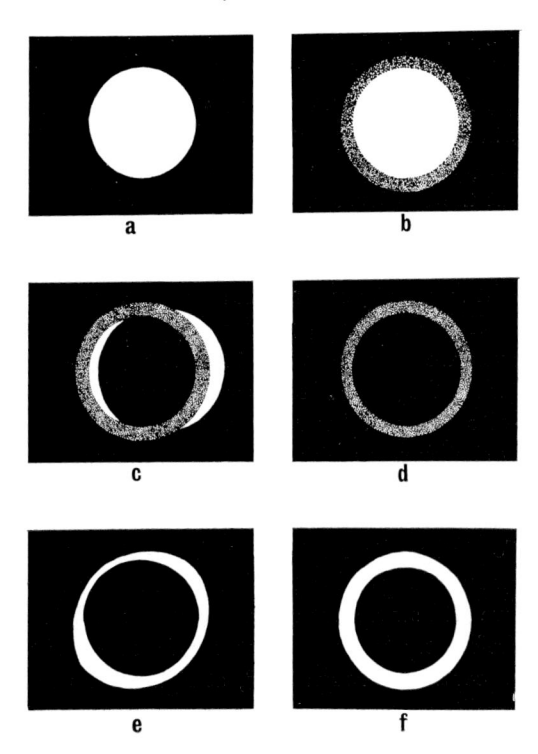

Fig. 169.

conjugate area of the diffraction plate, the image will be a faint circle due to the absorption seen in f. The image indicates an overlap of the edges of the conjugate area and the image of the edge of the annular opening in the diaphragm.

With eyepiece positioned, reduce the iris diaphragm of the lamp until the illumination area laps the edge of the zone being photographed. Further reduction would be harmful to the end product and would result in an over-emphasised "halo" around the subject.

There will always be variations in the method set out above due to the variation in make, style and optical system of the microscope. Nevertheless, it is important that a set pattern of adjustments, leading to a high quality photomicrograph, be adopted. Light leakage around the conjugate area will decrease the definition of the photomicrograph.

References

Cuckow, F. S. (1949). *J. Iron Steel Inst.* **161**, 49.
Francon, M. (1954). *Opt. Acta.* **1**, 50.
Wilska, A. (1954). *Mikroskopie*, **9**, 1–10.
Zernike, F. (1955). How I discovered phase contrast. *Science*, **121**, 345–349.

Polarization

In order to be able to discuss polarization and interference phenomena, it is necessary to form a simple picture of the nature of light. If we imagine a rope supported loosely in a horizontal position, we can transmit a "wave" along the rope by jerking one end quickly up and down. The wave travels along the rope to the far end, but the top itself remains in position. This travelling wave is therefore made up of a series of transverse displacements of short lengths of the rope and is known as a transverse wave. Before we apply this elementary analogy to the nature of light we may press it a little further. If we imagine that the rope passes through a small hole in a fence, the "wave" cannot be transmitted through the hole. If, however, we enlarge the hole into a slit, the wave will pass through the slit, but only when the plane of the wave is parallel to that of the slit.

Light consists of a series of transverse vibrations in space, which may be imaged to be very similar to a series of transverse waves passing along a loose length of rope. There is one very important difference however. The wave motion of light is not confined to one plane but occurs simultaneously: i.e. all round the rope. To picture this imagine that, for instance, the rope is "frozen" with a wave somewhere along its length. If the rope is now spun about its length as an axis, the wave will sweep out a circular shape, Fig. 171D. Reverting to light we may therefore try to form a picture of a series of transverse waves, which expand and contract through 360°, at right angles to the direction in which they are travelling.

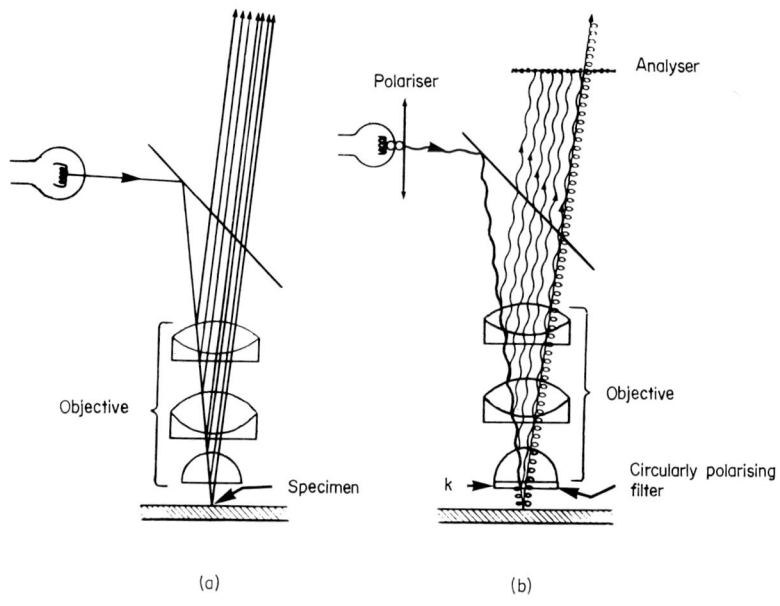

Fig. 170. Epi-objective.

If we now project a beam of such light through a substance (Fig. 171, A) which acts optically as the slit in the fence, all the vibrations which do not lie in the plane of the slit will be blocked, and the only light transmitted will be that in which the vibrations are parallel to the slit. Such an emergent beam is said to be plane-polarized because its vibrations are confined to one plane. If a second piece of similar material, such as an analyser, is placed in the path of the emergent plane-polarized beam the beam will be transmitted only if the plane of polarization of the second sheet of material is parallel with that of the first, B (Plate 17(c)). If it is perpendicular to it, no light will pass, C.

However, light waves from a given source, vibrating at the same velocity can interfere with each other. The result is that amplitudes are doubled and the light appears more intense. Two waves of light having the same wavelength and amplitude but one being half a wavelength out of phase with the other leads to zero amplitude throughout, zero intensity and so darkness. If two waves vibrate in planes at right angles to each other and are viewed along the direction of transmission, several points arise. Should the two waves vibrate with phase differences of a whole or half wavelength then the result is a straight line. But if the phase difference is a quarter or three quarters of a wavelength, the combination then produces the appearance of circular motion. The light is claimed to be "circularly" polarized. Should the phase difference be anything except an even or odd multiple of one quarter of a wavelength ($\lambda/4$) the motion will be along an

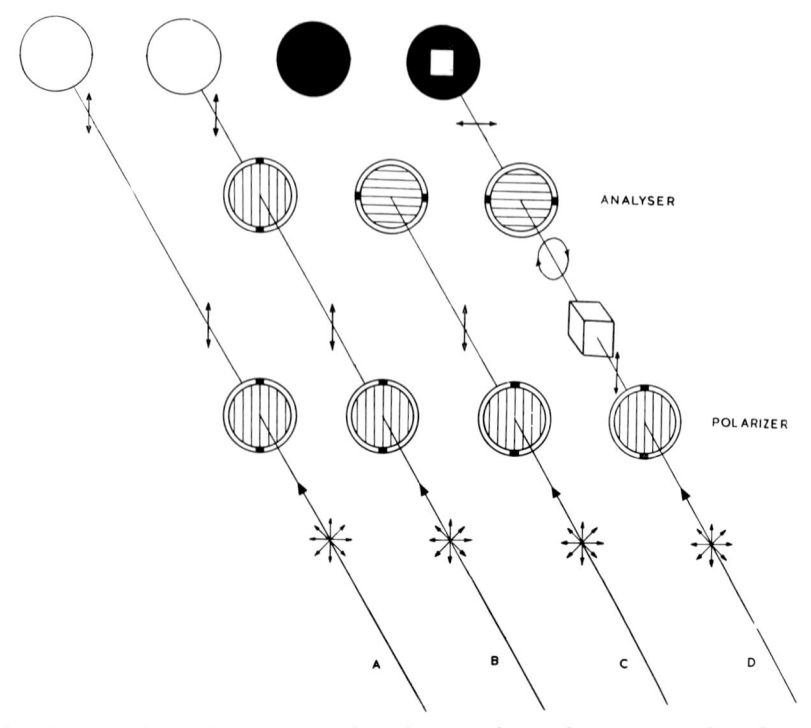

Fig. 171. A. Plane-Polarization (or linearly). B. Plane-Polarization (or linearly). C. Extinction. D. Polarized light (elliptically or circularly).

ellipse and the light is then said to be "elliptically" polarized. "Elliptically" or "circularly" polarized light indicates that the end point of the light vector does not follow that of a straight line, as in plane-polarized light (Fig. 171, A and B), but an ellipse or circle D (Fig. 172). Therefore doubly refractive material appears bright on a dark background, as indicated in Plates 17 and 35. Elliptically polarized light will be seen to change in intensity when viewed during the rotation of a Nicol prism (analyser) just as if it were a mixture of plane-polarized light. If a quarter wave is set in one particular position so that its axis is parallel to one of the axes of the ellipse, or at 45° to the axis of the plate, plane-polarized light is given which the Nicol prism will detect, see Plate 17(c).

There are several materials which may be used to produce polarized light. The Nicol prism is composed of two pieces of calcite Iceland spar, suitably cut, polished and mounted together with Canada balsam, Fig. 173, *a* and *b*. Iceland

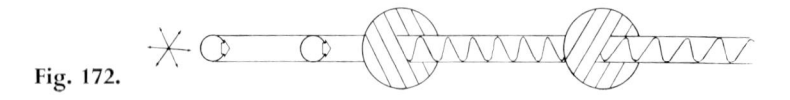

Fig. 172.

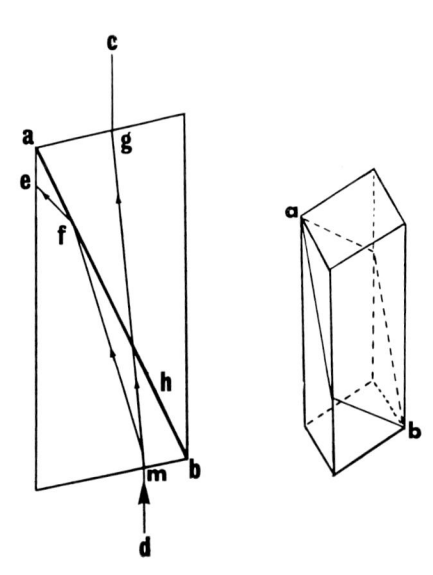

Fig. 173. The right figure shows two cut surfaces, a–b, cemented together with Canada balsam; the two end faces are ground down so as to convert an angle of 71° to one of 68°. Left of the figure shows a section of the Nicol where the plane of the paper is perpendicular to the join, a and b, with the vertical edges near parallel to the axis of rotation, when in position in the microscope. A ray of light entering at the lower surface at m is split in two. The ordinary ray is the more refracted and meets the film of balsam at f with an angle of incidence greater than the critical angle and is totally reflected to e as the refractive index of the calcite for the ordinary ray is greater than that of the balsam (1·54). None of its light emerges at the upper surface of the nicol. The extraordinary ray is less refracted, and takes the path m h g c, penetrating the film of balsam at h, because the angle of incidence is less than the critical angle.

spar is a doubly refracting material and a beam of light incident to its surface is split in two within the crystal. (Some singly refractive materials are gemstones, crystals of the cubic system, glass and plastic material.) These two beams travel at slightly different angles to each other, depending upon the relative disposition of the incident-beam, the face upon which it impinges and the direction of the optical axis of the crystal. But the important thing about these beams is that they are both plane-polarized.

The angle of the balsam cemented interface, between the two Nicol components of the completed prism, is arranged so that one of two beams is reflected to the side of the prism, where it is absorbed, while the other is allowed to proceed and emerge from the opposite face of the assembly to that which it entered (Fig. 173). The efficiency of a Nicol prism is very high, but it is difficult to get large pieces of optically perfect Iceland spar. The diameter of Nicol prisms is often rather small, commonly of the order of 1 cm.

The most convenient modern method of producing polarization is by the use of the synthetic material, Polaroid, which is available in large sheets and is quite thin. In order to use light in the production of a polarized image, polarizing elements must be fitted in the light train of the microscope, both for use in reflected and transmitted light. A Nicol prism is shown in position in Fig. 174a, but a disc of Polaroid may also be used, as indicated. When the plane-polarized light passes through the specimen, a number of changes may be produced in the nature of the emerging beam received by the objective. To examine these, a

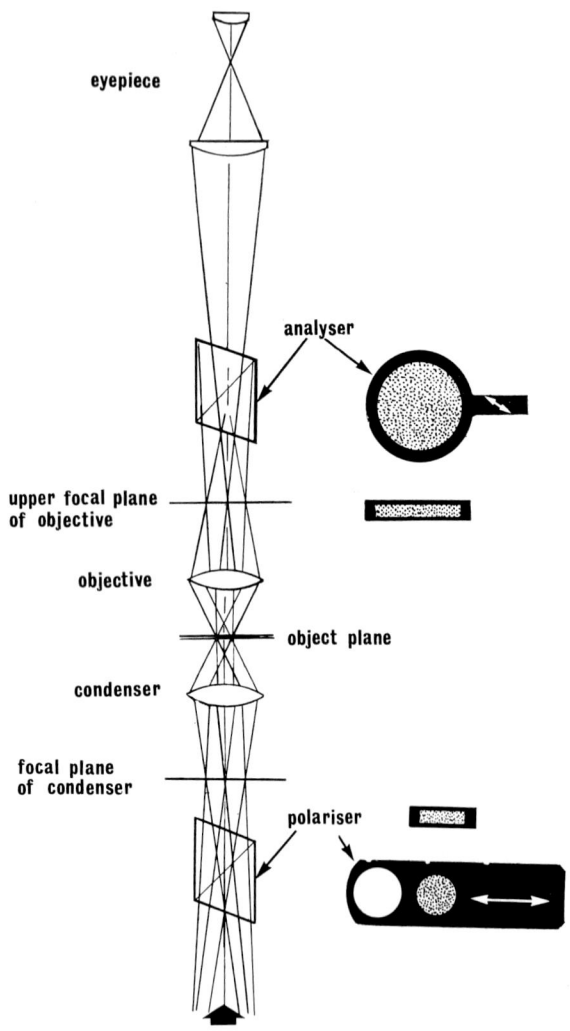

Fig. 174a. Optical layout of illumination by polarized light.

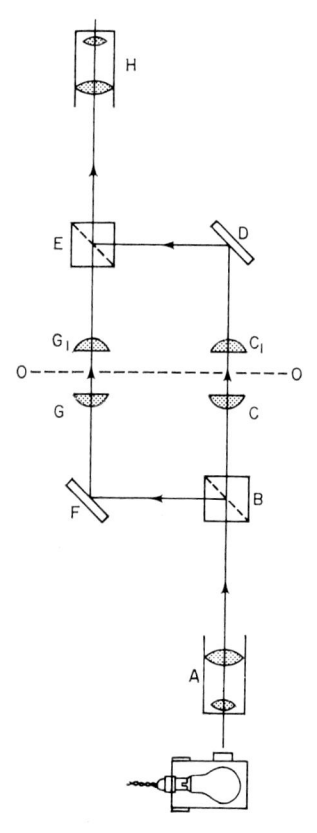

Fig. 174b. Round-the-square method of interference, after J. Philpot.

second polarizing element is mounted above the objective and within the microscope body tube, Fig. 174a. The only difference between this component, known as the analyser, and the lower polarizing element, is that the analyser is capable of rotation through 90° about the optical axis of the microscope, whereas the polarising element is usually fixed. There is a graduated quadrant with the analyser so that its degree of rotation may be observed and also it is usually provided with a click stop, or some similar means of establishing when it is "crossed". This condition occurs when the planes of polarization of the two polarizing elements are at right angles (Polaroid sheet does not create this condition) and corresponds to the position in which the minimum amount of light appears at the eyepiece (in the absence of a specimen on the stage). Figure 174a demonstrates light from a source having passed through two Nicol prisms or two Polaroids. The Nicol prisms can be rotated, changing the order of the light.

For photomicrography in this field, a graduated rotating stage is essential, and a slot at 45° to the fixed plane of polarization of the polarizer is often provided. Various accessories, such as quarter-wave plates and quartz wedges, may be inserted in this slot to facilitate interpretation of the nature of the subject material, Plates 77 and 78. Bausch and Lomb Polarizing Microscope LC and the Leitz Polarizing Microscope are recommended if a considerable amount of photomicrography in polarized light is to be carried out. There are also several body tubes equipped for work in this field which may be interchanged with conventional body tubes. Near infra-red = 1,200 nm can be used together with polarized light (incorporating Nicol prisms) and metals often show a more marked preference in reflecting power, producing improved contrast in the negative. Be careful when focusing as there is a slight change from visible to infra-red radiations. Polaroids are not so successful in this application.

When a crystal of an optically isotropic material is examined under the microscope with the analyser inserted and in the "crossed" position, the crystal will appear dark. This is due to the fact that the plane of polarization of the light transmitted by the polarizer is at right angles to the plane of polarization of the analyser and, since the specimen examined is isotropic, the beam of light undergoes no change in character between polarizer and analyser. The result is like that shown in Fig. 171 C. This condition will still be obtained if the specimen is rotated. Thus, if a transparent crystal appears uniformly dark under crossed Nicols, irrespective of the degree of rotation of the specimen, the crystal is probably isotropic.

Anisotropic crystals may be of either two or three refractive indices: material with three is known as biaxial and material with two indices is unaxial. These differing refractive indices mean that a beam of light incident to one of the crystals is divided into two or three beams within the crystal, each travelling in different directions and at different velocities. When these rays leave the upper face of the specimen under observation, the various beams combine and the result depends upon the precise nature of each of the emergent beams. Since the refracted beams within the crystal travel at different speeds, and since also the optical path length must depend upon the angle of each beam, it is clear that the emergent beams will not bear the same phase relationship to each other. The result is the introduction of a varying measure of ellipticity in the emergent beam and this is then "analysed" to give a series of patterns which bear a marked similarity to interference figures.

If an optically anisotropic crystal is examined under cross Nicols, the specimen will go black, or "extinguished", four times for each complete rotation of the analyser: that is, at intervals of 90° (Fig. 171 D).

If a wavelength of white light, projected through anisotropic material (colourless in tungsten light) is extinguished by interference, a colour, very beautiful and complementary to that of the extinguished wavelength, results. Anisotropy is restricted to substances which show different geometric dimensions in different

Table 21. *Colour scale (adapted from Quincke's table)*

Retardation for "D" line (microns)	Interference colours between crossed Nicols	Order
0·00	Black	
0·04	Iron-grey	
0·097	Lavender-grey	
0·158	Greyish-blue	
0·218	Clearer-grey	
0·234	Greenish-white	
0·259	White	
0·267	Yellowish-white	First
0·281	Straw yellow	
0·306	Light yellow	
0·332	Bright yellow	
0·430	Brownish-yellow	
0·505	Reddish-orange	
0·536	Red	
0·551	Deep red	
0·565	Purple	
0·575	Violet (the "sensitive violet")	
0·589	Indigo	
0·664	Blue (sky-blue)	
0·728	Greenish-blue	
0·826	Light green	Second
0·859	Yellow green	
0·910	Yellow	
0·948	Orange	
1·801	Violet-red	
1·128	Bluish-violet	
1·151	Indigo	
1·258	Greenish-blue	
1·334	Sea-green	
1·426	Greenish-yellow	Third
1·492	Flesh colour	
1·534	Carmine	
1·621	Dull purple	
1·652	Violet-grey	
1·682	Greyish-blue	Fourth
1·711	Dull sea-green	
1·744	Bluish-green	

directions of space. The exhibited colours are usually referred to in terms of Newton's colour scale. The lowest colours in the scale vary from grey to red, violet, blue, green, yellow and a number of other colours, most arresting to the eye and one's artistic taste. Seven orders can usually be distinguished, of which the second and third are the most brilliant, Plate 73, upper. From the fourth order the colours become more pale, until after the seventh they merge and become colourless. (For first order see Plates 77 and 78.)

The precise order of the colours observed may be determined by the introduction of a quartz wedge in the slot at 45° to the plane of polarization of the analyser. For the benefit of those who cannot undertake this technique, Plate 17(d) has been included. Table 21 on the previous page lists the first four orders of Newton's colour scale.

Gemstones can be photographed by using the polarising microscope and immersion techniques. Stones are submerged in a liquid, such as monobromonaphthaline (refractive index 1·549), contained in an optically flat glass cell (Fig. 152). By rotating the stage or analyser, stones can be examined under polarized light. The liquid prevents the various facets from reflecting light and enables the interior of the stones to be clearly seen. A doubly refractive stone will appear to be bright at times, but will darken at intervals of 90°. A single refracting stone will normally be dark during the complete rotation. If slight bright areas are seen in single refracting stones, it is said to be an anomolous polarizing effect.

Epi-objective. Objectives with built-in "Antiflex" (Carl Zeiss, Oberkochen) device are designed to produce images, rich in contrast, of opaque (reflecting) specimens originally lacking contrast. The action of these objectives is based on the polarized light principle. Scattered and reflected light, originating in the lens surfaces of the objective, is polarized by a polarizer and then extinguished by an analyser located transversely to it.

Figure 170 demonstrates the production of reflected light at the glass–air surfaces in an epi-objective. There are two light portions in the image plane: 1. Light reflected at the glass–air surfaces. 2. Light reflected at the specimen of *b*. The light can be plane polarized before it falls on the objective and is extinguished by an analyser. The light falling upon the specimen is circularly polarized by a crystal plate (*K*) and, when reflected, can pass through the analyser.

Epilan immersion objectives are achromatic and manufactured for incident-light work with oil and methylene iodide. These objectives are specially suitable for specimens of low reflectivity, including those of a mixture of low and high reflectivity. Methylene iodine immersion objectives generally produce the same results as corresponding conventional objectives for oil immersion.

The circularly polarizing filter located in front of the near lens element makes the objective unsuitable for certain measuring techniques, e.g. for polarized light analysis of the vibrational state of polarized reflected light.

Interference technique

Since the development of phase contrast techniques, further changes have been brought about in the methods of image formation within the microscope. Phase contrast is a form of interference where a beam of light is separated, or retarded, causing the image details to be surrounded with a very slight bright fringe or halo to give a second image line, one less sharp than the other. Interference technique provides an image free from such haloes.

Although phase contrast and interference techniques are comparatively new, a simple lens system was devised in 1862 in which a beam of light was split into two beams (one advanced) and then recombined to form an image in which interference effects could be observed. This technique proved to be a great success in the study of smooth surfaces and now enables irregularities to be measured in such a surface.

The fringe patterns can be likened to a map with contours. The depth or height can be determined by observing the corresponding fringe deformation. These patterns (Plates 30, 79 and 80) are easily interpreted: plane surfaces generate straight lines or fringes, while spheres are shown by circular patterns and cylinders show either elliptical or straight fringes. In a photographic record it is, however, difficult to discriminate between a ridge and a groove, although the difference may easily be determined by observing the movement of the fringes as the fine focusing control is racked gently up and down. Thus, although the height (or depth) may easily be calculated from a photomicrograph, it is necessary to specify whether the observed area lies above or below the remainder of the surface.

Refractive index determinations have been applied to liquids, Canada balsam and cadmium sulphide; the interference microscope to research on wood, cotton, rayon and synthetic fibres especially. Interference has also been used with polarized light (*Leitz-Mitt. Wiss. Tech.* 1971). Interference technique enables two transparent specimens, lying upon each other in a transparent medium, to be separately examined. Staining can be introduced though at the risk of destroying delicate cells. Interference techniques make possible the study of transparent specimens without staining or similar treatment. Interference colour fringe patterns assist in identification and a method for examining freeze-dried and stained tissues has been described (Lacy and Blundell, 1955).

Round-the-square. This method (Fig. 174b) may be regarded as a form of the Jamin interferometer with a microscope, or microscopes and is included here as a contrast. The light-source in this case is projected by a collimator A onto an obliquely inclined partial reflector B. The latter consists of two prisms separated by a thin partial reflecting layer of aluminium, silver or rhodium. From this composite prism the light is transmitted through the lens C, and having passed through the specimen plane O and lens C, is reflected at the mirror D on to another partial reflector E, identical with B. Thence the reflected light forms

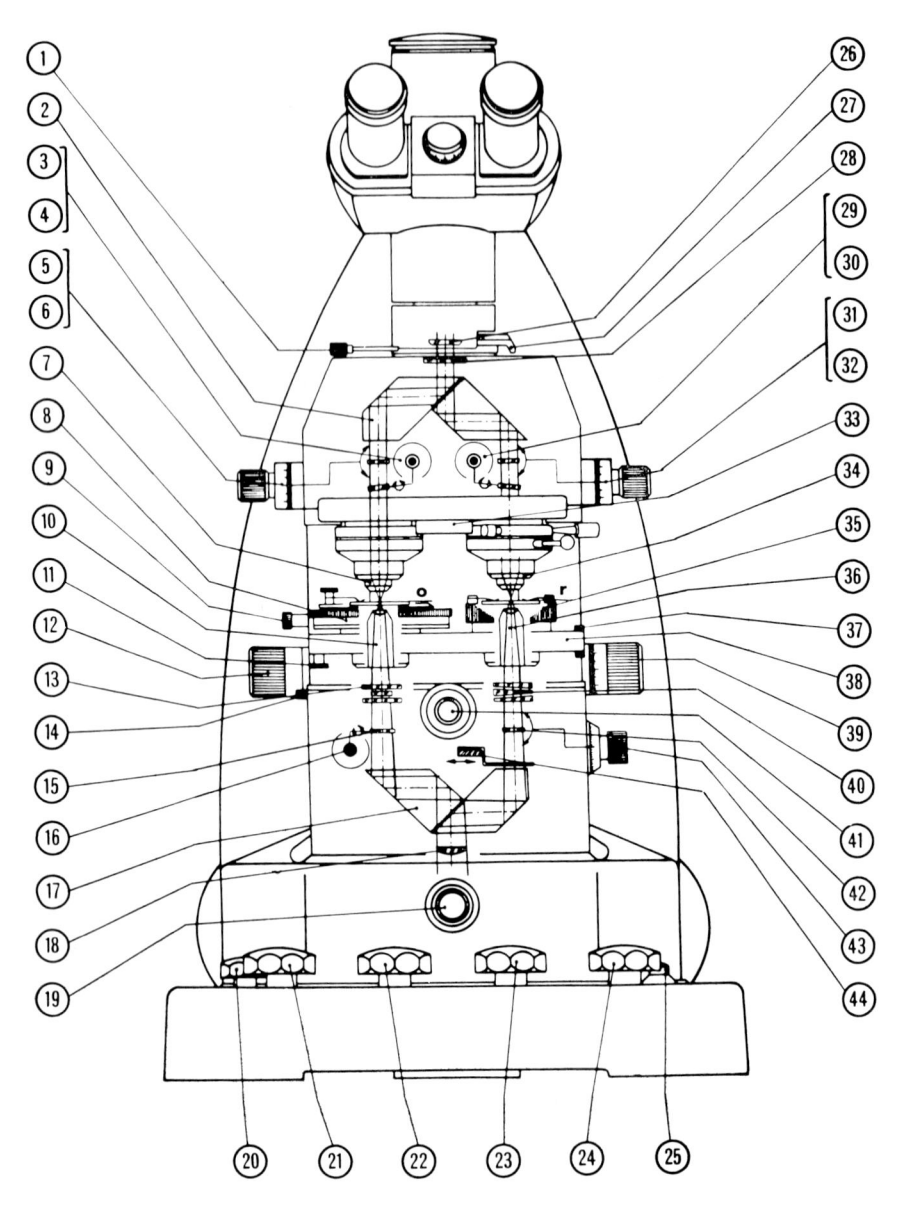

Fig. 175. Leitz "round-the-square" transmitted light interference microscope. 1. Knurled knob for inserting the analyser device, 28. 2. Beam-splitting prism II. 3. Pair of plane parallel plates (setting of the fringe direction). 4. Knurled knobs for adjustment of 3. 5. Pair of plane parallel plates (setting of the fringe width). 6. Knurled knobs for adjustment of 5. 7. Measuring objective, paired with 34. 8. Rotating stage for the measuring beam. 9. Screws for the mechanical adjustment of 8. 10. Measuring beam

"the odd beam" and continues onto the eyepiece H, the image forming in the normal manner. B, where part of the beam is bent after being reflected which shines onto the mirror surface of F, is bent at right angles and passes through lens G, specimen plane O and second lens G, to the partial reflector E. The transmitted light then blends with the odd beam to form an image at H. If no specimen is in position, and the media at the two specimen planes are similar, the system may be adjusted for any desired degree of interference. Then the introduction of a specimen modifies the phase relationship, and an interference image results.

The Leitz interference microscope, first announced in 1958, has been described by Horn (1958) and Grehn (1960) and later Scott (1971). The instrument is a double microscope with the "round-the-square" system of interference developed by Mach and Zehnder (Fig. 175), incorporating polarized light and TV image intensifier. Each light beam has its own microscope and the beams are interchangeable: one, the measuring beam, and the other the reference beam, placed 62 mm apart. The arrangement of plates and compensators can be followed in the figure.

The term "interference microscopy" is restricted to a beam of light which, having traversed the specimen is made to interfere with another beam of light which has been allowed to travel without any interference, Fig. 176.

A method of cyclic interference has been developed by F. H. Smith (Fig. 176). As we have seen, monochromatic light from the source is a parallel beam passing onto, and being divided by, the double prism, and emerging as two divergent beams, one to the left and one to the right. By virtue of the prisms B and B_1, they are plane-polarized at right angles one to the other. The beams are now deflected, becoming parallel to each other by prisms B and B_1, one beam passing through the specimen C. There is thus a difference in the optical paths of the two beams

---

condenser. 11. Clamping screw for the rotating stage 8. 12. Coarse and fine adjustment of 38. 13. Plane parallel plates of 14. 14. Step compensator (revolving). 15. Plane parallel plates (tilting axes vertical to each other). 16. Knurled knobs for adjustment of 15, 42. 17. Beam splitting prism I. 18. Intermediate objective. 19. Switch knob for polarizer. 20. Switch knob for grey filter set. 21. Switch knob for deflecting mixed light prism. 22. Rotating knob for aperture diaphragm. 23. Rotating knob for field stop. 24. Switch knob for selecting Hg-filters. 25. Igniter for Hg-lamp. 26. Analyser (foil, rotating and sliding). 27. Rotating lever of 26. 28. Coverglass for the upper prism housing. 29. Pair of plane parallel plates (setting of the fringe direction). 30. Knurled knobs for adjustment of 3, 29. 31. Pair of plane parallel plates (setting of the fringe width). 32. Knurled knobs for adjustment of 5, 31. 33. Objective change (slide with handle). 34. Reference objective. 35. Reference specimen carrier with knurled focusing ring. 36. Reference beam condenser. 37. Centering screws of 36. 38. Microscope stage. 39. Coarse and fine adjustment of 38. 40. Set for wedge compensator plates (3 plates). 41. Micrometer spindle with measuring scale (face screw for 40). 42. Analyser device. 43. Knurled knob for setting 42. 44. Sliding mask to shut the reference beam.

o Specimen plane. r Reference plane to o. (After R. G. Scott)

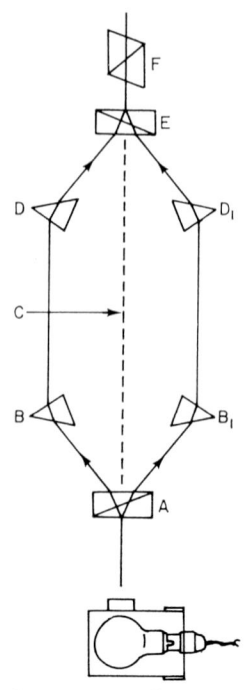

Fig. 176. Cyclic interference. After F. H. Smith.

owing to the object C. The beams are then combined by the prisms D and D_1, and the double prism E. A single beam, consisting of two superimposed perpendicular beams plane-polarized, emerges from this double image prism onto F. Rotation of the analysing prism F may then be used to vary the intensity of either of the polarized beams and in this way the condition for maximum contrast may be achieved. This condition is simply that the two beams be of equal intensity.

Interference technique is a means of measuring the refractive index and thickness of a specimen and the degree of contours. From the figure it can be seen that this is achieved by compensating one of a pair of exactly equal light paths for an addition to the other represented by the passage through the specimen. From the source the main light beam is split, and the two images of the source are superimposed, leaving the exit pupil as one body. The source images are adjusted by the optical system so that they are half a wavelength out of phase with one another, and cancel out. Now, if a specimen is inserted in one of the illuminating light beams, the adjustment will no longer be in accordance with the foregoing, and the object will be imaged as a bright field against a dark background. By adjusting the unaffected light beam, any part of the imaged specimen can be adjusted to make it invisible. The degree of adjustment needed represents the optical thickness in the specimen.

INTERFERENCE TECHNIQUE 207

It is possible to measure substances within the range of half wavelengths, brought about by the optical system. The whole of the specimen can travel through the half wavelength so that a series of parallel interference bands cross it, providing a complete picture of contour bands (see Fig. 177(b)). The image

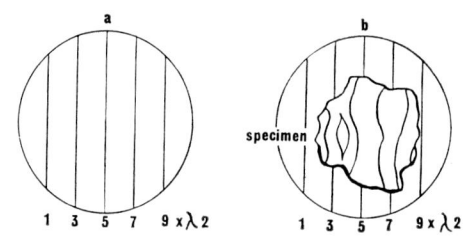

Fig. 177.

contrast is not, of course, solely due to refraction in the specimen, but is a form of polarized light. Colour contrasts can be seen between varying parts of the specimen field and such changes can be introduced by adjusting the system until the specimen reverts to Fig. 177(a).

Carl Zeiss, Oberkochen interference equipment is assisted by polarizing optical means (Fig. 178). Thereby, it is possible to measure slight phase differences with the utmost accuracy. The Zeiss interference microscope was developed on the basis of Jamin's and Lebedeff's ideas (in 1868 and 1930 respectively), with greatly improved optical systems for high magnifications.

The working principles of Carl Zeiss transmitted light interference microscope can be followed in Fig. 179. The light leaving the microscope lamp is linearly polarized (vibration direction in the normal position is east–west), then passes through the condenser and is split into two coherent sections by a beam-splitter†. The two sections propagate themselves parallel to and in a fixed distance from each other in the direction of the objective and while one passes through the specimen (observation and measuring beam), the other follows exactly the same way but without traversing the specimen. Both are linearly prolonged at right angles to each other.

After the beam combining system, the rays produce a uniformly polarizing wave whose components interfere after passing through the analyser. The beam-splitter and the beam combining system are suitably cut crystal plates with orientated vibration direction. One of these is located directly over the front lens of the condenser. The other is mounted in front of the front lens of the objective. The specimen lies between these two lenses in the region of the observation beam.

The optical lengths of the observation and comparison paths of rays, whether like or different to one another, result at times in definite darkening or coloured

---

† Beam-splitters, or semi-transparent mirrors, reflect part of the incident light and transmit the rest. The beam-splitter can be positioned anywhere in the beam path without interfering with the optical image.

values of the microscope image background, from which the specimen, differently darkened or coloured because of its optical thickness, stands out in contrast. (Plate 77 and 78.)

The difference in optical thickness of the specimen to its surrounding, or that

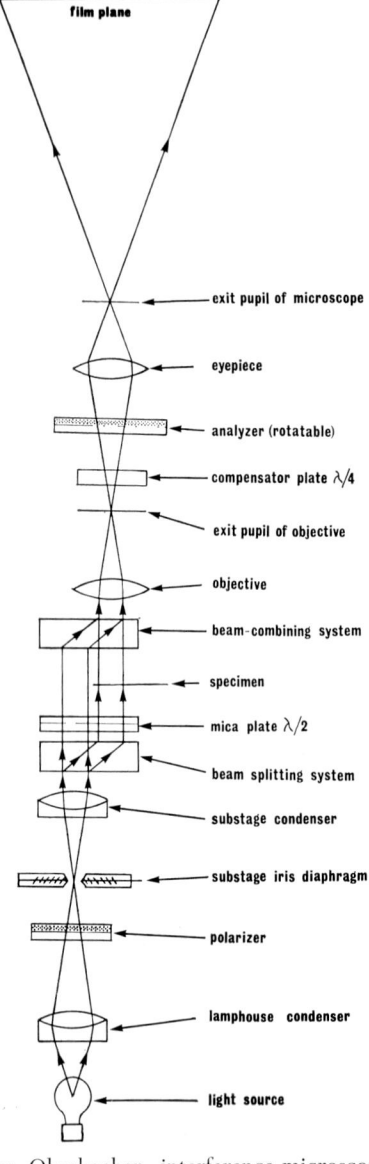

Fig. 178. Layout of Zeiss, Oberkochen, interference microscope and camera.

INTERFERENCE TECHNIQUE

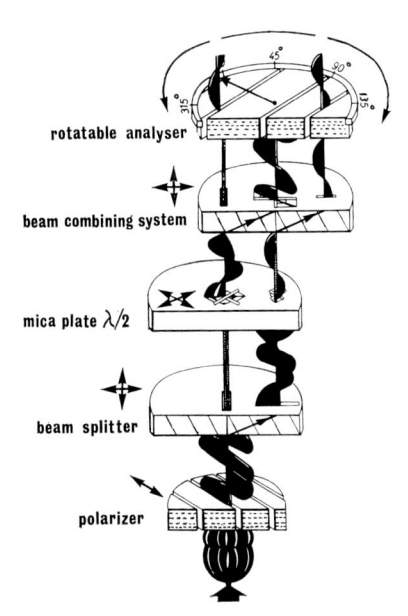

Fig. 179. Detailed section of Zeiss, Oberkochen, interference microscope.

of single area elements of the specimen to each other, can be measured with a compensator. In this case, starting out from a fixed (generally maximum) darkening value of the surrounding, the compensator is adjusted until the area in the specimen to be measured shows the same darkening or the same colour that the background previously had. For measurement of phase differences up to $\lambda/10$, $\lambda/20$ or $\lambda/30$, the Brace-Köhler type compensator is most practical (see 1 below). For phase differences up to 1λ Senarmont's compensator method is recommended for which a $\lambda 4$ disc is inserted in the slit underneath the analyser and the analyser rotated azimuthally until compensation is reached (see 2 below).

The rotation of the analyser and the phase differences to be measured are proportional to each other. For large phase differences up to 122λ, the rotary compensator with compensating calcite Iceland spar plates should be used according to Ehringhaus (see 3 below). The compensators can be inserted alternately in the tube slit provided for this purpose.

The distances between the observation and the comparison paths of rays in the specimen space are very large with respect to the scale numbers of the objectives and therefore merit special attention. The following are paired objectives-condensers forming part of the transmitted light interference equipment by Zeiss. They come to:

0·175 mm for the system POL Int.   I
0·054 mm for the system POL Int.   II
0·0546 mm for the system POL Int. III

These values determine the distance between the observation and comparison field, and at the same time the diameter of the observation surface free from overlapping, or the width of a fringe going through the specimen, accessible to measurement at each individual setting.

1. The compensator, according to Brace-Köhler, which serves for measuring the smallest phase differences, consists of an azimuthally rotatable, thin, birefringent lamina. The smallest rotation angles of this lamina can be taken and converted into phase differences with the aid of a calibration table that comes with the interference equipment.

2. In the compensation method according to Senarmont, which can be employed within the region of a single wavelength, a $\lambda/4$ disc orientated parallel to the polarizer is used. The compensation itself is effected by a rotation of the analyser directly proportional to the phase difference to be measured which can be read directly, up to $0 \cdot 1°$. The $\lambda/4$ disc can be replaced and must be adjusted to the wavelength used in the measurement.

For exact measurements of phase differences monochromatic light is necessary. This can be generated by interference filters, inserted in the microscope base and in front of the source. Generally light of wavelength $\lambda = 546$ nm is used. The interference continuous running filter monochromator (consisting of filter, calibrated frame and a holder for insertion in the light exit aperture of the microscope) is a valuable supplementary device. With this small piece of equipment any light wavelength in the visible spectrum can be isolated (Plate 44).

3. The basis of the Ehringhaus' compensation principle is the dependence of a phase difference in a birefringent object on its tilt position. This likewise can be read very accurately and translated directly into the proper phase difference.

Uniaxial interference. By way of contrast and as an example of the genius of the inventor and the flexibility of an optical system, Fig. 180 shows a compound microscope, with a means of dividing a single beam in two which then traverse the specimen and, after passing through it, cross and recombine. Thus optical path lengths differ slightly one from the other to produce an interference effect which can be viewed or projected onto a sensitive emulsion in the normal manner. The beam is reflected at the surface of the mirror B (vertical transmitted light bypassing the mirror can be used) and is directed through a polarizing screen C to a collimating lens D situated between the latter and a Wollaston prism $E - E_1$. The incident beam of light, now divided into two mutually divergent beams which pass through lens F and specimen plane G where they are brought to a point of focus, enters the objective on either side, as shown. These beams are polarized in planes at right angles one to the other. The beams go through lens F_1 and the double prism H and collimating lens I, which brings the two beams together. After passing through the Soleil Babinet compensator J, the joint beams now pass through the eyepiece K, which is fitted with a polarizing screen L.

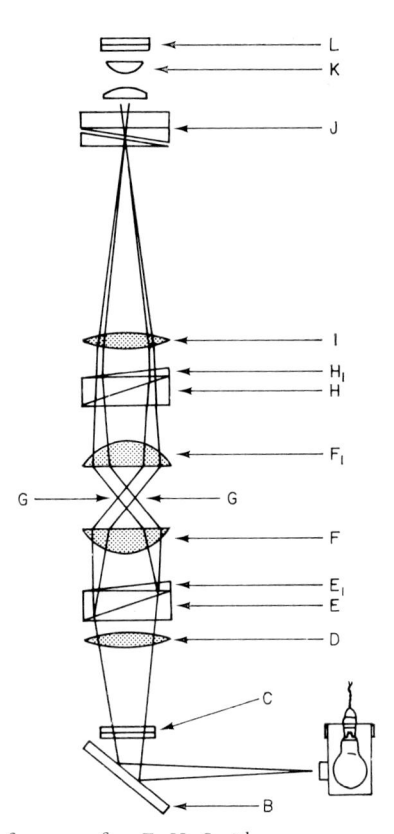

Fig. 180. Uniaxial interference, after F. H. Smith.

As already mentioned, the inclusion of a particular lens, condenser or prism, may introduce difficulties, and may cause the light beam to go "off course". In this particular case the additional prisms E_1 and H_1 have been added to maintain the correct relationship. The use of prisms E and H on their own introduces dispersion. The measurement of contours is then possible well within the range of half a wavelength in thickness. This is achieved by adjusting the specimen and by increasing compensation across the plane of the field, so that a series of parallel interference bands are brought to a position where they cross, and are displaced by the specimen. The depth of the specimen and its contours can be mapped as illustrated in Fig. 177(b).

Polarization interferometer after Nomarski. The Reichert Universal microscope makes provision for several changeable accessories including one most suitable for this application in interferometric techniques. This piece of apparatus has been designed to fit onto the microscope, and to save space in doing so. The new accessory detects changes from flatness as deviations of the diagonal systems of interference bands and can be measured within the range 0·01 to 1 micron. If the surface is uneven as, for example, a badly polished specimen, the order of magnitude of the "roughness" may be estimated. The measurement is carried out without any physical contact between specimen and objective. The specimen can be displaced without disturbing the interference image and without causing any damage.

Interference bands (Plates 54(b), 79 and 80) nearly always appear with the same contrast independent of specimen reflectivity. There is provision for varying the spacing of the interference bands and also the scale of the depth measurement. Changing to comparative bright-field photomicrography is achieved by rotating the analyser incorporating a low voltage lamp fitted with a monochromatic filter through 45°. If necessary, a sodium vapour lamp can be used.

References

Grehn, J. (1960). *Acta Histochem.* **9**, 204.
Horn, W. (1958). *Jb. Optik. Feinmech.* OVI-85.
Lacy, D. and Blundell, M. (1955). *J. Roy. Microsc. Soc.* **75**, 48.
Leitz-Mitt. Wiss. Tech. (1971). Vol. 11, **2**, 36–44.
Scott, R. G. (1971). E. Leitz (Instruments). Scientific and Technical Information. Vol. II, 2, 36–44.

The interferometer microscope

The optical system of the Interferometer Microscope was developed by Dr. J. Dyson in the Research laboratories of A.E.I. Ltd., and is now manufactured under licence (Brit. Pat. 676749). Figure 181 illustrates this optical system.

A function of the interferometer microscope (Fig. 182) enables quantitative measurements to be made of relative optical thickness in, say, waxes, plastics, synthetic fibres, glass, etc. In addition a large range of biological material can be investigated. The "halo" effect (already referred to) is absent in this method of recording because the phase plate does not restrict the aperture of the objective, the same areas being used by both interfering beams. Any section of a specimen can be made to give maximum contrast due to the variableness of the two beams of the relative phase.

An almost plane parallel glass plate A, of partially aluminised surfaces, intercepts the illuminating cone of light from the aplanatic condenser 1·3 N.A., seen at C. The cone of light is divided, one beam is directed and traverses the glass plate A and is brought to focus at the specimen K, while part of the beam is reflected by the top surfaces of A and then upwards by the lower surface of A. The newly formed beams traverse the specimen plane (see illustration), one, the "comparison" beam being reflected twice within the plates with a greater diameter than that of the field. (Beam-splitting systems: mirrors, semi-aluminised mirrors

THE INTERFEROMETER MICROSCOPE

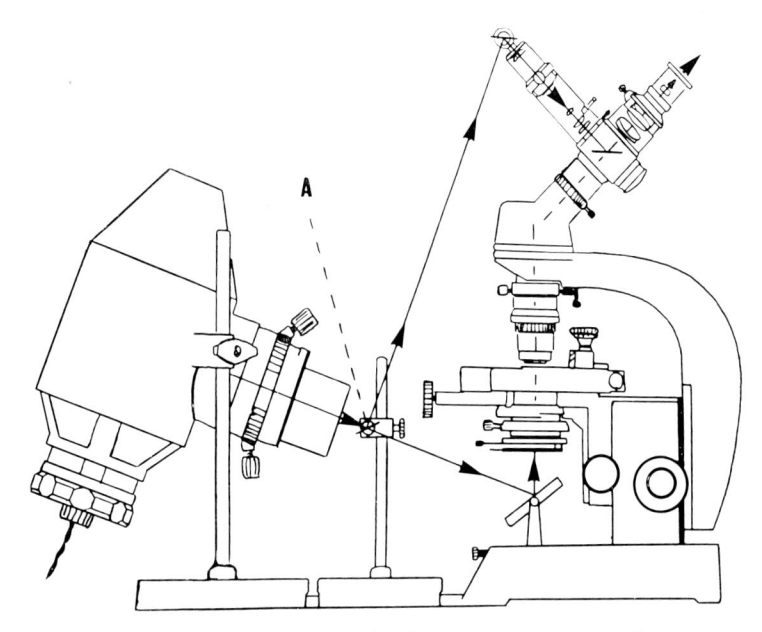

Fig. 181. Interferometer microscope with photometer eyepiece and mercury vapour lamp. *A*. Beam-splitter.

and prisms are arranged in the light train as a means of deflecting part of the light beam from its normal course. Semi-aluminised mirrors are incorporated into the illuminating indirect light system, when transmission is approximately 20%.)

Plate *B* is similar to that of *A*, its surfaces partially aluminised but, in addition, its upper surface has an axial opaque spot *D*, larger than the maximum field covered by the microscope. The function of this spot is to prevent direct light, and that from the comparison beam, entering the objective. The comparison beam has traversed the specimen and that which actually passes through the specimen is recognised as the illuminating beam and is internally reflected twice, while part of the comparison beam passes through the plate without being reflected. If the plates are correctly aligned the two beams emerge at the same time from the upper surface. The projected image of these coincident beams consists of the light source with specimen superimposed, and the light source only. Apart from the disturbances caused by the specimen, the phase difference between the two projections will be zero and the conditions such that interference will take place.

The plano-convex lens *G* is cemented to the top plate *B* (these operating as one unit), the spherical surface of this lens is fully silvered with the exception of a small axial aperture *F*. Due to reflection at *G* and again at the top surface of

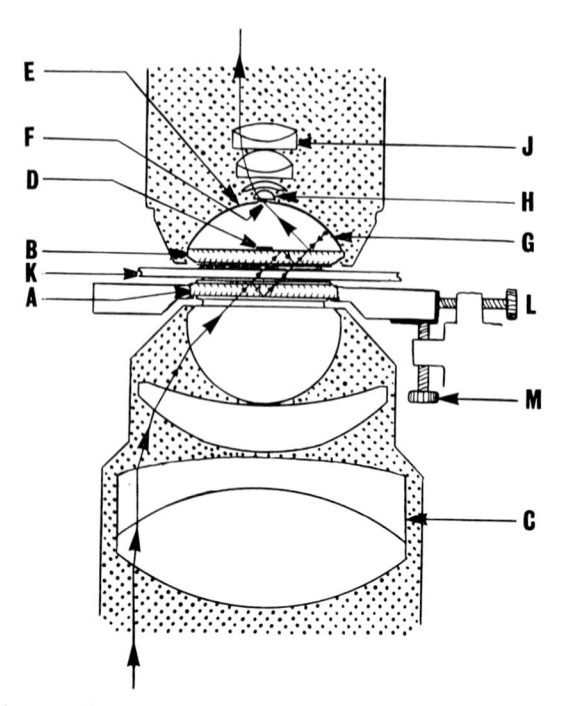

Fig. 182. A schematic lay-out of the optical system of the interferometer microscope. *A.* Lower nearly plane parallel plate. *B.* Upper nearly plane parallel plate. *C.* Aplanatic condenser. 1·3 N.A. *D.* Opaque spot. *E.* Fully silvered surface. *F.* Clear aperture. *G.* Plano-convex lens. *H.* Lower objective lens. *J.* Objective. *K.* Slide and specimen. *L.* Transversing adjustment. *M.* Levelling adjustment.

the plate, the two fields are imaged just below the unsilvered aperture at the top surface of *G*, and also beneath the lens *H* of the objective. The combined images are then transmitted by the objective.

The interferometer plates *A* and *B* are made with a 5 min wedge angle enabling the path difference between the two imaged fields to be varied accordingly. The image formation is controlled by horizontal movement of the interferometer by means of a micrometer adjusting screw, *L*, working through a 10:1 reduction lever. A wavelength difference is represented by some 120 graduated divisions of the micrometer drum.

To obtain the same path difference over the whole of the field the wedge axes must be parallel and the lens attached to the upper plate must be free to rotate as one unit. Levelling screws (*L M*), to enable the bottom plate *A* to be correctly aligned, bring about the conditions necessary for interference. The rotation of the upper plate *B* provides a uniform change of the path differences across the field of view, thus the wedge axes are no longer parallel. The field is then crossed

THE INTERFEROMETER MICROSCOPE

by parallel interference fringes. There is a choice of one of two objectives for this unique system: they are a ×40 of 0·85 N.A. and a ×95 of 1·3 N.A., both set in centering mounts complete with plates. The substage unit is specially constructed to house a particular condenser and plate with tilting and traversing movements.

There is a ×10 photometer eyepiece attachment with this microscope for the high precision measurement of optical path differences which eliminates the possibility of subjective errors in the visual determination of such differences. Optical path differences of high accuracy (1/300λ) within cells can be determined quite easily. The principle of the photometer eyepiece, C, can be followed in Fig. 183. A small silvered spot in the centre of a reflector J is perhaps the hub of the eyepiece system, being illuminated from either a compact source mercury vapour lamp or a tungsten filament lamp, incorporating a beam-splitter, A of Fig. 181. Careful alignment of these is vital. The intensity of the spot is controlled by the rotation of one of two polaroid filters situated between the upper mirror D and the reflector J. This filter is adjusted at F until the brightness of the comparison area lies midway between maximum and minimum brightness of the fringe system in the field of the microscope. When an area of the specimen is brought to match the spot by using the micrometer phase, the variation of intensity for a given rotation of the screw is maximum. Under these conditions

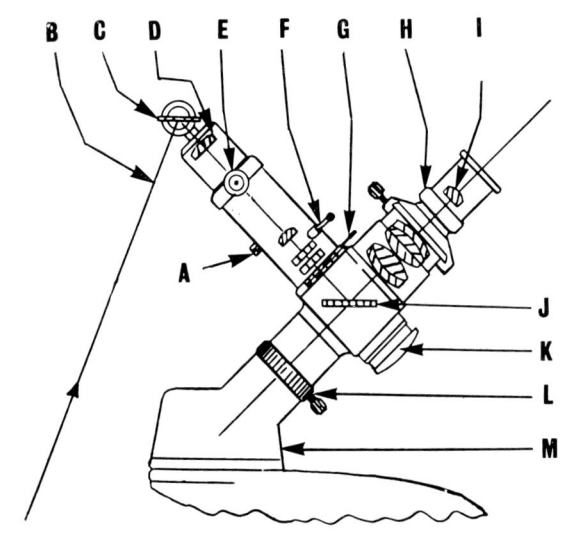

Fig. 183. A schematic layout of the photometer eyepiece for the interferometer microscope. *A.* Exit pupil focusing screw. *B.* Beam from beam-splitter. *C.* Upper mirror. *D.* Condenser lens. *E.* Control. *F.* Intensity control. *G.* Filter. *H.* Becke Klein lens attachment. *I.* Eyepiece. *J.* Reflector. *K.* Reflector retractor. *L.* Screwed body tube adapter. *M.* Inclined head.

maximum sensitivity is obtained and far greater accuracy is possible than by setting to maximum or minimum brightness. The reflector can be removed when not required. (When using a mercury vapour lamp as a source it is advisable to incorporate a green filter such as the Wratten 74, 77A + 58 or Ilford 807 filter. The Ilford 624 or 604 filter is recommended when the source is a filament lamp.)

The interferometer is intended to work satisfactorily with microscope slides up to 1·25 mm thick and 0·18 mm thickness cover glasses, and sizes up to 32 mm (1¼ in.) square may be used. However, the makers of this microscope do state that, when attempting to obtain very high accuracy (1/100λ–1/1000λ) imperfections in the surface of cover glasses due to optical inhomogeneities, which are approximately in focus with the specimen, may introduce an error. When these difficulties are encountered cover glasses must be specially selected (e.g. Chance resistance micro cover glasses).

In order to avoid inaccuracies due to unevenness in the slide and cover glass, homogeneous immersion is employed and to assist, an immersion fluid is used between the specimen and the upper and lower plates.

Reflecting microscope

As early as 1672, when both reflecting and refracting objectives began to be improved, Sir Isaac Newton described probably the first reflecting microscope to the Royal Society. The apparatus consisted of a concave ellipsoidal mirror and a diagonal flat. Considerable progress has been made since the first mirror objective.

The behaviour of polished metal surfaces in ultra-violet is interesting. Silver, which reflects about 95% of the incident light in the visible spectrum, reflects only 4% at wavelengths in the region of 316 nm, though for shorter wavelengths its reflecting power rises and fluctuates. A very thin film of metallic silver can be used as a filter to cut off visible light and transmit only ultra-violet. Nickel and magnolium and many other alloys are more effective than silver, while one of the most useful reflecting surfaces is made by depositing a thin film of aluminium on glass and quartz.

On the other hand, the spectral transmission of quartz is from 185 nm in the ultra-violet to approximately 3,000 nm in the infra-red (Fig. 184). This excludes

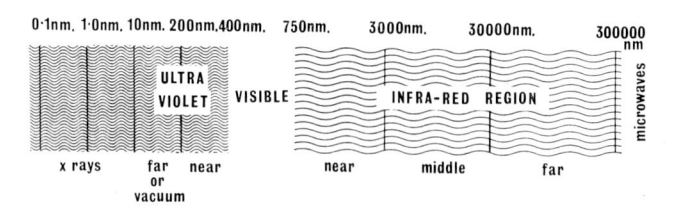

Fig. 184.

the use of quartz objectives in the vacuum ultra-violet where increased resolution as well as differential absorption and reflection effects can be of importance. We have already referred to the microscope mirror as being a lens of certain focal length, but this particular mirror reflects light only from source to the specimen and not a specimen image as does the mirror or catoptric objective. A mirror objective, made up of highly reflecting surfaces, has become a lens of great importance. It is extremely difficult to make and imperfections of the image are expressed with the increased curvature of the spherical surfaces. A series of spherical mirrors can exhibit spherical aberration and coma in a similar way to a simple lens. Because of this, optical systems are made up in groups, each lens surface having a less shallow curve until the computations of the lenses add up to one power.

A simple Schwartzschild-type reflecting objective, Fig. 185 (a) consists of two mirrors, *a* concave and *b* convex. The mirror produces an image of the specimen and this image is then magnified by the convex mirror *b* to produce a magnified image at *d*. Considerable diffraction can become evident and the image deterioration leads to a series of rings of decreasing intensity, identified as diffraction rings. As we have already seen, the size of these rings or disc determines the

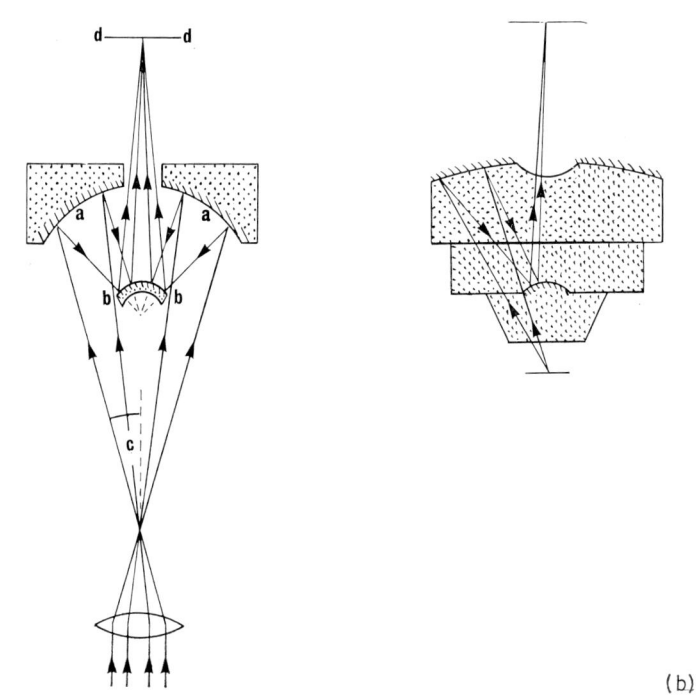

Fig. 185. (a) Path rays through a simple Schwartzschild reflecting mirror objective. (b) Solid quartz Schwartzschild reflecting objective (UV).

clarity of the projected image detail imaged by an objective. When two dots or lines are separated or resolved, the distance is $=0\cdot 5\lambda/\text{N.A.}$ If this is to be decreased the numerical aperture must be increased or a change made in wavelength. This could be done by introducing an ultra-violet source, or deep blue filter into the light train. On the other hand, if the angle indicated by c of the figure is increased, optical errors may arise. It is evident that a high numerical aperture is unlikely.

The error corrections so necessary in a mirror objective are an old problem and were first recorded by Schwartzschild (1905) when he used two components in his aplanatic telescope mirror objective which proved to be achromatic. Little attention was paid to this until, in 1932, Maksutov patented a fully reflecting catoptic microscope objective. The general design of this was, in fact, the reverse of the Schwartzschild lens system. After this, mirror objectives entered a new phase of development. In 1950 a reflecting objective was introduced by Ambrose for use in the near infra-red of the spectrum (Ambrose et al., 1950). With a conventional microscope, this covered a wide spectral range from approximately 240 nm to 1,000 nm. Today, the reflecting microscope is used to further research in many fields.

One of the functions of this particular method is the examination of photographic emulsions by ultra-violet and infra-red radiations, overcoming some of the limitations of the conventional refracting microscope. The applications of the reflecting microscope now cover phase contrast and interference techniques, polarized and unpolarized radiations from one end of the spectrum to the other.

In order to eliminate aberration, a quartz monochromat may have as many as seven lenses separated by air spaces. In vertical illumination in photomicrography, back reflection from the surfaces of these lenses may impair the image quality, more seriously so in ultra-violet than in visible light. Because of the possibility of such errors the reflecting microscope was introduced, with quartz and reflecting surfaces. Thus the use of quartz elements in objectives makes possible photomicrography with ultra-violet radiation as a source and makes a great improvement in the detail of the image as the resolving power of the optical system increases with decreasing wavelength. Generally, if ultra-violet objectives are used at a wavelength other than the one for which they were computed, the resolving power will deteriorate (Fig. 66).

Schwartzschild also produced a quartz reflecting objective (Fig. 185 (b)), made up of three quartz lenses, said to be N.A. 0·85, in which the space between the mirrors was filled with an aluminium deposit, and the entrance and exit surfaces were so arranged as to have the specimen image point at their centre of curvature. This and similar objectives have a high central obstruction and when used with immersion oil it is possible to increase the numerical aperture by a factor of 1·5. Over the years manufacturers such as Burch and Beck have reduced the obscuration ratio in their well known mirror objectives, in which the elements are

REFLECTING MICROSCOPE

aspheric and the N.A. 0·65 and the obscuration ratio 0·14. These objectives can cover a range from far ultra-violet to far infra-red.

Ealing Beck (formerly R. and J. Beck) no longer produce a reflecting microscope, though they still specialise in making reflecting objectives which mostly fit existing microscopes to enable them to be used as reflecting microscopes. These objectives (Figs 67 and 186) were primarily designed for operating over

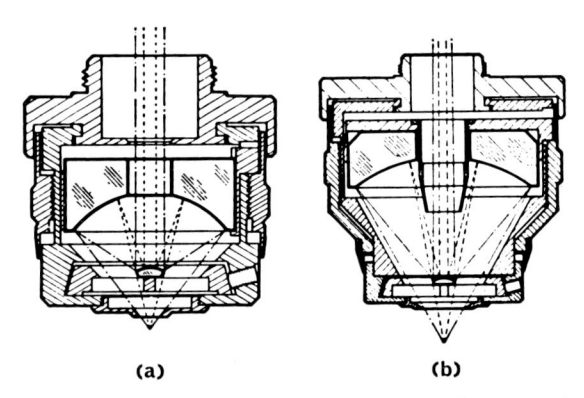

Fig. 186. (a) Ealing Beck. 3·5 mm Focal length. ×52 Magnification. 0·65 N.A. 3·5 mm Working distance. (b) Ealing Beck. 5·6 mm Focal length. ×36 Magnification. 0·50 N.A. 8·0 mm Working distance.

the full spectral range. They have such additional properties as extra long working distances and there is complete absence of chromatic aberration. The resolution is similar to the standard refracting objectives. The annular aperture, however, tends to enhance the contrast in the upper spatial frequency regions while degrading it in the mid spatial frequencies.

Table 22.   *Ealing Beck reflecting objectives*

Focus	13·0 mm	5·4 mm	3·5 mm	2·6 mm
Magnification:				
160 mm tube length	×15	×36	×52	×74
250 mm tube length				
Visual field of viewed object	1·2 mm	0·50 mm	0·34 mm	0·24 mm
Numerical aperture	0·28	0·50	0·65	0·65
% area of central obstruction	17·5%	12·5%	17·5%	12·5%
Working distance (approximation)	24 mm	8·0 mm	3·5 mm	2·5 mm
Length (from R.M.S. shoulder)	50·7 mm	53·8 mm	32·3 mm	32·3 mm
Maximum external diameter	44·7 mm	58·7 mm	38·1 mm	38·1 mm

The objective mirrors are adjusted by a moving collar with an engraved scale which represents the equivalent of the cover glass thickness. This moves against a second scale representing the tube length. By setting the two scales against each other for the particular combination of tube length and cover glass thickness aberrations are eliminated.

The exceptionally long working distance (Fig. 67) makes the objective especially valuable for work with furnaces or dewar vessels. It has the largest exit pupil in the range of four, so that it should be selected for investigations in the infrared where energy values are often low. It can be used with incident illumination. It is normally adjusted for 160 mm tube length and 0·18 mm cover glass thickness. It is even possible to adjust for any tube length between 80 mm and 230 mm including cover glass or furnace window thickness up to 3 mm.

A medium-power system (Fig. 186b) has resolving power equivalent to a 8 mm refracting objective. To obtain the high resolution and the long working distance, the diameter of the mirror systems has been increased and it is essential that the microscope with it is adequate in size. A collar permits correction to be made for a furnace window up to 2 mm thick, interposed between specimen and the objective, but if this thickness is exceeded by 0·25 mm monochromatic light must be used to avoid chromatic aberration.

The use of an aspheric reflecting mirror minimises the possibility of glare and other optical faults and, in addition, the reflecting system permits a far greater working distance than that of conventional lens systems. The Burch microscope has a working distance of approximately 12 mm as compared with 0·5 mm for a refracting objective of similar numerical aperture. The N.A. of one objective is 0·65 dry and the other, 0·95 dry: the latter increases to 1·4 when oil immersion is introduced. The photographic image is practically free from curvature of field owing to the aspheric surface of the mirrors.

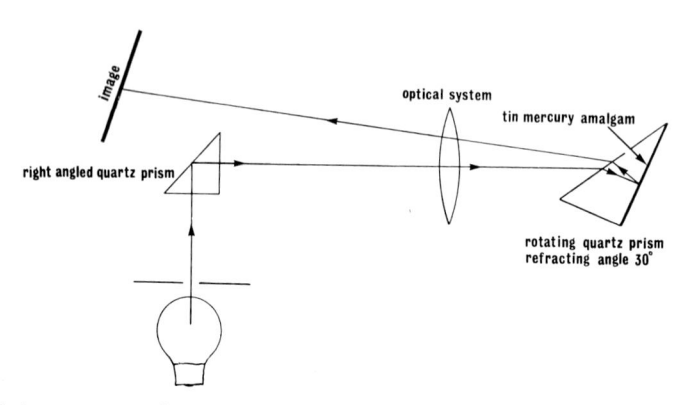

Fig. 187. Littrow monochromator.

The use of a cover glass over the specimen will, in some cases, cause defects in the photographic image when working with reflecting objectives because of the introduction of spherical aberration. To overcome this difficulty a specimen cover glass, the same thickness as the test slide, can be used when setting up the apparatus and focusing the field of view.

The illumination used with the Burch microscope can be either a tungsten lamp or a Littrow monochromator, in conjunction with a medium pressure mercury-arc lamp to give either white light or ultra-violet radiations respectively. A Littrow monochromator includes prisms, or hinged or pivoted mirror systems, Fig. 187. Photomicrographs are usually made *without* the eyepiece in order to avoid the possibility of aberration. The tube length of the Burch microscope differs from that of others and is operated at 34·5 cm. Normal optical filters should not be used with this apparatus. There are two points of focus in almost the same plane, one due to fluorescent light which radiates from the unstained areas of the specimen and the other due to transmitted ultra-violet radiation. For work in the ultra-violet, the visible must be filtered out with an absorbing filter (as discussed later).

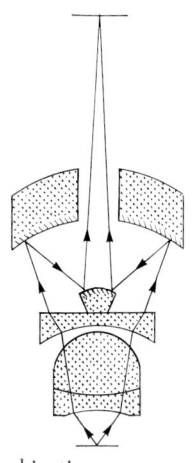

Fig. 188. Polaroid-grey reflecting objective.

The Polaroid-grey reflecting objective (Fig. 188) was designed by L. V. Foster of New York. Its focal length is 2·8 mm and N.A. 0·72 with a magnification of ×53. It has two reflecting surfaces of aluminium film evaporated on glass, a quartz fluorite doublet and a fluorite refracting component which reduces the residual aberration set up by the spherical surfaces. The two uppermost lenses are glass and the third is fluorite. The lower element is quartz and is fused to a fluorite lens. Any chromatic aberration which may appear through the use of a quartz cover glass is corrected. This unique objective has an occluded aperture of 0·21 and is operated with a ×4 negative-type eyepiece to give an overall magnification of ×212 at 160 mm tube length. For higher magnifications, the use of a ×10 quartz eyepiece of the Huygen type is recommended. This objective can be used in the ultra-violet or infra-red regions with a very slight change in the point of focus. The objective is absolutely free from chromatic aberration provided the specimen is mounted in air but, if mounted with a cover glass, an alteration in the optical path length is unavoidable. A mirror objective after Nomarski, ×40 N.A. 0·52 producing an image range from ×80 to ×500 is used with the efficient Reichert high temperature (1,800°C) microscope.

Los Alamos Scientific Laboratories, New Mexico, USA, have developed and patented a microscope capable of examining materials at a high temperature, making possible the observation of a wide variety of physical, chemical and metallurgical phenomena.

Photomicrography and cinémicrography can now be used throughout a complete phase of melt. For the first time the inspection at ultra-high temperatures of different crystal structures is possible. In addition, the Don Olson designed microscope has been used with the specimen at over 2,500°C in the investigation of phase equilibria, chemical reactions, sintering, polygonisation and grain boundary migrates.

The Berlyn Brixner optical company has assisted in designing an optical system which is complex, but which makes possible photomicrographs in the ultra-violet and infra-red radiations. A concave ellipsoidal first stage mirror, with a 10·2 cm (4 in.) working distance and N.A. 0·47, enables the specimen in the furnace to be examined at a long working distance. Approximately 8·7% of radiated light from the specimen is picked up by the ellipsoidal mirror and some 84% of this is reflected through a 13 mm (0·5 in.) thick plane parallel quartz window. Beyond this window is the heater and here an optical path is bent by a series of seven plane first-surface mirrors. The last of these is at the point of the second focus of the ellipsoidal mirror situated a considerable distance from its vertex. The final reflecting mirror receives a real image of the specimen surface at a magnification of ×30, which can be stepped up to as much as ×600 by the incorporation of the eyepiece system. Cameras using 35 mm and 12·7 × 10·2 cm (5 × 4 in.) film material can be used as well as ciné.

There is a split-tube resistance heater which by radiation heats objectives as large as 13 mm in diameter to beyond 2,500°C without affecting the optical system. The field covered by the optical system is in the region of 3·18 mm (0·125 in.) in diameter but it is possible to scan the whole specimen in the furnace.

It is claimed that nearly theoretical resolution can be obtained over the entire image range, and that spherical and chromatic aberrations are not present.

References

Ambrose, E. J., Elliott, A. and Temple, R. B. (1950). *J. Sci. Instrum.* **27**, 21.
Schwartzschild, K. (1905). "Theorie de Spiegelteleskope". Gottingen.

Infra-red photomicrography

The application of infra-red photomicrography has no advantage as far as resolving power goes for, as we have seen, resolving power is directly proportional to the wavelength employed. Therefore, with infra-red radiation the resolving power is not as fine as with ultra-violet radiation and visible transmission. However, much can be observed with the aid of infra-red radiation that hitherto was not visible. Figure 184 indicates the near infra-red, middle or conventional infra-red and far infra-red. Beyond this is microwaves. The eye is not sensitive to radiations of wavelength below 400 nm and beyond 750 nm. This can only be an approximation as there is no actual demarcation line but a gradual merging from one band to another. The region between 750 nm and approximately 40,000 nm is the area in which we are interested. The existence of this part of the spectrum was first demonstrated by Sir William Herschel (1738–1822).

As we have seen there are a number of quartz illuminants and objectives specially made for use in the spectral region of the infra-red. In contrast, there is always the standard optical system which, when the thickness of the glass lenses to the lenses of the eyepiece and substage condenser have been added, amounts to something in the order of 10–12 mm. Even so, such a thickness will not reduce the intensity of the infra-red radiation below 1,000 nm. If a quartz optical system is available the transmission will increase to approximately 4,000 nm.

Most opaque subjects absorb infra-red as such so there is less scope for expression in the field. On the other hand, some subjects, regarded as opaque, are in fact transparent to infra-red radiations. This is a great advantage to the plant pathologist and mycologist for infra-red can be used here to good effect.

An incandescent source produces a considerable amount of infra-red. This includes several of the widely used discharge lamps. Mercury vapour lamps and hydrogen discharge tubes, usually in quartz envelopes, are used in photomicrography as sources of both ultra-violet and infra-red. The XBO 150 W Xenon high pressure lamp presents strong lines in the short wave infra-red region at 800–1,000 nm (see Fig. 126). The Zirconium 100 W lamp, Fig. 122, also shows strong lines in the region of 750–940 nm.

Low and high intensity arcs and filament lamps are also efficient radiators in this region of the spectrum. As a rule the peak emission is in the region of 900 nm. Even so it is advisable to use an infra-red filter to control the output.

Infra-red photomicrography (Köhler, 1921), has proved to be invaluable to research in many scientific fields. Some alloy metals show a marked preference in reflecting power in the near and middle infra-red region of the spectrum to the visible region. Also, owing to the differential reflection coefficients of the various constituents of alloy, a greater degree of contrast is obtained in infra-red photomicrography (Fig. 189).

It is practically impossible to focus an infra-red spectrum: the position of the intensity is located but not the sharpness. Further, most of the infra-red spectra are continuous. So the chief advantage of mirror objectives over objectives containing glass elements is that they do not require focusing as such. If mirrors are focused, the infra-red spectrum will also be in focus to the farthest infra-red.

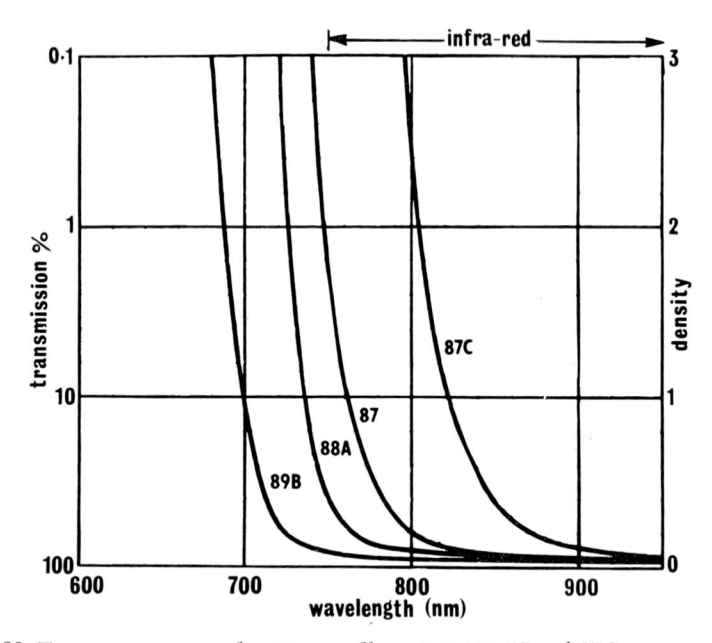

Fig. 189. Transmission curves for Wratten filters 89B 88A 87 and 87C.

Mirrors are cheaper to manufacture and can be used deep into the infra-red where even fluorite and rock salt absorb. The Wratten filter 87 (Table 25) transmission is from 720 nm or Ilford Tricolour Red, 610 nm and can be used for focusing purposes as the difference in focus between this line and infra-red radiation on the photographic material is approximately the same. A red filter can be used when imaging with an apochromatic objective, since this band is suitable for the colour correction of the system. Likewise a deep red filter can be used prior to infra-red radiation for focusing the substage condenser. The heat absorbing filter should be removed during the exposure lest it absorb infra-red radiation (Fig. 190).

More recently, the visualisation of bacteria within unstained living roots has been recorded in colour by Casida (1968). In a recent series of experiments by

INFRA-RED PHOTOMICROGRAPHY

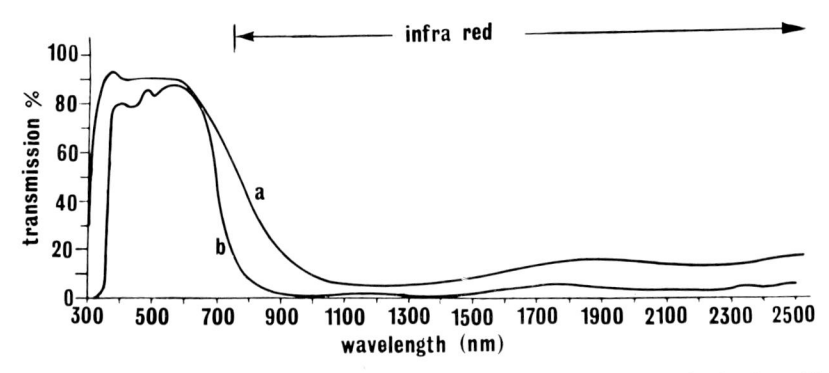

Fig. 190. (a) Transmission curve for heat absorbing filters K.G.1 2 mm thick. (b) Calflex 2 mm thick reflectance filter.

Farrell (1968) the sporulation of *Phytophthora infestans* photographed in colour through the open stomata of a leaf was carried out.

Specially coated colour film for use in the lower end of the infra-red has been used with success in recording many subjects without firstly having to use a stain. Staining techniques often distort the subject. Successful "false" colour recordings of colourless subjects can be made on 35 mm Ektachrome infra-red film, type 8443, supplied in 20 exposure lengths.

Infra-red colour film consists of three image layers sensitised to green, red and infra-red (Fig. 191) instead of blue, green and red used in conventional colour

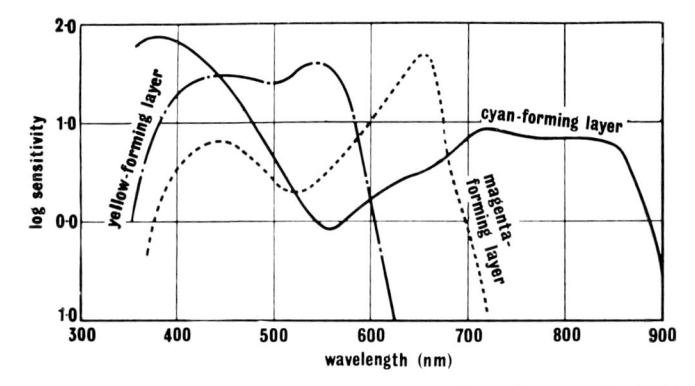

Fig. 191. Spectral sensitivity of Kodak 'Ektachrome' infra-red Aero Film 8443.

film layers. The curves above show the variation of log sensitivity with wavelength: the sensitivity is expressed as the reciprocal of ergs/cm^2. A filter such as a Wratten 87 (Table 24 and Fig. 189) or Ilford 207 + 813 is used to maintain a true colour balance during the exposure. Positive images of magenta and cyan appear in the red and infra-red, and a yellow positive image records the green sensitive layers respectively. Such colours can be termed "false", but at the same time they are a new means of identification. The following table demonstrates the use of additional filters for changing the colour balance.

Table 23.

Filter for 3400K	Colour shift
Kodak CC, cyan or green	From green to more magenta
Kodak CC, blue	From cyan to more red
Kodak CC, magenta or red	From blue to more yellow
Corning C.S. 1–59 (3966)	From red to more cyan

Neurohistological specimens stained by silver nitrate were photographed by Blair and Davies (1933) to record the cytoplasmic structure of ganglion cells (Plate 79), separated from background material, which could not be recorded in any other medium. Fowler and Harlow (1940), too, devoted their time and attention to photographing wood stained by neocyanine and obtained significant results.

It is not necessary to employ an infra-red filter when operating direct with an infra-red source. The range of infra-red transmission output limits the substances of optical interest, which are as follows

Glass	3,000 nm
Quartz	4,000 nm
Calcite	5,000 nm
Fluorite	9,000 nm
Rock salt	15,000 nm
Sylvine	23,000 nm

There is a wide choice of filter material absorbing ultra-violet and the visible spectrum and transmitting wavelengths from 700 nm upwards to 4,000 nm. Figure 135 demonstrates Balzers Filtraflex B–IR 1444 and B–IR 1437. Further filters by Kodak are listed in Tables 24 and 25 and Fig. 189.

When using a 250 W filament bulb the lamp must be enclosed in a lightproof housing with the aperture facing the microscope and be covered by an infra-red filter to eliminate all visible light. The spectral transmission of the filter should reach its maximum around 900 nm. Thus the maximum sensitivity of the emulsion will permit maximum image contrast (Fig. 192). In both cases the film reaches its maximum sensitivity in the region of 900 nm to 800 nm. Thereafter it quickly falls off into the visible region.

Early in 1940, E.M.I. Ltd. made an infra-red converter tube for use with a microscope (Fig. 193) in which an image is recorded on an anode (from Holst, 1934). The infra-red radiation is directed on to a cathode surface and photons

INFRA-RED PHOTOMICROGRAPHY

Table 24.

Wavelength nm	\multicolumn{4}{c}{Percentage transmission}			
	87	87C	88A	89B
700	—	—	—	11·2
10	—	—	—	32·4
20	—	—	—	57·6
30	—	—	7·4	69·1
40	0·10	—	32·8	77·6
750	2·19	—	56·3	83·1
60	7·95	—	69·2	85·0
70	17·4	—	74·2	86·1
80	31·6	—	77·6	87·0
90	43·7	—	79·7	87·7
800	53·8	0·32	81·4	88·1
10	61·7	3·20	82·6	88·4
20	69·2	8·90	83·7	88·6
30	74·1	17·8	84·7	88·8
40	77·7	28·2	85·5	89·0
850	81·4	41·0	86·1	89·2
60	84·0	53·8	86·6	89·4
70	85·4	61·6	87·2	89·6
80	86·8	69·2	87·5	89·8
90	87·8	74·1	87·8	89·9
900	88·4	78·5	88·0	90·0
10	88·8	81·5	88·2	90·1
20	89·1	83·6	88·4	90·2
30	89·1	85·1	88·6	90·3
40	89·1	86·0	88·8	90·4
950	89·1	87·0	89·0	90·5

and electrons are released which, through high voltage (4400) are in turn directed to an anode, forming a fluorescent image. After passing through the lens system the green fluorescent is imaged.

The figure demonstrates what is in fact a modified photo-electric cell, placed above the microscope tube. Here the cylindrical chamber-like cell made of Pyrex is approximately 5 cm in diameter and is the same length as apertures indicated. The end aperture facing the microscope consists of a semi-transparent silver–caesium oxide layer which acts as a cathode. A short distance beyond this is a glass plate coated with zinc ortho-silicate, known as Willemite, to which is attached a small meshed metallic grid, acting as an anode. Beyond this is the viewing eyepiece and camera.

An infra-red sensitive image converter is made by Zeiss, Jena. The infra-red image is projected on to a plate which converts the image to a fluorescent visible image on a screen as set out above. Photography then follows in the normal way.

A reflecting mirror objective was constructed by Elliott for use in the near infra-red region at a magnification of ×5. It consisted of an off-axis paraboloid, figured by evaporated selenium through a baffle on to a spherical mirror. This system, with an infra-red spectrometer and with polarised infra-red radiation, gave results of importance in the elucidation of the structure of synthetic polypeptides and fibrous proteins. An objective like this is often designed for a specific purpose and not all the possibilities in this field have been exploited. Unfortunately, such an objective would not fit a conventional microscope.

Further infra-red studies were made by Barer and others (1949) when they used a Burch microscope to examine single crystals and orientated fibres, which

Table 25.

Filter	Code No.	Transmission
Wratten	88	From 680 nm
Wratten	88A	From 729 nm
Wratten	87	From 740 nm
Ilford	207	From 740 nm
Ilford	207 + 813	From 740 nm
Muster Schmidt	K.G. 100	From 720 nm
Muster Schmidt	K.G. 500	From 760 nm
Schott and Gen.	B.G. 3 + B.G. 5	From 850 nm–1,000 nm
Interference (with X.B.O. 100 W source)	—	From 800 nm–1,080 nm
Balzers	B–IR 1437	From 800 nm–1,400 nm Peak 1,250 nm
Balzers	B–IR 1444	From 750 nm–1,500 nm Peak 1,050 nm
An opaque solution of iodine in carbon disulphide in an optically flat cell		From 1,000 nm–2,000 nm
Ilford infra-red dyed gelatine sandwiched between glass flats		From 1,000 nm–4,000 nm
Chance 2 mm thick containing nickel oxide	0.X5	From 1,000 nm–2,000 nm

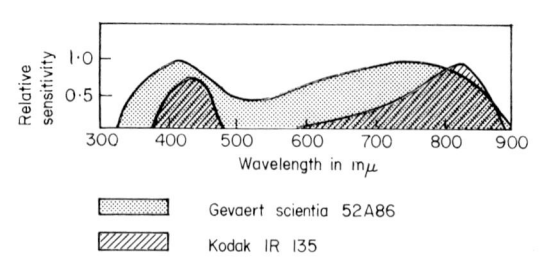

Fig. 192. Relative sensitivity curves of infra-red films.

INFRA-RED PHOTOMICROGRAPHY

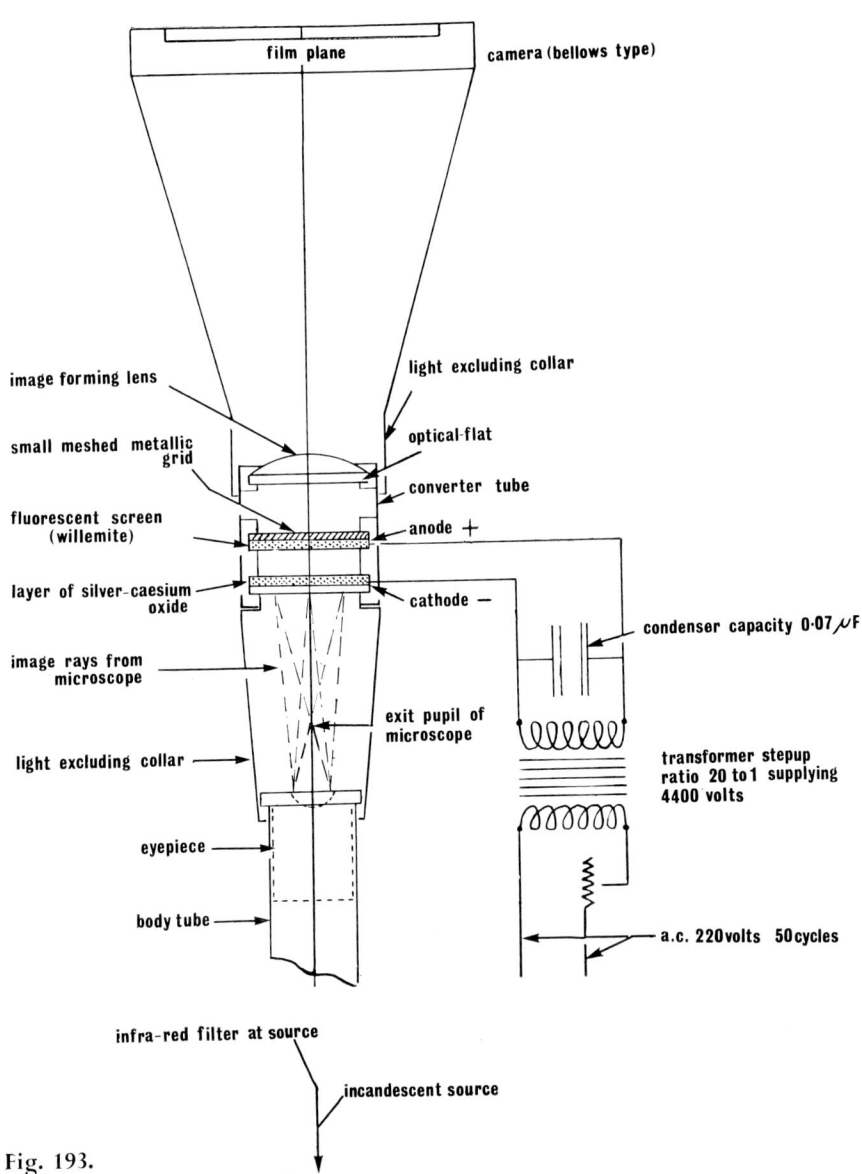

Fig. 193.

included subjects such as nerves and muscle fibres. Such studies could not be made with short wavelength but reflecting objectives make possible long wavelength infra-red microspectroscopy. The Burch microscope functions as a microilluminator enabling infra-red obsorption to be examined. Aberrations are not an important factor in this application, as they are when working in the ultraviolet. A spherical reflecting system of N.A. 0·65 was used with a plano–convex lens of silver chloride and thallium bromoiodide to increase the numerical aperture. The full radiation of the source is incident on the specimen and the radiation output from the microscope is fed into a monochromator and on to a radiation detector and thus on to the recording medium. It is still important to have optimum illumination of the microscope.

Barer (1950) continued his studies in the infra-red region at wavelengths up to about 1,000 nm. He employed a photo-electric image converter of conventional type giving an image on a fluorescent screen, as well as infra-red sensitive photographic materials. Photographic records were made of stained and unstained biological material with phase contrast, polarized and unpolarized infra-red illumination.

Radiations through the microscope extend sight by revealing what cannot be seen in conventional colour film and by the human eye. But, more importantly, a new dimension of interpretation becomes possible. For example, disease, stressed metal and botanical subjects can be recorded on film sensitive to near infra-red even though they appear normal to the unaided eye.

Infra-red colour film is an example of the use of false colour to enhance and display information not visible to the naked eye. False or unconventional colour is a form of multi-spectral manipulation. Interpretative image processing techniques and filters interposed in the path of radiations can be used to alter the colours, making possible a photomicrograph which emphasises hitherto unobserved features.

As a rule, penetration is far greater for black and white recordings in the infra-red than for colour but colour does express greater detail in the heavy stained areas. Further, colour serves a useful purpose when photographing sections heavily stained by haematoxylin and eosin, for here such colours are expressed as yellow and yellow/orange, in which great detail is seen. From this it does appear that this particular application is suited more to photographing red stained specimens, perhaps because of the extra transmission in the red region. Blue records very well, and is expressed as red, so detail can be seen in blue areas which prove to be difficult to record in black and white. False colour was first applied through filters when colour-blind men looked through them to discover what had been hidden. Infra-red recorded in colour does precisely this. The expected and conventional colours of a specimen are altered or replaced with other colours selected to emphasise desired information which would be invisible in conventional colour. Such illumination of colours is like varying the tonal contrast in black and white photomicrography in order to enhance certain images of interest and to minimise others.

The range of photographic material suitable for infra-red photomicrography is limited but adequate. A number of sizes of Kodak Class "R" high contrast maximum resolution emulsion (plates) are available and have a peak sensitivity from 740 nm to 850 nm. Emulsion (plates) known as Kodak Class "M" overlaps the transmissions of that of "I.R." and listed as being sensitive from 860 nm to a peak of 1,000 nm. A further emulsion, "Z†", continues deeper into the infra-red transmission and offers a coverage each side of 1,085 nm. I have imaged this material at 1,500 nm and found it satisfactory. As this material must be hypersensitised before being used it should be soaked in a solution of 4 cc of 0·880 sp. gr. ammonia in 100 cc of water for about a minute. Before using the negative material it should be evenly dried. (Kodak 2485, 125 ASA, is for black and white.)

Exposures for recording opaque subjects are usually long in this region of the spectrum: a period of 20–30 min can be expected. Experiment will help determine the exposure times since here the exposure meter may not give a reading. When using an infra-red converter (material exposed to fluorescent light) the exposure is generally short.

References

Ambrose, E. J., Elliott, A. and Templer, R. B. (1950). *J. Sci. Instrum.* **27**, 21.
Barer, R. (1950). *Photogr. J.* **90B**, 83.
Barer, R., Cole, A. and Thompson, H. W. (1949). *Nature London*, **163**, 198.
Blair, D. M. and Davies, F. (1933). *Lancet*, **225**, II, 801.
Casida, L. E. (1968). *Science*, **159**, 199.
Farrell, G. M. (1969). *Ann. Appl. Biol.* **63**, 265–275.
Holst, I. G. (1934). *Physica*, **1**, 297.

Ultra-violet photomicrography

Much has already been written about ultra-violet. Some further notes, however, may be of interest.

Unfiltered ultra-violet rays find industrial use today in the production and researches into photochemical changes of many kinds: the activation of Vitamin D in foods, medicine, polymerization of oils and varnishes, sterilization of water and many other liquids, durability tests on paints, dyes, rubber and other colloids, fabrics and papers, are among the most important. More widespread is the use of "filtered" ultra-violet radiations for fluorescence analysis, which affords a rapid and specific means of qualitative test in many industries. Immuno-fluorescent techniques and fluorescent photomicrography in the diagnostics of integumental and visceral forms of lupus erythematosus have been

†Available only in special circumstances.

reported from the Clinic for skin diseases of the Friedrich Schiller University, Jena (Jena Review, 1970).

The shortest light wavelength of the visible spectrum which it is possible to use with the microscope and the eye is in the region of 400 nm to 450 nm. We have seen already that the resolving power is the least distance between two lines produced by an objective of a given numerical aperture, with a given wavelength, recorded as two separate lines and can be expressed by the relation

$$\frac{0 \cdot 5 \lambda}{N.A.}$$

where λ is the wavelength and N.A. the numerical aperture of the objective. If we wish to improve the resolving power, to give greater detail, we can either increase the numerical aperture of the objective or decrease the transmitted wavelength of the source with which the optical system is used. As objectives have reached a point of producing maximum resolution, and there is little likelihood of any optical improvement, we are left with one variable, that of wavelength. Improved resolution can be achieved by incorporating a light filter in the light train, such a filter could be green (550 nm), blue-green (490 nm) and blue-violet (450 nm) (See Fig. 184). The limit of resolution for wavelength 550 nm (green) is 0·20 μ and, as a comparison in figures, wavelength 257 nm (ultra-violet) is 0·11 μ. As can be seen, (Plate 22) improvement in resolving power can be obtained by employing ultra-violet, reducing the value of λ. The eye has lost contact with light or radiations below 400 nm and so photographic emulsion takes over to record happenings below this region.

For most purposes the standard source of ultra-violet radiation is the high pressure mercury arc in a tube of fused quartz or other material, transparent to the active rays. The characteristic spectrum of the mercury arc shows lines down to 185 nm when generated in quartz. By interposing suitable interference filters, lines required for special purposes can be selected. For fluorescence work a filter is selected passing ultra-violet radiations or blue light. The mercury arc in quartz was first introduced by Hanovia, in 1905. It was a triumph of technical skills and serves man throughout the civilized world.

When employing ultra-violet radiations, depth of field shows the greatest improvement compared with focusing in the visible. This change in depth of sharpness is directly proportional to the wavelength of the source used. Therefore, even greater depth of field is found in the ultra-violet region of the spectrum, as opposed to that in the visual spectrum.

To photograph a specimen irradiated by ultra-violet, an optical system must be made of a material which freely transmits ultra-violet radiations. This means that the substage condenser, objective, and eyepiece, not overlooking microscope slide and cover glass, must be made of a material which permits radiation to pass unhindered. The glass components, included in a microscope for visual purposes, would absorb most of the ultra-violet transmissions. Normal apochromats can, as a makeshift, be used for wavelengths down to 320 nm but, owing to the high percentage of ultra-violet absorption through the lens system, the image quality is usually very low.

In dealing with the reflecting microscope we have already mentioned the optical value of quartz and that the limit of spectral transmission is down to 185 nm. The transmission of fluorite is even lower but, for various reasons, its properties tend to limit its use. Therefore, quartz is used for the optical systems of objectives, eyepieces, condensers, cover glass and microscope slides. As yet it has not been possible to produce fused quartz systems for the purpose of achromatizing high power ultra-violet objectives and these have been computed for a restricted wavelength. Such monochromats are listed on page 63 and their coverage demonstrated in Fig. 66. The performance of these monochromats deteriorates considerably if the wavelength is more than 2 nm from the figure for which they have been computed.

Quartz monochromatic objectives were first introduced at the turn of the century made by Zeiss, Jena for use in the far ultra-violet. Today, improved objectives are manufactured and computed to operate at 257 nm and tube length 160 mm. The Bausch and Lomb Optical Co. have developed an ultra-violet microscope which permits visual focusing by use of the new "ultrascope tube", made by the Radio Corporation of America. So it is now possible to focus the image and dispense with a substitute visual wavelength to estimate the region of the ultra-violet. The new ultra-violet photo-microscope has applications in such fields as pathology, tissue cell screening, bone marrow observations, and the study of petrography and synthetic fibres. Carl Zeiss, Oberkochen, have produced a microscope-camera (Fig. 194, Plate 59(a)) for use in the ultra-violet. They have also introduced Ultrafluars (not monochromats). These objectives are as follows:

×10   N.A. 0·20
×32   N.A. 0·40
×100   N.A. 0·85 and 1·25
×32   N.A. 0·40
×100   N.A. 0·85

The latter two are for use with phase contrast. The objective produces a complete coverage of the visible spectrum from inside the red band (Fig. 195). This is a remarkable spectral coverage which has been long awaited. Like many other objectives designed by Carl Zeiss, Oberkochen, they operate at a tube length of 160 mm together with a × 5 projection eyepiece computed with a projection distance of 45 cm using a quartz cover glass of thickness 0·35 mm. Further, an achromatic condenser of N.A. 0·8 for use in the ultra-violet forms

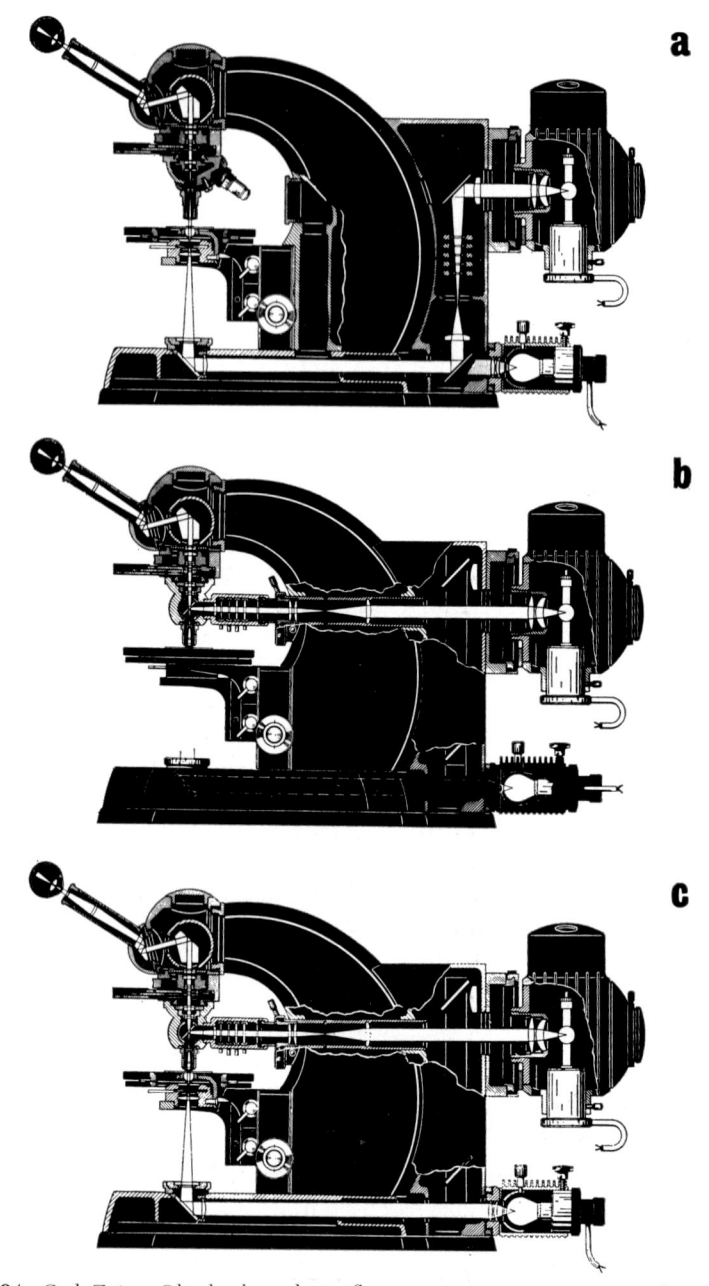

Fig. 194. Carl Zeiss, Oberkochen, large fluorescent microscope. (a) Path rays with transmitted light. (b) Path rays with vertical incident light. (c) Path rays with combined vertical incident light fluorescence excitation and transmitted light phase contrast illumination.

ULTRA-VIOLET PHOTOMICROGRAPHY

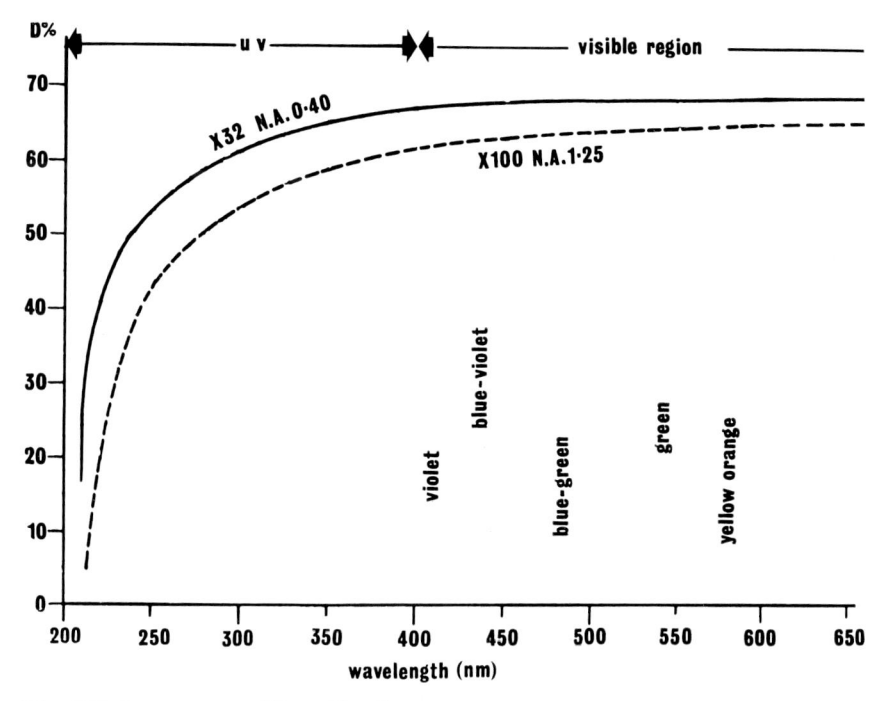

Fig. 195. Transmission of Zeiss Ultrafluars.

part of the combination of systems to be used with these objectives. At a wavelength of 280 nm its focal length is 6·3 mm and its magnification is × 29. Although it is a dry condenser, it may also be used as an immersion condenser (immersion medium has been mentioned). Its long working distance of 0·45 mm (with cover glass thickness of 0·35 mm) is sufficient even for a specimen slide 0·9 mm thick.

Mirror objectives can be used in the ultra-violet and do not present focusing difficulties. Unfortunately they have a restricted numerical aperture, perhaps not so noticeable in such a wavelength. Further, glass optical systems used within a mirror objective lessen the purpose of such an objective and this also applies to quartz monochromatic objectives. The microscope mirror should be replaced by a quartz total reflecting prism or by a surface-aluminized mirror of fused quartz.*

Immersion oil is unsuitable for use with ultra-violet but a mixture of glycerine and water has proved to be most satisfactory. Such a medium will transmit a radiation of approximately 275 nm. Adding water to glycerine enables the medium to satisfy the refractive index of fused quartz, though in a short time the water tends to evaporate and thus brings about a change in the refractive index and transmission wave. A non-evaporating fluid, like that obtained from

* Blazers UV mirrors down to 220 nm, transmission is 95%.

cane sugar (refractive index 1·451–1·535), together with glycerine (1·453) is more lasting and equally efficient. When crystals sometimes are seen in the fluid it is an indication that a replacement is necessary. Ultrafluars are to be used with an immersion fluid of undiluted glycerine, refractive index 1·453.

Far ultra-violet has been included in Fig. 184, but in fact very little work is carried out in this area and as a rule, catoptric objectives are used. Much experimental work was carried out by Johnson (1953) in the region of 200 nm to 100 nm with a spark discharge, with tin or zinc electrodes as a source with a fluorite monochromator. The electron microscope can cover this particular field with less physical fatigue and greater success in imaging the specimen. The most versatile Cambridge Stereoscan S4, scanning electron miscroscope, is capable of resolution in the order of 10 nm.

Consideration must be given to irradiation. This should be strictly monochromatic and of the desired wavelength in the ultra-violet region for which the objective is designed. A carbon arc lamp is satisfactory but unfortunately is not made today. HBO 200 W mercury high pressure lamp gives good results (see page 119) but this indicates that some light is absorbed. The zirconium arc high watt lamp is suitable and easy to operate. Heat absorbing filters should be used but care must be given to see that these do not absorb ultra-violet. The most intensive ultra-violet radiator is the H.T. cadmium spark, brought about by applying current passed through a transformer to about 5,000 V for sparking across 3 mm gap. This set up really requires a special piece of apparatus and occupies a great deal of bench space (Köhler, 1933).

Interference filter. A controlled output of radiation through the medium of filters is necessary and some are listed below. Filters, usually colour (Figs. 196a and 196b), blot out certain wavelengths mainly by interference. The advantage of interference over standard absorption filters lies in the wavelength being "screened out", or separated from the spectral range, which can be more precisely set and

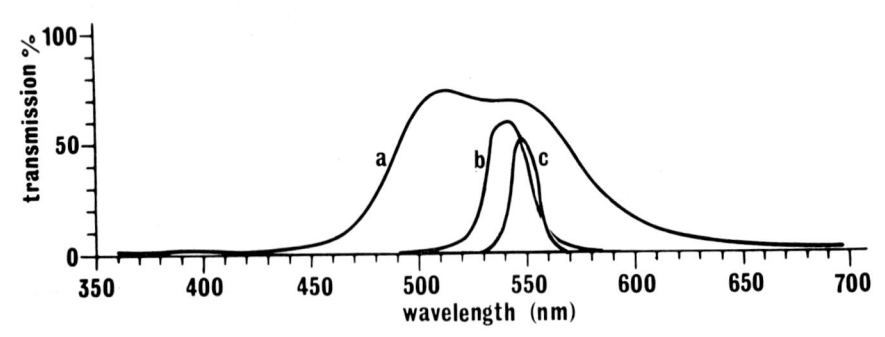

Fig. 196a. Interference filters. (a) Wide-band pass filter, green. (b) Interference wide-band filter, green 546 nm ± 5 nm. (c) Precision narrow-band pass filter, green 546 nm ± 2 nm.

ULTRA-VIOLET PHOTOMICROGRAPHY

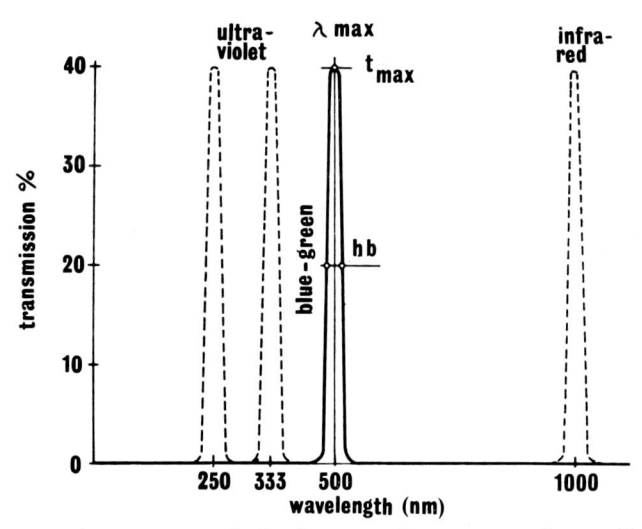

Fig. 196b. Typical transmission of Filtraflex monochromatic interference filter. Dotted lines represent the suppressed higher and lower transmission bands.

controlled. The wavelength transmission within the selective range becomes considerably greater.

Metal interference filters. Balzers produce the Fabry–Perot type metal interference filters under the trade name "Filtraflex". Basically, they consist of two semi-transparent metallic coated mirrors separated by an absorption free intermediate film. The thickness of the intermediate film determines the position within the spectrum of the transmission bands and the degree of background transmission. The coating on most semi-transparent mirrors is silver. The transmission value, with the width of the transmission bands depends largely on the optical characteristics of the metal film. The first metal film is the intermediate Dielectric layer, the second is coated on a glass substrate. To obtain optical symmetry and to protect the coated film from damage a further glass is added. Broad band interference filters are, as a rule, colour glasses, with a steep absorption ridge and used to exclude unwanted higher and lower transmission bands. Characteristics of typical monochromatic interference filters are shown in Fig. 196b. The dotted curves represent the suppressed higher and lower order transmission bands. (a) λ_{max} position within the spectrum of the peak transmission wavelength, i.e. the wavelength which has the highest transmission value. (b) T_{max} transmission value at the peak wavelength expressed in %. (c) Half band width, hb, = the transmission band width measured at a point equal to half the peak wavelength transmission. (d) Background transmission = zones of minimum or near zero transmission.

The 1/10th and 1/100th band width values applicable to Filtraflex metal-interference filters, define the extent to which they approach the monochromatic ideal. A range of interference filters is set out below.

Table 26

Filter	Code No.	Transmission
Interference (Schott and Gen.)	UG5/3 mm thick	280·4 nm
Interference (Schott and Gen.)	UG5/3 + WG5/4	312·6 + 313·2 nm
Interference (Schott and Gen.)	UG2/2 + BG12/2	366·3 + 365 nm
Corning	CS 7–39	365 nm
Corning	CS 7–54	313 nm
Balzers	R—UV1430	275 nm—350 nm
Balzers	B—20.1432	450 nm
Balzers	R—UV special 1428	250 nm
Liquid nickel sulphate heptahydrate 68 g in 100 cc water in cell of 60 mm	—	312·6 + 313·2 nm
Liquid picrinic acid 63 mg in 1000 cc water in cell of 10 mm	—	280·4 nm

Small cameras with an eyepiece fitting, specially made for photomicrography, are not really suitable in the ultra-violet since the beam-splitter would absorb a certain amount of the ultra-violet in the region of 320 nm and below. A camera of the bellows type should be used. Radiations leaving the microscope will be unhindered coming into contact with the emulsion. A light excluding collar around the eyepiece is necessary (Fig. 83 and Plate 27).

Photographic materials do not pose any particular problems, when used in the ultra-violet region of the spectrum. Contrasty materials are suggested as panchromatic material produces a flat image, and for this purpose Kodak ortho emulsion B.10, for scientific applications is sensitive down to 210 nm. Ilford high contrast emulsions, such as N40 and N4, are ideal. A method of increasing ultra-violet sensitivity is to bathe a plate (emulsion) with a fluorescent substance which transforms short waves into longer and can more readily penetrate the gelatine. Two Eastman organic chemicals, available from Kodak Ltd., Kirkby, Liverpool, are designed for this purpose. One is A.3177 ultra-violet sensitizing solution, and an alternative is 8269 ultra-violet sensitizer No. 2. After exposure, but before development (because these chemicals are impervious to water) in the A.3177 this must be removed in a solution of isopropyl or ethyl alcohol, the water in 8269 must be removed in cyclohexane. Emulsions so treated are sensitive to approximately 200 nm.

For colour expression, daylight colour film is usually preferred, which will minimize the exposure time. There is always the possibility that bleaching of the specimen may take place if exposed to radiations for long periods.

References

Jena Review (1970). **4**, 229.
Johnson, B. K. (1953). *J. Roy. Microsc. Soc.*, **73**, 24–29.
Köhler, A. (1933). *Naturwissenschaften*, **21**, 165.

In-line mirror monochromator

The Leitz new in-line mirror monochromator (Plate 38) for the ultra-violet, visible and near infra-red regions, represents a variation of the Wadsworth arrangement in which the optical axis of the emergent monochromatic beam forms the linear extension of the optical axis of the illuminating beam. The new monochromator is therefore highly suitable for inclusion on benches in a series of optical instruments arranged without deflecting the beam: a great advantage when the work has to be carried out on long, narrow benches. In addition, it covers the ultra-violet spectral range from 205 to 360 nm, which represents a considerable extension of the working range of the new monochromator compared with the previous high-intensity Leitz monochromator.

Figure 198 shows the diagram of the beam path of the monochromator.

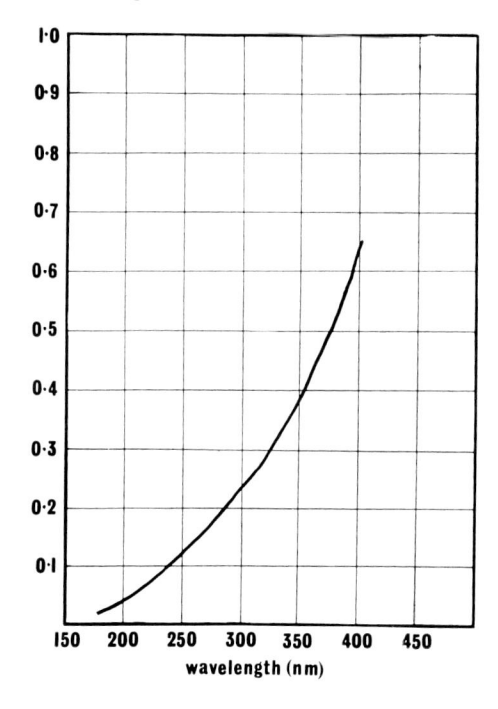

Fig. 197a. Spectral split width per unit of the drum scale, as a function of λ for quartz glass prism.

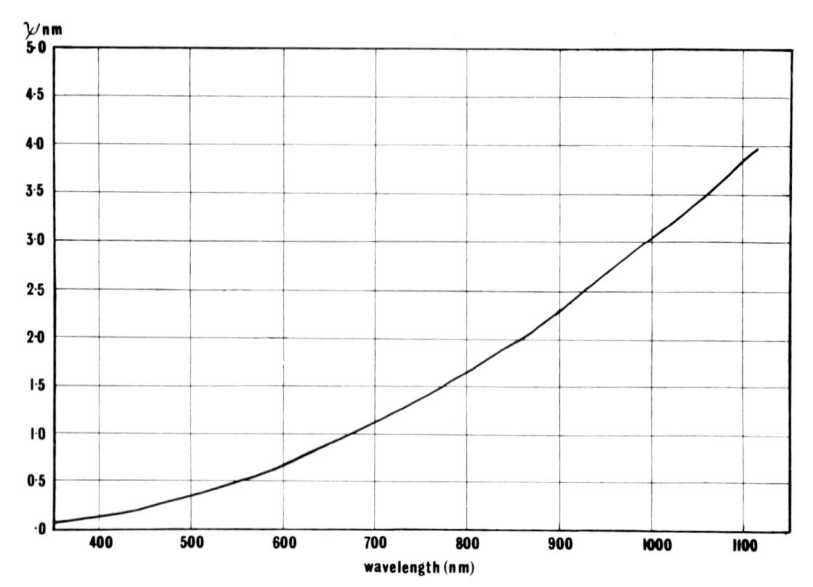

Fig. 197b. Spectral split width per unit of the drum scale, as a function of λ for quartz for F2 glass prism.

The monochromator proper (Fig. 198) consists of the entrance slit *ESp*, the surface-aluminized deflecting prism *U*, the collimator mirror *K* (focal length 200 mm), the two 30° dispersion prisms *P1* and *P2*, made of F2 glass or quartz glass (Spectrosil), with large mirror-coated short-side faces, mounted on a prism stage and interchangeable by tilting, and the exit slit *ASp*. The geometrical factor[4] of the monochromator is based on the formula

$$\frac{1 \cdot Q}{f} = 0.5 \text{ (cm}^2\text{)}$$

where l = the slit length
f = the focal length of the mirror } in cm
Q = the cross section of the beam near the prism in cm²

Beam path. The two component Suprasil quartz glass condenser *C*, incorporating an adjustable lens, forms an image of the light source *L* at *ESp*. The diverging rays emerging from *ESp* are deflected by the surface of the prism *U* on to the collimator mirror *K*, which in turn reflects them as a parallel beam to the dispersion prism *P1*. Here the beam is dispersed into parallel rays of different wavelengths, which strike the large mirror-coated short-side face of *P1* and, after being reflected, pass through *P1* again in the opposite direction. By means of the collimator mirror *K* and the second plane surface of the deflecting prism *U*, the monochromatized beam is transmitted to the exit slit *ASp*, where the concave mirror *K* forms an image of *ESp*. To prevent any obstruction by *U* of the rays transmitted to the prism *P1* and *P2*, the dispersion prisms are placed high enough for the parallel rays passing from K to $P1$ and $P2$ and back to just clear the top of U. $P1$ and $P2$ are suitably inclined for this purpose. An adjustable projection lens of Suprasil quartz glass in a screw mount is arranged beyond the exit slit.

The wavelength can be set by turning the wavelength drum (Plate 38a) which is coupled with the dispersion prisms by means of a cam, and interchanging the prisms simply by moving the lever through 180°. The wavelength drum has three scales:

(1) A uniformly graduated calibration scale with vernier, a scale from about 370 nm to 1100 nm for the glass prism, and a scale from 205 to 420 nm for the quartz glass prism.

(2) A shaft, projecting from the monochromator and coupled with the prism rotating mechanism, which can be connected to any recording instrument or to a synchronous motor.

(3) The connecting cable can transmit impulses for wavelength indication to the Rohde and Schwarz recorder.

In this combination the monochromtaor becomes part of a spectral recording instrument.

The entrance and exit slits are coupled to each other and can be adjusted by a slit drum; they can also be uncoupled. The width of the entrance slit can be read on the appropriate scale. The drum for adjustment of the slit has 90 scale divisions; with each division of the scale at 0·02 mm the largest slit width which can be set is therefore 1·8 mm. The slit length is 5 mm.

The spectral slit width is defined as the wavelength range $\Delta \lambda$ separated by the slit of mechanical width s (mm) and covering all the wavelength ranges, which transmit at least 50% of their total energy through the exit slit. Provision of a chart showing the spectral slit width per unit of the drum scale (i.e. $s_1 = 0·02$ mm) helps in using the monochromator. This quantity is called ψ and is a function of the wavelength.

If the number of divisions on the slit setting drum is N, then, generally, $[\Delta \lambda = N \times \psi)]$, with the mechanical slit width $s = 0·02 \times N$ (mm). For the Spectrosil quartz glass and F2 glass prism used in the in-line mirror monochromator the quantities ψ_{SiO_2} and ψ_{F_2} are represented as a function of λ in Figs. 197a and 197b. To determine the spectral slit width $\Delta \lambda$ for a given drum setting N, the ψ value for the particular wavelength setting must be derived from the curve for the prism used and multiplied by N, as in the above equation.

The standard light source equipment for the monochromator includes the lamp housing 250, which is easily removed after the clamping screw under the base plate of the monochromator has been loosened. It is suitable for all high pressure lamps such as the HBO 200 W and CS 150 W ultra-high pressure mercury vapour and the XBO 150 W ultra-high pressure xenon lamps and for the DL–100 deuterium lamp (Figs 199 and 200). The best form of illumination can be

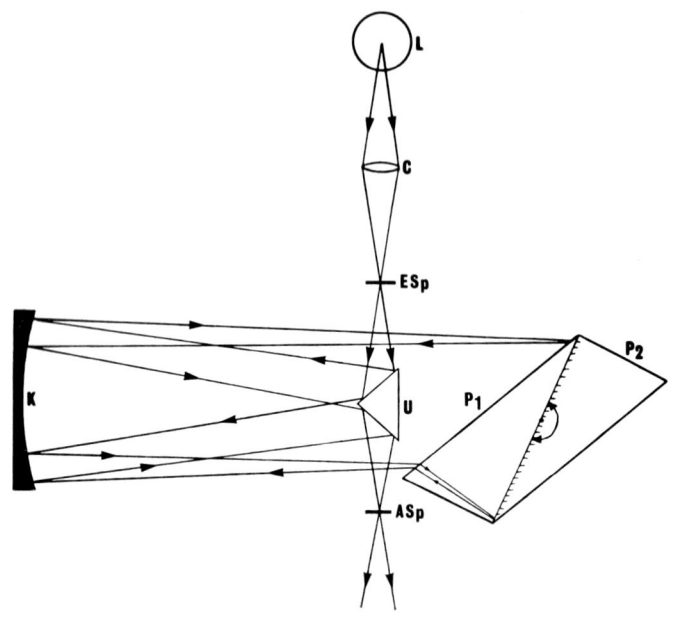

Fig. 198. Beam path diagram of the in-line mirror monochromator, after E. Leitz.

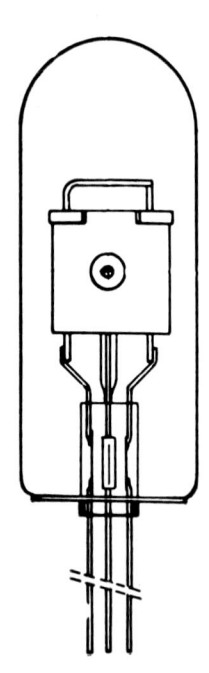

Fig. 199. Deuterium lamp.

IN-LINE MIRROR MONOCHROMATOR

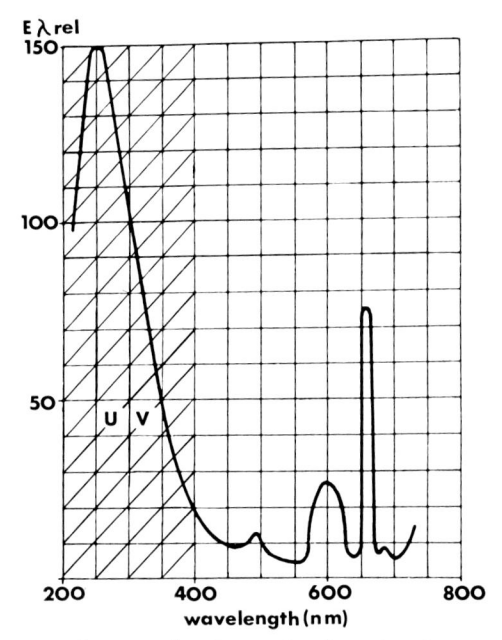

Fig. 200. Relative spectral energy distribution for deuterium lamp.

set quickly and reliably with the convenient centring devices for reflector, lamp and quartz glass condenser, all operated from outside. A deflecting mirror, which can be withdrawn from the beam path, makes a choice of UV light source or low-voltage filament lamp possible (see Fig. 201).

The light source for visible and short-wave infra-red rays is a 6 V 4·35 A low-voltage filament lamp with screw base. The ultra-violet light source is an air cooled deuterium lamp, type DL-100. The current for the deuterium lamp is taken from a special supply unit for connection to 220 V mains, providing a stabilized anode current (variation of the anode current less than $\pm 1\%$ with mains voltage fluctuations of $\pm 10\%$). (A special prospectus containing the description and instructions for use of this supply unit is available from Leitz.)

The deuterium lamp produces a continuous spectrum (Fig. 200) from approximately 200 nm, up to infra-red. Above 400 nm the spectrum contains moreover the Balmer-lines[†] and the strongest lines of the multiline-spectrum of deuterium. The filament current must maintain a constant intensity and is to be used on DC according to the circuit diagram illustrated in Fig. 202. Electrically and geometrically the deuterium lamp is identical to the hydrogen lamp, while absolute output is about 40% higher.

[†] Spectrum of simple element, hydrogen, Balmer 1885. Wavelengths of most intense lines expressed by $\lambda = Km^2/(m^2 - 2^2)$, K is a constant and m one of a series of numbers, 3, 4, 5, 6 and so on. This series of lines in the hydrogen spectrum is known as the Balmer lines, approximately 30 in all.

The very low working pressure of the mercury–helium lamp results in an absence of continuous radiation (Fig. 203). Thus the energy is dissipated in the well known mercury lines only. The outer bulb is made of quartz so that the

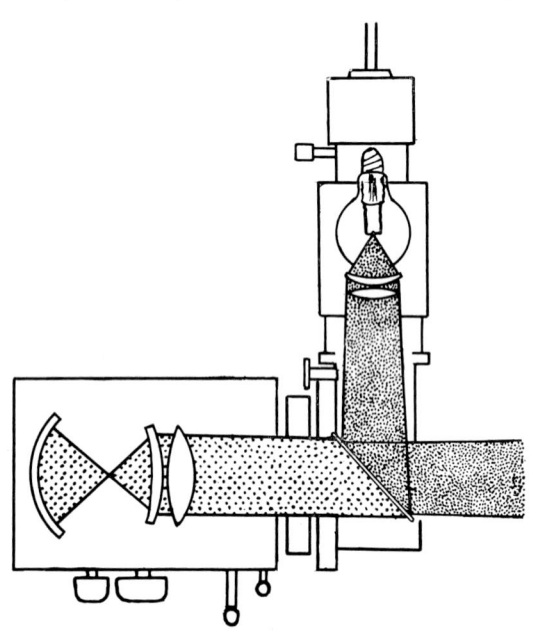

Fig. 201. Beam path in lamp housing 250, ultra-high pressure mercury lamp. With the deuterium lamp the reflector remains unused in view of the one-sided emergence of the beam.

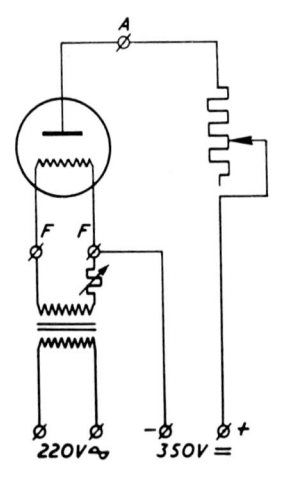

Fig. 202.

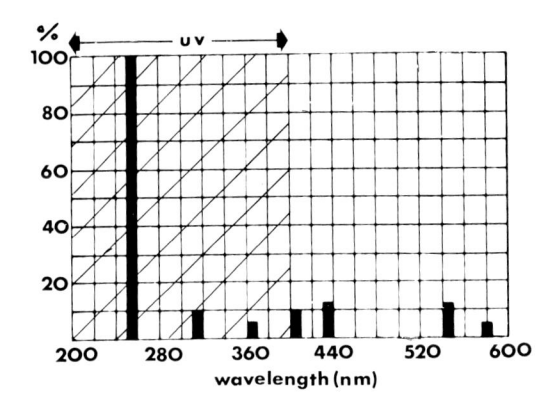

Fig. 203. Relative spectral energy distribution.

lines in the short-wave UV are transmitted too. The mercury lamp is also to be used on DC (see the circuit illustrated): the same circuitry as for the deuterium spectral lamp without any further adjustment.

If required, the monochromator can be supplied with only one light source either for visible or for UV radiation. When very high intensity of continuous UV radiation is essential the water-cooled hydrogen lamp, type WHS-200, is recommended. Operated on an anode current between 1 and 1·5 A, its intensity in the UV range is approximately 2·5–3 times that of the deuterium lamp, DL-100. A stabilized supply unit for connection to 220 V mains is available and the equipment includes a special lamp housing with a quartz glass condenser (f′ ≈ 30 mm).

The mercury–helium lamp, type HeQSB-60, is recommended. It is very convenient to handle because it has the same dimensions as the deuterium lamp DL-100 and can be exchanged readily for the latter in the normal lamp housing: here, too, the supply unit for the DL-100 is used. As the HeQSB-60 emits the lines of mercury and helium, it provides a large number of calibrating lines for the most important spectral regions.

Scattered-light filters. Together with the light passing through the exit slit, every monochromator emits a certain unavoidable proportion of rays of wavelength ranges other than that determined by the "spectral slit width". This scattered light, which is caused by reflections at the edges of prisms, the rims of mounts, internal mechanical parts, unavoidable optical flaws in the prism material (e.g. bubbles), or very small residual errors on the surface of optical elements, is kept to a minimum by the greatest possible care in manufacture. In most cases the unavoidable residue of scattered light can be reduced to a practically insignificant level by suitable filters inserted in the filter device in the lamp housing 250.

Monochromator accessories.

(a) By means of a rotating octagonal Bergmann prism† arranged in the beam path beyond the exit slit, it is possible to provide uniform illumination of a square area, e.g. in the object plane of a microscope arranged next to the monochromator, with a surface brightness greater by the factor

$$q = \frac{\text{height of exit slit}}{\text{width of exit slit}}$$

than when the same surface is illuminated by means of an enlarged slit image through a spherical lens. This attachment also includes an image forming lens between the exit slit and the rotating prism. For visible light this lens and the rotating prism can be made of glass, but for the UV range they must both be made of quartz glass.

(b) Two types of light-tight housing are supplied for attaching a photomultiplier directly behind the exit slit: One, for fitting photomultipliers by RCA, e.g. types 1P 21, 1P 28 etc. and a second for fitting a photomultiplier by Fernseh GmbH Darmstadt (e.g. types FS 9-A or VE 600).

(c) Connecting the photomultiplier is an electrical attachment StG 12 which is also used for the MPE single-beam microscope photometer. This apparatus does not contain a measuring instrument and a moving-spot galvanometer must therefore be used for measurements. If a galvanometer of this type is not available, Model 251J (sensitivity 2×10^{-9} A/scale division) by Messrs. Norma, (Fabrik Elektrischer Messgeräte GmbH., 6000 Frankfurt/Main, Osthafenplatz 6–8, West Germany) is recommended.

Applications. In quantitative optical processes and for many different methods of examination in physics, crystal optics, biology, medicine and technology—particularly in the large field covered by photomicrography—the use of monochromatic light is of special importance. Although colour filters, which may be of the liquid type in cells, or glass, gelatine, or interference filters, are very often adequate for the production of monochromatic light, in general they have a fairly wide transmittance range. But with increased monochromatism they transmit very little light owing to strong absorption. A high intensity monochromator offers, in addition to maximum monochromatism, the advantage of a continuous variation of the wavelength. It also provides the only easy means of separating monochromatic radiation in the spectral region between 200 and 300 nm since, even now, no convenient filters exist for the UV range below 300 nm (Fig. 184).

In photomicrography the colour of the background is of importance when recording stained specimens; contrast between different colours varies greatly and must be documented (Plate 72). By illuminating stained specimens with light from

† Heraeus Quarzschmeize GmbH, Hanau.

a strictly monochromatic source structures can be emphasized or suppressed as required. The in-line mirror monochromator is of special value in photomicrography in the UV range for, with the Leitz Micro Reflecting Optical Systems, it reveals a continuously variable wavelength and the structures of biological objects absorbing only UV light. The UV absorption of nucleic acids, proteins, lignines etc. are relevant examples. The quantitative analysis of such compounds can be carried out by spectrophotometric measurements of microscopic objects. In addition, a monochromator can be used for tests connected with the changes in biological tissues caused by UV radiation. A further field of application is fluorescence microscopy in UV light, of organic and inorganic substances.

In forensic photomicrography, practised by chemists, doctors and scientists of criminal investigation departments, the ability to obtain satisfactory monochromatic illumination down to the short wave ultra-violet range extends the scope considerably. Often the forgery of documents, postage stamps, seals, and cheques, or traces of various substances can be detected, or excluded, with certainty only by the use of ultra-violet.

In the arrangement of the spectrophotometers—whether single-beam or double-beam instruments, microspectrophotometers or fluorescence spectrophotometers—the monochromator always forms the principal unit: as for, example, the setting up of a fluorescence spectrophotometer even calls for two monochromators, one for the variation of the wavelength of the exciting lamp, the other for the analysis of the fluorescent radiation.

Biolaser system

Organism and environment are an inseparable whole and every change in environment ultimately influences the life of the organism. The environment is conditioned by the organisms it consists of, as well as chemical, physical and organic factors, *light*, temperature and water. Light, being one of the main sources of energy, dominates and promotes life. Micro-irradiation of living cells is carried out, first to study the effect of radiation on different parts of living cells and, secondly, to analyse the function of cellular constituents by selectively altering some of them. The method and principles carried out in this work includes micropuncture apparatus (laser illuminated microscope) ultra-violet source, specimen, adjustment of spot and phase contrast.

Micropuncture restricts the killing of one selected cell among many. So that the effect of the death of a particular cell upon the environment can be recorded. Living cells are attracted to the dying and finally remove it and thus we can see the phenomenon known as necrotaxis.

However, with the ruby 694 nm laser beam, micropuncture techniques by an adjustment of a micro-spot of 1 micron and larger can be applied. A commercially produced laser (biolaser system model 513), which can be used for micropuncture, is marketed by Laser Products TRG, of Control Data Corporation, New York.

To produce a beam diameter of 1 micron at the specimen, the output beam of the laser is focused on a pinhole, P1 by L1 of Fig. 204. The laser illuminated pinhole serves as the object for L2 which collimates the beam. The beam is reflected into the eyepiece by a multi-layer dielectric coated beam-splitter. The latter reflects a narrow band of the visible spectrum in the region of the laser output and transmits other wavelengths, thereby permitting simultaneous photomicrography and irradiation of the specimen. Micro-irradiation techniques can also be applied by the 513 Biolaser, when the specimen does not suffer cell structure damage, the image quality being recorded in greater detail.

The coupling (Fig. 204) is supplied with four different aperture sizes which provide the spot size ranging from 1–8 microns when used with a ×63 objective. Larger spot sizes can be produced by varying the objective or by removing the pinhole aperture. Alignment of the laser spot must be precise and is carried out by two easily accessible external dial adjustments. (Warning: Do not look at the source when aligning a laser beam, direct laser beam can permanently damage the eye.)

The biolaser attachment is fitted to the Leitz Ortholux microscope and has a typical sensitivity of 200 microvolts per joule. Bausch and Lomb eyepiece camera can be fitted with a dichroic filter to prevent scattered laser radiation from overlapping the film. Cinémicrography, time-lapse and close-circuit television techniques can also be applied.

The specifications are as follows:

- Laser head
  - Flux density on stage: Up to 10^4 joules/cm^2
  - Wavelength: 694 nm
  - Pulse length: 150 μ seconds
- Repetition rate
  - Uncooled: 1 per minute
  - Cooled: 1 pulse every 30 seconds
  - Power output: 0 to 300 millijoules
  - Power input: 230 joules (max)
  - Threshold: 150 joules
  - Effective beam diameter: 5 mm
  - Coating: Multilayer dielectric
- Power supply
  - Input voltage: 115/220 volts ac $\pm 10\%$ 50–60 cps, single phase
  - Trigger: Remote or panel pushbutton
  - Interconnecting cable: Line cord supplied to 103A laser Head to power supply

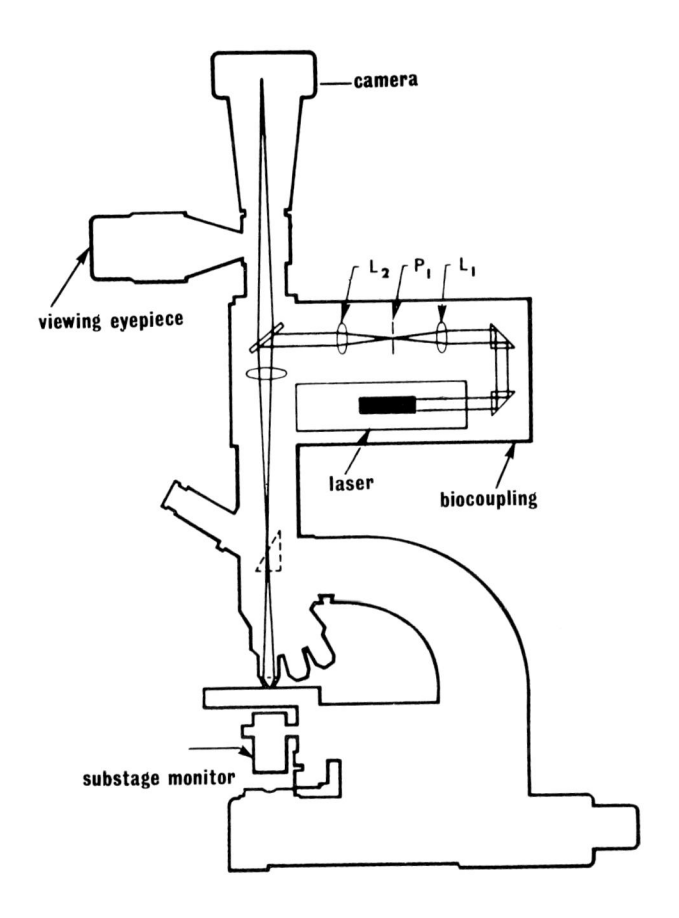

Fig. 204. Biolaser system after Laser Products TRG.

X-ray microscope and camera

This piece of apparatus made by General Electric X-Ray Corp., Milwaukee (Plate 60), is based on the shadow-projection method and is comprised of an objective, aperture, condenser, accelerating anode, electron gun, target, and associated vacuum pumps, separate power supplies and contact systems.

A narrow cone of electrons emerges from the electron gun and enters the condenser lens where it is collimated. This collimated beam enters the objective lens and is focused on to the target—a thin beryllium, usually coated with tungsten—X-rays (Fig. 205) are then generated from this tiny target area, and the rest of the apparatus is simply a shadow–projection microscope. A specimen placed near the X-ray source casts an enlarged image on a fluorescent screen or photographic emulsion. The complete electron gun assembly is movable to enable alignment with the condenser and objective lenses. Output is controlled by varying the grid potential, and the accelerating voltage on the anode is 20 kV. The aperture referred to in Fig. 205 is designed to keep the spherical aberration to a minimum and the function of the four apertures of different sizes is to limit and vary the size of the electron of three elements—two end elements at ground potential. Focusing is achieved by varying the potential on the centre element.

The objective lens employs the highly advantageous electrostatic focusing system. It has two lower elements, with apertures separated by an insulator, and a top grounded element which is also the target plate. This design places the target at the focal point of the lens and permits many sizes of specimen.

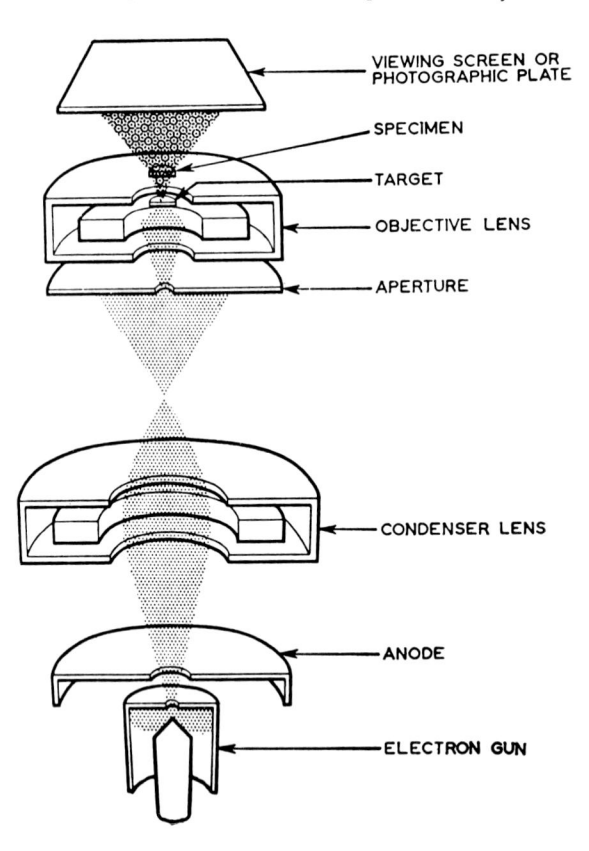

Fig. 205.

The centre element is at high negative potential. Varying this potential focuses the electron beam on the target.

In the centre of the target plate is a thin disc of beryllium plated on the bottom with tungsten or other materials. This is the target. It serves three functions:

(a) Intercepts electrons to produce X-rays.
(b) Acts as a window to release X-rays into atmosphere.
(c) Acts as a vacuum seal.

The X-ray microscope magnifies the various planes of the specimen by different amounts, each in focus. Thus stereoscopic viewing is possible simply by taking two separate photomicrographs and using the specimen manipulator for traversing the specimen between exposures.

The increased capabilities of the G.E. X-ray Microscope makes for a tremendously valuable laboratory tool. One of the most useful accessories for it is the Polaroid Land camera back. Operating with the microscope camera, this device permits a developed image to be obtained 60 sec after exposure (Plate 30). In addition to the speed of photographic operation direct viewing is possible through the lead-glass fluorescent screen. The camera extension can be adjusted to change the magnification and all settings are capable of reproduction (Plate 29).

X-radiography. This application of the microscope possesses certain advantages. One worthwhile feature is that this means of irradiation permits very high resolution, operating just under 100 nm. Although this resolution is an improvement over that with light, it does not surpass that of the electron microscope. Greater contrast and fine detail, coupled with resolution, and a far greater penetration of thick specimens is also possible. Materials of considerable depth can be viewed, especially useful with metallurgical and biological subjects. No special means of preparation of the specimen is necessary before photographing.

Contact radiography. The employment of almost any X-ray tube simplifies the process as no focusing is necessary. The specimen is placed in contact with a very fine-grained photographic emulsion, the X-ray beam being adjusted on the specimen (Fig. 206). The plate is processed and the X-ray image is then rephotographed through an optical system to improve the degree of enlargement. This second image can now be further enlarged to a third magnification. It is obvious that emulsions of a very fine grain size must be used so that the definition can be kept as high as possible, but even so it cannot possibly reach the standard which is produced by the use of an optical microscope. Some fine work has been carried out by Barclay (1947) and Mitchell (1950).

Reflection radiography. Another method is to employ curved surfaces to focus the X-ray beam (Fig. 207). Here a highly polished metal mirror will serve the purpose quite well or a curved crystal, but this projects a somewhat imperfect image.

Fig. 206.

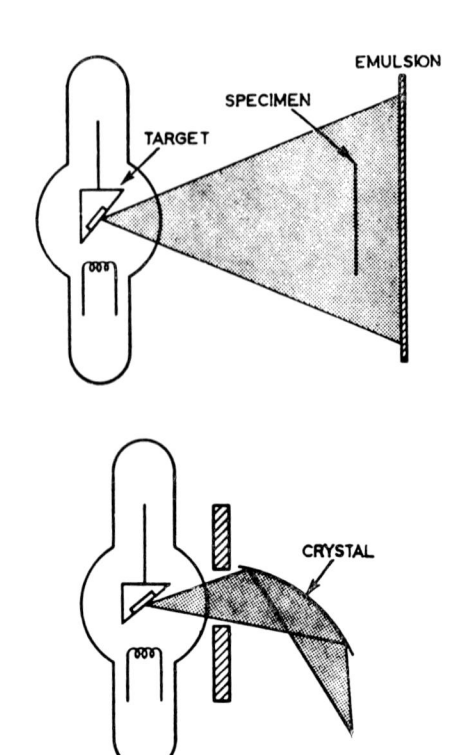

Fig. 207.

Mirrors, having a shallow curve, reflect rays at grazing incidence. The two mirrors are so arranged that one has its axis at right angles to the other. This arrangement of mirrors eliminates the possibility of astigmatism which a single mirror is prone to. This reflection X-ray microscope was developed by P. Kirkpatrick and A. V. Baez in 1951.

Shadow-projection radiography. Figure 208 shows an apparatus developed by

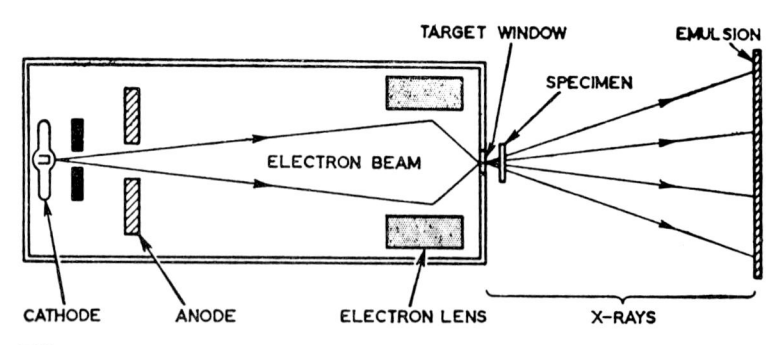

Fig. 208.

Cosslett and Nixon in 1953, from which some high quality microradiographs have been published. This method was made possible by using two magnetic lenses in series to form an image of the cathode on an area smaller than 1 sq. μ. The method adopted was to mount the specimen on a stage plate beyond the tube. Fitted to the stage was a micrometer adjustment which provided the necessary means to control accurate movements across the X-ray beam and also along the axis of the apparatus. The positioning of the specimen along the axis controlled the degree of the magnification of the projected image. An image of $\times 50$ was attained and by moving the specimen nearer to the source a higher magnification could be achieved. The final results have proved to be very successful and a high degree of resolution and depth has been achieved. The negative can now undergo considerable further enlargement without loss of detail.

Figure 209 illustrates another means of projection X-ray. Here the negative material can be enlarged to produce further magnification.

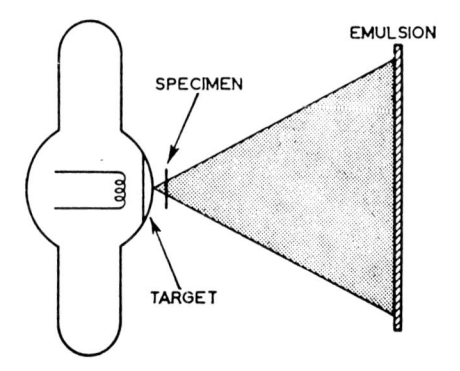

Fig. 209.

Reference

Kirkpatrick, P. and Baez, A. V. (1951). *J. Opt. Soc. Amer.*, **38**, 392.

Fluorescence illumination

This particular technique was not applied to microscopy until Lehmann, Heimsted and Reichert did so before World War I, although fluorescence had been present in animals, insects, rocks and many other substances since the beginning of time. Dr. Karl Reichert demonstrated his first microscope, specially made for the application of fluorescence illumination, before a meeting of scientists in 1911. Nevertheless it was ignored for many years. However, during the past 20 years interest has increased and a range of special microscopes for this technique has been developed in such sciences as histology, quantitative analysis, cytology and microbiology.

Fluorescence is the ability of material to absorb light of one frequency and to re-emit a different, usually longer, wavelength. Some substances continue to re-emit radiations for long periods after the external stimulus has ceased: this of course is phosphorescence. The light intensity can be measured with a fluorimeter.

Two methods induce fluorescence. First, direct ultra-violet radiations and, second, the introduction of a blue filter in the light train. In photomicrography ultra-violet radiations are used to study natural fluorescence, since these are principally absorbed in its production. (Care must be taken when working with intense ultra-violet radiations since the eyes can be damaged.) In addition to the natural fluorescence of an object a fluorescent material may be used to "stain" the specimen. Fluorspar and a range of specially prepared chemical reagents such as coriphosphine, auramine, acridine orange, berberine sulphate and rhodamine glow or fluoresce, stimulated by ultra-violet radiations, blue or deep blue light. Acridine orange is almost a password in some fields of research, due to the different fluorescent colours it assists in producing. These are stronger when the source is wavelength 400 nm to 440 nm where there are two strong peaks, violet and blue violet as set out in Fig. 184. Specimens can be immersed in the fluorescent impregnant (fluorochromes) for periods ranging from seconds to twenty minutes or more, according to the subject, before passing through the other routine processes and being placed on a microscope slide of glass or quartz. Fluorochromes are used in the same way as a chemical stain which will dye certain tissues, making identification easy. Stained thin transparent sections usually absorb light in the stained area, whereas fluorochromes re-emit light. Fluorescence is thus the opposite to absorption, where a thin stained specimen reduces the transmitted wavelength, and introduces a new way of seeing. This is evident when viewing beautiful fluorescent colours against a black background. The main fluorescence bands lie in the green, yellow and red regions.

Canada balsam, Venetian turpentine, styrax, equaral and sirax all have certain fluorescence properties and it is advisable, therefore, not to use these as mounting mediums. Methoxybenzene, Cargill's Immersion Oil and pure glycerine is a good substitute immersion fluid which does not fluoresce and has a high refractive index. There may be some loss of light when using the mirror to reflect the ultra-violet radiations from the source, but a quartz prism used instead of a mirror will retain the original output of ultra-violet radiation. If a prism is not available, the illumination can be vertical direct from beneath the substage condenser.

With ultra-violet as a source, the type of optical glass used in the condenser slightly affects the image brightness, and if ordinary crown glass is used, some ten per cent and more of the ultra-violet radiation becomes absorbed. The more expensive quartz condensers (Ealing Beck), however, do not absorb the ultra-violet, but transmit it. Apochromatic objectives and in particular, fluorite objectives, are to be preferred to those other than quartz.

No special microscope or optical system is really necessary in order to produce fluorescence photomicrography but the quality cannot be compared with the results from a microscope specially made for this task. If the microscope is old it is possible that some of the optical parts themselves may become fluorescent. Modern microscopes such as the fluorestar made by the American Optical Company provide an easy and efficient route to fluorescence photomicrography. The source distribution can be divided into three stages as seen in Fig. 210. An exciter filter, either glass or liquid, is positioned between the light source and the specimen, to selectively pass wavelengths through the microscope. They pass through the condenser into the specimen to simulate its specific fluorescent characteristic. A liquid filter, such as cupric tetaminosulphate, is recommended. What is more, it acts as a heat absorbing filter. A list of suitable filters for this application, illustrating their characteristic curves, is seen in Fig. 116.

Selective colour filters are used to isolate narrow boundaries in the spectrum. The light blue filter strengthens the effect of the heat absorbing filter and remains in the path of the illuminating light with fluorescence photomicrography. Schott and Gen. neutral density filters, N 0·1 and N 0·01, are designed to be as neutral as possible in the visible, permitting minimum transmittance in the ultra-violet spectral range, so that they likewise have a barrier filter character. When combined their neutral filter effect amounts to 0·001. The types of exciter filter employed which permit transmittance of the just visible blue and violet or the long wave ultra-violet rays can, of course, also be used to increase the resolving power of the microscope by using light of shorter wavelengths.

Selective filters include such as green and yellow-green which correct any colour reproduction shortcomings of the microscope optical system. They permit transmittance of the visible range of light radiation only and, if possible, bar the ultra-violet as well as the infra-red bands. Ultra-violet radiations, owing to their comparatively high energy, are dangerous to living organisms and have a bleaching effect if directed to histological sections too long. Infra-red radiation is very similar and it is undesirable for the study of microscopic preparations for long periods; more so when preparing for photomicrography. As all microscope lamps supplying a continuous spectrum give off more intensive radiation in the infra-red than in the visible range, thermal protection filters are necessary.

For fluorescence photomicrography a barrier filter placed between the microscope objective and emulsion will bar the transmission of undesirable exciting wavelengths (Fig. 117) and pass only fluorescing wavelengths re-emitted by the specimen. Barrier filters are used in various combinations and housed as indicated in Fig. 210. From the transmittance curves of fluorescence filters (Fig. 117) it will be seen that 41, 44, 47, 50 and 53 are sharp cutting

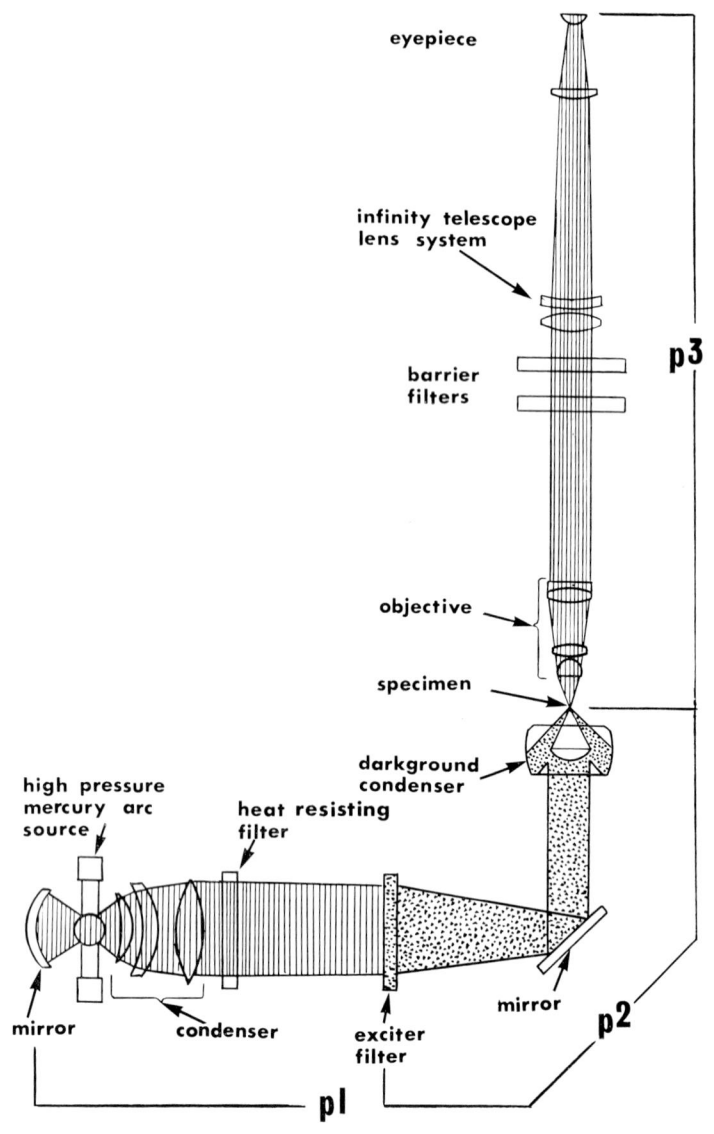

Fig. 210.

filters; that is, with a steep slope towards the short-wave region of the spectrum. The wavelength region in which the transmission of the filters operate is about 10% and is typical of the short cut. The barrier filter 65 is distinguished by its long passband, from approximately 350 nm to 650 nm, and has a gradual slope or cut towards the long-wave region and therefore it may be termed a "minus red filter". Its usefulness is appreciated when operating with filters 41, 44 or 47, when it reduces the unwanted red background. For instance, the barrier filter 41 has a transmittance of 0·1 and transmits approximately 10% of the incident radiation at a wavelength of about 410 nm (see Fig. 117). Below 410 nm hardly any exciting light or fluorescent light is allowed to pass. Thus, as indicated, filters 44, 47, 50 and 53 in turn allow less exciting light to pass, while 50 and 53 may fluoresce when directly exposed to this light and a red hue will then be seen over the entire projected image. In such cases barrier filters 44 and 47 should be used with either 50 or 53, the former absorbing the exciting light.

The following table lists some Schott and Gen. filters, arranged in suitable combinations, for use in fluorescence photomicrography when using the HBO 200 W mercury lamp.

Table 27

Barrier filter employed	Exciter filter employed in conjunction with barrier filter	Intermediate barrier filter employed	
		Barrier filter I	neutral glass
41	BG12/UG1		N.0·01
41/–65	UG5/UG1		50
44	BG3/UG1		53
44/–65	UG1		65
		Barrier filter II	neutral glass
47	BG12/UG5		N.0·1
47/–65	UG5		41
50	BG12/BG12		44
53	BG12		47

Neutral and grey filters are used in all cases when no other possibility exists to damp the available light. They are designed to weaken, as evenly as possible, the entire range of visible light from 400 nm to 750 nm (Fig. 65). Further, these filters are indispensable when working with vapour discharge

Table 28

Transmittance	100%	50%	25%	12%	6%	3%	1·5%	0·75%
Grey filter	—	50	50 + 50	12	12 + 50	3	3 + 50	3 + 50 + 50

lamps of high intensity, where it may be imperative to shield the "stained" specimen. Such "neutral glass filters" are tinted and therefore, to some extent, absorb light.

Schott and Gen. exciter filters are usually 32 mm in diameter and vary in thickness. Their curves can be followed from Fig. 116. The UG5 violet exciter filter (3 or 2 mm in thickness) has a passband from 230 nm to 420 nm and a short high peak from approximately 300 nm to 370 nm and, also, a side band beyond 680 nm. The Hg-lines at 365 nm are transmitted almost unattenuated, while only a small proportion of the energy of the lines at 404 nm pass. Paired with BG38 filter, this combination forms a filter when red is only attenuated.

BG3 filter, dark blue (2·5 or 4 mm in thickness) has a steep slope toward the long region and has a substantially higher transmission in the ultra-violet. Its passband extends from approximately 275 nm to 475 nm with a wide effective peak from about 320 nm to 400 nm. In addition, there are considerable sidebands in the near infra-red, about 700 nm and beyond. When used on its own in fluorescence, this filter background is a slight red. In addition to the mercury lines in the visible blue and violet it is high in the lines around 340 nm to 375 nm. This exciter filter produces a higher ultra-violet radiation than that of the BG12 filter, but a slightly less exciting blue. BG3 with the BG38 exciting filter excludes all red. Ilford's Micro 2 transmits from 450 nm to 510 nm and Ilford Tricolour blue transmits from 370 nm to 510 nm.

The UG1 exciter filter appears to be almost opaque and can be purchased in 3 mm and 1·5 mm thickness. It has a passband from inside 300 nm and is out at 400 nm, with a peak transmission at 360 nm, plus a shallow side band at 700 nm. Consequently it will transmit only the 365 nm group of intense mercury lines and its transmission ends with a steep slope at the border line between ultra-violet and the visible spectrum. The red transmission of the filter alone is so low that the fluorescent background will be black. The UG1 exciting filter and BG38 form a successful unit.

The BG38 exciter filter is light blue and in one thickness, 2·5 mm. This filter is distinguished from the others by its high transmission over a wide wavelength range with a sharp cut in the ultra-violet at approximately 290 nm. It also has a distinguishable gradual slope into the visible region as indicated in the figure. This filter is designed to reduce or suppress the red in the background. Its passband is long, 300 nm to 700 nm and a sharp cut peak up to 359 nm which flattens out to approximately 600 nm.

The BG12 is pure blue and is obtainable in 4 mm and 3 mm thickness. Its peak transmission is at approximately 400 nm and passband from 330 nm to just under 500 nm. Excitation of fluorescence is high, red is completely absorbed. When paired with BG38 exciter filter their transmission characteristics remain unchanged. All these filters fit into special filter carriers in the microscope seen in Fig. 194 and Plate 59a.

The Leitz BX fluorescence microscope also has built-in exciter filters which are not affected by the heat from the lamp. In some of the eyepieces, made by Leitz for this particular purpose, the filters are housed beneath the eye lens.

Success in fluorescence photomicrography depends, to a great extent, on the right combination of microscope, barrier filter, exciter filter and illuminator. There are many types of mercury-vapour lamps such as the high pressure 200 W mercury arc source which is rich in ultra-violet (Fig. 120) which also produces an intense brilliance providing maximum image brightness. The light intensity of these lamps is measured by the candle-power per square cm. Therefore, the d.c. carbon-arc, being 18,000 candles/cm², has a lower intrinsic brilliance than the HBO 200, which is 25,000 candles/cm². This and the HBO 150 mercury-vapour arc lamp is specially designed for photomicrography by Osram. (See page 119 for further information.)

The Turner filter by Gillett and Sibert (Fig. 211) has been designed to function with an iodine quartz light source. Transmission for this filter is

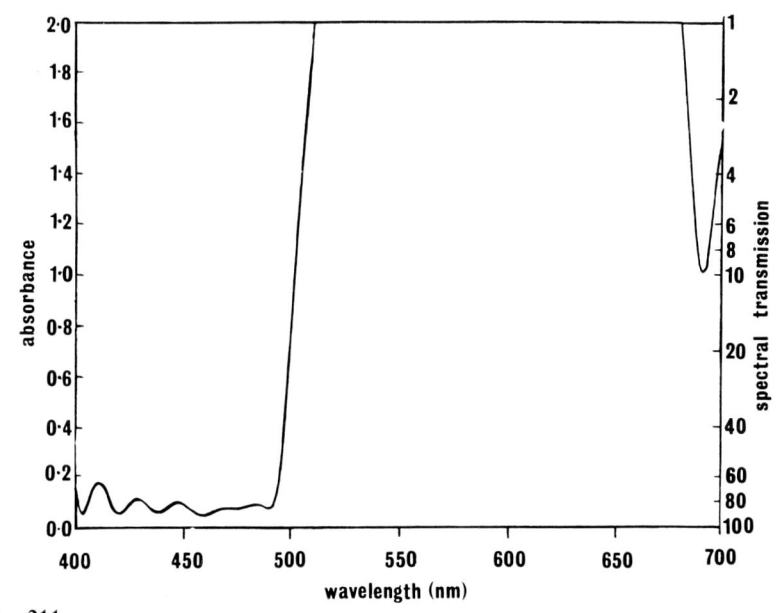

Fig. 211.

approximately 90% at a peak of 490–502 nm; thus producing "blue-light" fluorescence.

Incident light fluorescence. Until recently, incident activating light has found only a limited application, its use has been restricted to opaque specimens. However, it has been demonstrated that excitation with incident light can be carried out effectively with thin transparent specimens (Hauser, 1960; Grehn and Kornmann, 1965; Trapp, 1965; Ploem, 1965). In addition mounted transparent insect and pollen grains have been recorded by this method.

Ploem (1967) developed a vertical illuminator with interchangeable dichroic mirrors enabling the excitation light of different wavelengths for transparent specimens. We have already seen that with transmitted illumination there is often an absorbance of excitation light in the lower layers with thick specimens, and also an absorbance of fluorescent light in the top layers. This effect is not revealed when applying incident fluorescence light coupled with phase contrast via transmitted light.

A high pressure mercury lamp or high pressure xenon lamp can be used. The maxima in the ultra-violet and blue of the emission spectrum ranged from 365, 405 and 436 nm and were sufficiently spaced from the maximum at 546 nm, to prevent interference of the line with the fluorescence, excitation taking place with ultra-violet or blue light.

Two dichroic mirrors and an interference band filter were used for the selection of excitation light (Fig. 212). The dichroic mirrors (45°) have interference coatings with a very high reflectance of 90 to 95% for the excitation light and transmit 75 to 95% of the light of the wavelength. A relatively dark

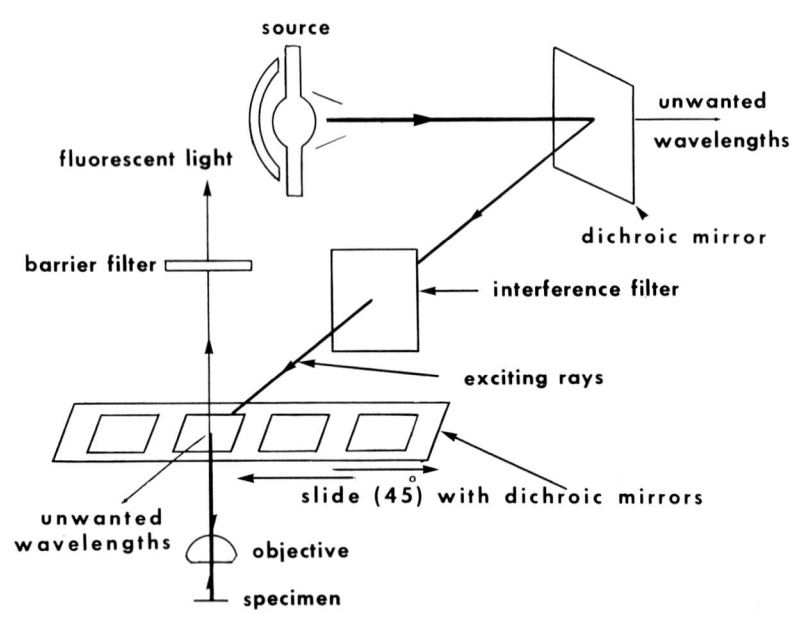

Fig. 212. The light path in incident illumination. The light emitted by the source is reflected by a dichroic mirror (45°), then passes an excitation filter and is finally reflected by a second dichroic mirror into the objective (after Ploem).

FLUORESCENCE ILLUMINATION

background can be obtained since the dichroic mirror also deflects the excitation light that is reflected by the glass surfaces of the objective and cover glass (Fig. 213). A barrier filter with low absorbance was sufficient to eliminate the lesser excitation light that penetrated the dichroic mirror and as a result only little auto-fluorescence is evident in the cover glass. Transmission curves of dichroic mirrors are illustrated in Figs. 214a and 214b.

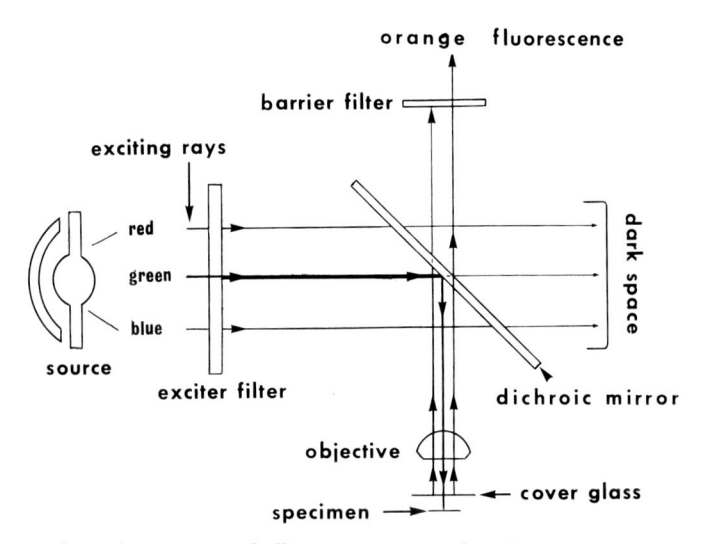

Fig. 213. Light path in a vertical illuminator equipped with a dichroic mirror (45°) with high reflectance for green light and high transmission for blue and red light.

For visual observation a much less intense light is necessary and for this purpose the H.P. 80 watt and the H.P. 135 watt (both manufactured by Philips) are most suited. The microscope should not be subjected to intense heat, as the metals would expand and contract and possibly damage the optical system.

Fluorescent tubes are low pressure mercury lamps operating to 1,000 volts. The inside of the circular tube is coated with a fluorescent material and so there is an improved light output in relation to the high pressure lamps. It is unfortunate that such a shaped tube is unsuitable for other than low power work. Even if the source output were concentrated into a small point source, it would be unsuitable for use with the compound microscope. Carl Zeiss, Oberkochen, market a very useful diffuse-light illuminator with transformer (Plate 67) which is well suited as an incident source for large specimens. It uses special fluorescent tubes, and the large reflector has an arm of double links so that it can be moved to the most favourable position. Highlights or shadows to

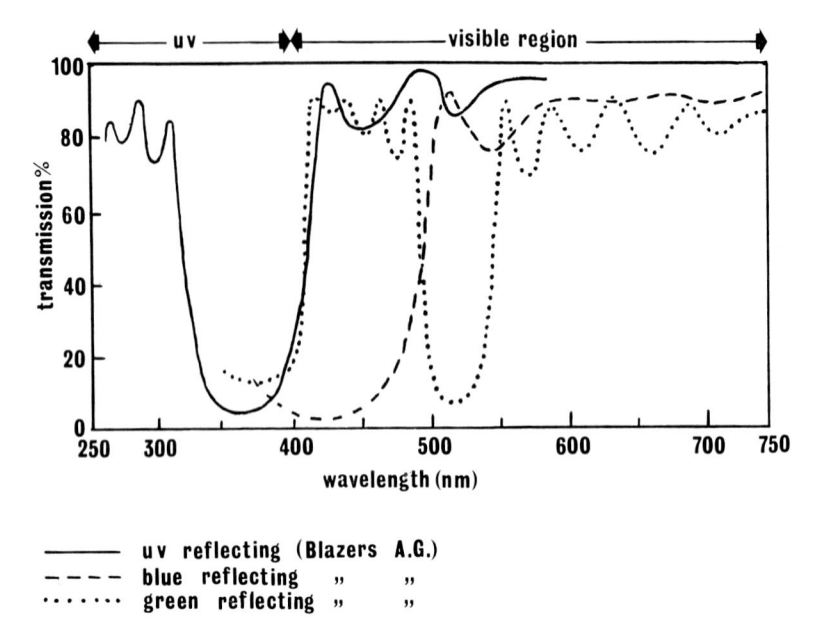

Fig. 214a. Transmission curves of dichroic mirrors.

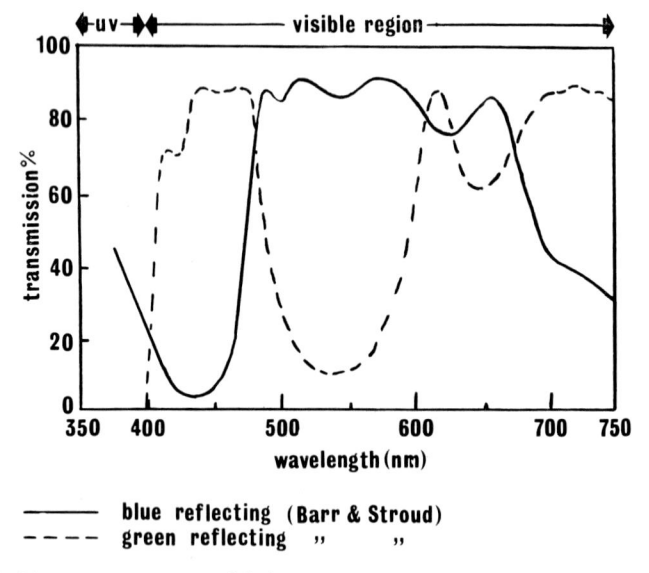

——— uv reflecting (Blazers A.G.)
– – – blue reflecting ,,   ,,
······ green reflecting ,,   ,,

Fig. 214b. Transmission curves of dichroic mirrors.

——— blue reflecting (Barr & Stroud)
– – – green reflecting ,,   ,,

increase the stereo impression can be achieved by means of an additional low-voltage illuminator. The fluorescent tubes of the diffuse light illuminator are distinguished by high light output and low consumption. Their rated life is in the region of 4,000 hr.

Two types of fluorescent tubes can be supplied:

Watts	Colour	Type	Rated life	Luminous flux
4	Daylight	F4T5/D	4,000 hr	121 m
4	Whitelight	F4T5/W	4,000 hr	1351 m

The type D (daylight) fluorescent tube is used to observe specimens in natural colours. The type W (white light) fluorescent tube emits a slightly brighter light.

The Ilford 805 filter is solely for absorbing ultra-violet and thus protects the eyes.

It is, of course, an advantage to have the microscope adjusted for Köhler illumination, though not over-illuminated. In fact this should be limited to the precise area, controlled by the substage condenser diaphragm. Excessive light or radiations for long periods will bleach the specimen.

Agfa CT 18 daylight colour film has proved more successful in this field. Film coated for artificial light is not really suitable. Orthochromatic film is more suited for black and white recordings.

References

Grehn, J. and H. Kornmann (1965). Kontrastfluoreszenz mit Opak-Illuminator. *Leitz-Mitt. Wiss. Tech.* **3**, 108.

Hauser, F. (1960). "Das Arbeiten mit auffallendem Licht in der microscope". Akademische Verlagsgesellschaft Geest and Portig K.G., Leipzig.

Ploem, J. S. (1965). Fluorescentiemicroskopie met opvallend licht gecombineerd met fasecontrastmicroskopie met doorvallend licht. Vergadering van de Nederlandse vereniging voor histochemie en cytochemie, Leiden, januari.

Ploem, J. S. (1967). "The Use of a Vertical Illuminator with Interchangeable Dichroic Mirrors for Fluorescence Microscopy with Incident Light". *Z. Wiss. Mikrosk. Mikroskop. Tech.* **68**(3), 129–142.

Trapp, L. (1967). Uber Lichtquellen und Filter für die Fluoreszenzmikroskopie und über die Auflichtfluoreszenzmethode bei Durchlichtpräparaten. *Acta Histochem. Suppl.* **7**, 327.

Darkground photomicrography

The application of darkground illumination began in 1856 when Whenham produced a paraboloid made of glass as an immersion condenser. It was shown that if parallel marginal rays of light enter a paraboloid reflector that they, now a cone of light, will converge to the focus of the lens (Fig. 215).

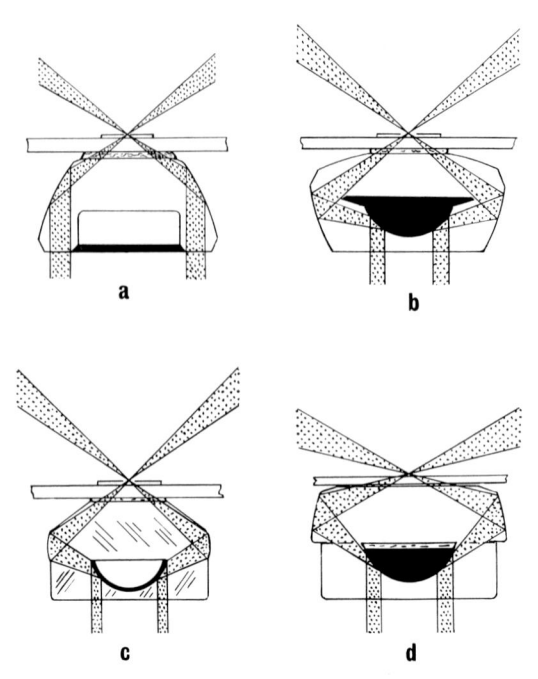

Fig. 215. Darkground condensers. (a) Parababoloid. (b) Leitz Bicentric N.A. 1·2. (c) Bausch and Lomb Cardioid N.A. 1·2. (d) Zeiss Luminous Spot N.A. 1·4.

Darkground illumination does not provide high resolution, nor even improve the resolving power of an optical system, but it does reveal objects clearly, delineated against a black background.

Most photomicrographs are taken under conditions of brightfield illumination. But subjects which have a refractive index, similar to that of the medium in which they are retained, are extremely difficult to observe. Since diffraction and reflection effects are most marked at the boundaries of such subjects, they may be rendered visible if the direct rays of light from the condenser are prevented from reaching the objective. The specimen is then illuminated by marginal rays, as shown in Fig. 216, a patch-stop being placed in the light train to leave a hollow cone of light so the specimen appears to be self-luminous against a dark background. This, darkground or darkfield illumination, depends for its efficiency upon accurate location and adjustment of a suitable objective and condenser (Fig. 216), a high-intensity light source, a microscope slide free from blemishes and a clean specimen.

It is false economy to wash and re-use microscope slides for this particular application. A slide which appears tolerably clean and well prepared when viewed under brightfield illumination often looks very dirty under darkground

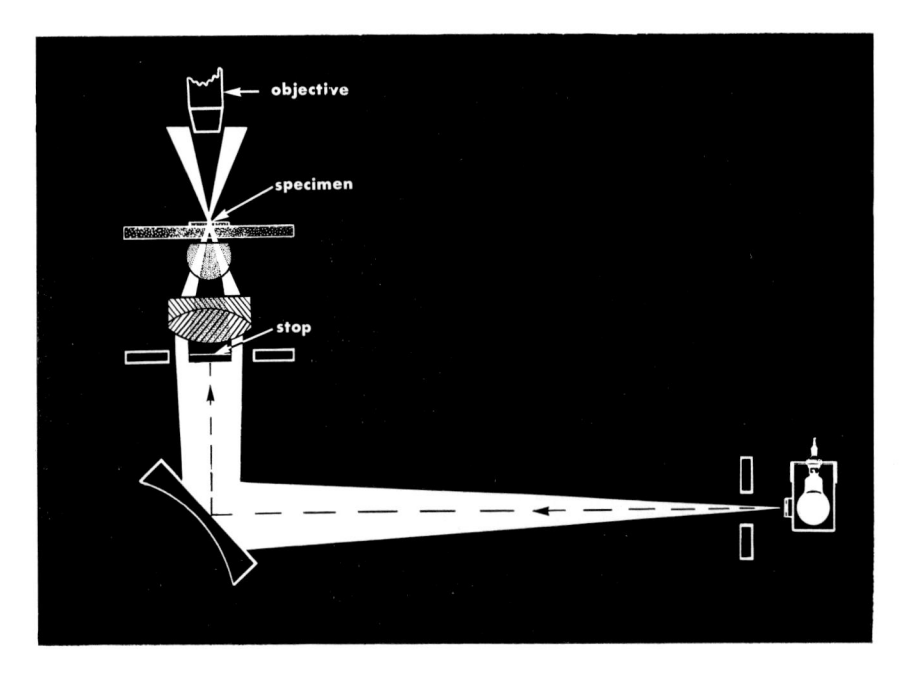

Fig. 216.

illumination. In fact, air bubbles and *minute* particles of foreign material may become so obvious that they confuse the details of the actual subject matter, and even Tyndall beam effects may be seen (Plate 35). If a thick microscope slide is used it may cause diffraction, so it is advisable to use those from 1·3 to 1·6 mm thick. If the slides are washed before use, great care should be taken to see that when in use they do not reveal Tyndall beam effects (page 274). Drying marks will also produce an unwanted pattern and perhaps interfere with the image of the spectrum. Clean slides should not be fingered, as the slightly greasy smears they make will give a hydrophobic surface (Plate 55). When cleaning glass and quartz slides and cells a *mild* application of an alkali substance (ethyl alcohol and hydrochloric) is satisfactory, but in time will etch these. Thereafter if used with darkground work, score marks will show as bright zones.

Before mentioning specially made condensers for use in darkground illumination, let us have a look at the possibilities of converting brightfield condensers into darkground condensers.

Consideration of the conditions necessary to illuminate a transparent specimen for darkground examination will show that the condenser must have a numerical aperture greater than that of the objective in use, and the central patch-stop placed in the condenser must block all the central rays which contribute to a numerical aperture equal to, or less than, that of the objective. If this is not so, and the condenser transmits any rays of smaller numerical aperture than those of the objective, then these will be able to enter the lower lens of the objective as direct transmitted light, thus ruining the photomicrograph. Carl Zeiss, Oberkochen, market several objectives specially manufactured for use in darkground. These are fitted with an iris diaphragm, which, when reduced, reduces the N.A. of the objective. The objectives are as follows.

Achromat	×100 N.A. 1·25	(oil)
Planachromat	×100 N.A. 1·25	(oil)
Planapochromat	×100 N.A. 1·3	(oil)
Apochromat	×40 N.A. 1·0	(oil)

A central expanding stop, such as the Traviss expanding stop (Fig. 217) or

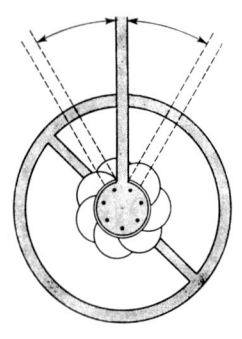

Fig. 217.

one of a set of fixed stops, may be placed below the condenser to block the central rays. Whichever type of patch-stop is used, care must be taken to see that the stop is accurately centred and arranged to be just large enough to obliterate all direct illumination of the specimen. Patch-stops can be made of black plastic or similar soft material and attached to the outer surface of the lower condenser lens, or even between the two lower elements of the condenser. By experiment one will arrive at a suitable size. The swing-out lens in the foot of the stand in microscopes with a built-in illuminator, should be removed from the beam path when darkground condensers are used.

A two lens Abbe brightfield condenser may be used with objectives which have a N.A. up to 0·40, but it is advisable to use an achromatic condenser with objectives of N.A. in the range 0·40–0·75. With objectives of a high numerical aperture, specially computed oil-immersion darkground condensers are required. (These are listed later on.) In an extreme case when the numerical aperture of an objective is too high it may be reduced by the introduction of a funnel stop (Figs. 218, 219) placed above the objective optical system (Fig. 219). Funnel stops appear to be restricted, but are still available through various

DARKGROUND PHOTOMICROGRAPHY

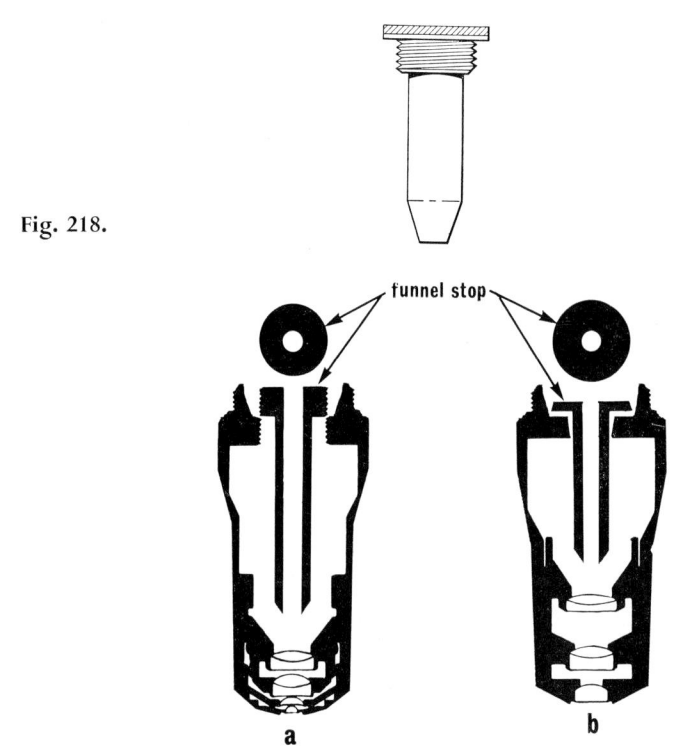

Fig. 218.

Fig. 219. Funnel stops for darkground illumination. (a) Screw-in type. (b) Drop-in type.

manufacturers. A darkground condenser, a beautiful piece of equipment which could be bought in the recent past, included a matching objective and funnel stop. In the case of an objective having a high back lens, a small funnel stop can be used.

The Abbe chromatic brightfield condenser, N.A. 1·2, is also ideal for this application of photomicrography and gives brilliant results with most objectives, especially dry objectives of N.A. 0·65. The insertion of an immersion medium between the top lens of the condenser and the under side of the microscope slide will often improve the brilliance of the image. It is suggested that microscope slides measuring 5·8 × 7·62 cm (2 × 3 in.) be used, because of the extra wide hollow cone of light which extends beyond the conventional microscope slide.

The Abbe chromatic brightfield condenser of N.A. 1·4 has three elements and its aperture allows a large stop of 22 mm diameter to be used between the two lower lens elements. It can be used successfully for the photographing of soap crystals, pond creatures and plants (Plate 17f), often quite difficult to record. Bacteria in a live state, fungi or minute colloidal particles, can be

K

recorded quite well if an objective of N.A. at least 0·85 is used with the condenser. An efficient darkground patch-stop, now in production (Fig. 220a) in the U.S.A., is designed to attach beneath the condenser or into the filter carrier. The stop is 22 mm in diameter and more suited for use in the higher magnification bracket. Patch-stops measuring 10 to 12 mm are necessary for small specimens.

One of the first commercial bicentric double reflecting darkground condensers was made by Leitz in 1908. Later H. Siedentopf of Zeiss designed an efficient cardioid reflecting darkground condenser (Fig. 215). The cardioid is a heart-

Fig. 220a.

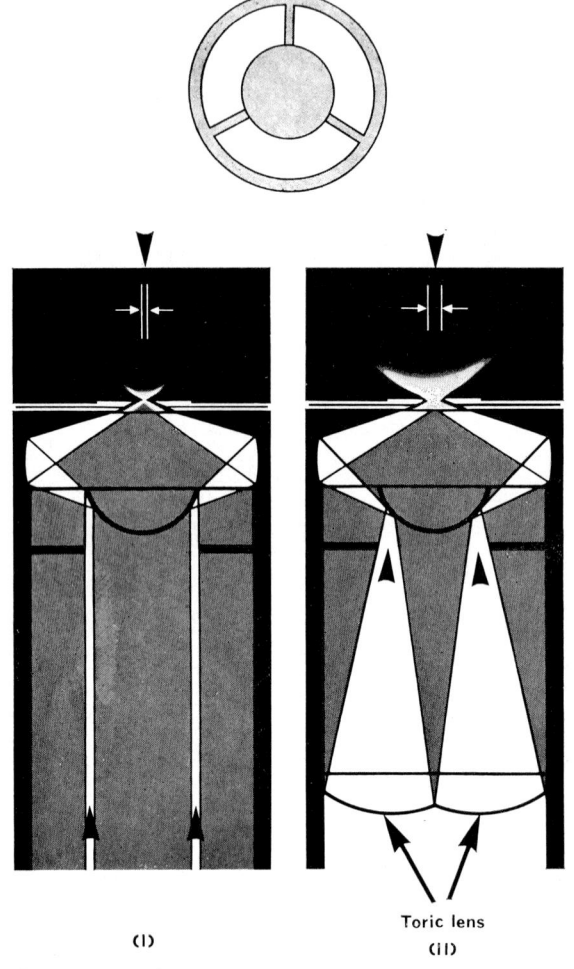

Fig. 220b. Reichert toric condenser. (I) Conventional darkground immersion condenser. (II) The new Reichert wide-angle immersion condenser carries an additional toric lens.

shaped curve with a normal cusp. The light rays, originally parallel to the axis of the optical system, strike the convex surface and are reflected to the concentric concave reflector, bringing the rays to an aplanatic focus at the specimen. The whole system is aplanatic anastigmatic and completely achromatic. Owing to the perfection of focusing achieved, fine details can also be recorded in photomicrographs of high magnifications. The aperture of this darkground condenser varies from N.A. 1·2 to 1·33 and its focal length is 3·6 mm, allowing objectives up to N.A. 1·05 to be used. This condenser is now sold by Bausch and Lomb and by Carl Zeiss, Oberkochen.

Another useful bicentric double reflecting darkground condenser is that of Leitz with a N.A. 0·80–0·95, and a focal length of 9·3 mm, usually used with objectives of N.A. 0·65 which would be 8 mm or 4 mm focal length. Leitz also market the bicentric double reflecting darkground condenser of N.A. 1·2, achieving a focal length of 3·6 mm to be used with achromatic and apochromatic oil-immersion objectives of N.A. 1·0 and N.A. 0·95. When high-power magnifications are required it is advisable to use a bicentric double reflecting darkground condenser of high numerical aperture to be incorporated with oil-immersion objectives fitted with built-in adjustable iris diaphragm to reduce the numerical aperture (Fig. 57 E).

The Watson darkground catoptric condenser is designed on the concentric principle. It is suitable for use with microscope slides 1·0 to 1·2 mm thick and oil must be used with ×100 or ×90 objectives whose aperture must be reduced to N.A. 1·0 by a funnel stop. However, the catoptric darkground condenser can be used with a ×40 or ×50 objective without a funnel stop. The condenser gives an intense black background and is easy to centre.

During the mid 1920's Zeiss improved their bicentric double reflecting darkground condenser which became known as the Leuchtbild Condenser (Fig. 215, d). This spot ring illuminator had an N.A. of 1·3 extending to N.A. 1·4 and those in use today are possibly the best in the field. At the same time Leitz produced a high N.A. cardioid darkground condenser which was said to be N.A. 1·4 to N.A. 1·5, later known as a "smear condenser". These high functioning condensers operate with immersion objectives, oil and water, but it is becoming more and more difficult to purchase them.

Toric darkground condenser. The Reichert new darkground condenser carries an additional toric lens (Fig. 220b) at the entryface which concentrates all the light from the source into the annular channel of the cardioid system. This makes it possible to illuminate a much wider specimen field than previously, without at the same time reducing the illumination intensity. The Reichert wide-field condenser can therefore be used for photomicrography incorporating a ×10 objective. This toric condenser can cover the entire magnification range from ×50 to ×1600, in addition the new condenser is made from UV transmitting glass so that it can be used in fluorescence photomicrography.

The alignment of the microscope must be carefully conducted for work by darkground technique. In fact it should follow Köhler's method of illumination. If a lamp with a very small source is being used it may be necessary to use an auxiliary condenser in the light-train to fill the whole of the condenser lenses. If this is so, preliminary adjustment of the mirror (if used) should be carried out with the auxiliary† condenser in position. The darkground condenser may now be inserted with the objective. The alignment of both should be checked. If the mirror expresses an incorrect tilt, or the condenser is not centred accurately, an unbalanced cone of light will result, and this will produce uneven illumination and an ill lit specimen (Fig. 221b).

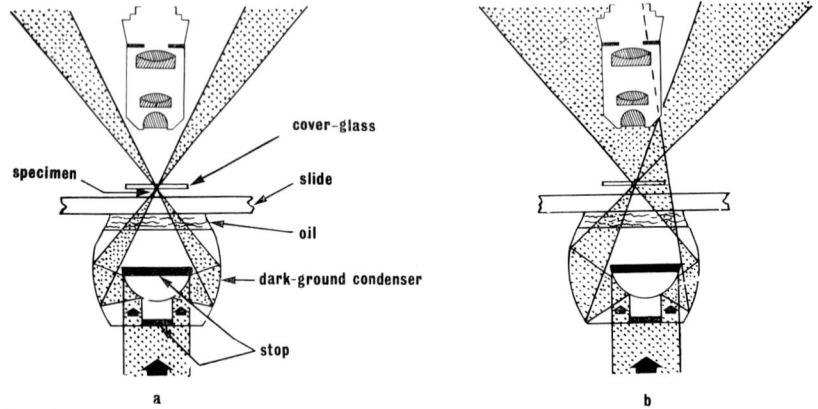

Fig. 221.

Since the adjustment is most exacting for high-power work, we will discuss the technique from this aspect. An ordinary condenser will have to be centred by inspection of the position of the cone of light which it provides, but most darkground condensers bear some sort of centering mark on their upper lens. This should be focused by the objective and centred by the darkground condenser centering screws. If the condenser is corrected for one thickness of the microscope slide it should be checked against the slide to be used. Some condensers are equipped with a correction device for slide thickness which must be checked too.

Assuming the light source has been correctly adjusted, and parallel light fills the centre of the darkground condenser, the next operation is to examine the projected cone of rays. A lower power objective and eyepiece is inserted into

† Auxiliary lens system. Such systems are made up of diverging or converging lenses, placed between the source and the substage condenser, bringing about a planned shift in the light source image. Such auxiliary optical systems are most useful with a free standing source.

DARKGROUND PHOTOMICROGRAPHY

the microscope and focusing is carried out on the specimen. The darkground condenser is moved in line with the optical axis of the microscope until a narrow bright cycle (Fig. 222c) is seen in the eyepiece. It is advisable to familiarize oneself with the condition of the light each side of the correct focus (a and b), to arrive quickly at the point of focus (c). Having located the narrowest bright cycle it must then be carefully centralized (Fig. 221a and b) by the adjustable screws so that the final position of the darkground condenser is accurate. Some very delicate final adjustments will almost certainly be required to ensure the best results.

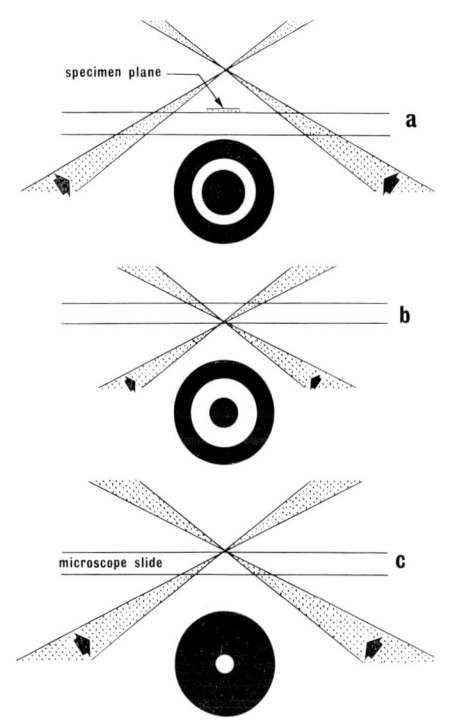

Fig. 222. Adjusting darkground condenser for height when viewing through the low-power optical system. (a) Condenser too high. (b) Condenser too low. (c) Condenser correct.

Incident darkground. Lower power incident light darkground illumination is a necessity and is taken care of by several optical manufacturers. There are two kinds of illuminator, the catoptric, of which a well-known form is that of Vickers (Fig. 223a), and the dioptric, such as is used with the Leitz Ultrapak microscope/camera. Zeiss, Jena, produce a small, space-saving illuminator. Figure 224 sets out a similar piece of apparatus. A parallel light beam passes around a

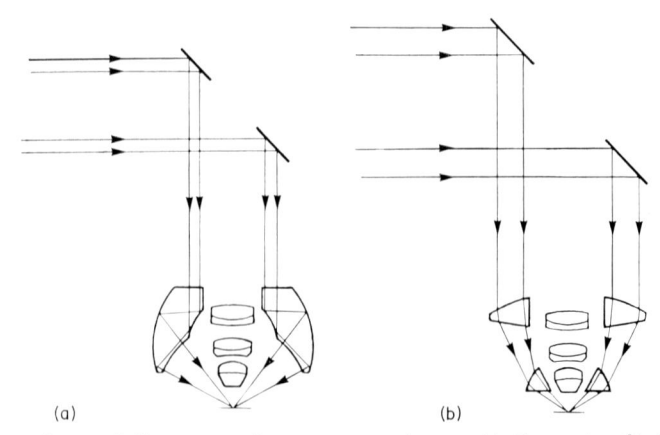

Fig. 223. Darkground illuminators for opaque specimens. (a) Catoptric. (b) Dioptric.

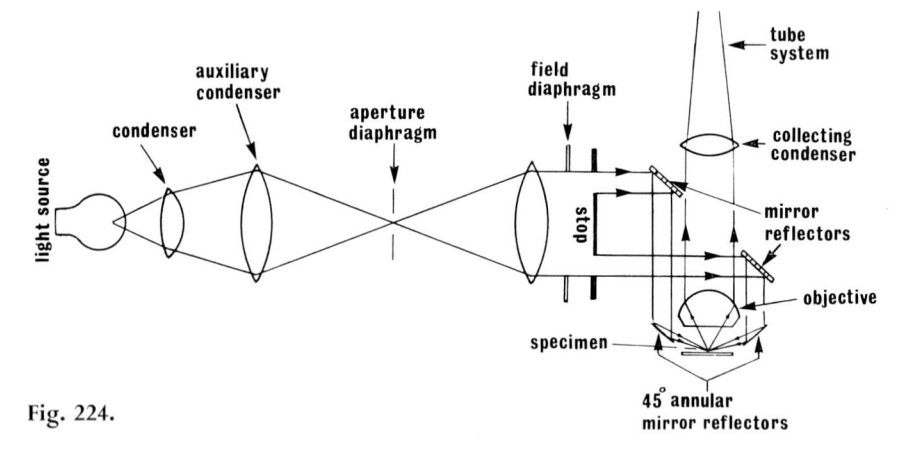

Fig. 224.

central patch-stop, is directed to an annular 45° mirror, which in turn directs the internal beam of light to a concave mirror condenser. This is responsible for directing the source into the specimen plane, but the source condenser iris diaphragm should be adjusted to provide a suitable cone of light, corresponding to the aperture of the mirror condenser. The light source should be centred correctly and adjusted, to provide an even distribution of light in the mirror condenser and in turn in the specimen plane, to avoid chromatic effects. Achromatic optical systems (Fig. 225), fitted with catoptric illuminators for incident light darkground applications, are now available from Vickers. They are as follows:

 ×10 N.A. 0·25 working distance 1·7 mm
 ×20 N.A. 0·50 working distance 1·4 mm
 ×40 N.A. 0·65 working distance 0·7 mm

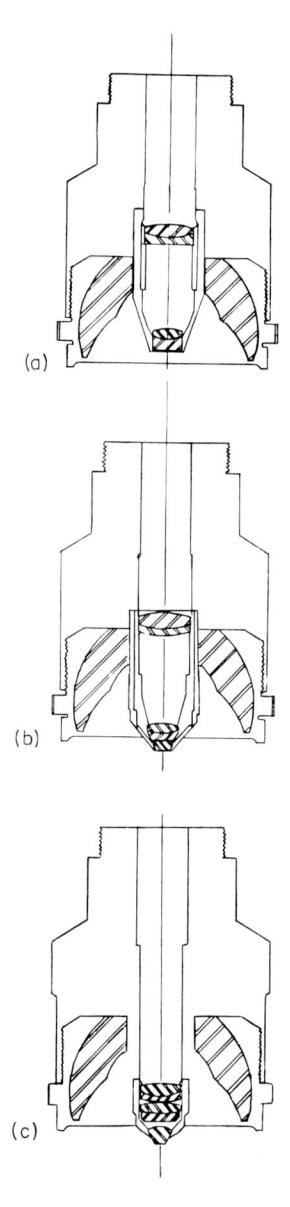

Fig. 225. Catoptric objectives (Vickers Ltd). (a) 16 mm achromatic darkground N.A. 0·25. (b) 8 mm achromatic darkground N.A. 0·50. (c) 4 mm achromatic darkground N.A. 0·65.

High-power incident light darkground illumination is necessary as a recording medium in some fields where opaque surfaces must be examined (Fig. 224). Great care should be exercised when adjusting the illuminator for photomicrography, all the rules for Köhler illumination apply.

The high intensity light source is centralized and is imaged on to the aperture diaphragm by a well adjusted condenser and auxiliary condenser. The parallel beam is now directed to a patch-stop positioned beyond a further condenser and the beam cone adjusted by an iris diaphragm placed beyond the collecting condenser. The patch-stop obstructing the main beam causes a hollow cone of rays to traverse the outer zone of the lens, and image, at a 45° annular mirror. From here the internal cone of rays travel down until intercepted by a mirror condenser adjusted to the objective in use. The mirror condenser redirects the light supply to the specimen, where it is imaged and reflected up through the objective, finally imaging in the film plane. As maximum light output is necessary the apertures should always operate at maximum capacity. The auxiliary lens must be adjusted to occupy maximum field diaphragm aperture.

Tyndall effect

The Tyndall beam (Plate 35a) resembles fluorescence in appearance, but actually the phenomenon is quite different. The light scattered in the Tyndall beam is polarized and the distribution of intensity among the wavelengths is determined to a great extent by the size of the particles.

The smallest particles of matter visible even under high magnification have a diameter of not less than 0·0001 mm. Colloid particles are much smaller than this and therefore cannot actually be viewed. They are, however, usually large enough to scatter light sideways, so that if a beam of light is passed through a colloidal solution the path of the light may become visible. This is known as the Tyndall effect, or Tyndall beam, and is not observed in true solutions. For the Tyndall effect to be observed in an aqueous colloidal solution, not only must the particles be of a suitable size relative to the wavelength of the light, but the refractive index of the material of the particles must differ greatly from that of water. It can be shown that the particles of starch, glue, etc. contain much loosely bound water, hence the refractive index does not differ sufficiently from that of water for the Tyndall effect to be seen. Particles of oils, or of inorganic substances, however, readily exhibit the effect. (Plate 35a.)

Lord Rayleigh showed that for electrical insulators the relationship between the wavelength λ of the light and the intensity I of scattering was

$$I = k(v^2/d^2\lambda^4)$$

where k is a constant, v is the volume of the particle, and d is the distance from the particle to the observer. Since the intensity is inversely proportional to λ^4, it is clear that radiations of short wavelength will be most intense in the scattered light; this accounts for the fact that many colourless or pale coloured solutions have a bluish Tyndall beam. The constant in Lord Rayleigh's equation contains a term involving the difference between the refractive indices of the particles and the dispersion medium. When these are equal there should be no scattering and this may account for the small Tyndall effect observed with lyophilic solutions and with some lyophobic solutions.

Brownian movement

The irregular motion of particles is an example of Brownian movement—first observed in 1827 by the English botanist Brown. He noticed that pollen grains, when suspended in water, continually perform haphazard zigzag movements which he was inclined to attribute to the agency of living organisms. Suspensions of other fine particles, such as those of gamboge and gum mastic, show the same erratic, sometimes river-like movement. Wiener realized that the cause lay in the bombardment of the particles by molecules of the solvent. The particles are, of course, being bombarded on all sides, but when by chance the collisions on one side of a particle out-balance those on the other, the particle will tend to move. It will rarely happen that the difference in "push" is great, so that the Brownian movement will not be shown in great contrast by large and heavy particles.

Schlieren photomicrography

A method used to record optical glass surfaces and non-homogeneous areas in thin transparent specimens, creating irregular refractions of light (Schardin, 1942), usually expressed in the form of streaks or striations. (Schlieren, being interpreted, is striations.) Schlieren methods can be used to locate the position of streaks and their size. It is a form of shadow projection illumination with a very small light source and narrow pencil beam of light: the edge zone of the specimen being darker than the surrounding area. The narrow pencil beam of light is redirected by the striations to a new position on the ground glass screen. The method is brought about by the introduction of a graticule with parallel transparent and opaque lines, positioned in line with the schlieren. The graticule and specimen are recorded on one exposure and the lines (schlieren) are imaged on the undulating specimen area.

Meyer-Arendt (1957) introduced a method whereby the source is directed through a slit 1/7 mm in width. The slit is imaged by the substage condenser into the objective and through the graticule positioned behind the objective and in front of the eyepiece. When the graticule lines are parallel to the slit, the shadows of the graticules are recorded with the surface of the specimen but when the graticule lines are perpendicular to the slit, the specimen is illuminated as if bright-field illumination had been applied—minus striations.

Figure 226 demonstrates Töpler's method for producing schlieren effects,

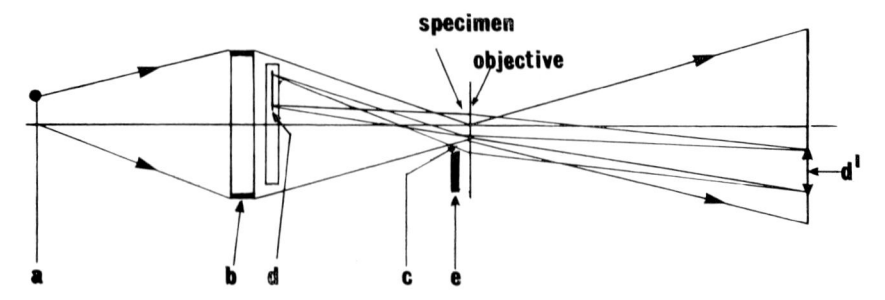

Fig. 226. Schlieren method after Töpler.

adopted by most. The area A is the light source and is directed by straight lines, as a rule, on one side. Some models have incorporated an adjustable narrow slit, positioned in the optical axis. The light source is first imaged by B, in the proximity of a condenser projected towards C, the objective. The condenser images the specimen with schlieren D in the image plane at D^1. In the plane C is an adjustable diaphragm ground to a fine edge, and in the same area E, is the schlieren placed parallel with the graticule lines of A.

The schlieren D changes the course of the narrow pencil beam, setting up refraction. If the deviated light passes D without being redirected, the image at D^1 records a zero schlieren effect. On the other hand, if E intersects the deviated light beam, then schlieren effect is imaged.

References

Schardin, H. (1942). *Ergeb. Exakt. Naturwiss.* **XX**, 303–439.
Von Jurgen Meyer-Arendt, J. (1957). *Mikroskopie*, **12**, 21.

Stereo-photomicrography

The principles of stereo-photomicrography ought first be understood. When viewing an object on a flat print, any impression of depth, or height, can only be given by shadows, or by tilting the specimen (Plate 11 and Fig. 115) to reveal depth through its structure. Shadows can easily be misinterpreted and a feature which appears to be in relief may in fact be a depression. Stereoscopic vision in photomicrography, depends upon seeing with both eyes, two pictures, correctly spaced, as one picture. It is largely a question of angles, as one eye sees slightly more of one side of an object than the other. Figure 227(a) illustrates the unaided eye viewing a small specimen at a distance of 250 mm. Stereoscopic

STEREO-PHOTOMICROGRAPHY

vision is retained in (b) where the eye is viewing the specimen through a simple magnifying lens. Higher magnification can be obtained by the introduction of a low-power stereomicroscope (Fig. 228).

Some animals and birds cannot see stereoscopically, particularly birds.

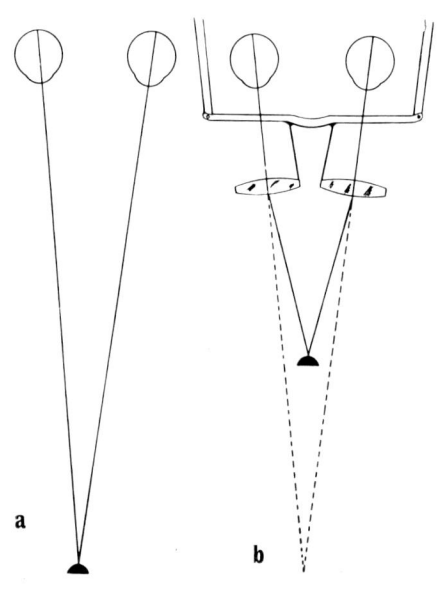

Fig. 227. (a) Stereoscopic vision, a near object viewed at a distance of ten inches. (b) The same object through a pair of magnifying lenses.

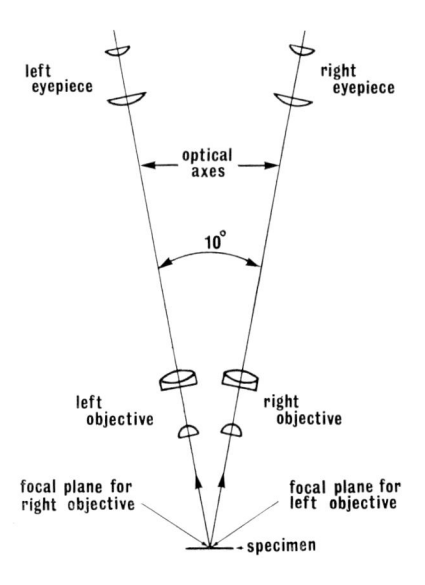

Fig. 228.

Most birds hold their heads on one side, either to see an enemy or when about to pick up their food. Most insects have compound eyes composed of minute facets which are believed to form images by a process known as mosaic vision. There are, of course, a few insects and pond creatures which have only one eye. In such cases the field of view covered is not seen stereoscopically.

A simple illustration indicates the practical value of stereoscopic vision. Imagine knocking a nail into a piece of wood while one eye is closed. How different and how very difficult this simple job now becomes and how often the fingers are hit. Using only one eye one cannot see round the nail, and it becomes difficult to judge distances because the ortho-stereoscopic effect vanishes. In other words, a person with normal stereoscopic vision cannot judge depth accurately with only one eye.

One of the most common forms of observation with a microscope is by using transmitted light, one objective and a single eyepiece. This monocular form projects a "flat" image without any sense of depth in a photomicrograph and made under such conditions, also lacks depth or stereoscopy, while viewing the specimen for any length of time is fatiguing. However, the single image projected by the binocular microscope objective (Plate 61) is divided by a beam-splitter system, providing images for both eyes. The optical system is so devised that both images are identical in field covered, form and magnification. Even so, there is no real stereoscopic effect and the images appear "flat".

The stereoscopic microscope is constructed differently from the monocular microscope (Fig. 228). Its design permits the production of stereoscopic images in which the magnification is the same in both negatives, but the images (prints), when viewed, have depth and perspective. In true stereoscopic vision or stereo-photomicrography, the left and right eyes, or prints, record a view of the specimen from slightly different points and therefore, in the case of the eye, the retinal images are slightly different. In a print one contains slightly more than the other. Thus it is essential in the construction of a stereoscopic microscope to present a left and right eye view, projected as separate images, to be viewed at the same time.

The two prints should be identical in tones and mounted so that corresponding points are about 6 cm (2·5 in.) apart and so that horizontal and vertical lines in each are in line, to avoid tilt. The mounted prints can be viewed with a stereoscope for study of the third dimension. Such prints could be viewed by fused direct observation. Another method for viewing pairs is to stand a piece of stiff card, quarto size, between the prints. Viewing takes place directly above the prints with forehead touching a board positioned between the eyes, as in Fig. 229. The right eye 250 mm (10 in.) from the print views the right print and the left eye the left. The prints can thus be fused together, with tridimensional depth. The brain "thinks" that A, B and C must be at a point where the converging images of vision meet at $A2$, $B2$ and $C2$.

STEREO-PHOTOMICROGRAPHY

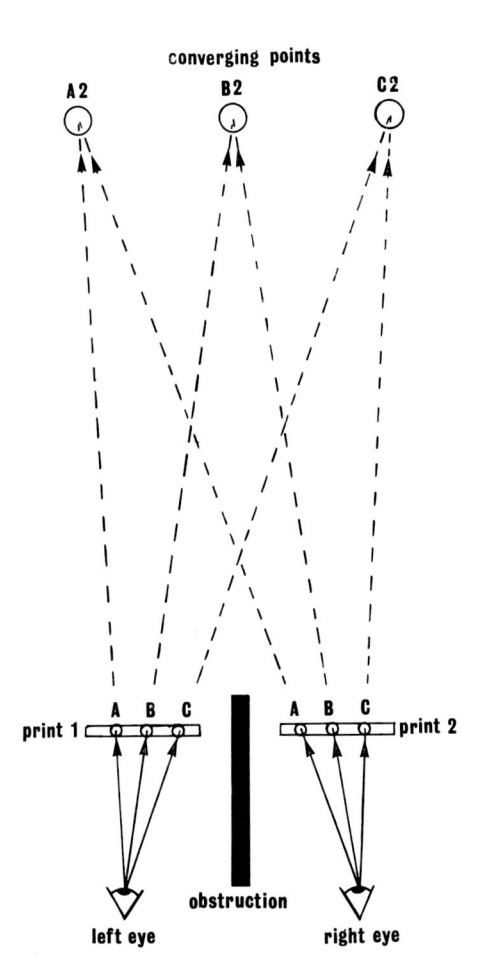

Fig. 229.

One of the earliest forms of obtaining stereo pairs through a monocular microscope was to divide the aperture of a single objective. The half lens method is based upon the Whenham system of stereos, using a D-shaped disc placed behind the top of an objective (Fig. 230). Successful stereo pairs can also be made by placing a half-stop over the front lens of the objective. As can be seen from the diagram only half of the optical system is used, and the numerical aparture is reduced by half. One exposure is made with the D-stop positioned over one half and a second negative made after rotating the annulus 180° or with the stop covering the other side. The straight side of the D-stop should run across the centre of the lens, at points equi-distant from the intersection of the horizontal with the vertical bisector. The specimen remains in

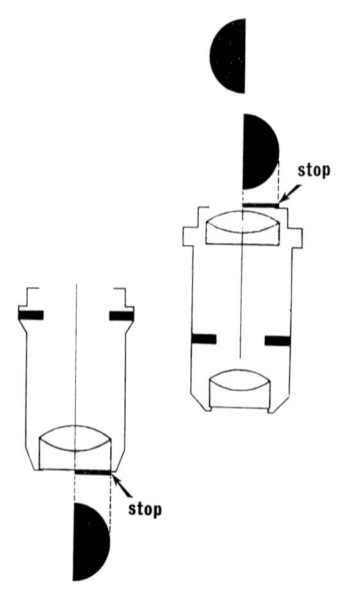

Fig. 230. Half lens method of making stereo pairs.

the same position for both exposures, as does the illumination. This simple arrangement produces satisfactory results but it is time consuming.

About 1900 a stereoscopic microscope was made based on the design of the Greenough microscope. This model (Fig. 228) consisted of two low-power objectives and two eyepieces, each in a separate microscope tube, at an angle of approximately 10°. Like Fig. 227 (a) and (b), the two optical axes intersect at the same point in the specimen and, when viewed, produce stereoscopy. Here the observer's eyes converge at the same angle as the optical axes and with an angle of 10° the specimen plane is 5° out of the vertical, perhaps more noticeable than when incorporating a tilting stage on which to view or photograph the specimen (Fig. 115).

The Greenough stereomicroscope is designed to use two lens systems arranged in a converging line, or formation, but when the first commercial stereomicroscope was manufactured it appeared with a large front lens, plus prisms. This became known as the "Porro system". The system provided an erect image. The microscope's mechanical arrangements allowed low magnifications only. If such a stereomicroscope were to achieve high to medium magnifications, the optical system would be brought nearer to the specimen. This could almost cause the two optical systems to be fused together: not at all practical in this instance. To help matters the angle of the two systems can be increased, which would affect the angle between the two viewing axes and also increase the stereoscopic effect. If the angle is approximately 15° the eyes must converge at a greater angle than normal vision and the specimen would be viewed from wide oblique angles. In overcoming these over-emphasized effects, a zoom system has been introduced into some optical systems, together with a large 100 mm f2 front lens of approximately 45 mm diameter. Figure 231

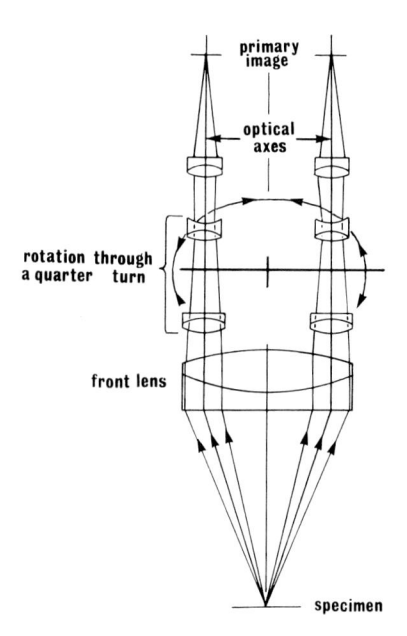

Fig. 231.

illustrates such an instrument. The telescope system rotated about the axes increases the magnification.

Stereoscopy further improved when an instrument was manufactured containing fewer optical elements, and axes which are not parallel. This apparatus had a large diameter front lens of flint, and crown glass to restrict aberrations. The 100 mm focal length front lens is the main objective to serve the two deviating prisms, as in Fig. 232. This lens system made for considerable depth of field and when the focused area was critical the two "wing" optical systems produced off-parallel axes. The eyepieces provided left and right hand views.

The Watson Zoom stereoscopic microscope was constructed without the large single front lens. Instead, it consisted of a paired optical system (Fig. 233) like the Greenough microscope. The front elements, or "pods", are housed in an interchangeable front collar. The lenses form the zoom objective and the movements are controlled by an outer rotating cam. Such an optical system has been designed to operate in a minimum and maximum position of magnification and is said to be free of aberrations. Primary magnifications are low

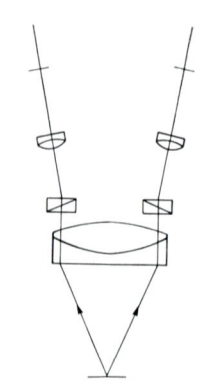

Fig. 232.

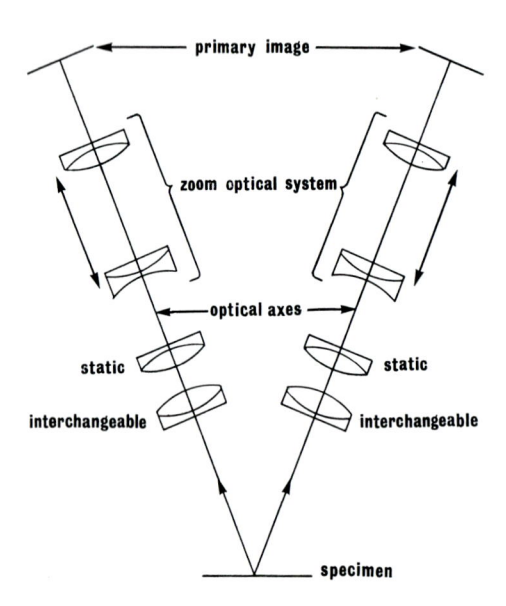

Fig. 233.

×1 to ×5 for standard optical systems, but can be increased by fitting an auxiliary lens system. Working distance, depth of field and magnification changes were similar to the previous method but unlike that seen in Fig. 228. The inclined eyepiece is over the parallel axes of the previous model. The eyepiece system ranges from ×10 to ×20 which assists in producing magnifications up to a medium range. This system does not suffer from the secondary spectrum sometimes created by a single lens.

Carl Zeiss, Oberkochen, produce a similar piece of equipment but the optical system is designed differently (Fig. 234). A front prism is used to further space the parallel optical axes. An erect image which is produced in keeping with

STEREO-PHOTOMICROGRAPHY

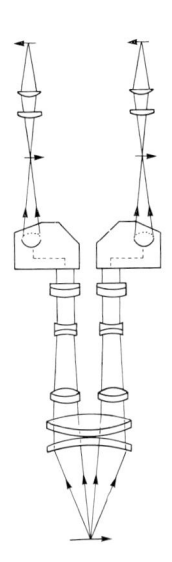

Fig. 234.

other models. The twin eyepiece tube is replaced by a single pillar camera attachment to take a Contax camera in the vertical position. Focusing takes place on a ground glass adaptor and it is necessary to make a separate exposure for each half negative. For this, the photo sliding tube is clamped in each end position respectively. When photographing with epi-illumination the iris diaphragm in the tube is closed to increase the depth of field. Working distance is about 80 mm.

A new microscope, the Stereomicroscope IV (Plate 67) produced by Carl Zeiss, Oberkochen, has a zoom magnification changer with a variable magnification factor of $\times 0.8$ to $\times 4.0$. Thus, with a pair of $\times 10$ eyepieces, magnification between $\times 8$ and $\times 40$ is obtained. By changing the paired eyepieces magnifications can go as high as $\times 200$. This model has greatly improved photographic facilities. The 35 mm camera is attached to one side of the eyepiece system and in no way interferes with the normal visual use of the microscope. Instead of the binocular tube a sliding photographic tube and 6.5×9 cm camera may be attached and either single or pairs of stereo-photomicrographs can be made. The 35 mm camera is attached by a beam splitter system to ensure successful stereo pairs. A special camera adaptor may be inserted, in place of the binocular tube, to house a 35 mm camera for stereo pairs side by side on one frame. The single front objective is interchangeable and may be replaced by a 63 mm macro-objective which, still in conjunction with the zoom system, will give higher quality photomicrographs.

Another method of producing stereo pairs is to incline the angle of the incident light. This is possible by the introduction of a single aperture mask below the substage condenser, as shown in Fig. 235. This means of illumination

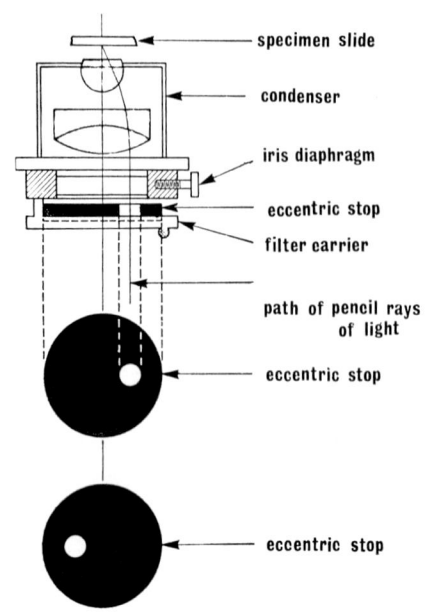

Fig. 235. Eccentric stop, as used with the substage condenser.

might be thought vertical because the rays of light pass through the substage and on through the objective in the normal manner but, instead of the normal axial rays, we use only oblique or marginal rays. The eccentric stop mask is housed in the filter carrier and is as near to the lower lens as possible. This form of oblique lighting must not be confused with top oblique lighting which is used when photographing an opaque subject. By using an eccentric stop, the pencil of light is directed to the edge of the condenser, which is operated at full aperture, as in Fig. 235. The light rays strike and penetrate the specimen at an angle, also seen in the illustration. After the first exposure the stop is turned round to the opposite side and the same procedure repeated. Various sizes of hole can be used according to the subject about to be photographed.

On occasions, when equipment allows, it is possible to use an eccentric stop with two holes which gives off two narrow light pencils running towards the axis of the condenser, but the instrument must be fitted with paired eyepieces. The rays of light are, of course, at an angle to the axis of the condenser. The angle of the pencil rays which strike the specimen is controlled by the positioning of the stop. The nearer the stop is to the centre of the substage condenser the more vertical is the angle of the pencil rays of light. With two apertures the two images can be exposed on the same negative, whereas with one aperture the images are made on separate emulsions.

Another means of producing stereo-photomicrographs of small subjects is to take two photographs from positions equally spaced from the perpendicular and to make two exposures as in other cases. For this method, however, it is advisable to use a ½ plate if two images cannot be accommodated on a ¼ plate. The procedure is quite simple. A line is drawn across the centre of the ground-glass screen and at equal distances on either side a mark is made, the overall distance being 2½ in. or 6 cm. This method cannot be employed unless the whole of the plate is illuminated and projected by the eyepiece, since two images must be made on the one emulsion.

An objective stop (Fig. 236) similar to that used with the condenser may be

Fig. 236. Eccentric stop, as used with the objective.

employed to direct a pencil beam of light after it leaves the objective. The eccentric stop must be small, approximately 21/2 mm, and fit over the back lens of the objective not farther than 1 mm from the lens. This provides a small circular aperture and ensures a pencil beam. The inner edge of the aperture should be as near as possible to the centre of the objective which will, of course, ensure the sharpness of the image. As applied to the substage eccentric stop, this too is turned 180° after making the first exposure. With this method it is not easy to turn the stop exactly 180°, in which case the stereoscopic effect is not secured.

The American Optical Company of New York have made stereo-photomicrography easy by the introduction of the A. O. Spencer Cycloptic stereoscopic microscope and camera. It has been specially designed for its job and is fitted with twin inclined eyepieces, positioned to house the special camera which gives a three dimensional photomicrograph. During visual observations the camera is mounted directly to the microscope in an "out of the way" position in readiness for instant use. With an easy forward movement the mounted 35 mm Graflex Stereocamera is placed into position over the eyepieces. The camera has special compensating prisms built into the adapter unit, thus rendering the camera parfocal with the microscope's optical system. No further adjustment is necessary. A full-size Graflex Stereoviewer, with built-in light source, is also marketed by this company.

Those who have Wild stereomicroscopes can easily photograph specimens with depth from two different positions and, by observing the resulting pair through a stereo-viewer, can examine a three-dimensional image of the specimen. There are two alternative methods of producing stereo-pairs.

One. The first exposure is taken down the right-hand tube of the stereomicroscope, the phototube is rotated 180° for the second exposure which is taken down the left-hand tube with the 35 mm camera.

Two. The two photomicrographs can be taken simultaneously on 35 mm film, so that they are side by side on one negative. This method is of course most useful for recording moving specimens. For this purpose a special stereo-phototube has been developed for use with the Wild M5 stereomicroscope. The light conditions in the intermediate image plane are particularly suitable for photography on account of the short exposure time; in addition, a larger area of the image can be included. Focusing is carried out via the reflex viewer of the camera and the exposure time is set on the camera shutter. The Wild electronic flash has been designed to operate with this application of photomicrography.

Negative material

The type of emulsion for making a photomicrograph is controlled by several factors; the colour of the stained preparation, the degree of contrast necessary to produce a grain free image, the colour of the light filter and the illuminating technique. The photomicrographer must also be able to differentiate between colours to which the photographic emulsion is equally sensitive and be able to accentuate one colour more than another. The wider the range of subjects handled the greater is the necessity for personal control over the photographic material. Under varied conditions, where no one emulsion suits two subjects alike, cut film or plates are recommended. Control must also be exercised in the use of colour filters (Plates 42 and 43). Sometimes it is advisable to use a chromatic emulsion such as seen in Plate 62. Here the specimen was stained only faintly so that it was lost in the background. (b) brings the subject to life and the resultant contrast retains the minute detail. (See also Plate 69).

Although panchromatic material can be used it is not really suitable because it produces a flat image and poses grain problems. Nor does it resolve as many lines per mm as the plates illustrated in this book. It is pointless buying the very best optical systems for resolving the finest lines if the negative materials used only resolve 45 lines per mm, compared with blue-green sensitive types which can resolve 100 to over 180 lines per mm. Agfa film (9 × 12 cm), medium speed is catalogued as resolving 1000 lines per mm and their colour film resolving 200 lines per mm. Naturally concern regarding the granularity of negative emulsions was inevitable when the use of 35 mm cameras became widespread. We now have negative material which, when used intelligently, will permit great degrees of enlargement without objectionable granularity becoming apparent. The minute negative illustrated in Fig. 2 was made some seventy years ago, the technical result of which (Plate 1), is far better than is often seen today in exhibition prints and reproductions in technical journals. Some workers have mastered the grain problem, while others search for some formula which will solve their problems though they seldom stay with one formula long enough to discover what it can offer.

Now grain in the negative, when examined under the microscope, is only of remote interest to the man whose sole concern is to produce good quality prints from objectionable granularity. The print as the end product is the thing that *matters*. My own negatives all receive the same treatment, namely two minutes in M.Q. type developer, and many photomicrographs even up to measuring 1 metre (3 ft) square have been made without granularity. This concern for fine grain is a question of general technique, and the presence or absence of grain in the finished picture is the product of several factors. It is easy to produce a printable negative and make a passable print of $\frac{1}{4}$ plate size or thereabouts from it; and that is just about as far as many workers get. Today many scientists are handling their own photomicrography and, provided they get an image of kinds, are not interested in high quality grainless prints. I have been given a print with the apology that "it was only taken on 35 mm film". To be candid, it was probably a proof of his own technical inability in print making or, getting to the root of the trouble, negative making. It should be possible to produce results of the highest quality at will, blending high quality photomicrography with real artistic vision and thus producing photomicrographs which are scientific and educational expressions.

Whatever the size of the negative with which one works, technical refinements inevitably combine to produce quality in the finished print, so the matter is not merely of concern to small negative format workers only. A factor, probably its importance is underestimated, is cleanliness. Dirty lenses, dust in the enlarger, or dirty finger marks on the optical system, produce flat results and grain is much more obvious in a flat print than in a print of correct contrast. Correct exposure must be given, and an exposure meter or table of exposures should always be intelligently used (especially with colour material). If any considerable degree of over-exposure is present the negative will develop a coarse, granular image, while under-exposure will invariably require printing on a more contrasty paper. This will reveal grain, perhaps more than a normal paper would do. It is only by standardising one's whole technique that one can obtain the fullest benefit from the materials which the manufacturers provide and this requires intelligent application of the facts, not credulous acceptance of involved mysterious formulae.

Processing plate and paper

The darkroom must be properly equipped. It is a waste of time taking all possible care with the microscope if darkroom essentials have been ignored. The darkroom, in which many hours are spent, should be spacious and well ventilated, the walls light in colour and the room completely "light-tight" when doors and windows are closed. The sink should be square so that the dishes rest on the flat bottom instead of rocking about, as they would in a domestic sink. Lead sinks are, of course, the best, and glass fibre sinks are very efficient. A porcelain sink will, however, serve the purpose. There must be hot and cold running water, a good waste-pipe and an adequate overflow pipe. The latter is essential, as the waste can quite easily become blocked if cotton-wool or a print is sucked over the waste hole. In addition, as a modification, I have made a perforated funnel stop which allows the water to flow away from the bottom of the sink at a rate equal to that of the in-flow. Syphon washing tanks are excellent for print-washing, etc. This type of tank can be left in the sink with the plug out, without fear of an overflow.

Safelights should be well arranged to give good service within a safe illuminating distance of the sink where processing takes place. When processing fast panchromatic plates, try to work without a safelight which, however carefully it is used, is likely to fog a fast panchromatic emulsion. If such an emulsion must be inspected do it after it has had several minutes' development, there will then be less likelihood of fogging. Use a pull-switch to eliminate the possibility of an electric shock.

Good cupboards and shelves are always very convenient and assist in keeping the room tidy, as do racks for housing various sizes of developing dish. Storage cupboards for photographic materials and microscope accessories should be in a dry, cool place and all opened packages of paper and plates should be stowed in light-tight drawers, or similar containers. Keep chemicals away from sensitive materials, perhaps on the floor. If bottles containing acids and developers must be stored in cupboards, attention should be given to the type of material stowed beneath them. When weighing-up chemicals, do so in a place away from all sensitive materials, taking great care not to blow or drop chemicals where any part of the microscope or similar instrument is likely to be. When the enlarger is in the same room as the sink it ought to be well away from the taps so that it is beyond the reach of splashing when the tap is left running. On no account should the microscope be where there is running water, or in a damp room, if corrosion of the metalwork is to be avoided. Always clean up after a day's or evening's work: cleanliness is one of the first and foremost attributes of good photography.

Printing

Although the types of printing paper are numerous, the question of the most suitable for reproduction need not cause anxiety. This is decided for us, as will be seen.

Papers vary in grain size, as do photographic plates, and in a way can help to control the final graininess of the picture. The glossy paper is devoid of grain: at least, it is not visible to the naked eye. Semi-matt papers show a little grain but not so much as matt papers, which are very rough. The final choice of paper is dependent upon the effect the worker is trying to achieve. In addition, the rough grained paper does help to camouflage a negative which is not quite as sharp as it ought to be, whereas if the same negative were printed on a glossy paper the defect would be seen at its worst. Some pictorial workers like a "fuzzy" print, whereas scientific workers must produce a print which is clear, sharp and full of detail. The recommended paper for such scientific work as photomicrography is undoubtedly glossy, with matt as second choice. Kentmere colour base paper is quite suitable for many subjects, especially exhibition prints. If the utmost care has been taken to produce a sharp grainless negative with fine definition, it leaves no choice but to use glossy paper if these finer points are to be seen to the best advantage.

Papers are, as a rule, made in three or four grades—soft, normal, contrasty and extra hard. The soft bromide papers will show a wide range of tones, from white to black. This paper has also a considerable latitude in exposure, but does not have the same latitude in development times as does a photographic plate. Normal paper gives less grey tones between white and full black, whereas a high contrast paper shows very few grey tones between black and white.

The final photomicrograph should show nearly all the detail as seen on the photographic plate. This can be ensured only by printing on the correct grade of paper. Test strips should be printed on the various grades of paper and finally placed side by side to assist the user in deciding which is the best, but without having too soft a picture or one that is too contrasty. The final print should have in it all the tones shown in the negative, ranging from a pure white to a dead black (neither a greyish-white nor a greenish-black).

The treatment of lantern slides is very similar to the printing process, the only difference being that lantern slides are transparent. When such slides are being made the image should not be made too big for the subject is highly magnified when seen on the screen.

Exhibition prints

The number of photomicrographs at photographic exhibitions is declining, perhaps owing to the lengthy period entailed in collecting, staining, mounting and finally photographing the subjects. Another reason for the lack of exhibition prints in this field may be the shortage of competent and knowledgeable judges. However, all the photomacrographs and photomicrographs in this book have been in British International Exhibitions.

The exhibitor should pay particular attention to presentation, thus showing his work to best advantage. The prints need not be very large: 76 × 76 mm (3 × 3 in.) is sufficient, with possibly whole plate or 30·5 × 38 cm (15 × 12 in.) as the limit. On rare occasions it is worth making a 40·5 × 40·5 cm (16 × 16 in.) print. It is not necessary in this field to fill the whole mount;

small prints submitted should, therefore, be mounted in some order of tonal values, the light prints at the top and the darker prints at the bottom. Twelve 76 × 76 mm (3 × 3 in.) prints fit nicely on a 51 × 40·5 cm (20 × 16 in.) mount and can be made to look very attractive. The top outside prints look best with subjects pointing inwards, and with a light background if possible. The bottom three prints should be the darkest and have a dark background. The centre three prints are best displayed if the image is upright.

The annotating should be done carefully and be consistent throughout. Presenting the scientific print in an attractive manner plays an important part in gaining the verdict "in". The print ought not to be retouched or defaced in any way. If it is, it is not a true record of the subject and ceases to be of scientific value. When a photomicrograph negative has fine lines very close together it is possible to make an enlargement from it, thereby giving greater visible detail in a higher magnification of the overall area. There is, however, a danger of making a print too large. Any details already clearly visible on the negative do not need further enlarging, as this only results in empty magnifications. Often a negative reveals such fine detail where two lines are so close together that it is necessary to enlarge in order to separate the lines on the print. This, then, is not a case where it can be said that an enlargement produces empty magnifications.

The aesthetic and pictorial application

Photomicrography takes us so much further than photography and enables us to express ourselves in a more intimate way. The very essence of photomicrography as a means of expression lies in the existence of contrasts between various subjects: their different surfaces, colours, shapes and lines, and their positioning, plus the uses of illumination.

Art. Because this application of photography is highly scientific there is perhaps a tendency to think of it as being without scope for artistic application or that art cannot be mixed with science. But a work of art, in any medium, is the deliberate creation of unity, and Nature through the microscope is one medium which provides us with plenty of scope for such unity. The application of art, while stimulating, is naturally difficult, however attractive the subject of a scientific recording may be. But by means of original and unusual treatment pictures can be created to catch the eye and the "record" photograph can be made to look quite attractive. The transforming of objects into artistic pictures has some extra appeal, no doubt due to the rareness of the subject. I have been exhibiting such pictures for over 25 years, and was perhaps among the first to do so. Photomicrography offers expression not in what the painter has already done but in what the painter cannot do. Can it be assumed that photomicrographs are abstract pictures? The term abstract is used too often to refer to pictorial photomicrographs of a pattern nature. Abstract photography is the

THE AESTHETIC AND PICTORIAL APPLICATION

making of pattern pictures which have aesthetic value, due to colours, shapes and arrangements, but which lack scientific value and meaning, and are non-factual. Such abstract patterns are either deliberately created and arranged or could, of course, consist of forms from factual material. This is sometimes referred to as subjective photography.

Aesthetics. One of the fundamental problems of aesthetics concerns the relationship between natural objects, considered beautiful, and the creations of man, whether in paint, stone, wood or metal, to which the same word is applied. Some of the photomicrographs in this book contain beautiful patterns, or rhythms, proportions which harmonise, recorded through the microscope. Unlike man-made patterns, they were *created* (Plate 34). They were not the result of an accident, nor made by measurement of composition and its related "true balance": they already exist for us to see and enjoy. It is entirely in the hands of the photomicrographer as to how the subject is presented. In scientific photography all should try to produce a picture which has a scientific message and, if possible, incorporates some aesthetic value. Though the latter is incidental to some, it is felt that it increases one's vision, and perhaps may be the means of a better understanding. There are some subjects which are purely scientific recordings and will always remain such, however much they are turned around or illuminated. With the introduction of better colour material, coupled with new and more reliable methods of reproducing colour, some publishing houses are using artistic photomicrographs, for instance, for book jackets.

Controlled arrangement. Most natural subjects, however small, are aesthetically perfect and lend themselves to pictorial treatment, either singly or collectively. The way they are displayed is controlled by one's own appreciation of nature's art. By skilful arrangement and illumination, commonplace objects provide great pictorial possibilities.

The arrangement of subjects such as insect eggs, antenna and scales of insects, often takes weeks to complete whereas such a subject as seeds or wet mounted weeds take only an hour or so. Each subject demands an individual approach and perhaps the appeal of pictorialism lies in the variety available. Personal expression involves arrangement (Plate 13, 17f, 39), background, illumination, view point, focal length of lens, magnification of the initial image and that of the final print, the field covered, depth of field, colour, detailed rendering of subject, photographic emulsion and, finally, tonal range of the finished print.

How does pictorialism function in photomicrography? Many people think that pictorial photography is limited to landscape, sea, cloud, mist and mountain photography, but this is not the case. Any subject will make a photograph. It is through the arrangement of lines and their placing within a certain format, illumination, and final darkroom treatment, that an attractive photograph is made (Fig. 237).

Composition. The composition is of equal importance to the photomicrographer

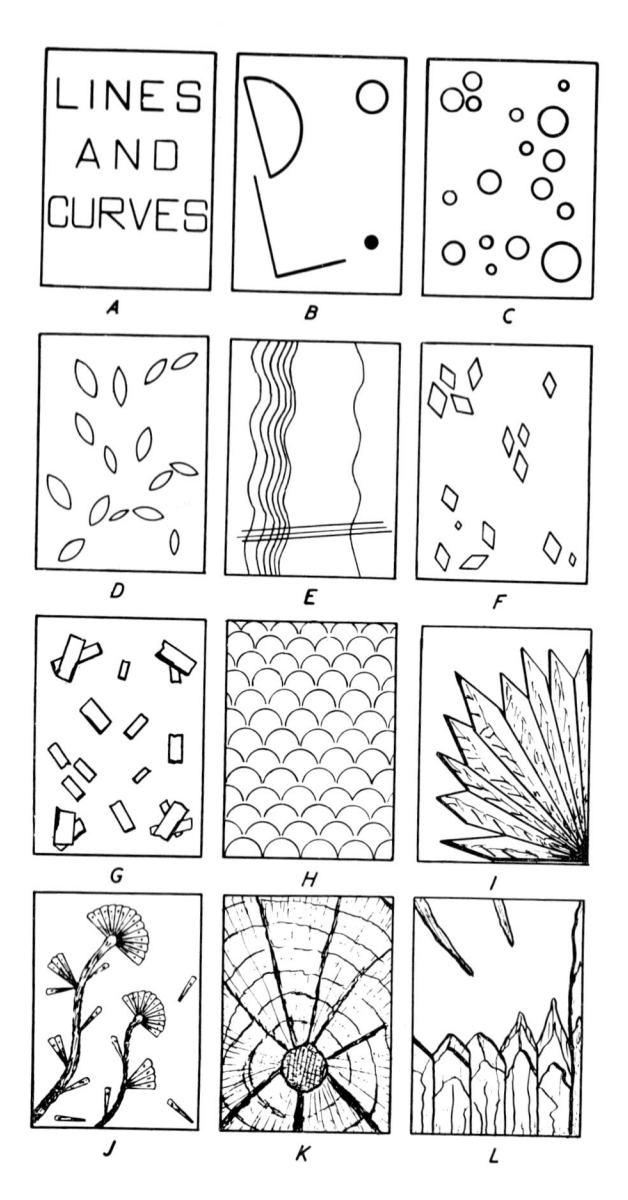

Fig. 237.

as to the painter. An artistic photomicrograph will reveal, underlying it, a definite plan not a haphazard jumble. Composition is a governing feature of this type of photography. It is something like technical craftsmanship, a means of expression, yet not the end: the combination of craftsmanship, composition and expression, make the whole. Take away any one of these parts and the picture becomes unbalanced. The skill of composition cannot be learned

THE AESTHETIC AND PICTORIAL APPLICATION

but comes of the imagination and sensitivity of the worker. Tutors can only inspire. Broadly speaking, the possibilities in this field are far more exacting, more rare, and more restricted, than in photography. The artist's personality will be stamped upon his work and what is conveyed to others will be his vision.

Lines of a repetitive nature like a mansard front of a house, or a few well-placed objects with flowing lines to add up to a pleasing composition are fields which intrigue the eye. Masses, like insect eggs or scales on the wing of a moth, must follow a particular pattern. If they do not the picture becomes monotonous. Sometimes it is possible to isolate, say, a single egg from the mass which should be well positioned away from the rest. On the other hand, if the shapes or particles of a mass have individual interest, one can usually accept them without being disturbed by such overcrowding.

Lines and dots are fundamental to composition in photomicrography. The microscope reveals that circles, ellipses, S shapes, crosses and hexangular patterns, form the most common shapes. The simplest are the symmetrical but composition here can be too obvious. Asymmetrical bodies are generally most interesting. The success depends largely upon the "correct" placing of lines and dots in the picture format, so that a wholeness is formed. Wholeness as such is not enough. There must be tension and diversity around and within that wholeness.

The simple drawings (Fig. 237), will, it is hoped, express the author's approach to pictorialism in photomicrography. At a glance we can read A, which consists of lines, the expression of which is understood. The same applies to B, but in addition to the letters there is a triangle which holds these together. This is brought about by the inclusion of the strong dot, which could be compared to a full stop in punctuation. The inclusion of a not so important circle at the top adds to the balance or design, and this is, shall we say, secondary to the full stop.

Moving to C there is a field of circles or spheres in which a triangle is concealed (follow from the top right to the bottom left and straight along to the lower right). Link these up and there is the framework. There is also a less spectacular triangle on the other side, and a small triangle at the top left.

Flowing lines always excite the eye and here in D are several, the most predominant being from top right to lower left. Often an S line is referred to as the "line beautiful". There are, of course, other constructional lines in this format.

Vertical lines add support to a picture, and curving lines such as seen in E have a rhythmic feeling about them. It is obvious that a glance at the top is enough to attract the eye and almost at once it follows the lines downwards until the "stop" is encountered. Here the horizontal lines cause the eye to rest, without which the eye would run right out of the picture. The single line on the right adds to the balance of the picture. Similar lines to these, incidentally, can be found in sections of wood and fibres.

There is no need to flood a picture with lines or shapes which follow the same pattern, as seen in F. A sparse field like this is sufficient to give the chemist a clear and precise picture of his crystals and these make a cross when all four corners are joined together. They are also interesting colonies. It is always difficult to know what to do with crystal shapes such as we see in G. A successful pattern is the forming of a strong "box" and the odd crystals are then held together. Never be afraid to leave blank spaces. It is a mistake to fill in with a little bit here and a little bit there. The blank spaces enhance the subject and attract attention to the right quarter. Sometimes the edge of the field will offer great possibilities.

Butterfly, moth and scales of fish, comprise the picture seen in H. This is an acceptable mass, which makes a whole. Take even one scale away and there would then be an incomplete picture.

The use of diagonal lines which form a fan, I always seems to give a feeling of movement and, a certain amount of satisfaction. From our earliest days we are intrigued by shapes, similar to those seen in I, and how important the lower right corner is. Many crystals give us patterns like this, and on occasions it is possible to complete the picture successfully by including the tops of a similar fan from the top, but if added here it would spoil the balance.

Curving lines tend to take the eye from one point of interest to another J. Curves are always more beautiful than straight lines, but a photomicrograph composed of curves alone would not be satisfactory. It is necessary to put in a straight line here and there to pull the picture together. In J there are two pleasing patterns—curves and fans. It is very important to add balance to the picture by the inclusion of some small related part, so as to form a coherent story, as seen here by the inclusion of the three frustules placed singly.

It is always a problem to know how to present a transverse section of a stem. It is not at all difficult to place the centre of the stem right in the centre of the picture area, but how unimaginative this can be. In K, the eye is directed and kept to the heavy circle, which represents the centre pith of the stem. The radiating lines contribute much to the picture as a whole. A similar pattern or quadrant could have been used as in I. It would allow a greater area of the perimeter to be included.

L includes verticals and diagonal lines. This gives a feeling of movement, while the verticals have an atmosphere of strength and height. Crystals frequently offer this fence-like pattern, but something has to be introduced into the picture for balance and this has been achieved here by the inclusion of the diagonals.

It is realised, of course, that it is impossible to find fields exactly as drawn, but these have been included simply as examples of analysis, and in the hope that all photomicrographer enthusiasts will benefit in some small way.

Colour composition. Not so many years ago the photographer crammed as many colours as he could into his photograph. This perhaps appeared to be the thing to do then but it jars with us, and the effect could very well be compared to many people talking loudly at the same time. Today we are working on different lines, those of colour selection, colour harmony, and colour contrast, in many cases controlled by the use of stains. We must also recognise colour stains which both dominate and subdue, and apply them to the finished print and transparency through the specimen. Simplicity and economy in colour is the best formula for producing an acceptable product (Plate 74). Nature is the best school in the world for colours and one can perhaps learn more about colours from the study of flowers and foliage than from anything else. When viewing green for long periods we gradually lose its purity or saturation unless red is also in sight, but by showing green with red, the blue and yellow rays are concentrated in one and the red rays in the other, and each appears at its full value. It is essential to understand the blending of colours, as well as the proportions of colour for it is very easy to spoil a good effect by an overdose of a certain colour.

Colour contrasts can be divided into three classes: contrast of tones; contrast of hue; and contrast of tone and hue, a combination of a tint and shade of the same colour. For instance, a light green and dark green give us a contrast in tone, but a combination of two distinct colours of the same tone such as a pale orange and pale blue will produce a contrast in hue. A combination of opposite colours, one light and the other dark, such as purple and light orange, will produce what is called a contrast in both tone and hue. In addition, black and yellow, brown and yellow, and brown and pink, unite to make a pleasing combination. As always, one must be cautious in the use of reds. Black is a good neutral and the lowest tone of all colours. Nevertheless, it can be extremely attractive to the eye if wisely used for wherever it is placed it will "tell". It can be used with rich colours, blending or breaking into what would otherwise be a monotonous picture. At the same time, black can be one of the best means of expression at the disposal of the photomicrographer. It can create a deliberate contrast and display shapes as well as you would wish to see. But as with all colours it should be used only sparingly.

Exaggerated colours should be avoided. Capture texture; display detail in the masses and fine lines; avoid complex colours; get to know what your light source and film will give you; use filters sparingly and aim towards perfection.

Colour versus black and white. There is perhaps a feeling that everything will soon be photographed on colour material but, in fact, this may not prove to be the case. Black and white photomicrography is firmly rooted as a recording medium and offers services which colour cannot hope to equal. Colour prints and transparencies which follow the monochrome pattern can be the most pleasing of all. This may, of course, be due to the fact that monochrome has always been with us. Kentmere tinted bromide paper will add interest in many cases. Autone tinted paper is similar but restricted in cut sizes.

The higher the magnification the less saturated becomes the colour, especially when using transmitted illumination. The incorporation of high-power objectives and a substage condenser increases the light intensity and reveals some colour objects as if they were transparent. Colour material does not give contrasts in delicate renderings as do the various black and white negative material. We have not yet reached the stage when various contrasts are available in colour film, as in the long-lived black and white film.

Colour reproduction. When photographing minute subjects on colour material a good quality image filling a large format is always better than a good quality image filling a small format. Any experienced photomicrographer can tell immediately he sees a reproduction taken on a 35 mm film. Colour photomicrographs on a 35 mm film can suffer from grain and quite often lack definition. Any subsequent enlargement is empty magnification and results in an all-round loss of fine detail. If the photomicrographer is to contribute in the light of his habitual "close-up" way of looking at things, he must always strive for quality production.

Essentials. Design legibility is of primary importance if the picture is to succeed at all. Confusion sometimes occurs through overcrowding: this can apply equally to similar and dissimilar shapes, for example, a picture with squares intermixed with, say, a delicate fern-like pattern.

Spacing of objects, or the grouping of objects, should be considered so that, if possible, a theme is formed.

Style and character. Each object should have an appeal of its own. Shapes have to be considered as individual units which will develop into a whole and be pleasing to the eye.

Cohesion. Our pictures must also show signs of cohesion. For a state of union of the particles of a mass to exist the picture must possess the following ingredients: *Climax*: a wholeness; *Contrast*: not only tonal values as such, but objects which exhibit differences must add up to make a wholeness; *Repetition*: characterized by the repeating of a strong object, not necessarily the same size or colour, but following the same shape.

Synclinal: a term which is rarely used—acceptable, supporting, evidence of sloping objects which fit into the whole and are usually found among crystals.

Choice of field. Sometimes photomicrographs come easily with little or no searching. On occasions attractive pictorial fields are found but cannot be used because they do not contain enough scientific information, or perhaps somewhere in the picture debris is to be seen, which would constitute an eyesore. When this happens it is advisable to make a new preparation. Exciting fields around the edges of most preparations can be found and as a rule these can be put to a good use.

THE AESTHETIC AND PICTORIAL APPLICATION 297

Format. Always try to fill the negative area and formulate the picture on a ground glass screen or viewing eyepiece, using these like a painter uses the whole canvas format. A secondary after-enlargement can be carried out if caution is exercised.

The filling of the negative area with a composed picture is usually brought about by using an extendable camera bellows (35 mm or $\frac{1}{4}$ plate), which can be moved to or from the eyepiece, but the distance of one particular bellows length rarely suits two fields. Some slight change usually has to be made in order to fill the format. The use of a 25 mm objective and a $\times 6$ eyepiece gives magnifications from approximately $\times 15$ to just over $\times 100$. This is achieved by extending the bellows to approximately 70 cm (Fig. 84). It is quite easy to vary the magnification between these two extremes without making any change in the optical system. This also helps to maintain a grip upon the exposure times, especially important with colour material.

It is unfortunate that a 35 mm camera attachment (Fig. 81) does not allow us the fine adjustment in image size equal to the adjustment of the 35 mm bellows type. Instead, the initial magnification is controlled mainly by changing the optical system, say, by using a series of objectives in conjunction with one particular eyepiece (Fig. 86). On the other hand, a specific objective could be used with various eyepieces to vary the magnifications (Fig. 85). The Exakta offers a "ground glass" on which to formulate the picture, whereas some other 35 mm cameras do not. In such cases the field photographed is viewed through the viewing eyepiece. The Leica bellows attachment does give a magnification of $\times 6$ which is a great help in the formation of a composed picture. The Nikon offers two 35 mm cameras attached to one head, one camera loaded with colour film and one with black and white film.

Opaque subjects. Subject and background should harmonise (Plate 74) if the picture is to be successful. The background is therefore almost as important as the subject. Newly laid green moth eggs contrast against old weathered timber and look very effective against the roughness of the wood. Smooth surfaces usually require diffused lighting, while on the other hand rough surfaces call for a strong low side lighting, referred to in this work as grazed illumination, Fig. 88 (h) and Plate 2. Every subject should show one predominant colour, which will impart the necessary balance. Full use must be made of the illuminant, so as to create highlights here and there which will separate subject from background. Where too many objects are disturbing it is best to illuminate the background more forcefully.

Optical limitations. In order to obtain greater depth of penetration (sharpness), comparatively long focal length objectives have to be used. These could range from 25 mm to 100 mm.

The use of these objectives does three things. It permits an increase in the depth of field in relation to magnification, decreases the resolution, increases the area of field covered and thus produces a low power magnification. A decision must be made as to how best these points can be applied when making a picture. Printing by enlargement from a negative of low magnification adjusts the depth of field in direct proportion to the secondary magnification, but since the negative image is a two-dimensional object, with virtually no depth, the image produced by enlargement also has no depth. Do not imagine from this that secondary enlargements cannot be undertaken, but remember to exercise caution. If a whole plate print puts over what you wish to convey, then further enlargement is not necessary.

Maintenance and care of the microscope

The microscope is a precision instrument and must be treated with respect. When not in use it should be covered up to prevent dust and fumes from affecting it. Dust can act as an abrasive and cause deterioration.

The optical parts of the eyepiece, objective and condenser should not be tampered with, nor should dust on any account be allowed to enter into the internal parts. Do not clean the optical parts unnecessarily. If the specimen image appears to have deteriorated in any way check the quality of the specimen, mounting medium and slide and try another objective and compare images. Out-of-focus specks appearing in the field of view are usually caused by dust or particles of wool from clothes which come into contact with the apparatus. Glass surfaces should never be touched with the fingers, as these will leave greasy smears. It is even possible that corrosive perspiration could damage the instrument if left for a long period.

Most moving parts within the microscope are protected by a permanent type of lubricant and, therefore, one should lubricate very carefully. External moving parts can be dusted with a fine dry camel hair brush. Carry the instrument with both hands to avoid any sudden jolting.

The lower lens of the objective should not be allowed to collide with the top of the cover glass or slide, as this would cause damage. Optical parts often become damaged because the operator has looked down the tube when focusing towards the specimen and, if he goes through the plane of focus unnoticed, a collision will occur. Great care should be taken when handling liquids, especially those which contain acids: after use make sure none is left on the microscope. Optical glass becomes tarnished very quickly if in contact with chemicals and chemical fumes which can cause oxidation on the lenses. They should be protected against this and properly cleaned before being put away. When not in use, eyepieces should always be kept in their cases.

As some optical parts are soft they are susceptible to scratching. Particles of grit can adhere to a handkerchief and be transferred to the instrument, if such a thing is used for cleaning purposes.

Lenses get particularly dirty when oil is used and, therefore, it is most important that the lens should be cleaned immediately before putting away. If this is neglected the oil will dry and become difficult to remove, and an even greater risk of damage to the delicate lens would ensue. A *little* xylol, or alcohol, is an excellent remedy to remove oil. Dust should be removed and not blown from one place to another. The U.N. Auto-Cleaner (Plate 63) is a midget vacuum cleaner with a brush of fine camel hair and is very useful. When removing the substage condenser a thick duster should always be placed beneath as a precautionary measure. If always looked after properly the microscope should never be in need of repair and, if one is to get the best from the instrument, it is necessary to keep all movable and optical parts perfectly adjusted.

Cinémicrography

The past few years have witnessed a multiplication of the number of amateurs making use of cinémicrography, and scientific laboratories in which cinémicrography is being carried out has increased. This technique is the best way of recording moving, expanding and retarding materials (Plates 62c, 64 and 65). Fusion methods in clinical microscopy are revealed in their full value: characterization and identification of fusible compounds, determination of phase changes, purity determination, qualitative analysis, quantitative analysis of various systems, study of mechanism of crystal growth, a study of other changes in the solid state such as recrystallization, grain growth and boundary migration and air–water interface. Usually when recording bacteria and some micro-organisms a thin layer of inoculated agar on a slide is overlaid with a cover glass or Teflon membrane, fastened to the slide with Vaspar or similar adhesive. The specimen thus sandwiched, is bound to be in focus even while still growing. Sometimes this technique hampers true growth and results can be non-representative.

Other techniques themselves include observation made during heating of a few milligrams of material on a "microscope slide". These observations are made during solidification of the melt, cooling and reactions to room temperature. Cinémicrography offers a complete optical crystallographic description of a given compound and may require any time from a few minutes to hours, or even days, in which to be completed. A hot stage is not absolutely necessary for many subjects, although it is extremely useful for more accurate determination of melting points. Griffin and George (Wembley) have developed a carefully designed hot stage microscope. The specimen is mounted at the top of a thermocouple of platinum–rhodium alloys which is used to register the sample temperature. Confirmation of a particular crystalline compound present, is usually obtainable by "freezing" the melt. The small thermal capacity of the sample enables cooling from 1700°C to 700°C in 0·5 second after the current is switched off. The "frozen" material can be readily examined and, in the case of metals, X-rays can be taken.

The advent of cinémicrography and the increasing use of tissue culture techniques in research, prompted the development of a tissue culture chamber by the Imperial Cancer Research Fund (Roberts and Trevan, 1961). The tissue culture chamber is now marketed by W. R. Prior and Company Ltd., of Bishops Stortford, and has been designed for the continuous filming of the specimen with microscope observation of cells and tissues in culture. Before cinémicrography can be embarked on, living cells from euthermic animals must be kept at a temperature which they are accustomed to before explantation so this chamber must be housed in a room or cabinet with an accurately controlled temperature. (See Prior cinémicroscope, page 30.) This chamber combines useful features from many of its predecessors but with some unique advantages. For example, the Prior chamber allows for prolonged observation and filming, and also replenishment and exchange of culture medium. The joint authors already referred to, have mastered this technique and have made continuous time-lapse films for over two months at intervals of 24 secs. When feeding or changing the culture medium, syringes are used to empty or fill the chamber by needle holes through the housing and silicone rubber rings. It is also possible to maintain a continuous gas-flow within the chamber. Cultures may be fed, removed or exchanged without interrupting viewing or cinémicrography.

The interior of the chamber communicates with the atmosphere by virtue of a small water seal positioned at the end of the gas exit line, giving a constant small difference in pressure. This is due to water in the seal between the inside and outside of the chamber, and it is interesting to note that the changes in barometric pressure do not affect the focus in any way, even over a long period of filming. Although the chamber was designed for use with the Prior Inverted Microscope (Fig. 238), it can also be used with a conventional erect microscope for the examination and filming of cells adhering to the top cover-glass.

The incubation chamber devised by Casida (1972) uses a microscope slide 38 × 76 mm (1·5 × 3 in.) through which a hole 1·9 cm (0·75 in.) in diameter has been drilled. The slide is placed in a petri dish with a thin covering of agar. The hole is covered with Teflon, overlaid with a cover-glass which has a microbial culture prepared as a smear, facing downwards, as is a cut mark on the face of the glass. This acts as a point of focus when filming. The drilled area of the chamber has the following layers over it, starting from the top: cover-glass (including the point for focusing purposes), smear, agar, and Teflon sheet. Microbial growth in the microculture chamber is unhindered and filming can take place as the organism grows.

Filming the growth of moulds or crystals growing on slides, such as illustrated in Fig. 264, does not present any difficulties for cinémicrography when time lapse is best suited. Yeast cells growing in gelatine in a petri dish, or in a thin layer of gelatine placed on a slide, can also be filmed through practically any microscope. A viewing eyepiece must be incorporated into the set-up. A

CINÉMICROGRAPHY

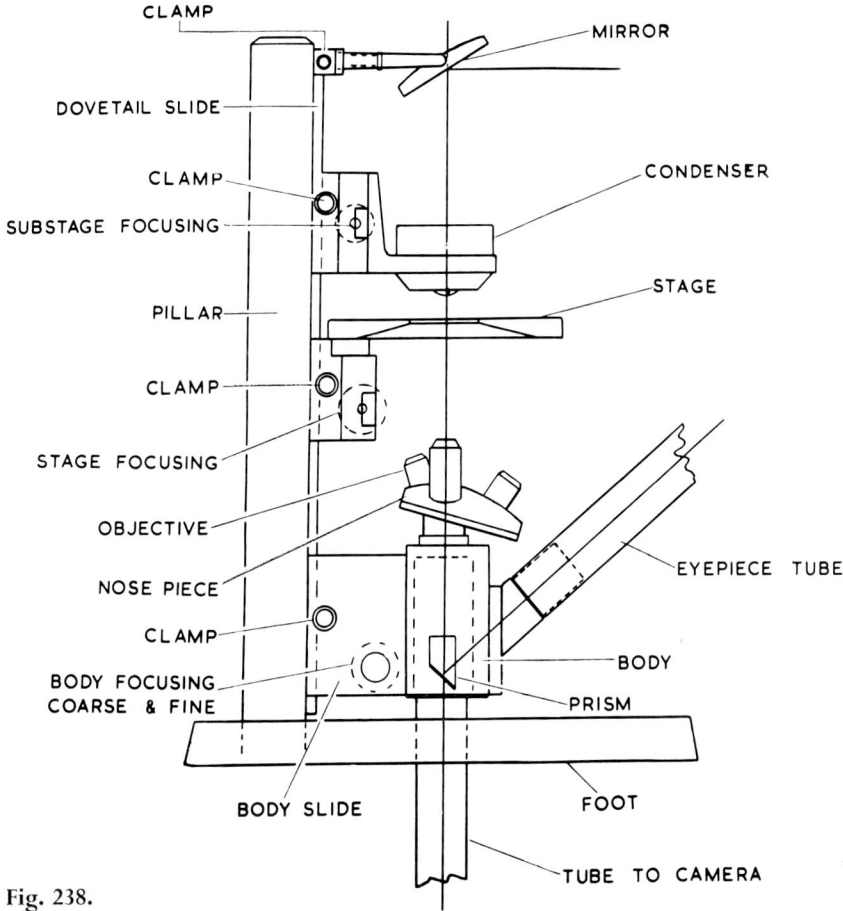

Fig. 238.

controlled room temperature is necessary for both specimen and the normal care exercised in all the departments of the microscope and camera. There are, of course, more subjects which lend themselves to being photographed when illuminated by transmitted light.

Cinémicrography offers much to further education, the filming of techniques and practical applications of photomicrography, interference, polarized light, phase contrast (to mention only a few), offer great possibilities and would offer more useful instruction to students than hitherto.

Filming opaque moving subjects (Plates 57 and 65) presents problems connected with illumination and depth of field. Moving insects, caterpillars hatching from their egg cases, demand an instantaneous exposure, such as coupled flash. Prolonged heat emitted from a constant source would obviously kill or damage the specimen or cause it to seek for cover. In such cases a weak pilot light is recommended for viewing purposes, making exposures with flash.

A strong continuous source can be used when filming metals stress (i.e. where a crack is followed) or melting rubber and fibres under different temperatures. A major problem connected with photographing a crack in metal whilst under pressure from hammer blows, is where to stand the microscope without vibration. Such dislocations usually commence slowly developing into a rapid conclusion and, to make recording easier, Prior manufacture a very strong supporting pillar with three pronged feet and an adjustable arm to house the ciné camera. With this arrangement the ciné camera can be placed close to the eyepiece, vertical, inclined or horizontal, without causing difficulties in operation. Photographing and following a crack developing into several inches in length often demands improvised techniques. Here the camera, simple microscope (less stage) and light source must operate as one unit. For this the author used the movable dovetailed top of a heavy lathe. A modified ex-R.A.F. G.45 16 mm camera gun was used. The film magazine housed 25 ft of film which was wound over by an independently operated motor. A viewing eyepiece situated in its original position assisted in following the zig-zag course of the crack while an additional pair of hands maintained good focus. A ×4 eyepiece was used at 180 mm tube-length and focused to give almost parallel rays to the ciné camera lens which was focused at infinity. This automatically increased the latitude permissible between camera and eyepiece, since the virtual image "seen" by the camera is at infinity. A "light-tight" tube was placed between the eyepiece and camera lens. A viewing eyepiece could have been inserted between these optical points had it been found necessary. Continuous even illumination was carried out by the use of an intensity lamp which moved with the "bench top". Such a ciné operation can be likened to photographing a rugged coastline from the air, covering it by a feature line-overlap, when the camera moves along over the subject. A ciné camera fitted with a zoom lens giving a short focus can assist in recording "close-up" images which, when projected, can be seen at a higher magnification.

The marine biologist will have problems peculiar to his subjects and demands. Here, free swimming creatures must be followed, and thus a two way (vertically and horizontally) means of focusing must be mastered while filming. As a rule, the illumination has to penetrate a glass container and here the inverted microscope offers much over erect models.

The integral substage illuminator (*Beck*). This is an extremely useful piece of apparatus for this application of photography. The illuminator is a complete lighting unit replacing the mirror, and being fixed at the base of the mirror housing. This long awaited piece of apparatus has a permanently adjusted lamphouse, housing a 6 V 15 W pre-focused bulb from which phase contrast, polarized light techniques can be applied. The illuminator is fitted with an iris

CINÉMICROGRAPHY

diaphragm which is imaged by the optical system of the substage condenser, on to the object, achieving Köhler conditions of illumination.

The Vinten 16 mm Scientific Ciné Camera. W. Vinten Ltd. produce several 16 mm ciné cameras which enable continuous and time lapse studies, including high speed, producing 250 pictures per sec. The Mk. 1A camera is constructed of aluminium alloy die casting and comprises a camera body and a detachable magazine fitted with a re-loadable cassette to accommodate 50 ft of double perforated 16 mm film. A magazine which will accept 200 ft of film wound on a daylight loading spool is also available. The camera houses an electric motor drive which permits a free flow of film without the drawbacks of manually wound cameras. The camera can operate at speeds of 8, 16, or 24 pictures per sec; in addition to this, there is incorporated a single-frame mechanism triggered by short pulses of direct current connected to a separate control circuit in the camera. The camera is operated by a direct-current of low voltage and can be fitted with either 12 V or 24 V D.C. motor. A tripod bush is provided in the case of the camera body; a special adjustable mounting for housing the camera is also available, and this fits the Vinten camera support and stand (Plate 66). An event marker is supplied as an extra. It is a small retractable pointer which projects into the camera film gate, recorded as a small square in the corner of each frame, and is brought into operation by an electro-magnet positioned inside the camera body. When using this camera for cinémicrography a viewing or watching eyepiece is used (Fig. 239a). This is quite complete and is screwed to the

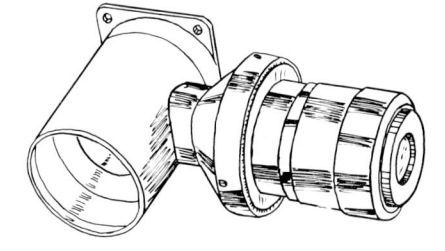

Fig. 239a.

camera, replacing the original camera lens. It is permanently focused and no other adjustment is necessary in obtaining a sharp image on the film, beyond the focusing of the microscope. There is a fine graticule within the eyepiece which shows the exact centre of the frame and the actual area of the recorded picture. There need be no anxiety about parallax, for the total area of the microscope image shown is the same as that recorded. The eyepiece attachment is fitted with a beam-splitter which enables 80% of the available light to be recorded by the camera and 20% is directed to the viewing eyepiece. The specimen image, of course, can be viewed while filming.

In addition to the mains power unit there is an Interval-meter, a mechanical device for feeding impulses in order to obtain single exposures at accurately controlled intervals. A cam mounted on a minute spindle operates a microswitch, which in turn controls the time intervals. A set of cams is provided giving periods of 2, 5, 15, 30 and 60 sec and also 2, 5, 15 and 30 min. Time exposures can range from 1/50 sec and, to provide further assistance in poor light conditions, synchronized flash contacts are included.

Vinten Mark 2 camera. This camera has an optically worked stainless steel shutter giving a true reflex image, which is also seen during the shutter closed part of the cycle. Like the Mk 1A it is fitted with a button switch when hand held but is electrically controlled for cinémicrography. The viewing eyepiece can also be used and, as illustrated, is fitted with a light trap collar which fits freely around the microscope eyepiece. There is little fear of the Vinten ciné camera breaking down—it has been run under all conditions for indefinite periods, as at the Division of Experimental Biology and Virology of the Imperial Cancer Research Fund, where some 12,000 ft of film were successfully exposed at time-lapse speeds without defects.

The Vinten 16 mm high speed camera is designed to operate at between 100 and 250 pictures per sec. This means that, with the film in a standard projector working at 16 p.p.s. the actual time taken for a physical event to occur can be magnified by any figure between six and fifteen. Consequently, slow-motion can often render the smallest and most unobtrusive of movements. A small neon lamp, housed in the camera body, records a time base along one edge of the film. Provision has also been made for data or identification marks to appear on any required frame. The camera can be used in very low temperatures owing to an enclosed thermostatically controlled heater.

Cooke, Troughton and Simms have produced specially designed time lapse and ciné equipment, incorporating the Vinten 16 mm ciné camera with their standard microscope. The time between exposures ranges from $\frac{1}{4}$ sec to 1 hr. In addition to this, individual manual exposures can be made as well as 8, 16 and 24 pictures per sec. The exposure times can be set on the electronic unit, ranging from 1/25 sec to 1 sec. Cooke, Troughton and Simms also market an automatic integrated timer, coupled to the Kodak Colorsnap Camera. This is a simple camera, which relatively unskilled hands can operate, incorporating 35 mm film of high film speed. The automatic integrating timer and time lapse equipment was developed in conjunction with Dr. R. Barer and Dr. J. Underwood of the Department of Human Anatomy, Oxford University.

Wild Heerbrugg Ltd. (Heerbrugg, Switzerland) produce a comprehensive outfit for ciné and time lapse recordings. The equipment is shown in Plate 66, and includes friction drive for raising the camera with a motion of over 51 mm (2 in.), an anti-vibration plate (obviating the transmission of vibrations from the bench to floor to the microscope), ciné camera, microscope, camera stand (bench model), Wild Universal microscope lamp incorporating xenon or mercury vapour high pressure source. In addition, a low voltage 6 V 30 W lamp is available. Fitted directly in front of the lamp-house lens is an electro-magnetic shutter which automatically prevents the light from reaching the specimen during the intervals between exposures. This is a safeguard against distorting or killing the delicate subject by a strong light and perhaps heat, especially over a long period.

The shutter is controlled from a pulse transmitter and can operate at single pulses, giving short exposures to $4\frac{1}{2}$ sec: but when free running 2 frames a sec are taken, with an exposure time of 1/25 sec. The pulse transmitter is connected direct to the mains and provides three pulses, synchronized to each other. The sequence is as follows: the first pulse switches the lamp on, the second pulse opens the shutter and the third sets the camera in motion. The rest period is preceded by the reverse cycle which prepares for the next exposure. The number of exposures or frames is recorded on a frame counter built into the transmitter. The automatic sending of pulses (that is, the intervals between pictures) ranges from 1·5 and 1000 sec, with an average projection rate of 20 frames per sec.

With the control gear Wilds have produced a special attachment for ciné-micrography which enables the operator to have greater control over the filming: incorporating framing, illumination, focus and placing of specimen to produce a more useful and intelligible end product. The special attachment (Fig. 239b)

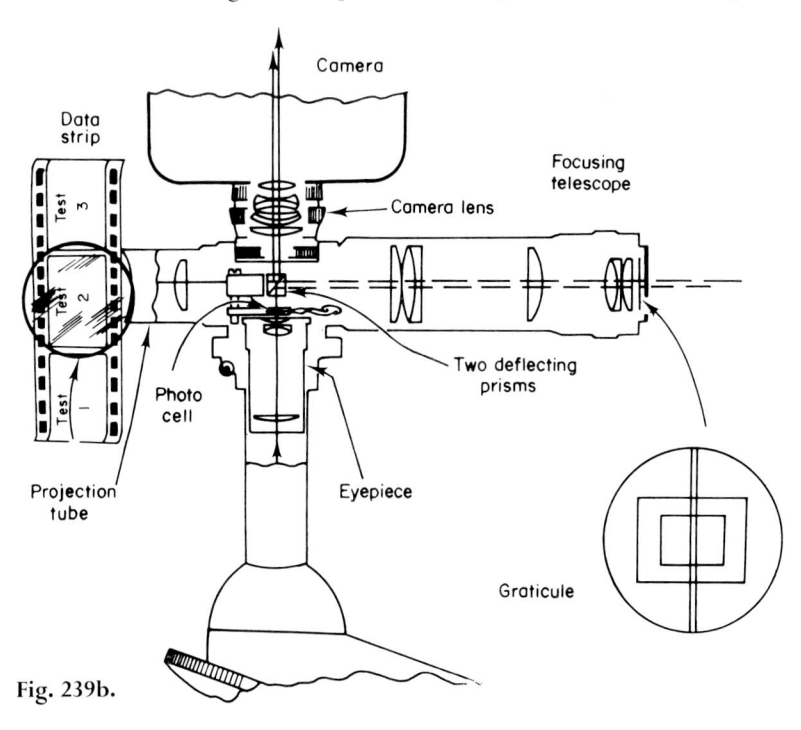

Fig. 239b.

consists of a focusing telescope, two deflecting prisms transmitting 95% of the projected light to the camera and 5% to the focusing telescope. The attachment also allows for 50% each way when using low level illumination, but the shot is always taken with the 95:5 prism. There is also a built-in photocell and a projection tube.

When using the attachment the camera, complete with lens, preferably the 50 mm, focuses on infinity. A 75 mm lens does not give a complete coverage per frame, and a light cutting off at all corners can occur. The camera lens shunts almost up to the microscope eyepiece. The Vinten Mark 1A, Bolex, Kodak Special or a similar camera can also be used like this. The projection tube enables necessary data or identifying marks, arrows or numbers to be filmed at the same time as the specimen, thus making scientific evaluation and recognition of the relevant footage simpler. It is useful for editing too. The film does not end when the last exposure has been made, for it has to be clearly annotated. Editing a film is an important feature, and is not merely sticking pieces of film together, but a putting together of facts clearly presented.

The focusing telescope is provided with a graticule of the cross hair type, plus two rectangular areas. The larger rectangle is for 16 mm film when the $f = 50$ mm lens is used on the motion picture camera. The inner rectangle is used with a $f = 75$ mm lens. The built-in photocell enables the illumination to be monitored via a mirror galvanometer and, when in position, the photocell clicks into the "out" position, thus leaving a through way for the light rays to reach the film. An electric motor specially designed for use with the Bolex H 16 ciné camera is connected to the 8 p.p.s. shaft of the camera. The Vinten Mk. 1A 16 mm scientific ciné camera can also be electrically powered.

A wall bracket support (Fig. 239c) is a useful addition and offers much to those working in scientific films or research where space is specially allocated to cinémicrography. Here there is no fear of vibration caused by the running camera and there is added support when manually winding and loading.

Wild automatic camera. This is a motor-driven camera using 35 mm film. The camera is used with any microscope of normal tube diameter or with any stereomicroscope of tube diameter 33 mm and has a vibration-free electromagnetic shutter which is operated from a control unit.

Robot camera (Telford Products Ltd., London). This is in great demand in research, and is proving to be an essential piece of recording apparatus. Star II offers 55 exposures 24×24 mm, while the spring motor will make 18 to 50 successive exposures per winding. The camera has a fully synchronized shutter speeded at $\frac{1}{4}$ to 1/500 sec.

The Pelco (Pelling and Cross Ltd.) 12 V time-lapse controller, accumulator operated, was originally designed for emitting a power pulse to an actuator, for making exposures at regulated intervals with a ciné camera. This unique piece of apparatus will switch power from the accumulator at intervals of 1 sec,

CINÉMICROGRAPHY

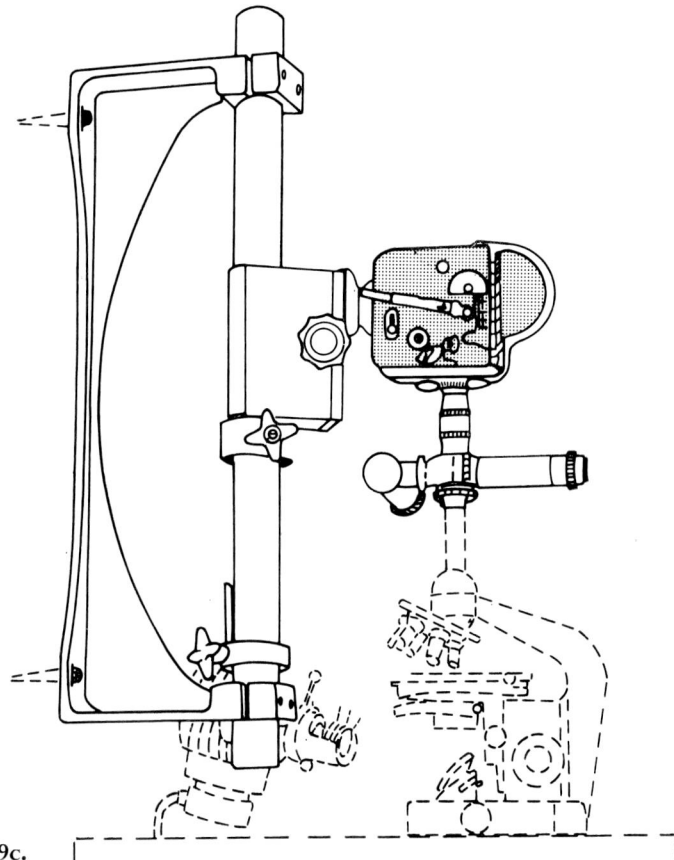

Fig. 239c.

the pulse being adjustable in duration from 100 to 500 ms. With the addition of an extra fitting the standard controller can be adjusted to give further intervals. This controller is also available for AC mains single-phase 220/250 V emitting a pulse of 8 V DC at any chosen interval between 1 sec and 10 sec. The supply from the controller can be connected directly to the Pelco/Bolex H.16 Camera Actuator Mk. III which is secured to the camera. The controller and actuator together will make the exposure of 1/25 sec and wind the film to a maximum camera capacity of 4,000 frames at any pre-set time interval.

The Prior Electromechanical Timelapse and Camera Control Unit for the Vinten Mk. 3 Scientific Camera and other 16 mm cinécameras is a compact piece of apparatus taking up very little room. This unit will operate an electrically driven ciné camera under conditions of continuous running, single shot and time exposures, either by manual control or automatic control, at predetermined

L*

intervals. The design of these units has been improved by using interchangeable plug-in parts for ease of servicing and flexibility in the range of automatic cycle times. Used with a Vinten Mk. 3 Scientific Camera the control unit provides the following features:

(1) Low voltage power supply to operate the camera for continuous running, single shot and time exposures.

(2) A stabilized electronic exposure control circuit giving time exposures of $\frac{1}{4}$, $\frac{1}{2}$ and 1 sec to 9 sec in 1 sec steps.

(3) A frame counter which may be reset to zero, operated by an impulse from the camera shutter, recording the number of frames actually exposed.

(4) A relay operated by the same shutter impulse provides, to outlet sockets on the front panel of the instrument, a circuit to operate an electronic flash when the shutter is open and a circuit to operate a light shutter when the camera is in use with a high intensity light source.

(5) Operation from the control panel of the camera event marker.

(6) Automatic cycle time control. When used on automatic control to take pictures at predetermined intervals, the beginning of each camera exposure is arranged to coincide with the beginning of each cycle time period so that exposures may be adjusted without altering the interval between each picture. The cycle time control will operate the camera automatically at intervals ranging from 1 to 9999 sec, adjustable in 1 sec steps with digital readout on the front panel of the instrument. (A Bolex camera can be used with this unit by the addition of a motor drive attachment.)

Prior cinémicroscope. This is an inverted microscope housed in a transportable self-containing cabinet (Fig. 240 and Plate 66c) and provides for accurate control of the temperature of the specimen, permitting recordings by continuous or time-lapse 16 mm cinémicrography of growth in any time. The instrument has been developed, after much research, to the specifications of the Imperial Cancer Research Fund (Pybus, 1960 and Pybus, *et al.*, 1961). The complete instrument (Plate 66c) is pleasing in appearance and of up-to-date design.

The cabinet housing the microscope is of sufficient size to allow easy manipulation of microscope during filming and general attending to the specimen within the Prior Chamber (already referred to in this section). The front and top of the cabinet has an air spaced double wall of perspex and the rest of the cabinet is thermally insulated. The temperature inside is maintained by forced circulation of warmed air passing along the cavities, round the back and sides of the cabinet, through the inner compartment via heaters controlled by an adjustable thermostat and fan. No cold air can enter the cabinet. Spent air is extracted at an even rate, the temperature of the interior is controlled to within $\frac{1}{2}°C$ of the required temperature.

Exposure timing is controlled by selector switches, giving any exposure between $\frac{1}{2}$–19 sec. The cycle time controls are graduated in a like manner but the

CINÉMICROGRAPHY

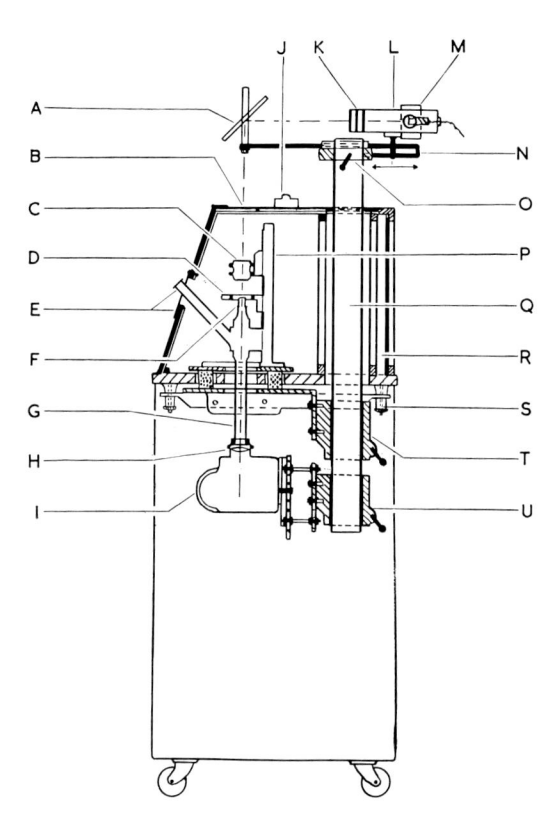

Fig. 240. Prior I.C.R.F. inverted microscope unit. (a) Reflecting mirror. (b) Transparent screen. (c) Condenser. (d) Stage. (e) Eyepiece. (f) Objective. (g) Camera tube. (h) Camera lens. (i) Camera. (j) Thermo control. (k) Illuminator filter holders. (l) Illuminator. (m) Cooling cowl. (n) Illuminator adjustment arm. (o) Illuminator locking lever. (p) Support pillar for adjusting optical elements. (q) Main support column. (r) Air circulation cavity. (s) Suspension bush. (t) Main pillar adjustment control. (u) Camera adjustment control.

timing can be regulated between 1–99 sec. Other time controls can be fitted if required. The cold cathode tube, a resistance capacitor timing circuit with neon stabilized power supply, controls the timing which is not affected by mains fluctuations. A relay in the control unit completes the camera shutter solenoid circuit for clockwork camera motors, or supplies a 1/10 sec pulse for an electric motor drive camera.

The light source is situated at the top of the cabinet and consists of a five filament strand, high intensity lamp, 48 W 6 V, forming a 2·5 mm spot. The source is housed in a Prior High Intensity microscope lamp. This is fitted with an iris diaphragm and lamp condenser. Two arms, extending horizontally from the front of the housing, support a mirror in an adjustable bracket which diverts the light vertically through the perspex top, on to the substage of the inverted microscope.

The instrument is a Prior inverted microscope (Fig. 238) with its eyepiece protruding through the front of the cabinet (Fig. 240) and insulated from the outside by a large rubber washer. The bottom of the tube houses a beam-splitter prism which diverts 20% of the light into the viewing eyepiece and at the same time allows 80% to travel through to the ciné camera. The microscope, lamp and camera are mounted on a vertical $2\frac{1}{2}$ in. steel column. In addition to this the microscope is mounted on a $\frac{1}{2}$ in. steel plate which is connected to the main column by a massive right-angle clamping collar. The camera is secured in a similar way, having full adjustment for height and alignment. The flat microscope mounting is attached to the main body of the cabinet by four anti-vibration mountings, ensuring optical alignment irrespective of any movement or vibration of the main cabinet.

The camera can be almost any make but Bolex or Vinten cameras are normally recommended. A correcting lens in the microscope camera tube ensures that the image is in focus in the film plane as, simultaneously, the projected image is in focus in the viewing eyepiece. Being positioned outside the cabinet, the camera can be adjusted without interfering with the specimen or filming.

The timer controls are positioned on the front left panel, this includes exposure time, cycle time and switch for manual or automatic operation. The front right panel consists of volt meter, main switch, rheostat light and heater control, and switches, mains pilot light and booster heater switch.

For an excellent work on this subject, *Cinémicrography In Cell Biology*, edited by George G. Rose, Academic Press, is recommended.

References

Casida, L. E. (1972). Internal scanning photomicrography of microbiol cell populations. *Appl. Microbiol.* **23**, 190–192.
Pybus, W. (1960). *J. Roy. Microsc. Soc.* **79**, 369.
Pybus, W., Roberts, D. C. and Trevan, D. J. (1961). *J. Roy. Microsc. Soc.* **80**, 89–95.
Roberts, D. C. and Trevan, D. J. (1961). *J. Roy. Microsc. Soc.* **79**, 361–366.

Some common faults in photomicrography

From time to time, the author is invited by one or another of the photographic or scientific bodies concerned with photomicrography to judge prints, to lecture on the subject, or to answer correspondents' questions. Frequently, when judging prints, one sees the same faults occurring and invariably in answering questions after lectures it is evident that the audiences are seeking the same remedies for these faults. The following notes, prepared over a fairly long period, indicate the basic recurring faults. They are described here, in the hope that readers whose work includes photomicrography may benefit.

(1) *Uneven illumination.* The form and application of lighting in illuminating the specimen are the two most important factors in photomicrography. (a) Uneven illumination can sometimes be traced to the substage condenser being too high or too low. (b) Or the trouble is due to the light source not being in alignment with the axis of the lamphouse condenser. (c) Or a little more care in positioning the microscope mirror is necessary. (d) Or when photographing opaque specimens with only one light, a white reflector should be introduced.

(2) *Diffraction rings.* (a) These are visible as a series of concentric lines around a specimen, specks of dust, etc. and are caused by the interaction between minute structural detail in the specimen and the wave-front of the incident light. (b) The rings can be caused by the substage iris diaphragm being closed too far.

(3) *Light bar marks across prints.* (a) Probably images of the filaments of the light source and occur when the wrong type of bulb is used. (b) The substage condenser is badly adjusted.

(4) *Bright spot on the print and always in the same place.* (a) May be due to internal reflections within the microscope tube. (b) It could be a flare spot from one of the elements. These faults can be prevented if the tube is lined with thin black paper. A wide draw-tube (50 mm) is recommended.

(5) *Overall unsharp image.* (a) Vibration or glare from the substage condenser which would flood the back lens of the objective may be the cause. (b) Or the body tube sliding slowly downwards. (c) Astigmatism results in the formation of two images, and if so, the final print shows an overall "fuzz".

(6) *Low contrast and poor definition.* (a) This can usually be traced to spherical aberration. (b) Old negative stocks.

(7) *Negative unsharp at one end.* (a) The optical system may be out of alignment. (b) The specimen slide is not lying flat. (c) Refraction from wet areas.

(8) *Unsharp edge zone.* (a) This can be caused by using a tube of the wrong length, or a cover glass that is too thick. (b) An ill-adjusted substage condenser will also produce this fault. (c) The objective may exhibit field curvature.

(9) *Lack of depth of field in opaque subjects.* (a) The introduction of a behind-the-lens iris diaphragm will allow for a much reduced aperture, giving a greater depth of penetration.

(10) *Print lacking detail in dark areas.* (a) The introduction of a correct filter (controlled by the colour of the stained specimen) would improve this.

(11) *Foreign marks, usually unsharp.* (a) These are caused by untidy or dirty specimens. (b) Dust on optical systems.

(12) *Poor resolution when photographing transverse sections.* (a) This fault may be traced to the objective: an apochromatic or fluorite is recommended. The

higher the N.A. the greater is the number of lines resolved. The introduction of a small stop to the substage iris diaphragm reduces the resolving power. Green and blue filters increase the resolving power.

(13) *Dark, uneven, unsharp spots on the print*. (a) These are usually due to dust on the optical system. To check this, turn the eyepiece and objective while watching the image on the ground glass screen. Dust specks will be seen to turn as the lenses are rotated.

(14) *Refraction around the image*. (a) Occurs when photographing wet specimens as they are drying. The specimen must be completely covered with water. (b) Chromatic aberration also produces this fault due to colour fringes which sometimes form around the image. (c) Mounting medium and glass refract colour light rays, varying the point of focus.

(15) *Void magnification*. (a) This is a common fault. The maximum useful magnification is approximately 1,000 times the N.A. of the objective used. To go beyond this often causes a breakdown of the image, especially if fine structural detail is involved. Always work below the limit.

(16) *Unsuitable negative material*. (a) It is advisable to use material which gives the highest resolution. Panchromatic material does not resolve as many lines per mm as ortho.

(17) *Print lacking definition*. (a) Check the quality of the specimen, cover glass and space between specimen and cover glass. (b) Oil or finger marks on the lower lens of the objective will also cause this fault.

(18) *Yellow or amber overall hue seen over the transparency*. (a) Maybe old mounting medium is the cause.

(19) *Indistinct lines with a blur in place of a clear line*. (a) Perhaps an overlarge cone of rays projected from the condenser.

(20) *Dark circular area around the print*. (a) When the substage condenser is lowered in order to illuminate a wider field, it will only poorly illuminate the outer area of the recorded image. (b) If the source is too bright, often the substage is lowered (not a good practice), resulting in uneven illumination.

(21) *Colour transparency too dark*. (a) Due to under-exposure, which means that a lower step number or change in shutter speed should have been used.

(22) *Colour transparency too light*. (a) Perhaps the exposure was too long, a shorter exposure and a higher step number should have been used.

(23) *Sharp black foreign marks on the print*. (a) As a rule these are sharper than the image proper, and are caused by dust on the negative material before and during exposure.

(24) *Sharp white foreign marks on the print*. (a) Due to dust on the negative or enlarger elements during exposure. Again the unwanted marks are sharper than the subject matter and appear to be more important.

(25) *Colour transparency exhibiting a pronounced yellow cast*. (a) It is possible that the light source was not adjusted to the colour balance (°K) of the film.

(26) *Colour transparency exhibiting a pronounced blue cast.* (a) Light source and film not adjusted to one another as in (25).

(27) *Slight yellow cast in colour transparency.* (a) Colour temperature of light source slightly too low.

(28) *Slight blue cast in colour transparency.* (a) Colour temperature slightly too high. A change in correction filter is necessary.

(29) *Colour transparency exhibiting greenish cast.* (a) May be due to a long storage time after exposure.

(30) *Colour casts on the margin and at times seen in the background.* (a) These could be colour borders to the image of the iris diaphragm, if the condenser is not colour corrected.

(31) *Colour transparency too dark in spite of correct exposure time.* (a) Perhaps due to old film stock, hence the loss in sensitivity.

(32) *Structural details of the specimen exhibit colour margins, (blue, red, orange, etc.) at the periphery of the image.* (a) Chromatic aberration. (b) Or chromatic magnification difference through the incorrect eyepiece. (c) Perhaps the various coloured rays of white light are very differently refracted, because lenses work as prisms in this respect: that is to say, they separate the light into the spectrum colours. (d) Geometric errors are caused by the curved form of the lens surfaces, the thickness and distances of the single lenses and the diaphragm introduced. The most important errors like this in microscope objectives are spherical abberation in the axis; spherical aberration in the main ray (coma or error contrary to the sine law); and curvature of the image. (e) Further, an error, which in microscope objectives is of importance, namely, the alteration of the spherical aberration according to the colour, called sphero-chromasy. All these errors have to be reduced so as not visibly to disturb the formation of the image.

(33) *Presentation.* It is important to put one's work over in a successful manner, and very few applicants seem to know how to present their work to the best advantage. Full scientific names, magnification, scale lines and data about the subject should all be given. A line drawing is always a welcome addition, as this provides further information which is welcomed by those outside this field. Last, glossy paper is best for scientific work.

Hot Stages

Carl Zeiss, Oberkochen, have perfected two heating stages suitably made to fit a standard microscope, Plate 63. One model for temperatures

30°–60°C (86°–140°F)

and another, simpler model for temperatures from

35°–43°C (95°–101·4°F).

Both stages are suitable for investigation into metals, organic chemistry and synthetic products.

A brass plate with a 9 cm central heating area is attached to the microscope stage, with permanently fitted contact shoes for cables connecting it to the heating device. There is a built-in control thermometer and magnet for setting the desired temperature of the specimen, either on a slide or in a vessel. The maximum time required for heating up the specimen is 15 min. A plexiglass cover encloses the whole stage while observations and photomicrography can be carried out without obstruction. The set heating values cannot be exceeded, even with prolonged observation periods, unregulated fluctuations of temperature do not occur. Up to 40°C the accuracy of the temperature is $\pm 5°C$. At 60°C the maximum deviation from the set value is $\pm 1·5°C$. The temperature constancy of the heating stage at the specimen is 0·2°C after heating up.

Leitz, Ortholux is a popular microscope for the investigation of temperature-sensitive organic and inorganic subject matter in transmitted and incident light. Four different heating and cooling stages are available for this microscope. For high power magnification it is advisable to use the Leitz special heating objectives, which allow for maximum aperture of N.A. 0·60 giving a maximum magnification of ×600. (See also *Jena Review* (1972) **2**, 75-83.)

Graticules

The term "graticule" is familiar to those practising photomicrography but it may not be to those who are not connected with optics. Graticules may best be defined as being the measuring scales or marks, placed in the focal plane of an optical system, for the purpose of determining the size, distance, direction or number of the objects which are viewed or recorded coincidently with the scale itself. Generally they are on glass and it is usual to refer to the complete discs or plates, with the measuring scales or marks on them, as "graticules". The word is a highly convenient general term, which comprises not only certain descriptive expressions for individual appliances, such as rectiles, cross-lines, sighting scales, stage micrometers, eyepiece micrometers, haemacytometers, but also various non-descriptive and loosely worded terms.

An important point, germane to all reproduction by photographic lenses, relates to the limits of accuracy set by the lenses themselves. An ideal objective lens would be one which produces an equally sharp image on a flat plate over the whole field it covers (Fig. 241), at whatever aperture the lens is used and with accurately equal magnification or reduction over the whole of the field. Every lens computer knows, however, that these and various other conditions cannot be rigidly complied with and that actually an objective represents a compromise between them. For any ordinary purpose the conditions are indeed amply fulfilled in the best existing lenses—no small tribute to the lens computers' skill—but graticule

GRATICULES

work and similar scale work bring out the limitations. Thus it is difficult by photographic means to secure lines of a fineness less than 0·010 mm (1/2500 in.) over more than a comparatively narrow angular field. With great care lines of a fineness of 0·051 mm (1/500 in.) can be got with the best objectives over a moderate angular field. And lines as fine as 0·020 mm (1/1250 in.) wide from original diamond rulings, have been produced.

eyepiece scale

stage micrometer

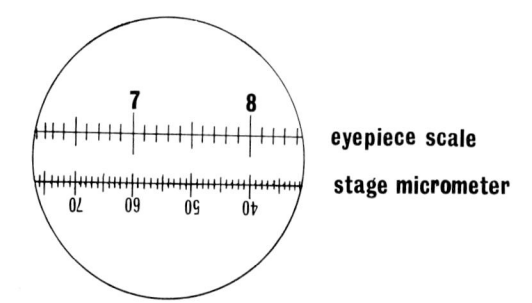

Fig. 241.

The majority of graticules are required for use with transmitted light and it is customary for them to have opaque black lines on glass, probably because they reproduce photographically. Another type is made by depositing a highly reflective coating of metal on the glass. The ruling cuts the metal and exposes the surface of the glass slide. With incident illumination, light is reflected from the metal surface and the graduations appear black. This type of graticule can also be used with transmitted light when the scale will appear white on a darkground.

Of course the thickness of a line is all important. A line 0·025 mm (1/1000 in.) wide viewed with the unaided eye at a normal viewing distance of 250 mm, subtends an angle of approximately 20 sec. An opaque line seen against a white background will appear rather thinner than it actually is, as the white light encroaches on the two edges of the line, by irradiation of the eye. If the line is too narrow the line appears faint and indistinct and is troublesome to the eye. With the unaided eye a line of 0·051 mm (1/500 in.) can just be seen but several lines placed close together would appear as a mass.

For eyepiece graticules the magnification of the eye lens has to be taken into consideration, but the best rule for comfort is not to have the lines narrower than needful, and with eyepieces of moderate power, ×5 to ×8, opaque lines 0·013 mm (1/2000 in.) will be found sufficiently fine. In micrometer scales divided into tenths or twentieths of a millimetre, the width of the lines may reasonably be reduced to 0·005 mm (1/5000 in.).

Calibration of eyepiece micrometer graticules is achieved by comparing the various characters of the particular eyepiece pattern with the standard scale micrometer placed on the microscope stage. The observed image of the stage micrometer graticule is aligned with the eyepiece micrometer graticule to cause some lines of both scales to coincide. In Fig. 79 the divisions of an eyepiece micrometer have been aligned with the divisions of a stage micrometer. By comparison 10 divisions of the eyepiece scale represent 20 divisions of the stage micrometer, thus one division of the eyepiece scale represents 20 microns.

The Ramsden-type eyepiece is usually used when measuring for particle size (Fig. 242). The Kellner × 10 is also a positive type eyepiece which will accom-

(a)

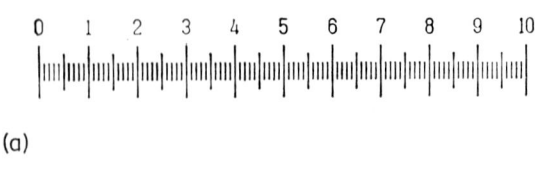

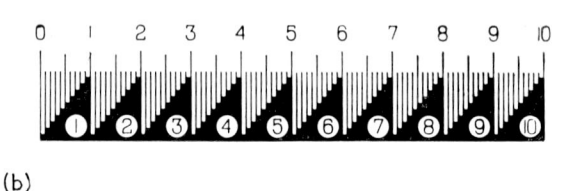

(b)

Fig. 242. (a) 10 mm eyepiece micrometer with 100 divisions. (b) Step micrometer 10 mm divided into 100 parts of microns.

modate a graticule in the focal plane so that it appears superimposed on the projected specimen image when the specimen features can be measured on the photomicrograph and ground glass screen.

Graticules Ltd. (Tonbridge, Kent) market a Huygen eyepiece† specially made to house a graticule. It allows the graticule image to be brought into sharp focus when it is superimposed on the specimen image, a difficulty often experienced with a fixed focus eyepiece. The graticule is held by a screw-on retaining ring and is easily removed for cleaning or replacement by another pattern. They also prepare a wide range of eyepiece and slide graticules; 40 divisions on the graticule equal 0·2 mm on the stage micrometer; the value of each division on the graticule is then $0·2/40 = 0·005$ mm (5 microns).

The Vickers Screw Micrometer Eyepiece (Plate 26) is unique in that a fine line is made to traverse the field of view by means of an accurate screw. The drum controlling the movement is divided into 100 parts, each representing 0·01 mm, displacement of the line. The fixed divisions provided on a graticule in the field of view indicate whole revolutions of the drum and are not intended for the purpose of direct measurement. Evaluation is made by comparison with

† Zeiss, Jena, also make a Huygen micrometer No. 3 eyepiece.

GRATICULES

the stage micrometer, as with a fixed graticule eyepiece. The tube which fits into the microscope is tapered to prevent rotation of the draw-tube.

Graticules Ltd. list a very wide variety of eyepiece graticules. Figures 242 and 243 demonstrate the micrometers supplied in the following range:

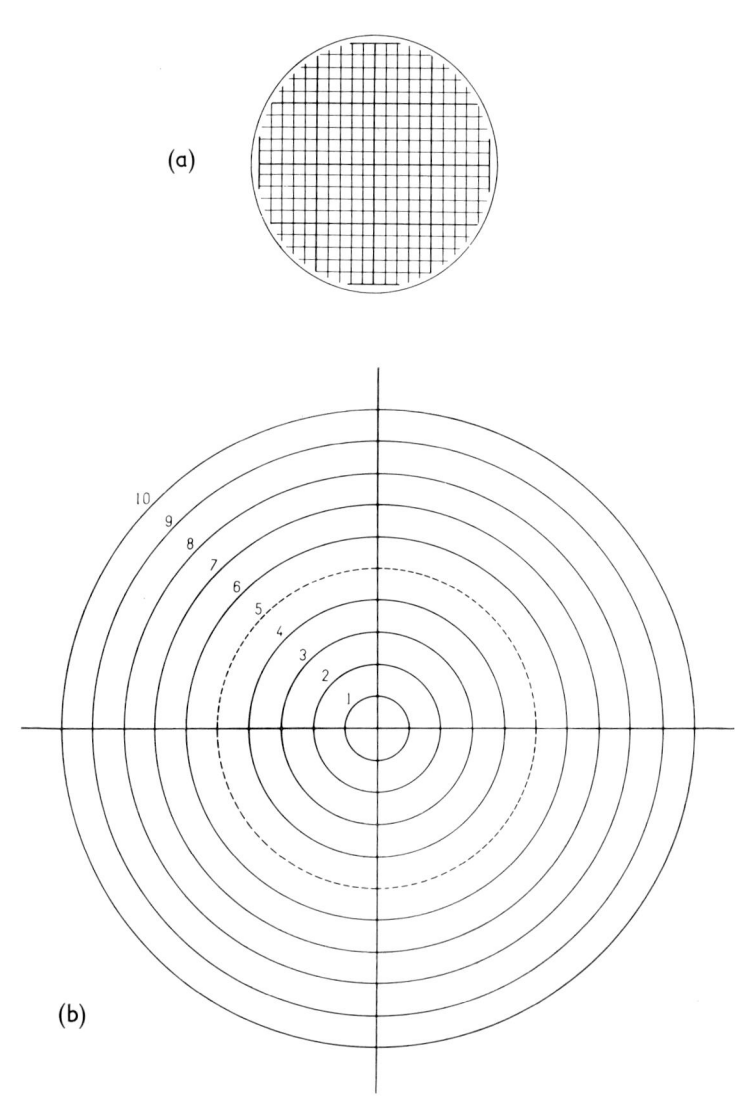

Fig. 243. (a) Net graticule for measuring conoscopic angles (polarized light). (b) Concentric circles for measuring grain size.

10 mm	in 100 parts of	100 microns
5 mm	in 50 parts of	100 microns
5 mm	in 100 parts of	50 microns
2·54 mm	in 100 parts of	25·4 microns
2 mm	in 100 parts of	20 microns
1 mm	in 100 parts of	10 microns
10 mm	in 200 parts of	50 microns

This type of selection goes for most graticules including combined vertical and horizontal scales, crossline, grids, chessboard, gauge lines, concentric, protractors, step micrometers, keyway circles, and particle size analysis of various types.

The England or relocation finder (Graticules Ltd.) is a precision device which enables quick and accurate relocation of fields. The specimen, mounted in the normal manner can be replaced with the England finder, a section of which is illustrated in Fig. 244. It consists of a glass slide 76 × 25 mm (3 × 1 in.)

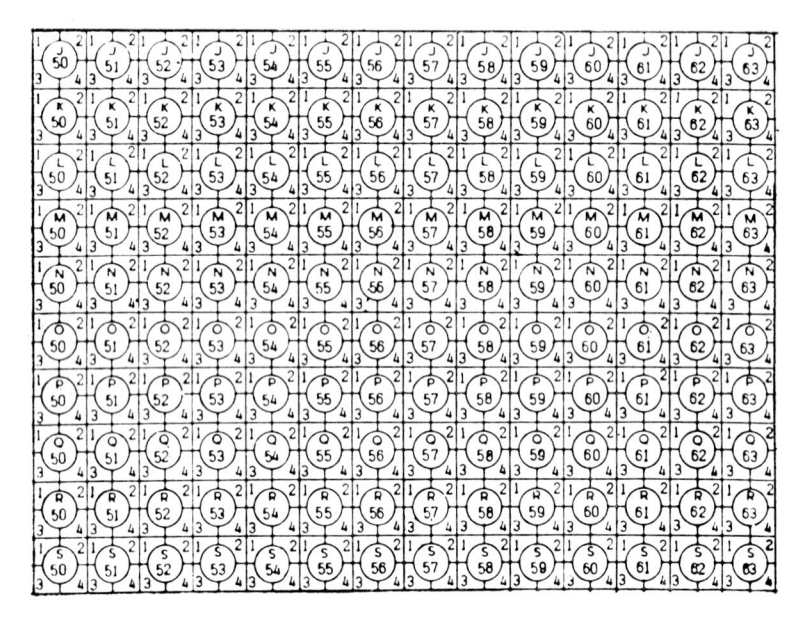

Fig. 244.

marked with a square grid at 1 mm intervals. Each square contains a centre ring bearing reference letter and number, the remainder of the square being subdivided into four segments numbered 1–4. Reference numbers run horizontally 1–75, and letters vertically A–Z (omitting I). The main location edge is the bottom of the slide which is used with either the left or right vertical edge of the

GRATICULES

slide, according to the fixed stops of the microscope stage; all three locating edges being marked with arrow heads. For reference purposes it should be read as indicated in Fig. 245, Q6.3 and S5.

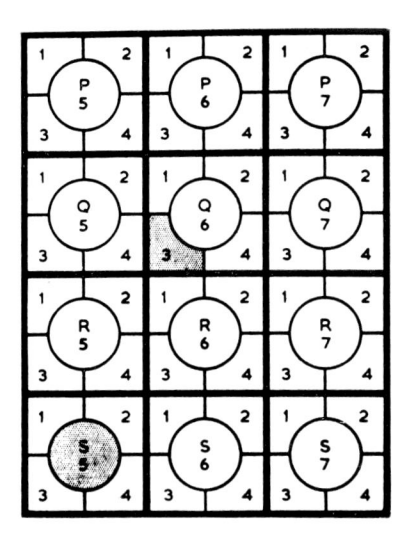

Fig. 245.

The relocation slide is of particular interest to the biologist, pathologist, botanist and medical photomicrographer. A small label (Fig. 246) on the specimen slide can be used to record the necessary evidence for relocation of any required area. This will not disturb the normal filing of the slide.

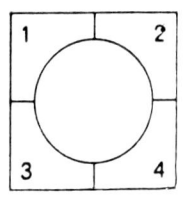

Fig. 246.

The Whipple Disc is a very useful graticule and easy to insert. It is divided into four squares, each further divided into 25 smaller squares. The sides of the larger squares are calibrated to coincide with the stage micrometer for the particular optical system in use. This graticule may be used in one of three ways: the image can be projected on to a white card; or viewed through the eyepiece in the normal manner; or a photomicrograph can be taken to include the calibrated squares on the negative.

The Portion graticule, Fig. 247, is suitable for a wider range of applications than the Patterson, Fig. 248. The right-hand side of the Portion counting graticule

320   PHOTOMICROGRAPHY

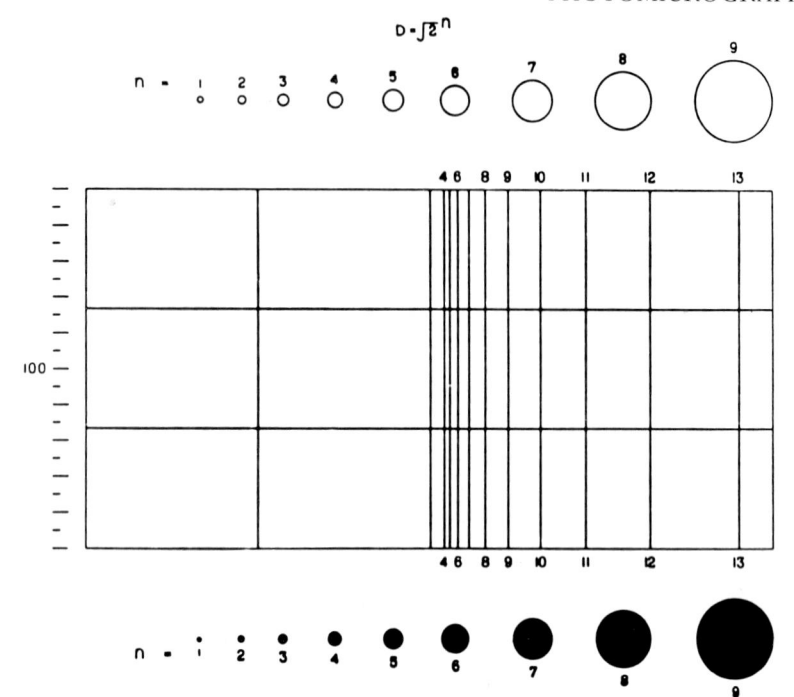

Fig. 247. The Portion graticule for particle size measurement.

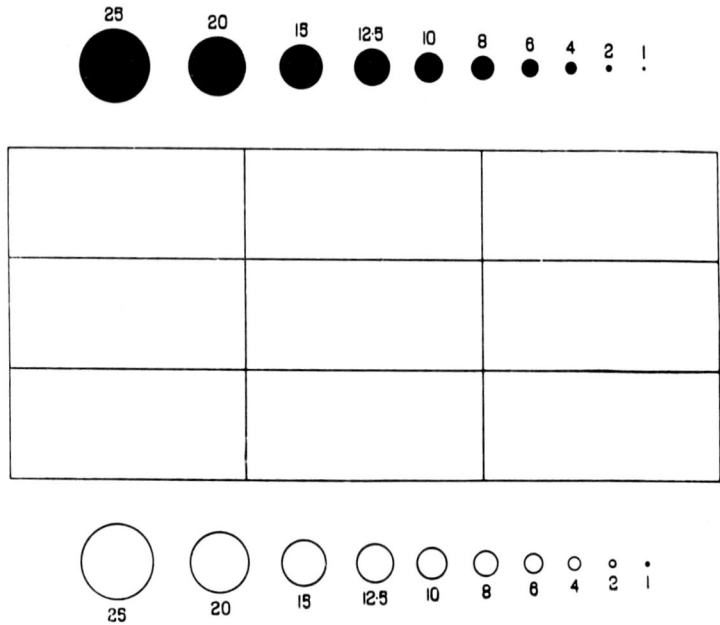

Fig. 248. The Patterson graticule.

GRATICULES

is divided vertically. The diameter of the globes and circles increases in geometrical procession. The Fairs series of graticules, Fig. 249, consists of three eyepiece discs, available in three sizes 18, 21 and 24 mm diameter and are suitable for use with all microscopes. The graticules constitute a counting chamber and a range of globes and circles numbered from 1–16, 16–45, and 32–128. Size

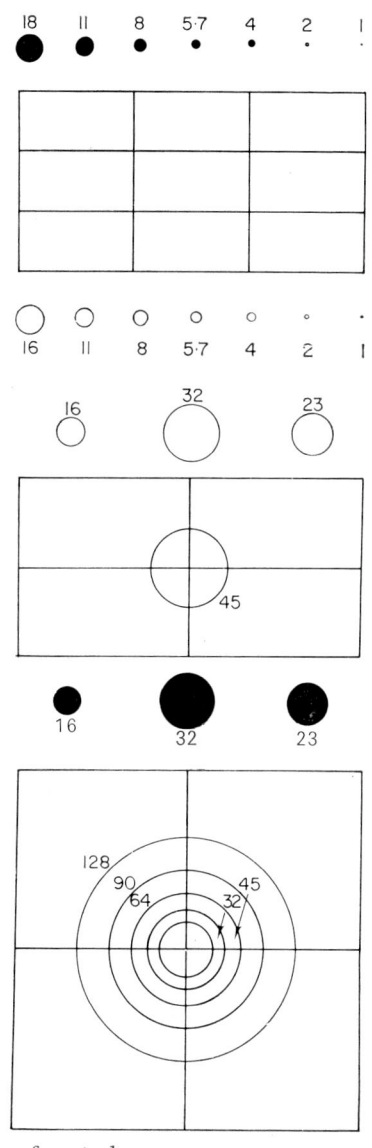

Fig. 249. The Fairs series of graticules.

progression of the globes and circles is in proportion to the square root. The simplified globe and circle graticule, Fig. 250, is for use when only approximate measurements are required.

The Carl Zeiss, Oberkochen, integrating eyepiece is a device for assisting quantitative and surface analysis. The eyepiece is fitted with a graticule I (Fig. 251)

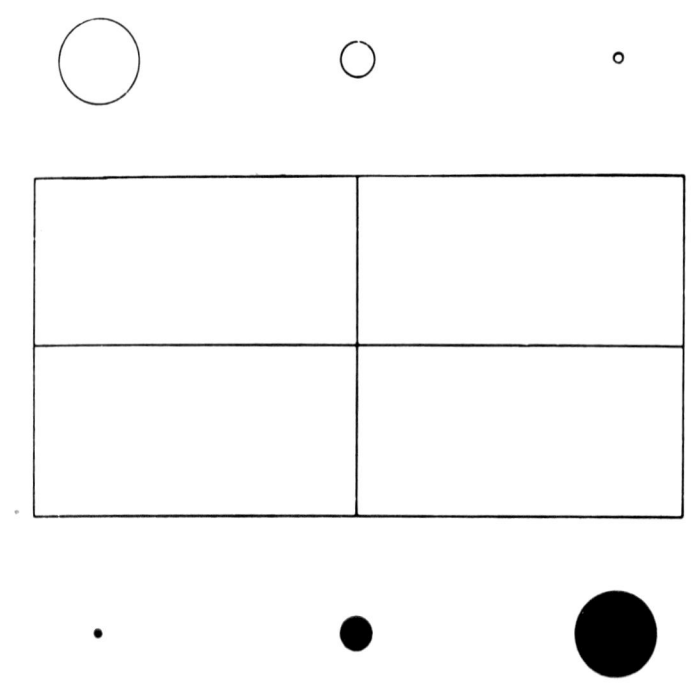

Fig. 250. Simplified globe and circle graticule.

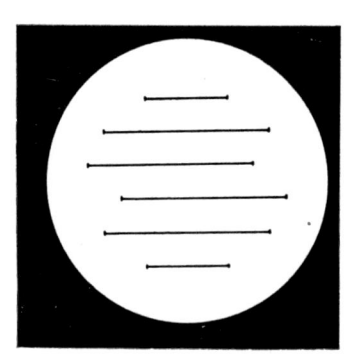

Fig. 251. Zeiss integrating eyepiece graticule I, test point graduation.

GRATICULES

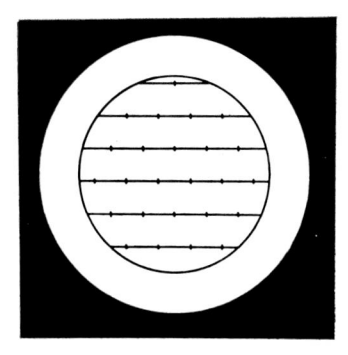

Fig. 252. Zeiss integrating eyepiece graticule II, length graduation.

and graticule II (Fig. 252). The former contains a network of 25 points, arranged within a circle for counting with the lines in sharp focus. This field is smaller than the field of view; field of view number 18 and magnification ×8.

The test-point graduations serve for determining the quantity (volume percents) of individual constituents in a heterogeneous material (metal alloys, rocks, biological organisms, crystals, etc.).

Any optical section through the material to be examined is covered with the point lattice. Every point of the lattice which accidently lies above a particular component (Fig. 253) is counted as a "hit" for that component. To obtain the

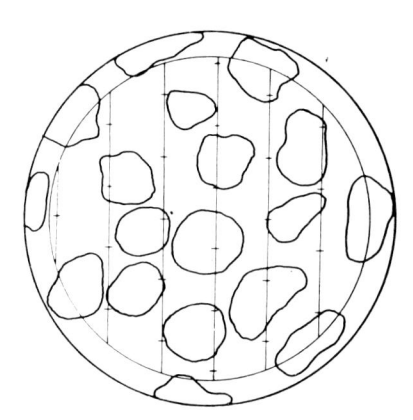

Fig. 253. Point counting graticule (Zeiss).

desired precision, the counts are repeated, either on different parts of the specimen with unchanged orientation of the graticule, or on the same place of the specimen but with a change in the orientation of the graticule (rotation of the eyepiece). After the count is completed, the hits for each kind of component are summed up. The ratio of the resulting sum for each kind of component to the total sum of hits is computed. It represents the volume share of the respective component.

Integrating eyepiece with graticule II (length graduation) contains a graticule also housed in a ×8 Kpl. (compensating flat field) eyepiece, focused on the number of parallel lines of definite length (Fig. 252). Sections as thin as possible are made through the specimen to be measured. In these, the boundaries of all bodies or cavities, whose surface is to be determined in the total volume, appear as curved or straight lines. All points which these lines intersect with the graticule lines are counted. The surface is directly proportional to the body volume V and to the mean value p_m of the point numbers from many tests, inversely proportional to the total I of all measured distances. The length of each of the middle four amounts to 1/5 of the total length, that of the upper and lower lines each of 1/10 of the total length I. Through the parallel arrangement of the lines, all separated by the same distance a, the fluctuations in the composition of the test specimen are best equalized. In special cases one has to rely on the magnitude of the constant amount a.

Carl Zeiss, Oberkochen, market special eyepiece micrometer discs, housed in turrets, manufactured for assessing and counting particle size distribution. Figure 254, 1–4 indicates some disc patterns obtainable. The eyepiece accepts the revolving discs which serve for a convenient disposition and rapid exchange

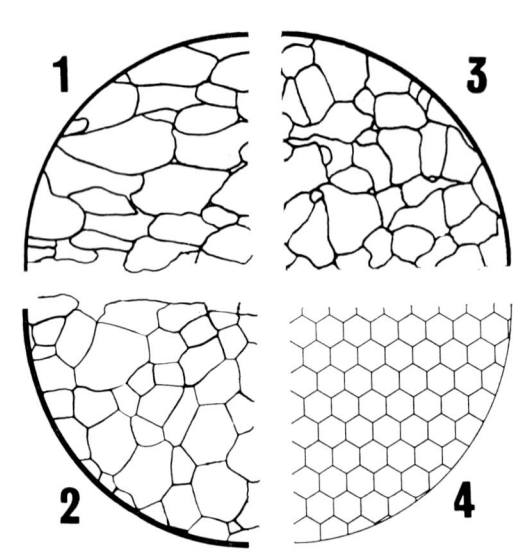

Fig. 254. Zeiss particle size distribution. (1) Type 'c' micrometer-disc turret "ferrite grain" size 2:1 and 4:1. (2) Type 'd' micrometer-disc turret "ASTM-E 112 untwinned grains." (3) Type 'b' "austenite-ferrite" micrometer-disc turret. (4) Type 'a' "honeycomb net" micrometer-disc turret.

of several micrometer discs in the intermediate image plane. This is an advantage for evaluating clusters of particles in connection with standard series, for measuring size and determining volume. Each turret contains either one revolving disc, or two, arranged one above the other, with six or twelve graticules, respectively. In another position, the light passes freely through the discs.

Figure 254, No. 1. This double turret is identical in design to the turret. The only difference is it covers the grain size ratio 2:1 and 4:1. The micrometer-disc turret reflects all the different conditions of the Steel and Iron Standard 1510-61 in conjunction with the turret No. 4 of the figure.

Figure 254, No. 2. On this turret the grain-size samples are arranged according to ASTM-E 112† with the whole numbers 0–4 or, respectively, 7 and 8 as well as the half numbers 4·5–5·5. Variation in the initial magnifications of the objectives thus allow the ASTM numbers 2–10 to be covered. The distribution of grain-sizes has been particularly adopted for the austenite grain sizes, but it may also be used in connection with isometric grain-size systems.

Figure 254, No. 3. Each of the two revolving discs holds six micrometer discs with austenite–ferrite grain-size in a ratio of 1:1 conforming to the Steel and Iron Standard 1510-61. The discs bear the data for the Steel and Iron Tests Nos. 2 to 7. In conjunction with Carl Zeiss, Oberkochen, epiplan objectives it is possible to cover the numbers 0 to 9. The micrometer-disc turrets Nos. 4 and 1 together give the overall requirement of the above standards and should therefore be treated as one unit.

Figure 254, No. 4. The micrometer discs built into this double turret conform to ASTM-E 19-46. The eleven discs correspond to the ASTM numbers 0–10. With the aid of the appropriate objectives, the ASTM numbers 2–12 can be covered. The honeycomb nets are extremely convenient for estimating particle sizes with relative radii of $\sqrt{2}:1$ and $2:1$. One opening contains a micrometer disc with 10 mm/100 mm and four 10 cm/128 graduations.

Figure 255. Integrating micrometer-disc turrent. With this, sets of points arranged in a square are superimposed on the microscope image according to the principle of point analysis. Any overlapping of a point with a component is counted as a hit. The ratio of the number of hits to the total number of points gives the partial volume or the percentage by volume. The revolving disc has seven openings. For optimum geometric adaptation to the specimen, integrating discs are mounted in four of these openings. The discs have the same base length 25, 100 and 400 points respectively. In addition, the large clusters of particles are centrally subdivided into 25, 100 and 400 points, such small particles can be uniformly counted with the particle nets. If the co-ordinate motions counted with the mechanical stages are made parallel to the sides of the counting discs, then the specimens can be completely covered with the nets.

† American Society for Testing Materials.

Carl Zeiss, Oberkochen, relocation finder (Fig. 256) is divided into three squares, A, B and C, each containing 30 × 30 smaller squares, engraved on 27 × 76 mm slides. The 0·75 × 0·75 mm squares are figured from 1 to 900 and further subdivided into 3 × 3 squares. The slide, together with the specimen, need only be exchanged for the relocation finder to give an accurate reading of the position of the stage.

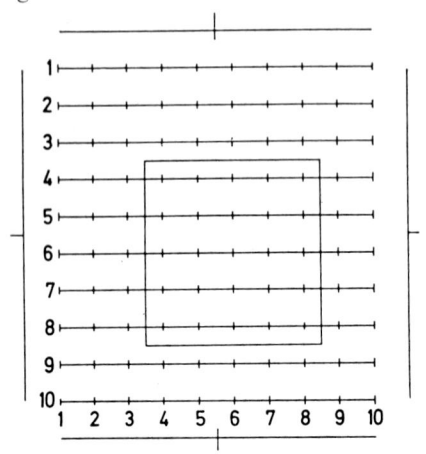

Fig. 255. Zeiss integrating micrometer-disc turret.

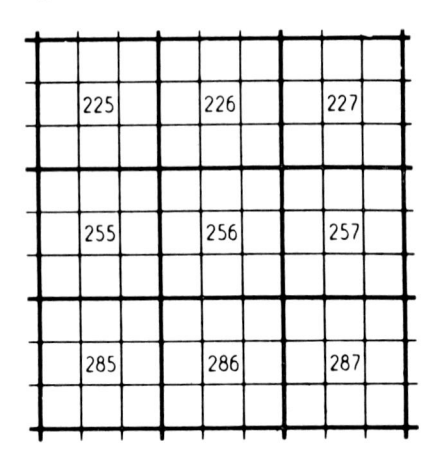

Fig. 256.

Mounting and staining

One of the ways in which one can improve one's knowledge of microscopy is to acquire facility in the preparation and mounting of specimens. Prepared slides, expertly cut, stained and mounted, can be purchased. Slides of this type are invaluable to the newcomer to photomicrography, since any defects in the final photomicrograph may be associated with an error in microscope technique rather than in the preparation of the specimen. Apart, however, from the satisfaction which results from the production of good-quality slides, it is often necessary for the working photomicrographer to be able to prepare suitable specimens.

Fixing and hardening. The nature of the specimen and the observations required must decide the type of slide to be prepared. Small animals and other specimens are sometimes mounted whole. Fluids such as blood may simply be smeared on to a slide. When the contents of a cell are to be studied and the structure of the cell is unimportant, the specimen may simply be squashed under a cover glass. But the majority of specimens which require examination by transmitted light are so thick that they must be cut into thin sections which must not distort the structure. Often it is necessary to stain the section selectively to render structural details visible.

The majority of tissues, such as wood, skin, hair, leaf and stem, are usually examined in this way. Most of these tissues are too soft to be cut without introducing distortion. Such tissues must, therefore, be fixed. The aim is to kill and harden the tissue without damaging or distorting it in any way. These twin objectives are not easily realised. There are several fixatives, each with its own characteristics. The most commonly used fixative is 70% alcohol, but it causes shrinkage of fine cell structures. Acetic acid is also used for fixing specimens and is more rapid than alcohol. In general, however, it does not harden as well as the 70% alcohol. However, a mixture of 99% of the 70% alcohol with 1% of glacial acetic acid is a very efficient fixative, especially for nuclei and chromosomes. Ethyl alcohol is most suited to animal and plant tissues and will fix in a comparatively short time, 2–3 hr.

Chromic acid is suitable for fixing plant and animal tissues and algae. The subject matter must be small, as this medium does not penetrate as quickly as others requiring 18–24 hr. However, tissues stain very well after being fixed with this reagent.

Formaldehyde is one of the most popular fixatives and is used in most laboratories. Unfortunately, it requires a fixing time of about 48 hr. It is used mainly for animal and histological material, and is not suitable for use with water, chromic acid or potassium dichromate. Used carefully it produces very little shrinkage in the tissue. If subsequent staining is to be conducted with haematoxylin, very good results are obtained with a formaldehyde fixative. A solution of 70% alcohol and 30% formalin (itself a 40% solution of formaldehyde) is a very good general fixative for plant and histological sections, which is capable of fixing within 15 min.

Tenker–formol is a very good fixative for vertebrate specimens. It contains a mixture of potassium dichromate and formaldehyde and fixes in 24 hr. An equal time of washing followed by a rinse in mercuric chloride is necessary.

If mercuric chloride has been used it is necessary to wash the specimen thoroughly in water and then soak it in an aqueous iodine solution, followed by further soaking in 70% alcohol. Failure to observe this washing procedure may result in the development of crystals in the finished slide.

After fixing, tissues must be thoroughly washed in order to remove all trace of the fixative. Failing this, the section may not stain evenly and reagents left in the section may later crystallise out. When alcohol is used as a fixative no after washing is necessary, but with the alcohol–acetic method the specimen should be washed out with a 70% solution of alcohol.

For preserving purposes tissues may be bottled in 70% alcohol or formaldehyde. If they are to be stored in the latter, make sure the section is thoroughly washed or staining may appear.

A simple fixing procedure may be summarised as follows:

Animal Tissue. (a) For very rapid fixation of small pieces of fresh tissue, where some degree of damage is acceptable:

(1) Drop into hot water at 90°C for 1 min.

(2) Transfer to a solution of 70% alcohol for 1 min.

(b) For general use:

(1) Cut tissue into pieces not larger than $1 \times 1 \times \frac{1}{2}$ cm.

(2) Transfer to formol–saline (strong)[a] for 24 hr.

(3) Transfer to alcohol (50, 70 or 90%), avoid water.

Plant Tissue.[†] (a) For rapid fixation of small pieces or sections of fresh tissue and algae:

(1) Formalin–alcohol[b] for 15 min.

(2) Alcohol (70%) for 2 min.

(b) For general use:

(1) Cut tissue into pieces not larger than $1 \times 1 \times \frac{1}{2}$ cm.

(2) Soak in chrome–acetic for 24 hr.

(3) Wash in running water for 12 hr.

Sectioning. This is best done with a microtome, but reasonably thin and uniform sections can be prepared free-hand with a razor blade, or even with dissecting scissors. Thin sections can be cut from a variety of subjects, such as hair, fibre, biological preparations, textiles, animal and human tissues and botanical material.

A section must be of uniform thinness. The cutting action with a very sharp knife must therefore be smooth. Any unevenness in the thickness of the specimen will lead to difficulties later on, since staining will give a deeper colouration in the thick regions and the variation in surface height may result in out-of-focus bands across the field of view. Thus, in making a microtome, rigidity and smooth

---

[†] H. Alan Peacock.

[a] To make "strong formol-saline" use sodium chloride solution (0·9% aq.) as the diluent for formaldehyde (10%).

[b] Alcohol (70%), 100 cc; formaldehyde (40%) 6 cc.

movement of the knife must be ensured. For facility in the preparation of specimens a microtome is necessary. It consists of a rigid frame supporting both knife and specimen, with means for traversing the knife across the specimen with a device for advancing the specimen as required. It is usually necessary to provide a supporting material, even with fixed and hardened specimens. Of course, durable specimens can be effectively frozen. The Reichert O.M.P. microtome is adapted for this purpose: a jet of carbon dioxide is directed on to the specimen to freeze it, which usually suffices to hold the specimen to the stage as well. More delicate material must be embedded in a suitable supporting medium, such as paraffin wax. The Reichert O.M.E. is an excellent example of a modern microtome suitable for handling embedded specimens. When the irregular surface material has been removed, either the O.M.P. or O.M.E. permits the production of a succession of slices with a minimum thickness of some 20 microns (1 micron = 10^{-3} mm).

A newly designed Leitz Cryostat stainless microtome is incorporated and is specially designed for operating at low temperatures. A large work plate houses the electronics for controlling the temperature regulation of the cooling chamber and the specimen stage. Furthermore, it contains a pneumatic device for the automatic section transport. The air and cooling compressors are installed in the left lower part of the cryostat. All controls are easily accessible from the outside and neither during the cutting nor for the transport of the specimen is it necessary to operate manually in the cooling chamber. (Its technical features are: specimen thickness, between 1 and 25 μm; cooling chamber temperature, -5 to $-30°C$; specimen stage temperature, -5 to $-30°C$ (independent of the cooling chamber temperature); automatic heating for rapid temperature changes of the specimen stage; automatic thawing for rapid removal of the specimen block; built-in specimen stretcher; automatic specimen feed and section transport.)

The author has made his own microtome, using one of the old "cut-throat" types of razor. Provided that the specimen and blade are rigidly supported and the razor is very sharp, extremely thin sections can be cut from a wide range of materials.

The majority of soft specimens can be supported in paraffin wax, tallow, lard or similar material, but there is much to be said for the freezing method. This may be used not only to make the specimen rigid but to freeze it to its support. If carbon dioxide is sprayed on the specimen it will quickly become covered with a thin film of ice. This treatment can be repeated should the specimen show any signs of thawing out. Alternatively, a block of solid carbon dioxide may be placed on the stage to which it will freeze and the top then levelled off to take the specimen. With this method there is the disadvantage that the specimen is supported by a block of carbon dioxide and there is some loss of rigidity.

If the specimen is to be frozen to the stage it may be washed in water after being soaked in 70% alcohol. Should the specimen require embedding in wax, it must first be dehydrated in 95% alcohol and then cleared of alcohol in a medium such as xylene. Figure 257 illustrates my home-made microtome. It is easy to make and very easy to manipulate, and produces thin sections which can be cut from objects frozen to the stage or held in wax, etc. The various parts of the microtome are shown in Fig. 258. *A* is made from a piece of hardwood, 1 in.

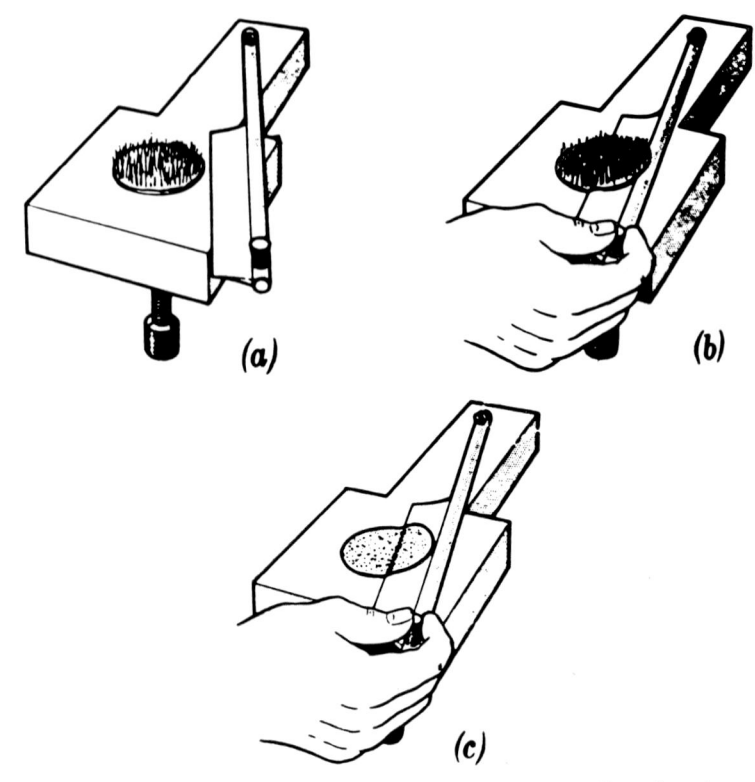

Fig. 257. Microtome (a) Hair or fibre before initial cut. (b) Levelling the subject which can be supported in wax or firm substance. (c) Making the thin section.

in thickness. A hardwood is necessary to resist wear and distortion and to clean easily. The microtome is fixed to a table or bench top L by a clamp K which is fixed to A by two or more screws N. The inner cylinder B fits tightly into the hole B_1, the top of the cylinder being fixed 12·7 mm ($\frac{1}{2}$ in.) below the top surface of A. The cylinder B is tapped to receive a fine-threaded screw F passing through it, with a knurled head at D. The other end of the thread has a conical retaining spigot E on which the stage E_1 is free to rotate. The specimen may be supported on the stage and advanced enough for each section by rotation of D. The blade M is fastened by a small screw G through the hole G_1 and a small grip

MOUNTING AND STAINING

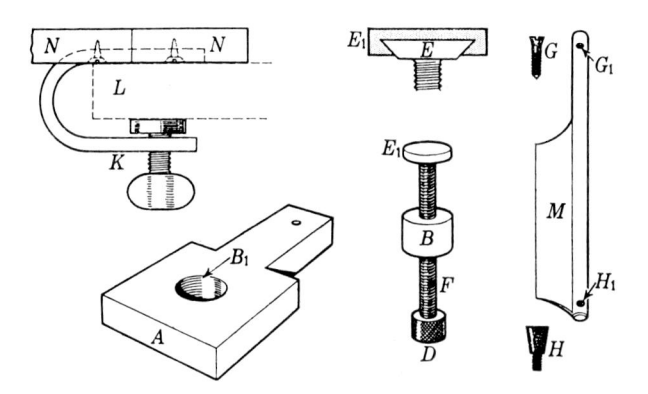

Fig. 258a. Component parts of microtome.

H fastened at H_1 enables the blade to be swung to and fro. Figure 257 illustrates the microtome in use: (a) showing the subject immediately after mounting; (b) illustrating the initial cut which leaves a flat even surface; (c) showing the first thin slice being cut. Immediately the required slice has been cut it should be placed in a beaker of water, from which it may be floated on to a slide.

Larger sections may require preliminary dissection, so that, in addition to a microtome, dissecting scissors are invaluable. A large range of suitable instruments is necessary. A complete set of dissecting instruments consists of: microtome razor; diamond pencil; scalpels of various blade sizes; Borrodaile's needle; scissors, straight and curved; forceps, straight, blunt and sharp; forceps, curved; cornet forceps; needles, straight and curved in handles; seeker, straight and curved; blow-pipe; and section lifter. Beck, Leitz and Zeiss supply dissecting microscopes and dissecting arms. The design and construction of these instruments give perfect stability, together with ample access for the hands and support for the wrists and arms. The lenses are carried on a swinging arm which is focused by the usual rack-and-pinion movement, providing a vertical travel of 76 mm (3 in.). A plane mirror and opal glass disc are provided, and the instruments are well finished and easily cleaned.

Staining. It is an advantage to be familiar with some of the generally used stains. As long ago as 1770 J. Hill tried to stain sections of timber: it is believed that cochineal was used. In an attempt to stain living organs Ehrenberg fed them either powdered indigo or carmine, hoping to examine the various digestive systems. Haematoxylin has had a long and useful run for it was first manufactured in 1840 but it was not until some time later (1865) that Bomer first succeeded in using it. Haematoxylin and carmine is still in use and over the years has proved to be indispensable. To some, this is the only stain. Botanists Goppert and Cohn were the first to use natural aligarine in their studies but since that time a synthetic stain known as aligarine has been introduced.

One of the first to use stains in plant histology was Hartig, in 1845, followed by Perkins who, around 1856, introduced several stains. It was probably Gerlach who first applied stains to thin sections of the human body. It is on record that overnight he left some tissues in a dilute solution of carmine and was surprised to find this had stained his preparation. An iodine solution, common to most, was introduced by Lugol and used in the technique of staining bacteria by the bacteriologist Gram, hence the long-lived Gram stain. Leishman in 1901 prepared a polychromed methylene blue, added eosin and dissolved the resulting precipitate in methyl alcohol. Today this is one of the most important stains for clinical and malariological work.

There are five uses for stains:
(a) To make cell structure and organisms visible.
(b) To clarify variations in cell structure; a means of identification.
(c) To reveal their chemical nature.
(d) To influence the growth of organisms, particularly in bacteriology.
(e) To induce fluorescence in unpigmented plant and animal tissues and cells.

Some staining procedures are highly complex and their discussion would be out of place in this book. It is therefore suggested that "Elementary Microtechnique" by H. Alan Peacock, and "Biology Staining Schedules" by R. R. Fowell be to hand. A list of the stains commonly employed is given in Tables 29 and 30.

If the specimen to be stained is embedded in wax for sectioning, the wax must be removed with a solvent such as xylene and then be removed with alcohol. The specimen must be washed afterwards with water if the subsequent stain is a water soluble type.

When the section has been satisfactorily stained, all water must be removed before the section can be infiltrated and mounted with Canada balsam or any similar medium. With ethyl alcohol a series of solutions of increasing strength will minimize the risk of damage by diffusion currents. Cellosolve is an efficient and rapid dehydrating agent which will displace the water from an average section in approximately 1 minute. The dehydrating agent must be removed by a clearing agent such as benzene, cedarwood oil, clove oil or xylene which will also render the section transparent. The whole series of operations may be readily conducted in a line of watch glasses, so that the section is simply moved along in a steady and logical sequence to the mounted specimen (Fig. 258b). "Synthetic Dyes in Biology, Medicine, and Chemistry" by E. Gurr, Academic Press, is a volume which ought to be to hand.

Mounting

The purpose of mounting the specimen is to enable it to be handled for examination and photography. Such mounts may be either temporary or permanent. Saline solution, glycerine and distilled water are suitable for mounting tissues in

Table 29 *Botanical stains*

Stain	Solvent	Staining Reactions	
Aniline hydrochloride	Water	Yellow	—
Bismarck brown	70 per cent alcohol	—	Brown
Delafield's haematoxylin	Weak alcohol	Yellow	Purple–blue
Erythrosine	Lactophenol	Yellow–Orange (dense protoplasmic structures stain red)	—
Gentian violet	Water or 20 per cent alcohol	Violet	—
*Iodine	Water / Starch	Deep yellow / Oxford Blue	Pale yellow / —
Light green	Clove oil	—	Green
Phloroglucinol	95 per cent alcohol	Red	—
Safranin	50 per cent alcohol	Red	—
Schulze's soln	Water	Yellow	Violet

* This must first be dissolved in a small quantity of strong potassium iodine solution.

Table 30 *Zoological stains*

Stain	Solvent	Staining Reactions	
Borax carmine	50 per cent alcohol	Red	—
Ehrlich's haematoxylin	70 per cent alcohol	Purple	Pink
Eosin	90 per cent alcohol	—	—
Leishman's stain	Methyl alcohol	Purple–blue (oxyphil granules red)	Pale blue
Methyl green	Water	Deep green	Pale green

a temporary manner. There is a large range of media suitable for the preparation of permanent mounts; this includes hydrax, naphrax, pleurax, lactophenol, gum styrax, realgar, clarite, Canada balsam, euparal, diaphane and glycerine jelly. There are many variations in the technique of preparing permanent mounts so it is possible to give only a brief general account in this book.

Fig. 258b. (1) Wash. (2) 75% alcohol, 25% water. (3) 50% alcohol, 50% water. (4) 25% alcohol, 75% water. (5) 100% water. (6) Canada balsam on rod and specimen about to be laid on the mounting medium. (7) Placing cover glass on the specimen. (8) The mounted specimen. (9) Spring clip to hold cover glass for at least 48 hr.

MOUNTING   335

Manipulative technique and cleanliness are very important in the preparation of mounted sections. In particular, dust and fibres from cleaning materials must not be allowed to spoil the specimen. The slides generally used are 77 × 25 mm (3 × 1 in.) and the object is, as a rule, placed in the centre. The cover glass rests on the mounting medium, which is allowed to spread out to the edge of it and set (Fig. 259). On occasions it is necessary to rest the cover glass on a ring,

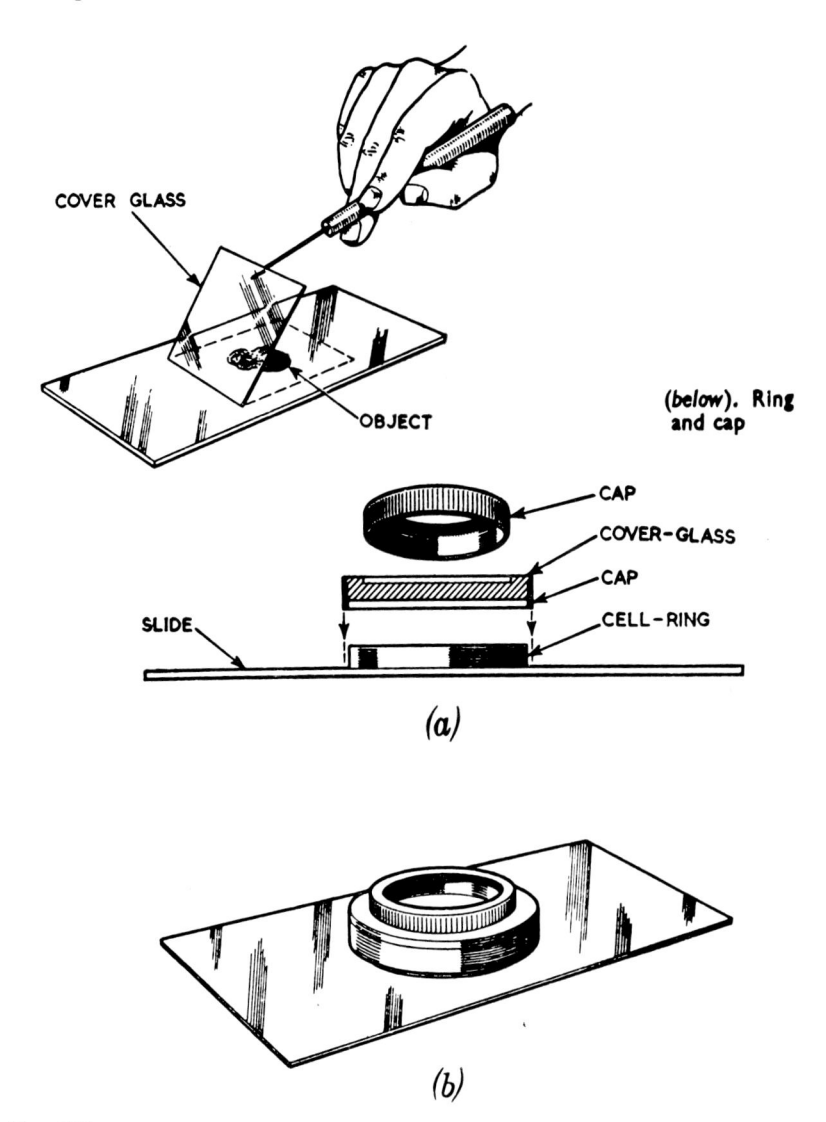

Fig. 259.

owing to the thickness of the specimen (Fig. 259a and b). Cover glasses are made round, square and rectangular, and are manufactured in several thicknesses, as mentioned earlier. Thick cover glasses, made from crown glass, are used for low-power and visual work, but for high-power photomicrography unannealed glass is the best. These cover glasses vary in thickness and are also very brittle.

First, dust both slide and cover glass with a camel hair brush placing a drop of mountant on the clean slide. With a flat spoon and camel hair brush (Fig. 260),

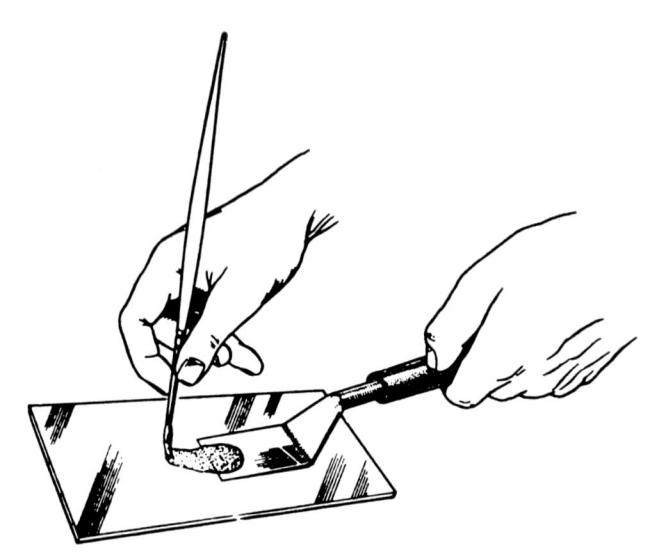

Fig. 260. Transferring a very thin section to slide.

or by holding one corner with tweezers, place the specimen in position. Put a cover glass on one edge and gently lower it by means of a mounted needle, as shown in Fig. 259, and allow it to settle. Do not press it down. If this operation is hurried, air bubbles may be trapped beneath the cover glass and subsequent attempts at remounting may harm the specimen. If the slide is gently warmed over a small flame, the bubbles are more rapidly dispersed. At this stage a spring clip may be used as in 9 of Fig. 258b. The mounting medium may take several days to set and if there is a small area near the edge of the cover glass into which the mounting medium has not penetrated, this can be filled later on. When the edges are finally dry the slide can be ringed with gold size and finished off. If it is not absolutely dry when the gold size is applied, the slide is almost certain to be spoilt by the later development of bubbles within the mounting medium.

If the specimen is thick it is advisable to introduce a built-up cell on which to rest the cover glass. There are several ways of doing this, as shown in Fig. 261. Square frames can be cut from millboard (as in a) but there may be difficulty if

MOUNTING

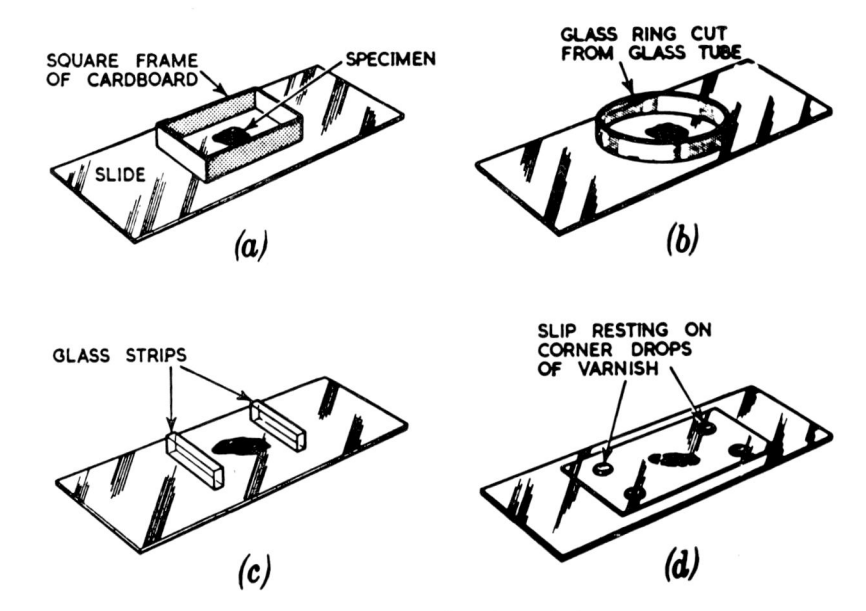

Fig. 261. Various methods of raising cover glass above thick specimen or opaque object.

the mounting medium is absorbed by the millboard. Thin strips cut from cover glasses can be used to support the cover glass (as in c) and to permit a thick layer of mounting medium to be placed on the slide, so that the final mount will be air-tight. Aluminium cell rings are available in three diameters, 15·9 mm ($\frac{5}{8}$ in.), 19·1 mm ($\frac{3}{4}$ in.) and 22·5 mm ($\frac{7}{8}$ in), and of thickness 1·59 mm ($\frac{1}{16}$ in.), 1·06 mm ($\frac{1}{24}$ in.) and 0·76 mm ($\frac{1}{32}$ in.).

When prepared cells are not to hand several layers of varnish, laid successively in rings around the specimen, may be used to support the cover glass. Alternatively, a short length of glass tube may be used (as shown in b). Such rings, however, are not easily prepared. Teflon rings, held in position with high vacuum grease, can be used when photographing temporarily mounted specimens when a cover glass is not necessary.

The "corner post" method shown in Fig. 261 (d) is sometimes used. Drops of varnish are placed on the slide where they will support the cover glass. However, it is difficult to arrange for all the blobs to be of equal height, and this may result in a crack in the cover glass. The crack will probably run over the area occupied by the specimen.

Fresh algae, which is easily damaged by the pressure of a cover glass, is usually mounted within a cell in glycerine jelly. Such cells are ringed with a cap and sealed off as a permanent slide. Care must be taken when the cover glass is sealed to the top of the cell wall for any trace of glycerine or water at the top edge of the wall will later result in a leak. Varnish is probably better than gold size for making this type of seal.

There are many subjects which, by their nature, are mounted in a fluid medium. Then one must run a ring of cement round the specimen. The following are very good cements:

(1) One part Canada balsam, one part paraffin wax. Melt and warm gently together until golden, apply warm.

(2) Tolu balsam cement. Two parts tolu balsam, one part Canada balsam, two parts shellac, and chloroform to bring to a syrupy consistency.

Most commercially prepared slides are ringed to provide an attractive finish and a more lasting slide. In addition to the cements mentioned above Brunswick black is very good, in both its application and appearance. If large numbers of slides are to be ringed, it is advisable to make or buy a turntable for the purpose.

It is not always necessary to make a permanent and mounted slide, especially as it is now possible to obtain a photographic record. In a way, the negative may replace the slide and it is always to hand for inspection and the production of prints. The preparation of temporary mounts is therefore increasing in the laboratory. If the specimen cannot easily be handled (a section of pickled cauliflower, for example), a very good method is to mount it in water and place a cover glass over the top, as in Fig. 266. The specimen should be well covered so that there is no possibility of spurious reflections or rapid drying.

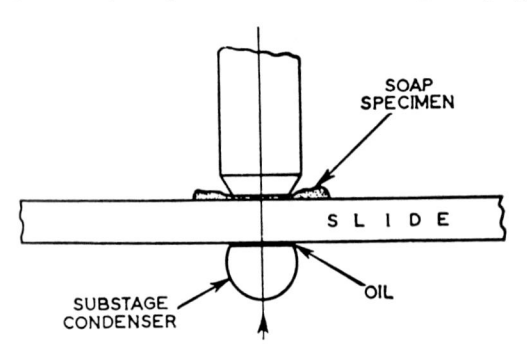

Fig. 262.

Considerable difficulty often arises when trying to find a suitable mounting medium as happened when trying to photograph shaving-soap cream and similar products (Fig. 262). Various media and mixtures were used and finally a mixture of equal parts by volume of diethylene glycol mono–ethyl–ether and ethylene glycol mono–ethyl–ether, with a refractive index of 1·42 at 18°C, was found to give the best differentiation in crystal structure. With oil-immersion technique, the oil is sandwiched between the lower lens of the objective and the top of the cover glass, and between the substage lens and the slide (Fig. 262).

The "hanging drop" method shown in Fig. 263 is particularly useful when

MOUNTING

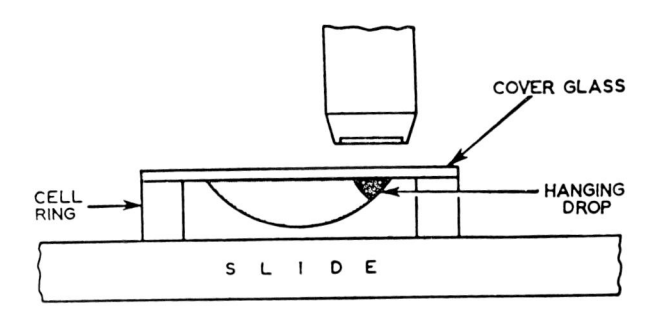

Fig. 263.

photographing emulsions. It makes the study and recording of the globular structure possible and comparisons can be made. Bacteria can also be photographed like this. The method is a temporary one and if the subject is left any length of time during photography evaporation takes place round the edges. The method adopted is quite simple and yet efficient. A thin glass rod is placed in a bottle of the emulsion, withdrawn and allowed to drain, leaving just enough to make a drop on a cover glass. This is then turned upside down and placed on a cell ring, as shown in Fig. 263.

Moulds can be grown on slide cultures (Fig. 264), enabling photomicrographs to be obtained. Mould spores can be collected in most places where there are foodstuffs. A simple method of collecting airborne spores is to expose a slide containing agar-czapek medium in a suitable position in store, larder, factory or hospital, for approximately 30 min. They will come into contact with the tacky surface of the medium from which they will grow (as shown in Fig. 264 a and b) if maintained at a temperature of approximately 25°C for about three days. When the growth is at its peak, carefully remove the agar, leaving the fungus growth (as in Fig. 264 c and Plate 7). Permanent stained slides can also be made and these give greater contrast when photographed, but some realism is lost and damage caused to the sporing heads.

Detel, a product used as a paint ingredient, can also be used in the laboratory as a mounting medium. There are some specimens (Plates 7f and 22d) which do not mix the many known and popular media, and Detel may often be used in such cases. Detel also has a very useful refractive index of 1·510.

Opaque specimens are comparatively easy to mount, there being very little preparative work, and such subjects, if semi-permanent, can be mounted in Canada balsam to prevent slipping. Moth and butterfly eggs (Plates 13 and 31) and many such subjects can easily be transferred with a leaf and fixed on a slide and successfully photographed many times during a period of days. Plate 57 of the butterfly's eggs clearly illustrates this point.

Some subjects, e.g. blood cells, do not lend themselves to the same treatment as tissues. In such cases it is usual to make a very thin smear along the slide, as in

M*

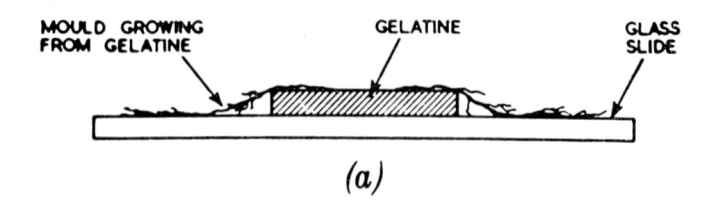

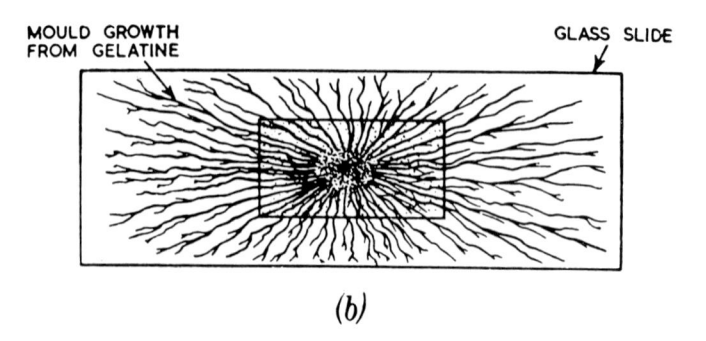

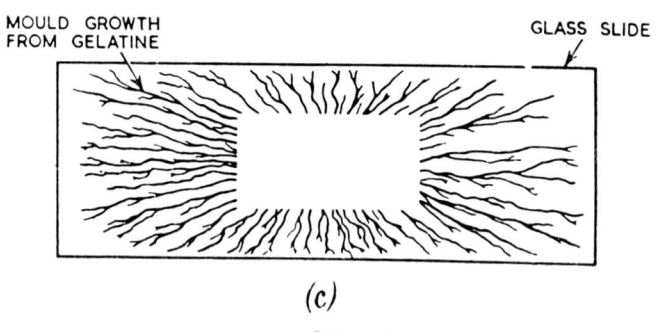

Fig. 264. Slide culture

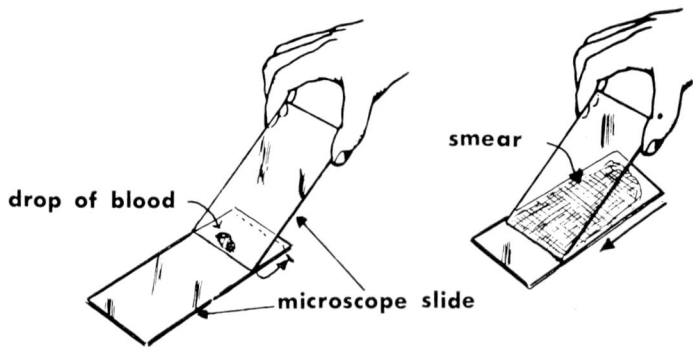

Fig. 265.

MOUNTING

Fig. 215 (a) and (b). High-power objectives are used and, as a rule, counts must be made. A small drop of blood is placed in the centre line of a slide about 1 to 2 cm from one end. The spreading slide is placed at an angle of about 45° to the slide and then moved back to make contact with the drop. The drop should spread out quickly along the line of contact of the spreader with the slide. The moment this occurs, the film should be spread by a rapid, smooth, forward movement of the spreader. The drop of blood should be of such a size that the film is about 3 cm in length (Fig. 265). It is essential that the slide used as a spreader should have an absolutely smooth edge and should be narrower in breadth than the slide on which the film is to be made, so that the edges of the film may be readily examined. If the edge of the spreader is rough the result is a film with ragged tails, containing many leucocytes.

The faster a film is spread the thicker and more even it will be. A common mistake is to make films too thin. The ideal thickness is such that there is some overlap of red cells throughout much of the film's length, with separation and lack of distortion towards the tail of the film and the leucocytes easily recognizable throughout its length.

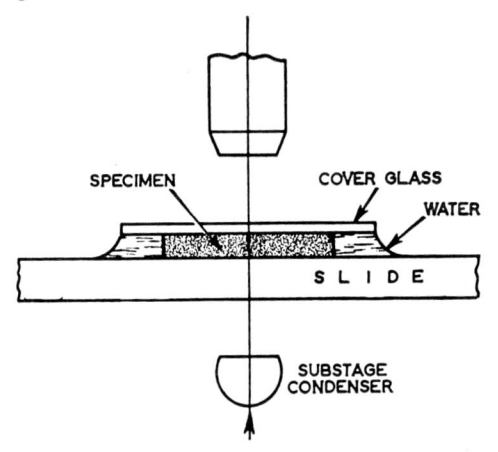

Fig. 266.

Objectives designed for homogeneous immersion may be used with covered or uncovered specimens, such as blood smears. The cover glass thickness is naturally of no importance with these systems but it is different with some present-day immersion objectives, in which the principle of homogeneity has been abandoned. This point should be remembered if immersion objectives computed for covered specimens are also employed for viewing smears which are left uncovered to save trouble. This is normally the case in the examination of blood and bacteria smears. While the slight degradation of the image thus produced may still be tolerable for routine work, the trouble of covering the specimen should definitely not be dismissed if full use of the high performance of the objective is to be made. It is true that the lack of a cover glass can be made up by immersion oil of a higher refractive index—n_D 20°C = C = 1·524 instead of 1·515 (Fig. 40)—or by increasing the tube length, but both methods are problematic. Some wet specimens must be photographed in a natural state: the best method is shown in Fig. 266. Figure 267 demonstrates prepared slides

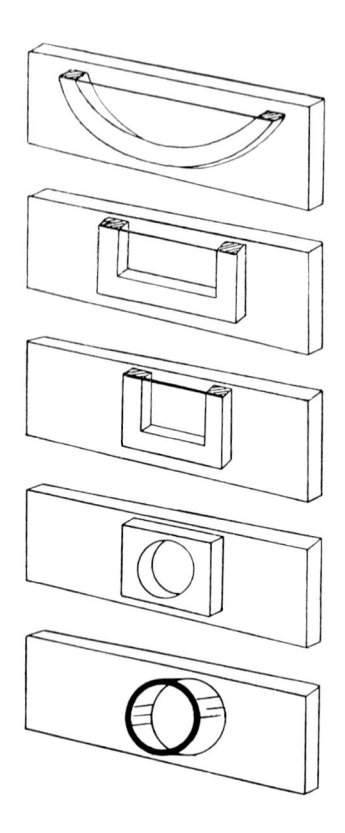

Fig. 267.

specially made for containing fluids. (For stains, chemicals, waxes, equipment for general pathology: Raymond A. Lamb, Ealing Road, Wembley.)

After the specimen is mounted it should be labelled before storing in the slide cabinet. The labels should be small and clearly written data about the subject, method of fixation, stain, date of preparation and the maker's name. It is interesting to buy an old prepared slide from a shop and find this information on it, and often startling to find that it may be well over sixty years old and still as good as the day it was mounted.

Some useful mounting media. Canada balsam. A yellowish resin, obtained from Canadian fir. Air bubbles rarely appear in this medium as they are automatically absorbed by the balsam.

Euparal. A very useful mounting medium for stained blood smears, starches and insect sections and particularly suitable for use with high-power objectives. It is a mixture of menthol, camphor, oil of eucalyptus, salol, paraldehyde and gum sandarac. Alcohol should not be mixed with or allowed to come into contact with Euparal, since it may cause bleaching. Dry within two or three days, when a covering of shellac can be applied.

Hydrax. This originated in the United States and requires more time and technique in mounting. After a liberal drop of Hydrax has been placed on the slide it should be left for 3–4 hr to allow evaporation to take place. During final mounting, the slide must be maintained at a constant temperature of 100°C, and should remain at this temperature for nearly 1 hr after mounting is complete. This is a drawback but the medium suffers a reduction in its refractive index. The dry slide is treated in the normal manner.

Methylene iodide. Most suitable for mounting semi-permanent preparations.

Pleurax. Useful in research and industry since it saves time when mounting specimens. These can be transferred direct from a 95% solution of alcohol to Pleurax. It has a high refractive index.

Diaphane. With its comparatively low refractive index, Diaphane is suitable for stained sections, insects and stained blood smears, and can be used directly after alcohol without any intermediate treatment. This quick-drying mountant can be used with nearly all stains with great success.

Glycerine jelly. Gelatine one part (by weight) and distilled water six parts (by weight) are left to digest for over 2 hr, after which seven parts (by weight) glycerine are added and thereafter 1 gm of pure thymol to every 100 gm of mixture. Warm, filter through filter paper, and store. Research laboratories make particular use of this medium in mounting drugs. The slide is warmed, the specimen placed in position and the jelly applied last. The keeping properties of this medium are equal to those of Canada balsam.

Detel. A paint product with many laboratory uses. It has a good dispersive quality and a useful refractive index (1·510). It is a useful mounting medium for starches, crystals, pollen grains and powders, is colourless and takes two to three days to set.

A.L.P. 1. Immersion oil. A non-drying mountant particularly useful for slides of the hanging drop type, when bacteria can be photographed alive. It is a transparent medium of refractive index 1·524 at 20°C. (It is available from Messrs. Cooke, Troughton and Simms).

Realgar. This very useful mountant is unequalled for finely marked diatoms, revealing far more detail than any other medium. It is not easy to handle. The dry specimen is placed on the slide with a piece of clean realgar alongside it. The slide is gradually heated until the realgar melts, when a cover glass is applied. When dry, the usual after-treatment can be carried out.

Green damar. This is preferred by some to the more common Canada balsam. The gum is dissolved in chemically pure benzene, after which it is diluted in resin. Most stained preparations mount and keep well in this translucent medium.

Henry green method. Green damar plays an important part in this method. It is mixed with cumar gum to make a really good dispersing medium.

Xylol. Its place in photomicrography is as a solvent and clearing agent.

Additional stains and routine methods.

Bismarck brown (aqueous). Stain for $\frac{1}{2}$–1 min. in a 0·5% solution of Bismarck Brown in distilled water, rinse and dry.

Ehrlich's haematoxylin. Stain for 20 min and wash in running water until the tissues appear blue. Counter-stain for $\frac{1}{2}$–1 min in eosin (yellow) and rinse in distilled water, passing quickly through the alcohol baths. Alternatively one may counter-stain for $\frac{1}{2}$ min in 0·5% aqueous Biebrick scarlet, wash quickly with 95% alcohol and finally pass on to absolute alcohol and into xylene, resulting in blue nuclei and yellowish-pink cytoplasm.

Van Gieson's stain. This enables rapid differentiation to be achieved between muscle and collagen. Stain for a period of 4 min, rinse in running water and pass straight on to the alcohols. Collagen fibres stain red, muscle and epithelia, yellow.

Malachite green. A superb stain to demonstrate nuclei and can be used with haematoxylin and eosin.

Masson's trichrome method. This brilliant technique, though not as simple as the Van Gieson method, can be relied upon once it is mastered. Histological and micro-anatomical sections 5–6 microns in thickness respond best. I. R. Baker has modified this stain as follows. Soak for 1 hr in 4% aqueous iron alum. Stain for 1 hr in Heidenhain's haematoxylin and differentiate in alcoholic picric acid until collagen fibres and cytoplasm are completely "decolorized", while nuclei still retain the stain. Wash for 5 min in running water. Stain the muscle and cytoplasm of the epithelial cells pink in xylidine ponceau and rinse in distilled water. The specimen is now placed in 1% aqueous phosphomolybdic acid and rinsed; the collagen fibres are stained in light green which occupies about 2 min, during which time the tissues are rinsed in distilled water for short periods and examined under the microscope. When staining is complete, pass the tissue quickly through an alcohol bath and into xylene. The final section should give blackish nuclei, green collagen fibres, and the muscle and cytoplasm of the epithelial cells, etc. pink.

Histological, pathological and zoological method. Volkensky's method is perhaps a little complicated, but nevertheless gives a brilliant tricolour finish. The chromatin is a different colour to the mitochondria and cytoplasm. Fix the section in

Mercuric chloride	5 gm
Potassium dichromate	2·5 gm
Sodium sulphate	1 gm
Distilled water	100 ml

for 6 hr and when ready for use add 1 ml of formaldehyde solution (approximately 40%) to each 20 ml of this fluid. Postchrome for two days in a saturated aqueous solution of potassium dichromate at 30°C and wash overnight in slowly running water. Sections should be cut 3 or 4 microns in thickness and put on a slide with water, the edges dried off and flooded with a 6% solution of acid fuchsin in aniline water. Place on a hot plate for 3 min, after which the slide should be left to cool for about 5 min. Lightly wash off excess stain with distilled water and differentiate with aurantia, 0·25% in 70% alcohol, until the cytoplasm is pale yellow while the mitochondria are still bright red. Rinse, and pass on into 1% aqueous phosphomolybdic acid for 4 min, rinse and stain for 4 min in Volkensky's methylene violet. Differentiate in tannin orange until only nuclei are blue and cytoplasm is yellow. Follow by a rinse in distilled water and lightly blot. Dip momentarily in 90% alcohol and transfer to absolute alcohol for 1 min. Pass section through xylene and into the mounting medium, Canada balsam. The final effect is

Mitochondria	red
Chromatin	blue
Cytoplasm	yellow.

Blood staining. Prepare a smear in the usual way (Fig. 256), allow to dry, fix in solvent methyl alcohol, B.D.H. Stain smears with Jenner's stain (Jenner's stain powder 0·5 gm, solvent methyl alcohol 100 ml) for 3 min in a moist Petri dish. Wash off with distilled water until pink.

Janus green. A 0·01% solution should be used for mitochondria (and lepidosomes). At extreme dilutions (0·002%) it is specific for mitochondria.

Methods for plant chromosomes. Navashin's fluid and iron haematoxylin. Fix for 1–2 days in Navashin's fluid (chromic acid, 1% aqueous solution 15 ml; glacial acetic acid 1 ml) and add 0·25 ml of a 40% solution of formaldehyde. Wash overnight in slowly running water and stain paraffin sections with iron haematoxylin, as follows: flood the slide with water and soak for 1 hr in 4% iron alum, rinse in water, and stain in Heidenhain's haematoxylin (dissolve 0·5 gm haematoxylin in 10 ml of 96% alcohol and add 90 ml distilled water), rinse and differentiate in 2·5% iron alum solution until only the chromosomes are stained. Wash for 5 min in running water and pass through alcohols and xylene and mount in Canada balsam.

Aceto-carmine. This is most suitable for the preparation of non-permanent preparations of chromosomes. Simply dip the section in a drop of aceto-carmine†, lay on a slide and cover with a cover glass. Chromosomes are stained red.

Some of these staining methods are described in a very useful booklet specially prepared for microscopical use by The British Drug Houses Ltd., Chemicals Group. I have used them all and found them most satisfactory.

Fluorochromes. The fundamental work on fluorochromes was carried out by Dr. H. C. Max Haitinger, of Vienna, who has also introduced many new stains and increased the staining intensity of existing fluorochromes. These stains, which cover a wide range of colours, provide a means of differentiating tissues as well as detecting tiny quantities of infiltrations or deposit. Furthermore, fluorochromes can be used for histological and bacteriological examination.

Narcotizing. It is often necessary to arrest all movement when photographing fast swimming pond creatures, without disturbing them in any way. The following has been proved to be very useful in this

2% solution hydrochlorate of cocaine	3 parts
Alcohol	1 part
Water	6 parts

Replica technique

For making impressions and then photographing them is not widely practised and yet such a technique is extremely revealing and rewarding. Epidermal impressions of leaves on living plants, the upper and lower surfaces of leaves and petals can be recorded through this medium. This technique does not cause damage to the living plant, in fact a series of impressions can be made over a period of time, making possible examinations hitherto difficult to record. Often the replica furnishes the photomicrographer with far greater detail than when photographing the original, and so the technique offers much to industry and research. A replica of a 360° surface can be made and recorded on one photomicrograph, this also applies to the inside of such a surface. Places inaccessible to the microscope can now be pictured in detail.

The technique, which used ICI Silcoset‡ 105 RTV Silicone rubbers, is both simple and highly effective. One teaspoon of liquid Silcoset 150 is mixed with a couple of drops of Curing Agent "D" and applied to the surface of the specimen. After 4 min at room temperature the liquid cures to a resilient rubber pad which can be easily peeled from the specimen without damaging it in any way. Liquid Silcoset has a very low surface tension and it flows into the minutest epidermal cavity. Also it contains no solvents and cures with virtually no shrinkage.

† Saturate boiling 45% acetic acid with carmine, cool, then filter.
‡ Ciba (A.R.L.) Ltd., Duxford, Cambridge.

These properties ensure it takes an accurate impression of minute details.

To examine or photograph the surface, the replica must be coated with a number of applications of nail varnish. Micrex† can be used in place of nail varnish but takes 24 hr to dry, whereas nail varnish dries in a few seconds. The thin dried layer of varnish is peeled off the surface of the Silcoset and is placed lower side up on a microscope slide, which is gently warmed so that the thin layer settles and sticks to the slide. The replica can then be shadowed with aluminium to give contrast and depth, two necessary elements.

Some terms used in photomicrography

Abbe test plate. A test slide with a regular array of intersecting lines ruled through a very thin layer of silver, protected by a cover glass. The lines are used in the determination of spherical and chromatic aberrations present in an objective.
Aberration, chromatic. The inability of a lens to bring light of different colours to a common focus.
Aberration, spherical. The inability of a lens to bring the marginal and central rays to a common focus.
Aberration, zonal. Spherical aberration cannot be corrected entirely and there are usually some zones with a very slightly different focus from the rest. This residual aberration is called zonal aberration.
Absorption bands. These occur when "white" light passes through an absorbent specimen, creating a reduction in the transmitted light, thus the gaps in the spectrum are referred to as absorption bands.
Achromatic. Partially corrected for chromatic and spherical aberrations.
Achromatic objective. An objective corrected for chromatic aberration in two colours (yellow–green) and for spherical aberration in one colour.
Alignment. Mutual coincidence of optical and mechanical components along a common axis.
Anisotropic effect. When rotating the specimen between crossed polarizers, a change in light intensity is observed.
Apertometer. A device for determining the numerical aperture of objectives.
Aperture, angular. The angle of the cone of light which, proceeding from a sharply focused specimen point, is received by the objective.
Measured by n. sin u. it is called numerical aperture.
Aperture, numerical. The product of the refractive index of the medium between objective and specimen and the sine of half the angular aperture of the objective. The higher the N.A. the greater the resolving power of the objective.
Aplanatic. A term used to describe a substage condenser corrected for spherical aberration.

† T. Gerrard and Co., East Preston, Sussex.

Apochromatic objective. An objective which is corrected for chromatic aberration in three colours and for spherical aberration in two colours. These objectives are required for the most exacting work.

Apochromats, semi-. Lenses having one or more fluorite components and having correction between that of the achromat and the more expensive apochromat.

Aspheric. Applied to a lens which forms an image in a parabolic surface rather than in a spherical surface. This reduction in spherical aberration is achieved by grinding the margin of one lens surface more than the centre.

Aspherical. Curved surface, not having the shape of a sphere.

Body tube. The tube which carries the draw-tube at its upper end and the objective, or objective holder, at its lower end. The focusing is accomplished by moving this tube up and down with fine and coarse adjustments.

Brightfield illumination. Common type of illumination applied in ordinary microscopy. Specimens appear dark against a bright background. The converse of darkground illumination.

Bull's eye. An auxiliary plano-convex condensing lens.

Coarse adjustment. The adjustment which moves the body or stage of a microscope relatively large distances.

Compound microscope. An instrument with more than one lens system, such as an objective and an eyepiece.

Condenser. The condenser focuses the light from the source on to the specimen.

Condenser, substage. Directs the light on to the specimen and into the objective. Many types are available, from the simple Abbe to the complex aplanatic-achromatic suitable for work at the highest magnification.

Correction collar. A collar fitted to an objective whereby the optimum cover glass thickness may be varied.

Cover glass. A very thin glass plate, cut in circles, squares or rectangles, used to cover the specimen, and made in various thicknesses. Thick cover glasses are suitable only for low-power work; the usual thickness for work at high-power is 0·18 mm and 0·17 mm.

Critical angle. The angle of incidence beyond which light will not pass into a medium of lower optical density.

Definition. The faithfulness with which the optical system magnifies and reproduces the specimen details. The brilliance, clarity, distinctness and sharpness of the photomicrograph image.

Depth of field. The axial distance through an object which may simultaneously be brought into focus in the image. In microscopy it is usually very small. The depth of focus of an objective decreases with increasing N.A. For an objective of $\frac{1}{2}$ in. focal length it is approximately 0·0008 in.

Depth of focus. Depth of focus can be termed the focal range an image is visible as a clear line or point imaged in the film plane of the camera.

Diaphragm, iris. A circular opening of variable aperture.

SOME TERMS USED IN PHOTOMICROGRAPHY

Diffraction. Interaction between minute structural detail in the specimen and the wave-fronts of the incident light produces diffraction phenomena. If the substage iris diaphragm is not opened sufficiently, spurious diffraction phenomena appear in and around the image.

Dispersion. The separation of light of different colours on passing from one medium into another of different optical density.

Draw tube. Adjustable tube situated in the top of the body tube.

Dry objective. Microscope objectives designed to be used dry, without immersion liquid. The $\times 40$ objective is quite frequently referred to as the "high dry" objective: the $\times 10$ "low dry" objective.

Empty magnification. High magnification which increases image size without enhancing resolution of specimen detail. The limit of effective magnification should be in the range of 500 to 1000 times the numerical aperture of the objective, e.g. $\times 43$ objective N.A. 0·55; effective maximum magnification = $1000 \times 0·55 = \times 550$.

Equivalent focal length. Usually used to designate the true focal length of a microscope objective.

Exit pupil. The point on the axis above the eyepiece where all the principal rays of light intersect. This can be found by holding a ground glass at about 7·4 mm. above the eye-lens of the Huygen eyepiece, or 23·6 mm. above the eye-lens of a wide field eyepiece.

Eyepiece. An optical system which magnifies the primary image of the objective forming a virtual image at a distance of 10 in. from the eye.

Field. The area of the specimen which is in view at one time.

Field diaphragm. A diaphragm limiting the field of view. In a properly adjusted and equipped illuminator, it is the iris diaphragm governing the size of the illuminating beam presented to the specimen.

Field of view. The visible area seen through the microscope when a specimen is in focus. It is usually expressed in mm diameter and can be determined by focusing onto a finely graduated, transparent millimeter scale on the microscope stage. The field of view varies directly with the resultant magnifications—the greater the magnification, the smaller the field of view.

Filter. A transparent sheet of colour material which absorbs some part or parts of the spectrum and transmits others.

Fine adjustment. The fine-focusing mechanism of the microscope, enabling small adjusting movements to be made.

Focal length. When a parallel beam of light is brought to a focus, the distance of the image from the nearest focal plane within the lens is the focal length of the lens.

Focal plane shutter. A shutter with a slit in it which moves across the face of the emulsion in the focal plane of the lens.

Graticule. A ruled disc placed in the focal plane of the eyepiece; a means of measuring.

Holoscopic eyepiece. An eyepiece manufactured by Messrs. Watson and Sons and suitable for achromatic and apochromatic objectives.
Glare. Unfavourable light scattered by a specimen, spoiling image detail. Scattered or strayed light within a microscope system generally caused by improper use of diaphragms and condensers.
Homal eyepiece. A compensating eyepiece made by Zeiss; it provides a wide flat field.
Huygen eyepiece. Consists of two plano-convex lenses with a diaphragm between them.
Illumination, critical. An image of the source is focused on the subject.
Illumination, darkground. A means of illumination whereby the specimen is made to stand out as a brightly lit body against a black background. No direct light can enter objective.
Illumination, Köhler. An image of the light-source is formed at the substage condenser iris.
Immersion objective. An objective computed to operate with a thin film of oil or water between the front lens and specimen or cover glass. As a rule cedarwood oil is used because its refractive index is similar to that of the crown glass commonly used. An oil-immersion system is sometimes called a homogeneous system.
Immersion oil. This has a set refractive index, dispersion and viscosity characteristics for use with high power "oil immersion objectives".
Iris diaphragm. An assembly of thin leaves controllable by a lever to produce variable-sized openings. It is generally associated with microscope and illuminator condensers of intermediate and advanced types.
Light. The human eye can see clearly radiation of 400 to 800 nm and is most sensitive at 555 nm, yellow-green light. Light is radiant energy of above wavelengths which upon reaching the retina of the eye stimulate nerve impulses to produce the sensation of vision. White light is composed of a mixture of coloured light of different wavelengths. When a specimen is too transparent to be seen effectively, it may be stained—then it can be seen by the colour image formed as the dye absorbs certain wavelengths of light and transmits the others to the eye.
Magnification. The ratio of the apparent linear size of an object projected through the microscope (the virtual image) to the size of the object as it appears to the unaided eye at a distance of 250 mm. In photomicrography the ratio is usually expressed in terms of "diameters", "power", "$\times$" or "times", e.g. $\times 180$. The compound microscope has two separate lens systems. The one nearest the specimen magnifies the specimen a definite initial amount. The other lens system, the eyepiece, further magnifies this image (real image), so that the resultant image seen by eye (virtual image) has a magnification approximately equal to the product of the two systems. The primary magnifications of objectives and eyepieces are engraved on each part.

Mechanical stage. A microscope stage or attachment which enables controlled traversing of a slide to be accomplished in two directions at right angles to each other.

Mechanical tube-length. The length of the body tube from the lower flange of the thread which houses the objective to the upper end of the draw-tube. Usually 160, 170 or 180 mm.

Meniscus lens. A concavo-convex lens, such as a spectacle lens, can be used as a supplementary lens in ordinary photography.

Micrograph. A free-hand drawing of the image seen through the microscope.

Micrometer eyepiece. An eyepiece fitted with a graticule in the plane of the field diaphragm. The graticule is usually divided into 100 equal divisions.

Micrometer, stage. A microscope slide 3 × 1 in., bearing a 1 or 2 mm scale divided into 0·01 mm spaces. Used to determine magnification.

Microphotograph. A minute image, on an emulsion, of a large object.

Microscope. See Compound Microscope.

Microscope, simple. A single magnifying glass mounted in a simple stand. When placed close to the eye a magnified virtual image is seen.

Monochromatic light. Light of one colour, incorporating a narrow band of the spectrum.

Nosepiece. A fitting attached to the lower end of the microscope body tube to carry from two to five interchangeable objectives.

Numerical aperture. See Aperture, Numerical.

Objective. The lens system used to form the primary image.

Oblique light. A narrow beam of light directed at the specimen at an angle to the optical axis. This method, particularly with transmitted light, must be used with care.

Oil immersion. See Immersion Objective.

Optics. The science dealing with the properties of light and vision.

Optical axis. The principal axis of a lens or lens system, through the centres of curvature of the faces.

Optical bench. A bench (usually of metal) upon which optical components can be moved without disturbing the alignment of the optical axis. Some microscopes are built upon an optical bench.

Optical filters. All glass or laminated gelatin filters coloured or neutral used to modify the light source. The light from a tungsten filament bulb is yellow and is usually rendered whiter by inserting a blue filter to absorb the unwanted excess red.

Optical glass. A very high quality glass made especially for use in scientific instruments. Good microscope lenses and prisms are made from such glass having specific refractive index and dispersion values.

Paraboloid condenser. A condenser specially made for darkground work, having a parabolic reflecting surface.

Parfocal. When objectives may be interchanged without altering the focus of the image they are said to be parfocal.

Periplanatic eyepiece. An eyepiece which produces a wide flat image. It is fitted with a doublet eye-lens.

Phase contrast. A discovery of F. Zernike which enables slight differences in optical density or path length to be seen as differences of contrast.

Photomacrograph. An enlarged image of an object often produced without the aid of the microscope.

Photomicrograph. An enlarged image of a minute object usually photographed through a microscope.

Ramsden eyepiece. An eyepiece consisting of two plano-convex lenses with their flat sides nearest the objective. This system is known as the positive type because the focal plane is outside the system.

Real image. The image formed in space (aerial image) by a system of lenses. Its presence can be viewed only by the insertion of a receiving screen, ground glass or projection screen.

Refraction. A beam of light passing through an interface between media of differing refractive indices suffers a change of direction, or refraction.

Refractive index. The ratio of the sine of the angle of incidence and the sine of the angle of refraction when a beam of white light passes from air into a given transparent medium.

Resolving power. The ability of a lens to separate the images of closely spaced lines. It is directly proportional to the N.A. of the objective and increases with a decrease in the wavelength of the light used.

Simple microscope. Single magnifying lens, usually operative with bellows or extension tubes.

Slide microscope. The slide of white glass measuring 3×1 in. used for mounting the specimen.

Spectrum. The radiation over a given range of wavelengths.

Stage micrometer. See Micrometer, Stage.

Stage, microscope. The platform upon which the microscope slide is placed for inspection. See Mechanical Stage.

Supplementary lens. A lens of meniscus type, often placed in front of a camera lens to reduce its effective focal length and produce a larger image of the subject. Used in photomacrography.

Tube-length, optical. The distance from the upper focal plane of the objective to the lower focal plane of the eyepiece.

Vertical incident illumination. Light is projected down through the objective in use on to the specimen and reflected back.

Vertical transmitted illumination. A system in which the light is projected through the specimen from a substage condenser.

Virtual image. The apparent size and position of the object specimen. This image (not a real image) seen with the microscope appears to be as far away as a book is when reading. This distance is generally agreed to be 250 mm.

Widefield eyepiece. An eyepiece fitted with an achromatic doublet eye-lens having the plane side of the lower lens nearest to the objective.

Working distance. The distance between the front lens of the mount of the objective, when the microscope is focused on a specimen preparation, and the top of the cover glass. The greater the primary or initial magnification of the objective the smaller the working distance.

Plates

1. The death of Lord Nelson, ×80.   361
2.A   (a), (b) Penicillin mould growing on agar.
       (c), (d) Bacteria.
       (e), (f) *Staphylococcus aureus*, ×14,400.   362
2.B   Diagrams of technique used in 2.A (a), (b), (c), (d).   363
3. (a) Corrosion on the inside of a lead tube, ×10.
    (b) Watch chain, simple microscope.   365
    (c) Micro lens, iris diaphragm and lens mount.
4. (a) Mite in flour, ×120.
    (b) Section of skin of electric Catfish, ×120.
    (c) Transverse section stem of Clematis, ×36.
    (d) Condensation droplets, ×120.
    (e) Radiolarian ooze, ×180.
    (f) Yttrium Platinocyanide crystals, ×150.   366
5. Carl Zeiss, Oberkochen, research microscope.   367
6. (a)–(e) The surface of a highly polished cigarette lighter, ×36.   369
7. (a) *Subtilis bacillus*, ×1,500.
    (b) Stomata, tobacco leaf, ×900.
    (c) *Aspergillus amstelodami*, ×360.
    (d) Transverse section of stem of tobacco, ×180.
    (e) Seed of paullini, ×15.
    (f) Pollen of lily, ×300.   370
8. Airy disc simulation.   371
9.A–E   The relationship between substage iris diaphragm aperture and the final image of the specimen, ×400.   373
10. (a) Chain of Conidia of hop mildew, ×360.
     (b) Bacteria *Escherichia coli*, ×1,200.
     (c) Flowering *Aspergillus terreus*, ×360.
     (d) Bacteria *Escherichia coli* normal cells, ×4,500.
     (e) *Fusarium solani*, ×360.
     (f) Bacteria *Escherichia coli* after treatment with an antibiotic, ×4,500.   374
11. (a) Objective with adjustable collar.
     (b) Diffraction around the edge of a specimen.
     (c) Tilting stage.   375

12. (a) Rabbit embryo normal palate.
    (b) Rabbit embryo cleft palate. ... 377
13. (a) Eggs of Large Cabbage White butterfly, *Pieris brassicae*, $\times 39$. ... 378
13. (b) Eggs of Common House fly, *Musca domestica*, $\times 72$. ... 379
14. 1–16 A series of exposures made through the eggs of a rabbit. ... 381
15. (a) Transverse section stem showing cell structure, $\times 38 \cdot 4$.
    (b) Transverse section stem showing cell structure, $\times 80$.
    (c) Enlarged from (a) to $\times 80$.
    (d) As for (a) but taken with an 8 mm objective, $\times 80$.
    (e and f) Bacteria *Salmonella typhi*, $\times 424$ and $\times 864$. ... 382
16. Hand made wooden microscope slides. ... 383
17. (a and b) Colony of Vorticalla, $\times 96$.
    (c and d) Transverse section of Olivine, plane polarized light and polarized light, $\times 18$.
    (e and f) Group of diatoms, transmitted light and darkground illumination, $\times 60$. ... 385
18. (a) Newspaper print, $\times 25$.
    (b) Newspaper print, $\times 80$.
    (c) Newspaper print, $\times 100$. ... 386
19. Magni optical changer, 16 corrected optical elements. ... 387
20. Standard optical system for Bausch and Lomb microscope. ... 389
21. (a) Corrosion on the inside of a pressurized dispenser.
    (b) Corrosion on the inside of a pressurized dispenser, $\times 8$.
    (c) Corrosion on the inside of a pressurized dispenser but with grazed illumination, $\times 8$.
    (d) Egg of Pine moth, *Hyloicus pinastri*, on pine needle, $\times 16$. ... 390
22. (a) Pollen of hazel, $\times 720$.
    (b) Pollen of poppy, $\times 720$.
    (c) Penicillin *Brevicompaction*, $\times 510$.
    (d) Glass wool.
    (e) Colony of bacteria growing on agar, $\times 21$.
    (f) Colony of bacteria growing on agar, $\times 21$ exposed to ultra-violet. ... 391
23. Vickers M15C microscope and camera. ... 393
24. Comparison eyepiece of Zeiss, Jena. ... 394
25. Left: pinewood, $\times 27$. Right: deciduous wood, $\times 27$ taken on Zeiss, Jena, comparison eyepiece. ... 395
26. (a) Vickers' micrometer eyepiece.
    (b) Vickers' screw micrometer eyepiece. ... 397

PLATES

27. Linhof Universal Accessory Stand.   398
28. (a) Damaged human hair, ×250 taken with 25 mm objective.
    (b) Damaged human hair, ×250 taken with 8 mm objective.
    (c) Polythene, ×80.   399
29. (a) Palate of Whelk, ×140 transmitted light.
    (b) Palate of Whelk, ×140 phase contrast.
    (c) X-ray of leg of mouse.   401
30. (a) Aluminium–tin alloy, ×100.
    (b) Scratched chromium-plated surface, ×180.
    (c) Aluminium–tin alloy, ×100 X-ray.
    (d) Scratched chromium-plated surface by interferogram, ×180.   402
31. (a) Egg of Rare Swallowtail moth, *Aurapteryx sambucaria*, ×48.
    (b) Fossil Promicroeras, ×6·4.
    (c) Ungraded sand, ×16.   403
32. Saw of Saw-fly (4 photomicrographs).   405
33. Guinea-pig siamese twins.   406
34. (a) Skin of eel, ×320.
    (b) *Licmophora flabellata*, ×480.   407
35. (a) Tyndall beam, ×54.
    (b) Foot of fly, ×48.
    (c) Rusted surface of metal after treatment, ×24.
    (d) Foot of Sheep tick, ×54.
    (e) Scales of eel under polarized light, ×18.
    (f) Surface of aluminium sheet after treatment with chromate-oxidation process.   409
36. (a) Beck versalite lamp.
    (b) Beck integral substage illuminator.   410
37. (a) Wine crystals, ×161 transmitted light.
    (b) Wine crystals, ×161 source out of vertical.
    (c) Wheat rust on wheat leaf, ×210.
    (d) Table salt, ×24·5.   411
38. (a), (b) Leitz in-line mirror monochromator. Courtesy of E. Leitz.   413
39. Pinheads and one actual size.   414
40. (a) Flea of human, *Pulex irritans*, ×49.
    (b) Water flea, *Daphnia Pulex*, ×42.   415
    (c) How not to do it: flea of human, *Pulex irritans*, ×49.   416
41. Line-overlap in the laboratory.   417

42. (a) Application of filters to photomicrography, stem of
    clematis, × 38·5.
    (b) Dissolution rate of aspirin and glycine crystals, × 10·5.   419
43. Application of filters. Molar tooth, × 12.   420
44. Continuous running filter of Carl Zeiss, Oberkochen.   421
45. Leitz Orthoplan microscope with fully automatic head.   423
46. Nikon Multiphot 4 × 5 bellows camera and diascopic
    illuminator.   424
47. Three Bausch and Lomb laboratory cameras.   425
48. (a) Rabbit embryo and double gall bladder.
    (b) Wedge glass container.   427
49. Particles in broth, × 24.   428
50. (a) Longitudinal section upper surface of precipitation band.
    (b) Transverse section of precipitation band.
    (c), (d) Ouchterlony plate.   429
51. Micro-gel precipitation bands in small glass capillary tubes.   431
52. (a) Rabbit blastocysts, × 9·9.
    (b) 14 day rat embryo, × 5·4.   432
53. Deformed rabbit blastocysts, × 36.   433
54. (a) Rabbit eggs as seen by the eye and camera.
    (b) Interferogram of echeloned surface, × 160.   435
55. Gas bubbles at the solid–liquid interface, × 55.   436
56. Defective bottle tops.   437
57. The hatching of Large cabbage white caterpillars.   439
58. Larva of water beetle, *Dytiscus marginalis*, centre, × 2.   440
59. (a) Zeiss, Oberkochen, large fluorescent microscope and camera.   441
59. (b) Vickers 35 mm camera, M15C microscope with projection
    base and J35 exposure unit.   443
60. Series of rat teeth taken on the simple microscope, × 40.   446
61. Wild, M20 binocular microscope.   447
62. (a and b) Testis of rat, × 117.
    (c) Crystal growth, × 180.   448
63. (a) Carl Zeiss, Oberkochen, hot stage.
    (b) U.N. Auto cleaner.   449
64. Crystal growth, × 77.   451
65. Larva of Great water beetle, *Dytiscus marginalis*, × 1·4.   452

PLATES

66. Three photographs of ciné apparatus.	453
67. Carl Zeiss, Oberkochen, Stereomicroscope IV.	455
68. Optical system within Bausch and Lomb objective.	456
69. Mast cell degranulation, ×800.	457

Colour plates

70. (i) Knobbed hairs on the edge of petal of Field spurrey, *Spergula arvensis*, ×190. (ii) Raft of gnat eggs showing larvae inside, ×110.	459
71. (i) Aspirin crystals, ×50. Demonstrating optical staining. (ii) Aspirin crystals, ×50. Demonstrating optical staining.	461
72. (i) Transverse section stem of clematis, ×25. Demonstrating optical staining. (ii) Vertical section of human lip, ×10.	463
73. (i) Fused Phloidian crystals, ×100. Polarized transmitted light. (ii) Machine embroidery, ×16. Reflected light and reflector.	465
74. (i) Eggs of *Attacus edwardsii*, ×16. (ii) Amphioxus, ×7. Unfiltered transmitted light.	467
75. (i) Human jejunum, ×190, stained with azane. (ii) Cancer pagurus L. (Common crab), Chitinous tendon. Congo red in polarized light, ×30.	469
76. (i) Frog's blood, fluorochromed, ×700. (ii) Particles of asbestos in sputum, ×750.	471
77. Transmitted light interference photomicrography in cytological research.	473
78. Transmitted light interference showing living nerve cell of the mesocerebral ganglion of the pulmonate gastropod *Helix pomatia*. L.	475
79. Reflected light interference showing the surface of a ground steel plate with sodium light fringes.	477
80. (i) Reflected light interference of the surface of germanium wafer polished to an angle of 2° to the top surface. (ii) An interferogram of the edge of a film of glue on a clear plate glass.	479

Plate 1. The death of Lord Nelson. This recording was made from the micro-negative seen in Fig. 2. The micro-negative was photographed through the microscope, using an objective and eyepiece. The original micro-negative is just under 2 mm × a little more than 1 mm, bringing the print to approximately ×80. Would your negatives enlarge to this dimension?

Plate 2A. (a), (b) Penicillium mould growing on agar. Often difficulties arise in producing a clear defined image; (b) was taken without further focusing, but a change in the direction of the lamps was made as seen in the illustration. (c), (d) Extremely small colony of bacteria recorded by two techniques. Although (c) is critically sharp it looks woolly. A change in the direction of the lamps has brought about a difference in the recorded image. The method can be followed by the illustrations. (e), (f) Bacteria Staphylococcus aureus, "Penicillin resistant strain". Normal cells untreated, ×14,400 (electron-micrograph). In (f) cells exposed to 10 μg/ml Cloxacillin (penicillin). The high resolving power of the electron microscope reveals that the bacteria has been destroyed, ×14,400.

Plate 2B. Methods of illumination. Diagrams of technique used in 2A.

Plate 1.

362 PHOTOMICROGRAPHY

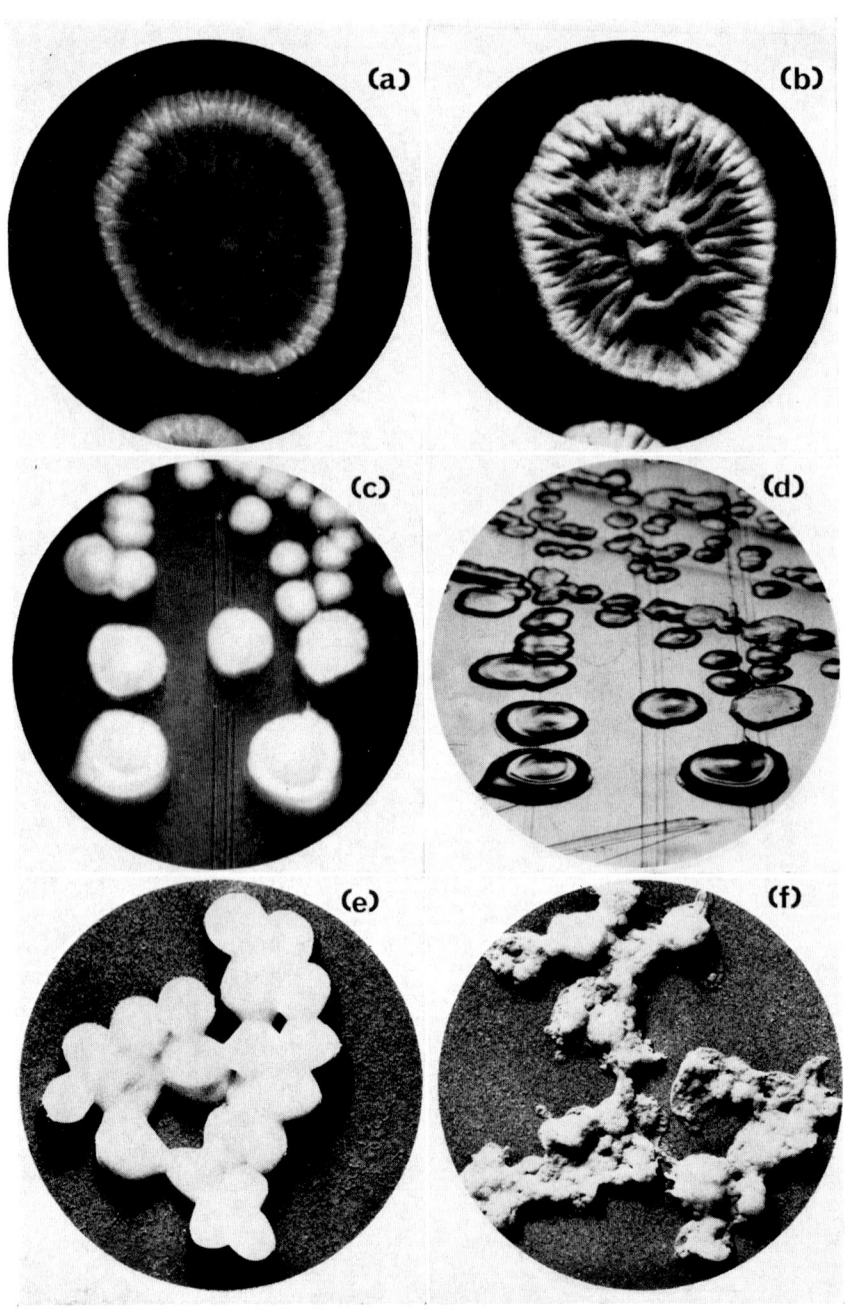

Plate 2A.

Plate 2B.

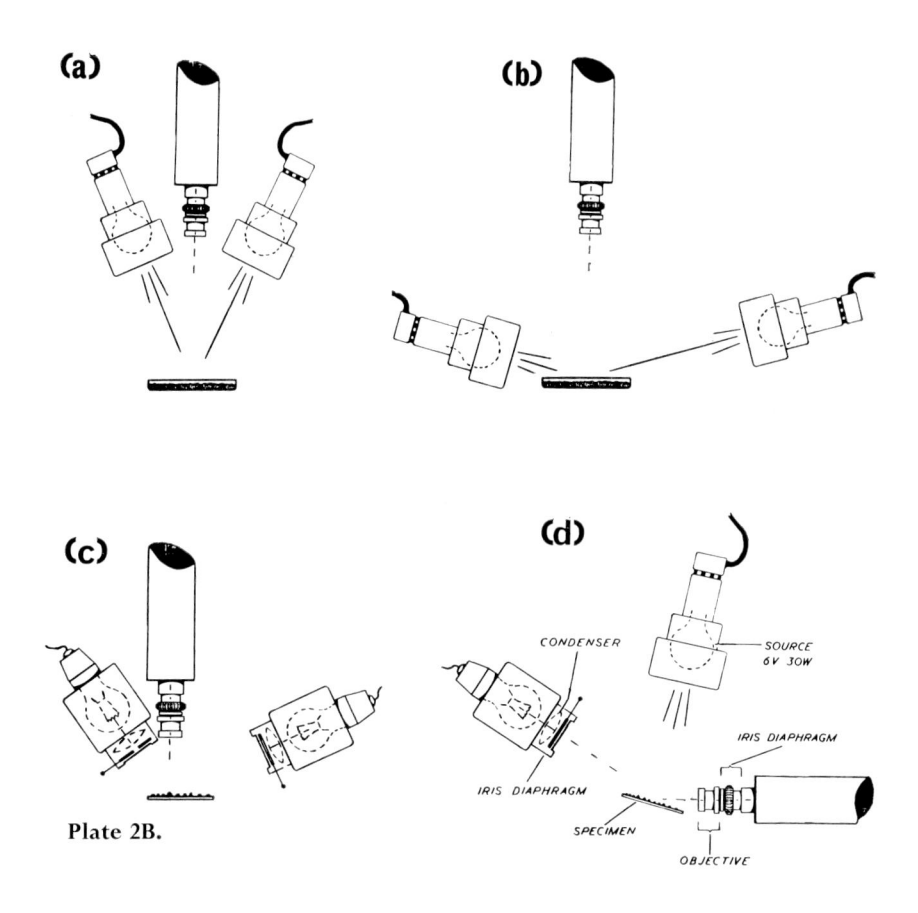

Plate 3. Corrosion on the inside of a lead tube, ×10. (a) Photographed by using a 50 mm objective attached to a ¼ plate camera. Two 6V 30W lamps were used to illuminate the subject. *e* shows the outside enamel, *f* corroded edge, *c* indicates a fracture on the inside of the neck of the tube, *a* was cut with scissors when opening the tube. The objective mount has a camera thread at the base and the standard microscope thread which takes the lens and iris diaphragm indicated by the arrow. (b) The value of a "simple microscope", note the recorded detail.

Plate 4. (a) Mite found in flour, ×120. (b) Section of skin of Electric catfish showing the parasitic Ciliate, Icthyophthirius *in situ*, ×144. (1) Cells of electric organ. (2) Epidermal cells. (3) Wall of vesicle containing extremely small parasites. (4) Vacuole. (5) Cross section of limb of horseshoe shaped nucleus. (6) Cytoplasmic inclusion. Stain: Short's iron haematoxylin, alcian green and eosin. (c) T/S Stem of clematis, ×36. (d) Fine condensation droplets photographed on the inside of glass lid. Illumination according to Fig. 88. *D*. Magnification ×48, Total ×120. (e) Radiolarian ooze, ×90, Total ×180. Two lamps used to illuminate the subject. (f) Yttrium platinocyanide crystals, ×150. Dry mount, transmitted light.

Plate 5. Carl Zeiss, Oberkochen, research microscope can be fitted with a camera. (Courtesy of Carl Zeiss, Oberkochen.)

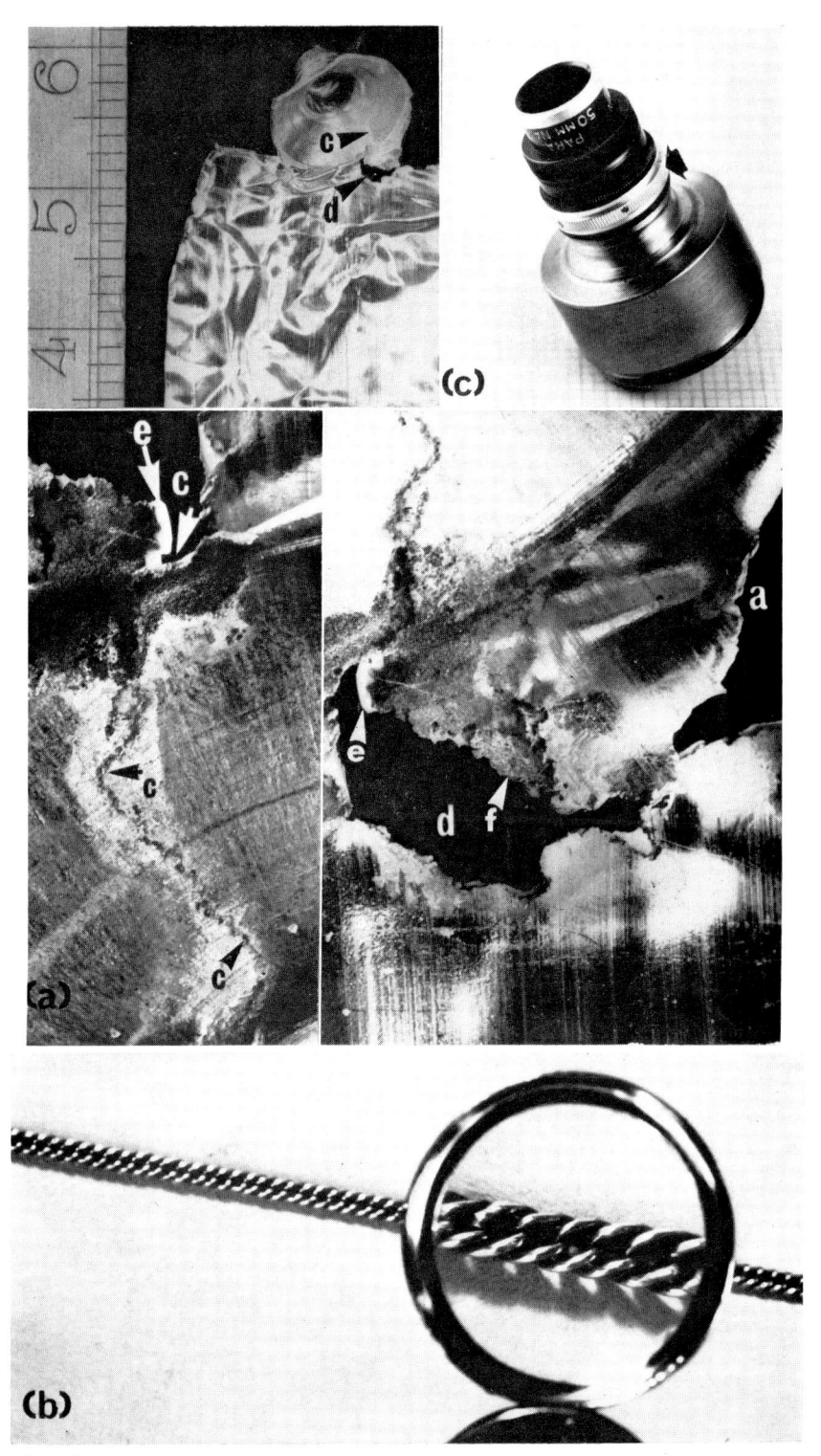

Plate 3.

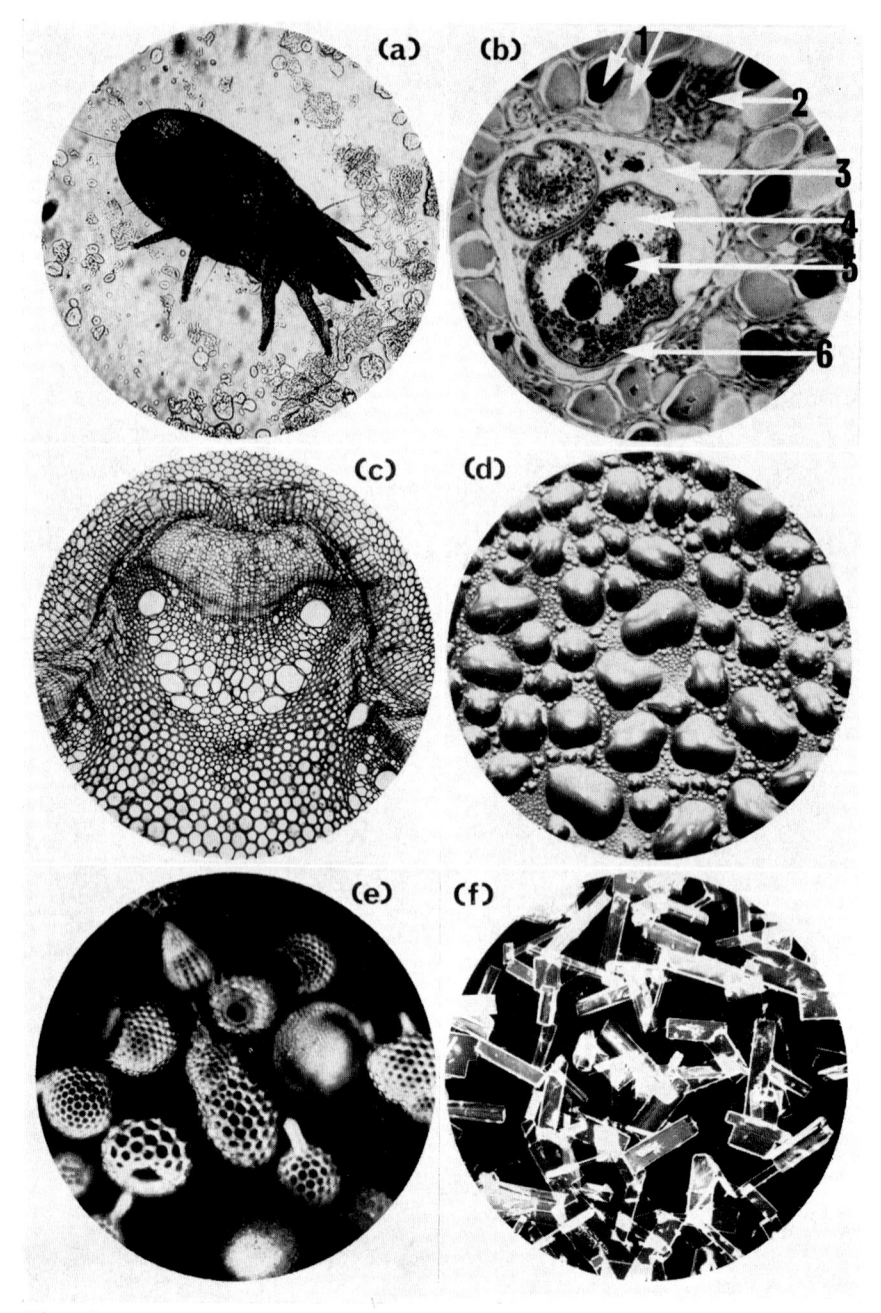

Plate 4.

Plate 5.

Plate 6. The surface of a highly polished cigarette lighter, ×36. (a) After coating, 6 sec exposure. (b) Coarse polishing, 6 sec exposure. (c) Fine polishing, 60 sec exposure. (d) Polished with lanolin, exposure 20 min. (e) Polished with lanolin and swan down, exposure 30 min.

Plate 7. (a) *Subtilis bacillus*, ×1,500. The objective was used at N.A. 1·37. Non-drying immersion oil with refractive index = 1·524 was used between substage condenser and underside of slide and between objective and top of slide. (b) Stomata, tobacco leaf, ×900. Data the same as for (a) with the exception of a shorter bellows extension. Optical system was the same. The specimen was tilted 4° for the exposure showing considerable depth in the stomata. (See Fig. 114.) Note the diffraction phenomenon is not present in either of these high-power reproductions. (c) *Aspergillus amstelodami*, ×360. Unstained specimen growing on agar on a microscope slide (Figs 1 and 264). No cover-glass used. (d) Transverse section stem of tobacco, ×180. Photographed with shadowed transmitted light. Specimen lies on tilting stage, photographed with 7° tilt to the left and then 7° tilt to the right (Fig. 115). Here the subject has a degree of modelling; compare with a similar section in Plate 4 (c). (e) Seed of paullini, ×15. Low-power objective used together with incident light illumination, one light at near vertical plus a reflector on the right. These and similar subjects are things of great beauty and not seen other than in this way. (f) Pollen of lily, ×300. This photomicrograph demonstrates the use of both vertical transmitted illumination and near vertical incident light from a 6V 30W lamp. The former light provided the silhouette while the latter enabled the rough texture to be recorded.

Plate 8. Airy disc simulation. (From "Microphotography", Chapman and Hall Ltd.)

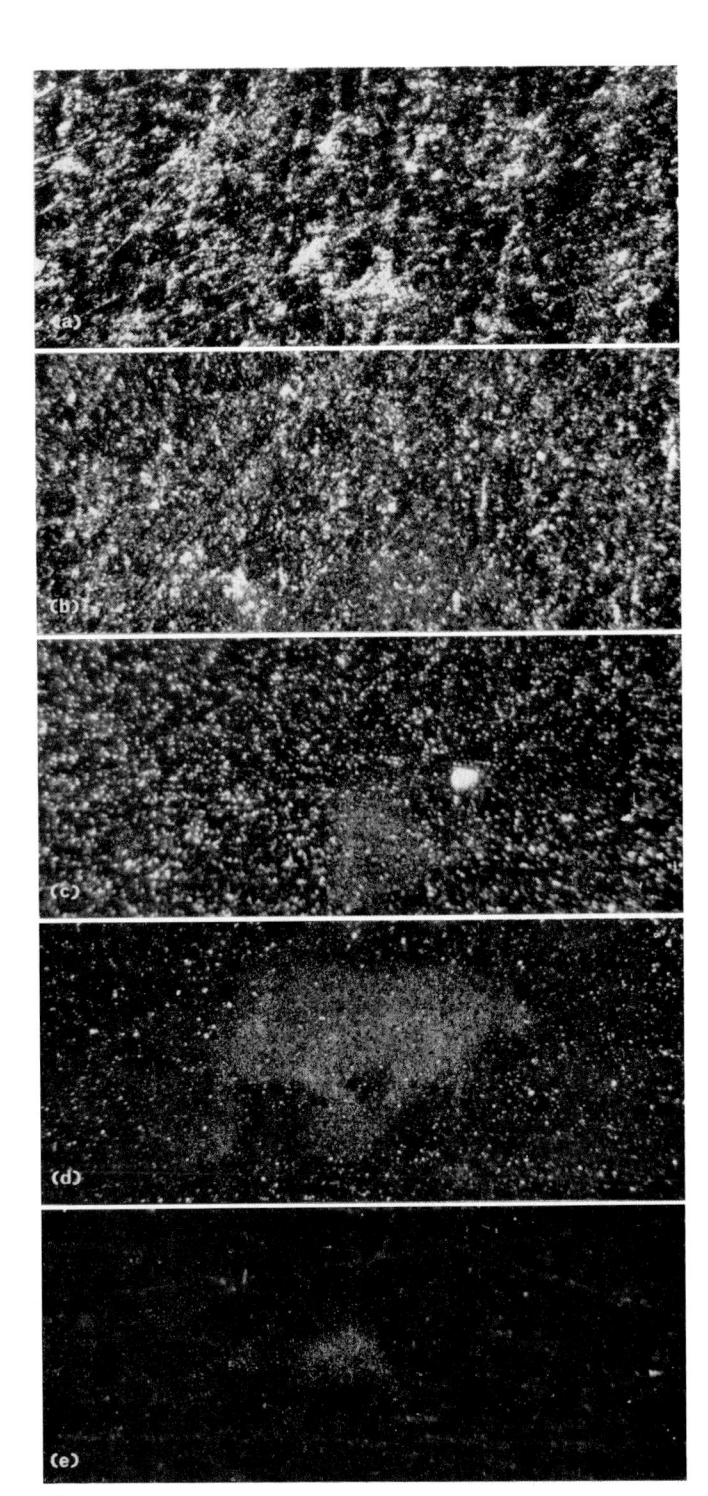

Plate 6.

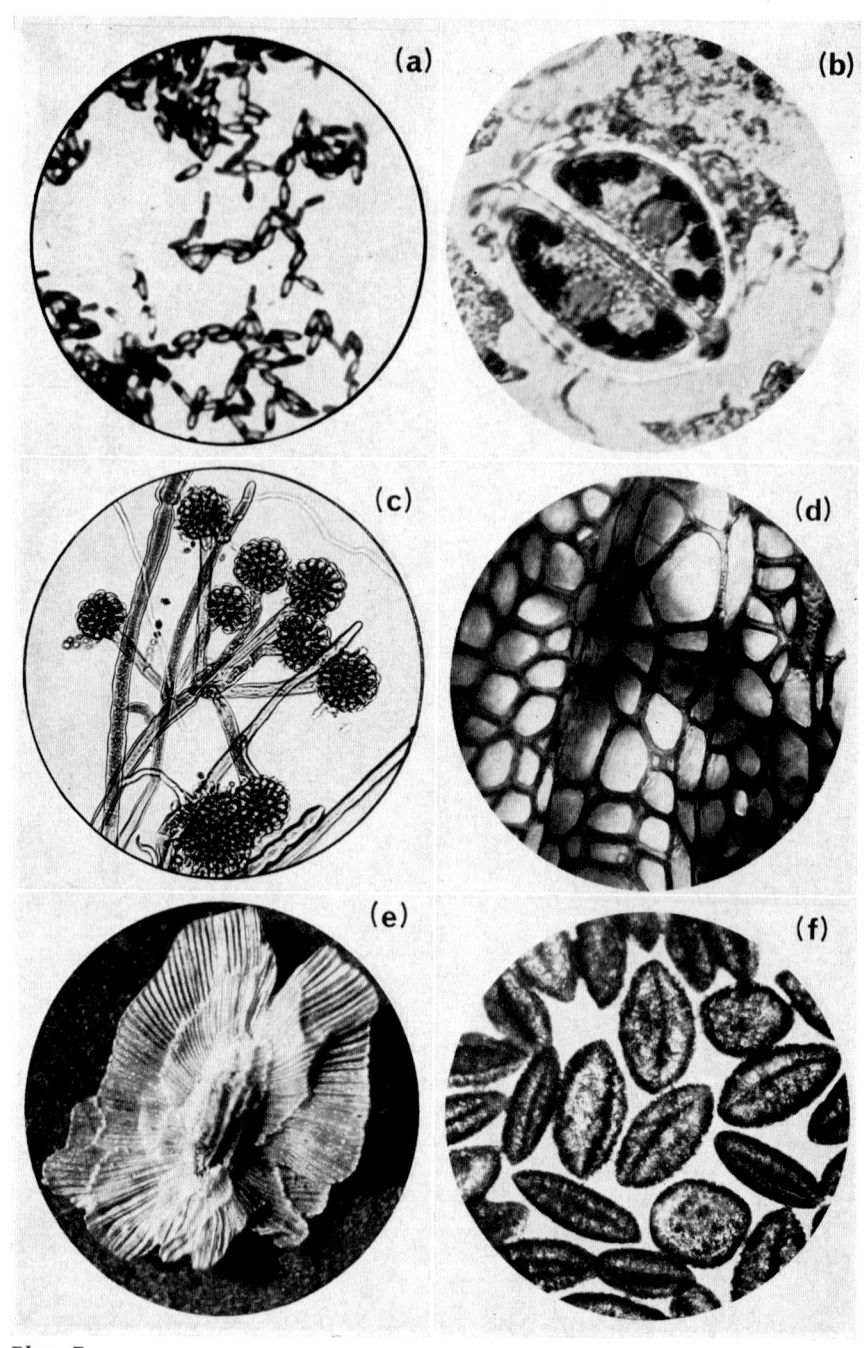

Plate 7.

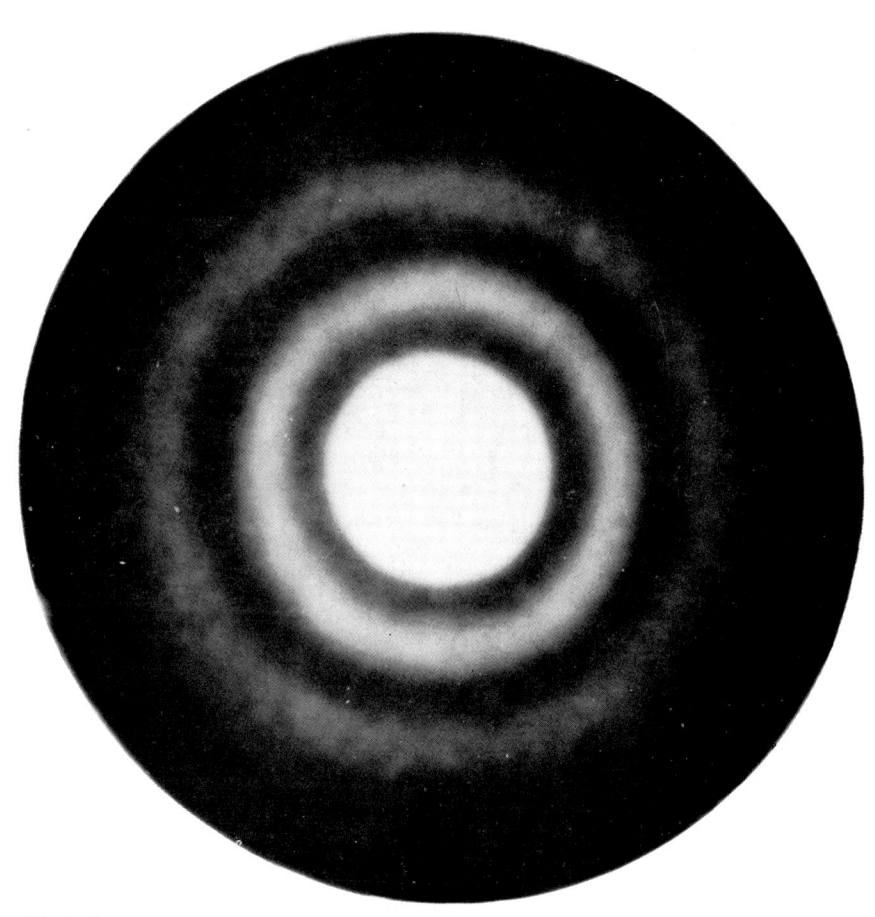

Plate 8.

Plate 9. The relationship between substage iris diaphragm aperture and the final image of the specimen.   A. Aperture too large resulting in lens glare.   B. Correct aperture producing a sharp image of the mould chain, $\times 400$.   C. Aperture too small producing diffraction rings around the image proper and in the area around the image.   D. Enlargement from C.   E. A non-image area enlarged.

Plate 10.   (a), (c), (e) Airborne fungi, $\times 360$.

This series were grown on Sabourand's malt agar. The agar was placed on a series of microscope slides, seeded and left to grow. After full growth the agar was carefully removed, leaving some mould on the glass slide as indicated in Fig. 264. Photographed by transmitted light. Initial magnification $\times 240$.   (a) Chain of conidia of hop mildew.   (c) Flowering *Aspergillus terreus*.   (e) *Fusarium solani*.

(b), (d), (f) Bacteria *Escherichia coli*.

(b) Normal cells taken with the optical microscope, initial magnification $\times 1,200$ enlarged here to $\times 1,800$. The specimen was stained by methylviolet and photographed on Ilford chromatic emulsion. This photomicrograph is approximately the highest useful magnification which can be produced by the optical microscope with oil-immersion objective and substage condenser.   (d) Normal cells taken by the electron-microscope. The specimen was unstained and the initial magnification was $\times 1,740$, seen here at a magnification of $\times 4,500$.   (f) Cells of the same culture of *Escherichia coli*, $\times 4,500$, after the cells had been exposed to ampicillin (10 μg/ml for 40 min). There is no doubt left in one's mind regarding the limits of the optical system of the microscope and the value of the electron-microscope, and ampicillin.

Plate 11.   (a) Objective. An adjustable graduated correction collar for use with cover-glasses of various thicknesses. (See page 56.)   (b) Diffraction around the edge of the specimen 1–6 (Fig. 46).   (c) Tilting stage (Fig. 114).

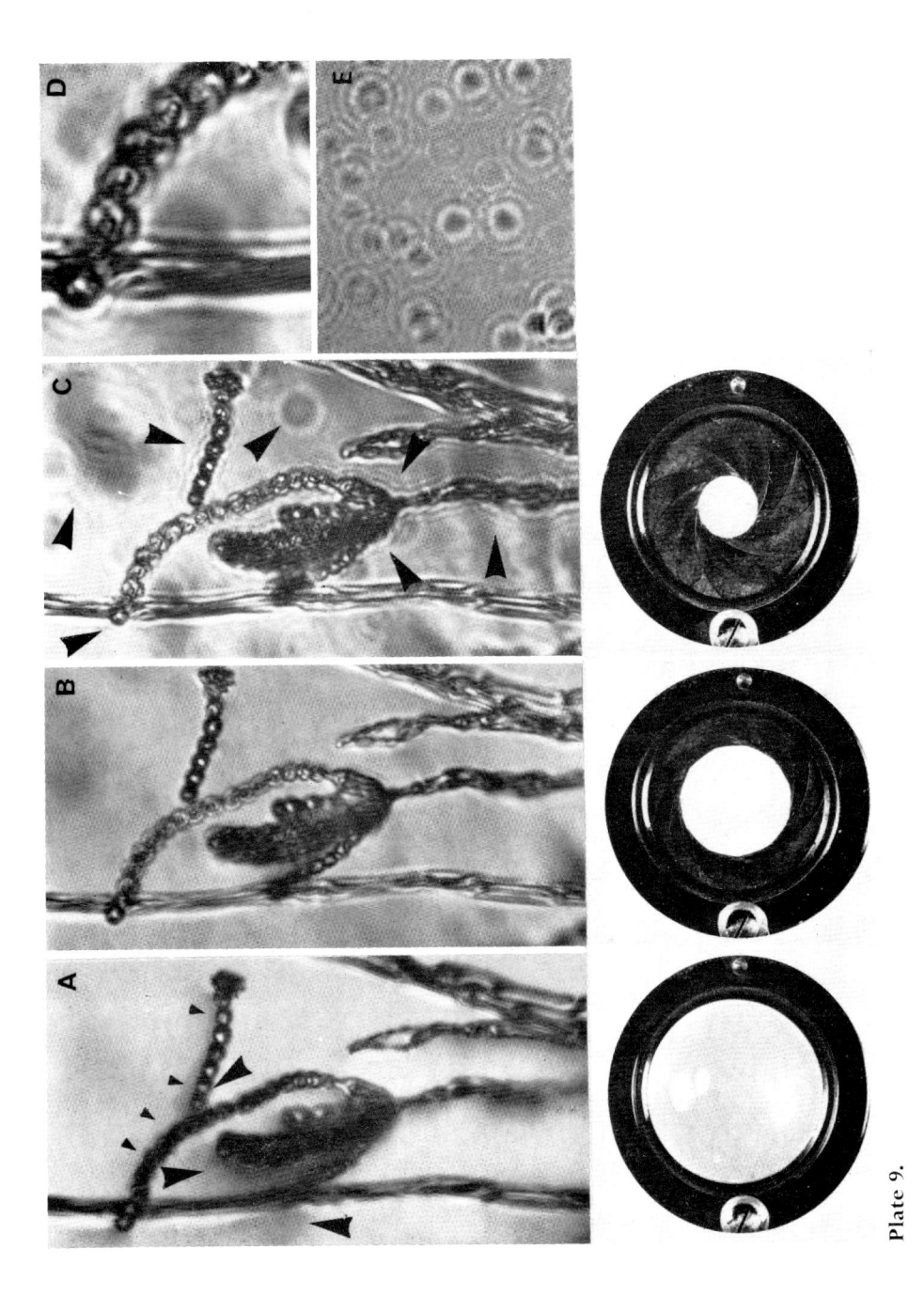

PLATES

373

Plate 9.

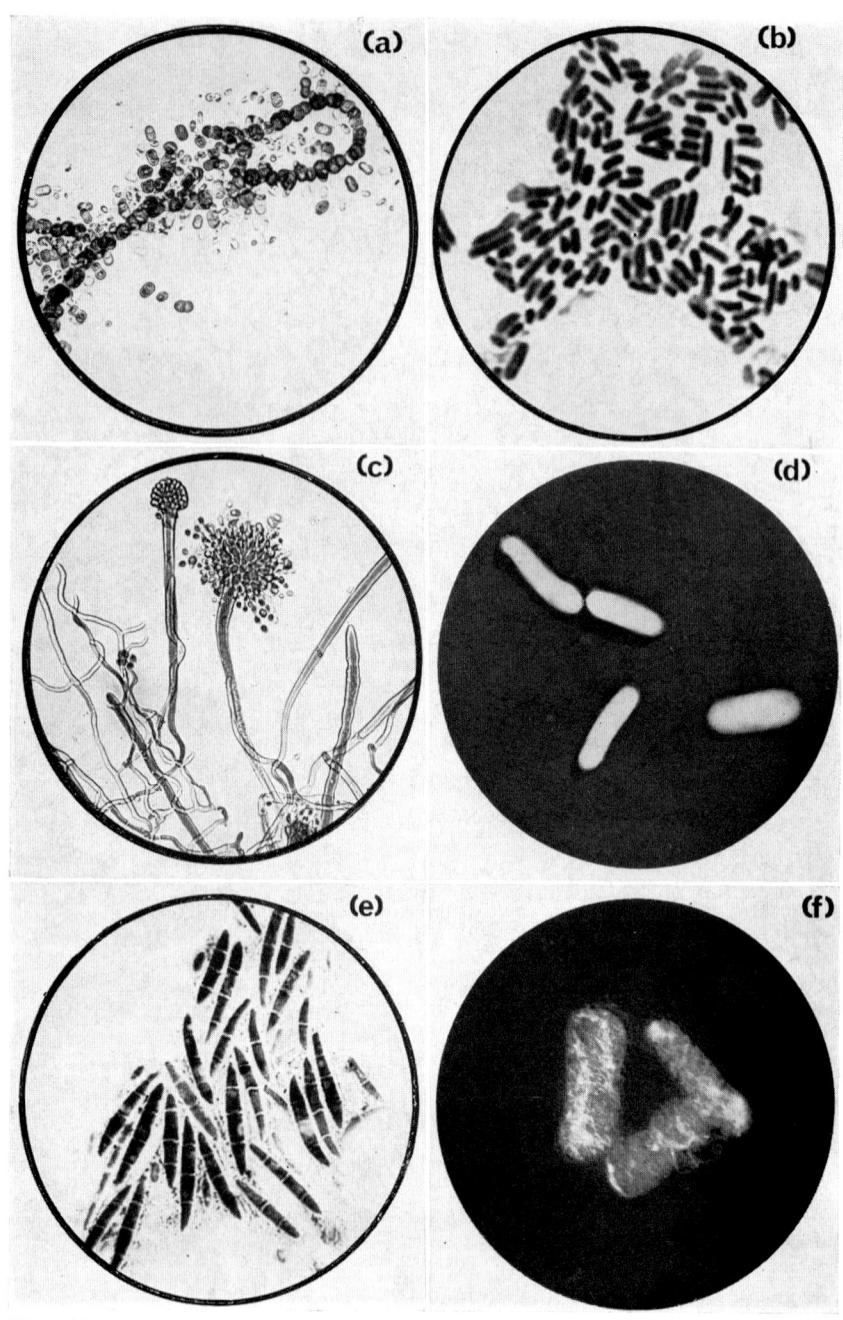

Plate 10.

PLATES

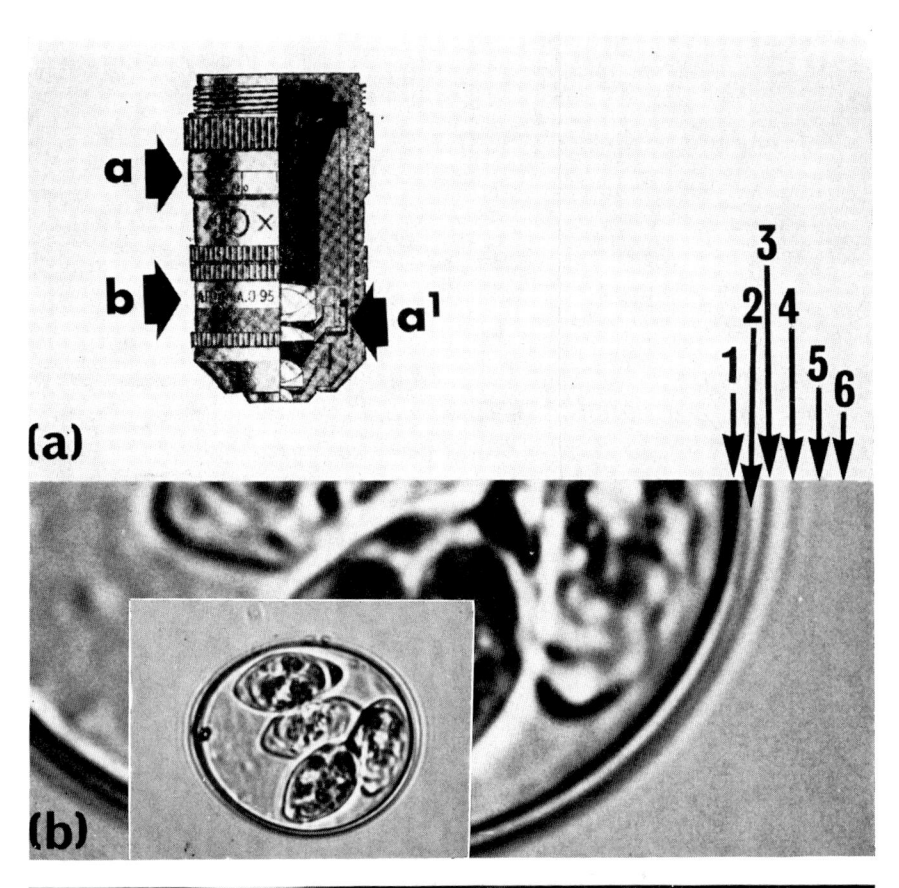

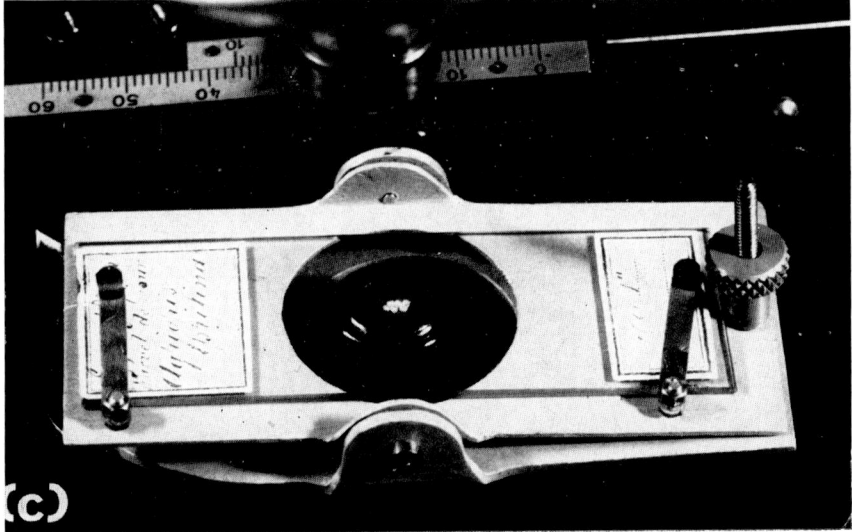

Plate 11.

Plate 12. (a) Rabbit embryo, normal palate.   (b) Rabbit embryo taken from a normal animal on a normal diet, showing cleft palate.   (c) Life size.

Photographed by using a 50 mm objective without an eyepiece. Two 6V 30W lamps were used to illuminate the subjects.

Plate 13a. Eggs of Large Cabbage White butterfly, $\times 65$. The dark tops, better seen top right, and the dark spot in the centre, are an indication that the caterpillar is about to free itself. This lovely specimen was illuminated with one intensity lamp and one reflector, the shadows play an important part here, giving a feeling of depth. (See Plate 57.)

Plate 13b. Eggs of Common House fly, $\times 90$. Such a common subject and yet how often is it photographed? The fine detail seen in each egg is of course hidden to the naked eye. It took me four years to perfect the photo-technique and capture this natural wonder.

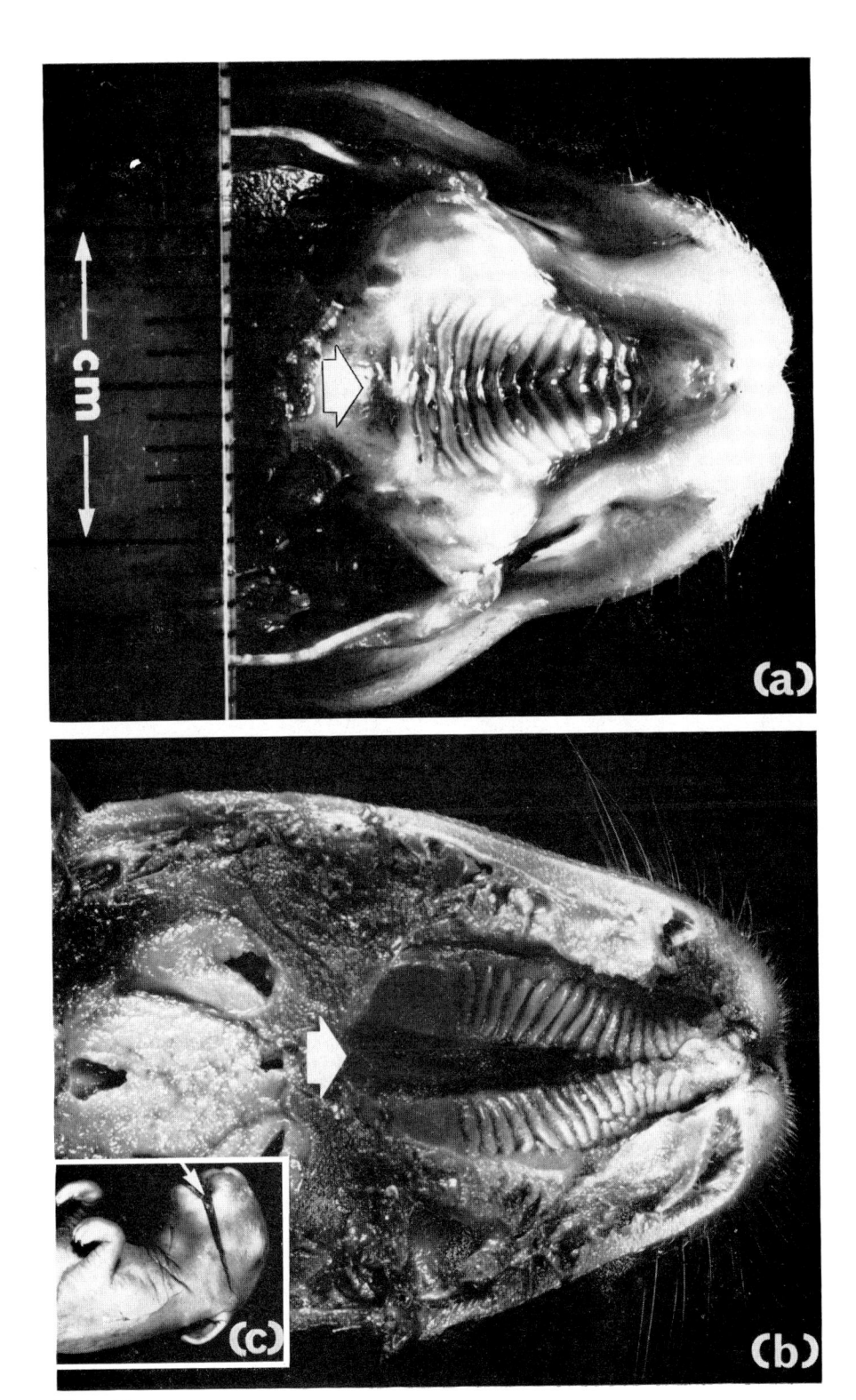

Plate 12.

(a)

Plate 13.

Plate 13.

Plate 14. A series of exposures made through the eggs of a rabbit. (See Fig. 54.)

Plate 15. Transverse section of stem showing cell structure.
(a) ×42·4   16 mm objective, ×4 eyepiece, bellows extension 23 cm
(b) ×80   16 mm objective, ×4 eyepiece, bellows extension 48 cm
(c) ×80   enlarged from (a)
(d) ×80   8 mm objective, ×4 eyepiece, bellows extension 23 cm
Photomicrographic data   (e)   (f)

Subject	Bacteria *Salmonella typhi*	
Mounted	Smear	
Magnification	×424	×864
Date	May 67	May 67
Objective	Apo. dry, ×40, 4 mm, N.A. 0·85	Apo. oil, ×90, 2 mm, N.A. 1·40
Eyepiece	×8	×8
Tube Length	170 mm	170 mm
Substage Condenser	Complete, dry	Complete, oil
Illumination	Transmitted	Transmitted
Iris Diaphragm Aperture	Maximum	Maximum
Camera Bellows Extension	40 cm	40 cm
Filter	Green	Green
Stain		
Emulsion	Ilford Chromatic	Ilford Chromatic

Plate 16. Hand made wooden microscope slides.

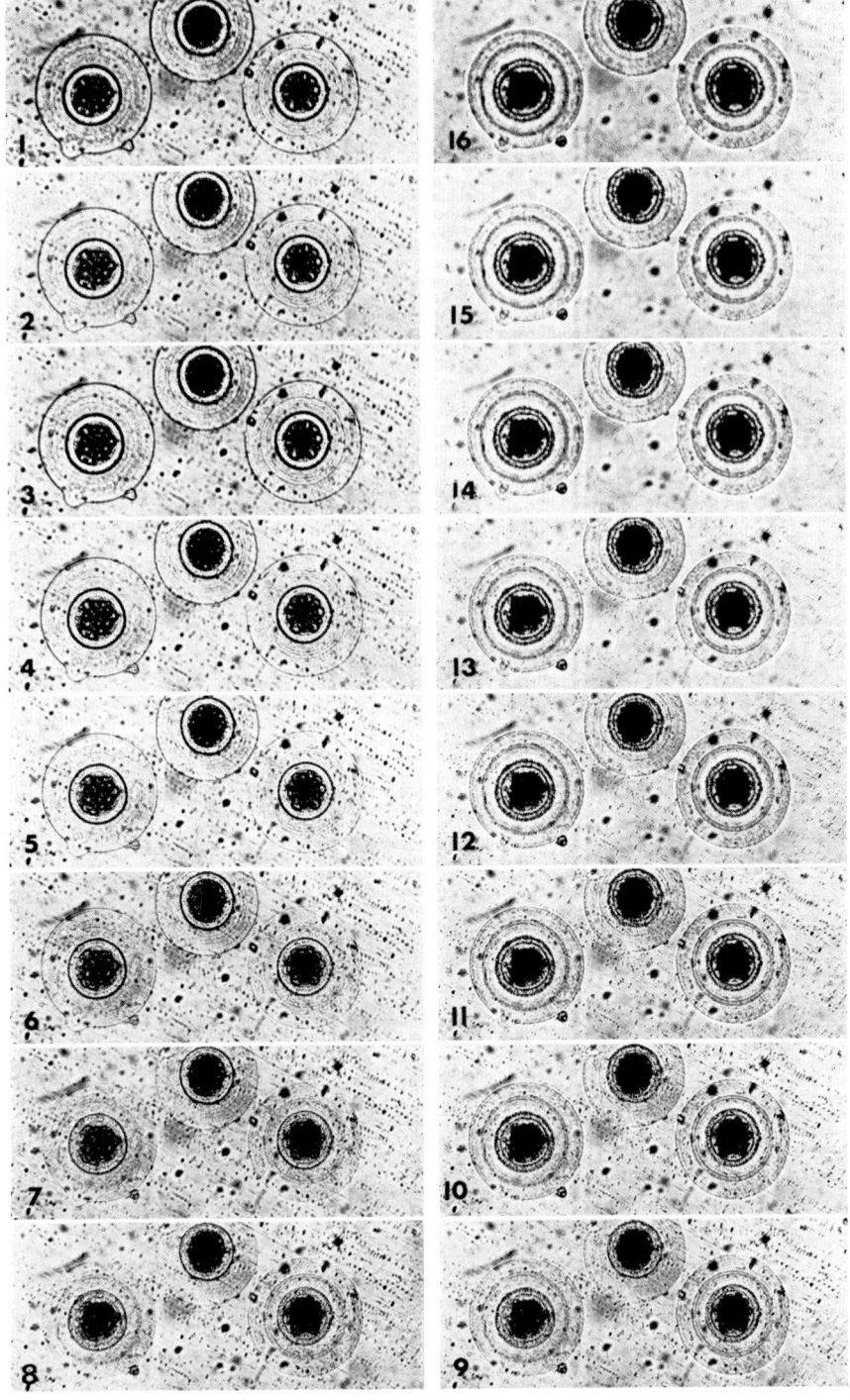

Plate 14.

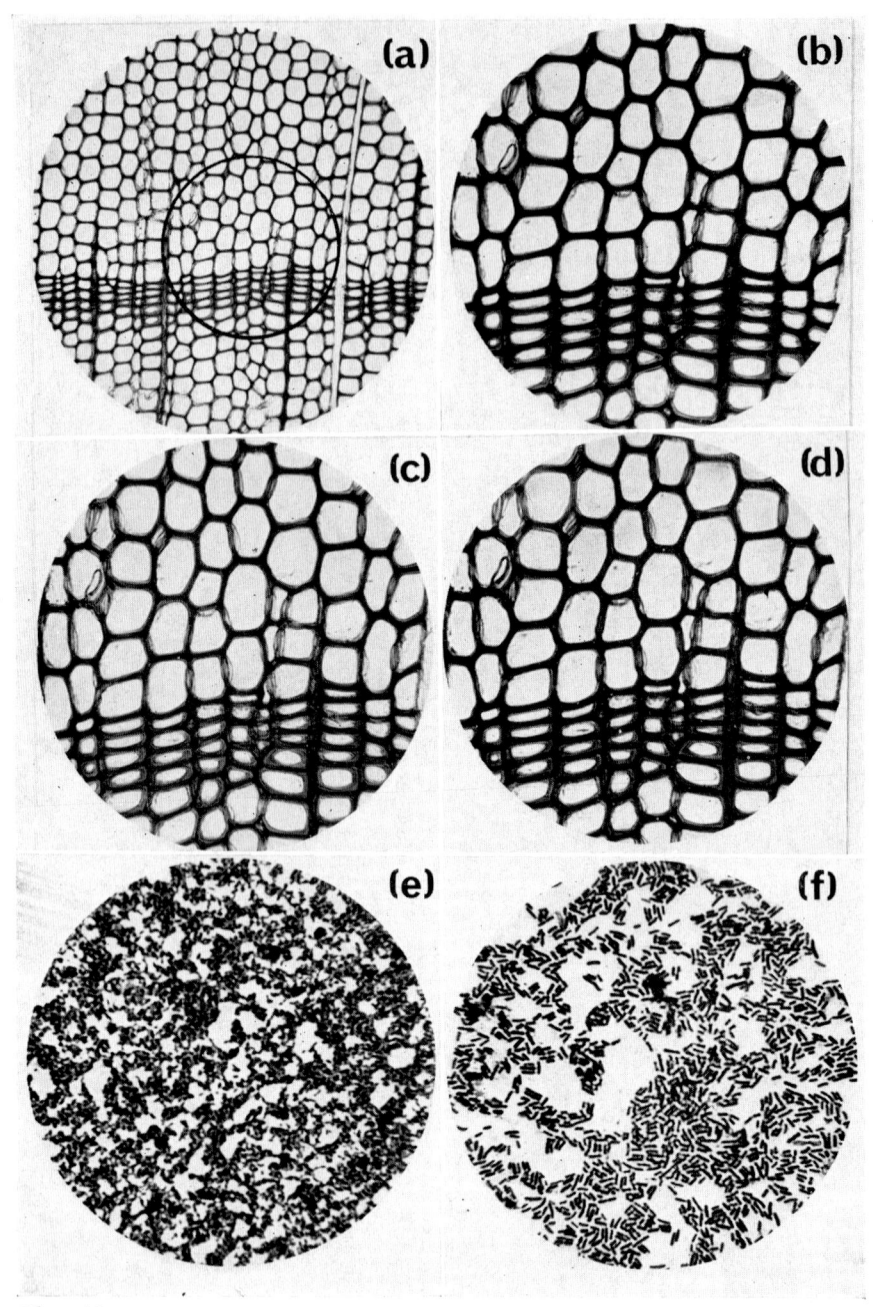

Plate 15.

Plate 16.

Plate 17. (a), (b) Colony of Vorticella, ×42, total ×96. Attached to their retractile stems, in water in a hollow ground slide (not narcotized). (c), (d) T/S Olivine, ×18. (c) Recorded by plane-polarized light. (d) The same field seen under polarized light, showing form and zoning. Form when idiomorphic, shows hexagons which are irregular but symmetrical about two of the diameters. Irregular cracks frequent. The line drawing illustrates the various colours as set out in the following order. 1. Royal blue; 2. Black; 3. Dark brown; 4. Olive green; 5. Mauve; 6. Pea green; 7. Yellow; 8. Brown; 9. Light brown; 10. Blue to purple; 11. Pink; 12. Blue-green; 13. Mid-green.

(e), (f) Group of Diatoms, ×21, total ×160. These minute algae, Heliopeltametric, are commonly found in both fresh and salt water. Because of their delicate and regular markings, certain species can be used to test the resolving power of an objective. Careful treatment of the negative material prevents distortion of the very fine detail. Transmitted light (e) does not do justice to this beautiful subject. Darkground illumination (f) reveals certain features that can scarcely be seen by transmitted light. It proved to be very difficult to judge the size of dark stop in order to achieve a sharp rendering of the outer diatoms.

Plate 18. Newspaper print and paper structure photographed by grazed illumination and one reflector. (a) ×25, (b) ×80, (c) demonstrates the paper fibre, ×100.

Plate 19. Magni-optical changer, housing 16 corrected optical elements, see page 65. (Courtesy of American Optical Co.)

PLATES

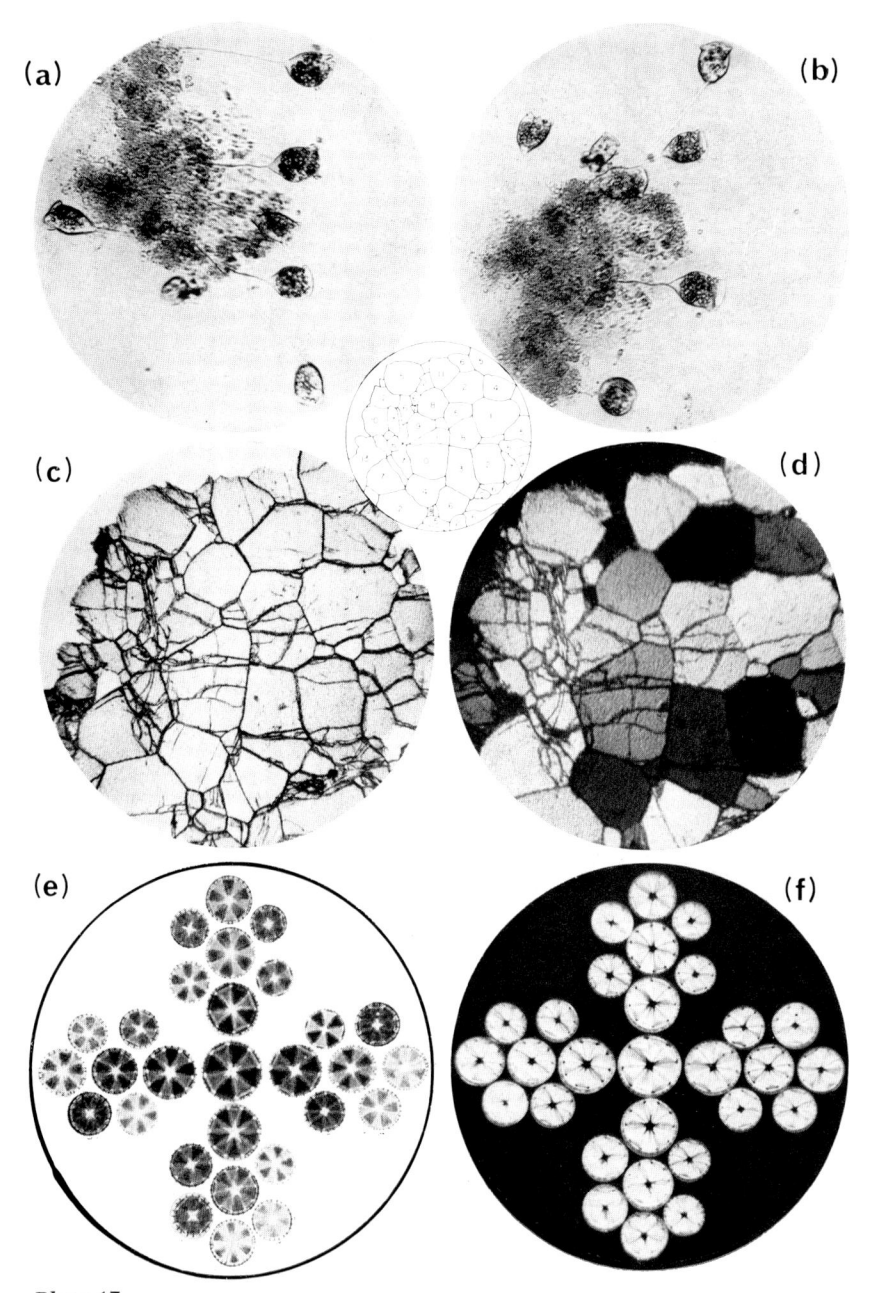

(a) (b)
(c) (d)
(e) (f)

Plate 17.

Plate 18.

PLATES 387

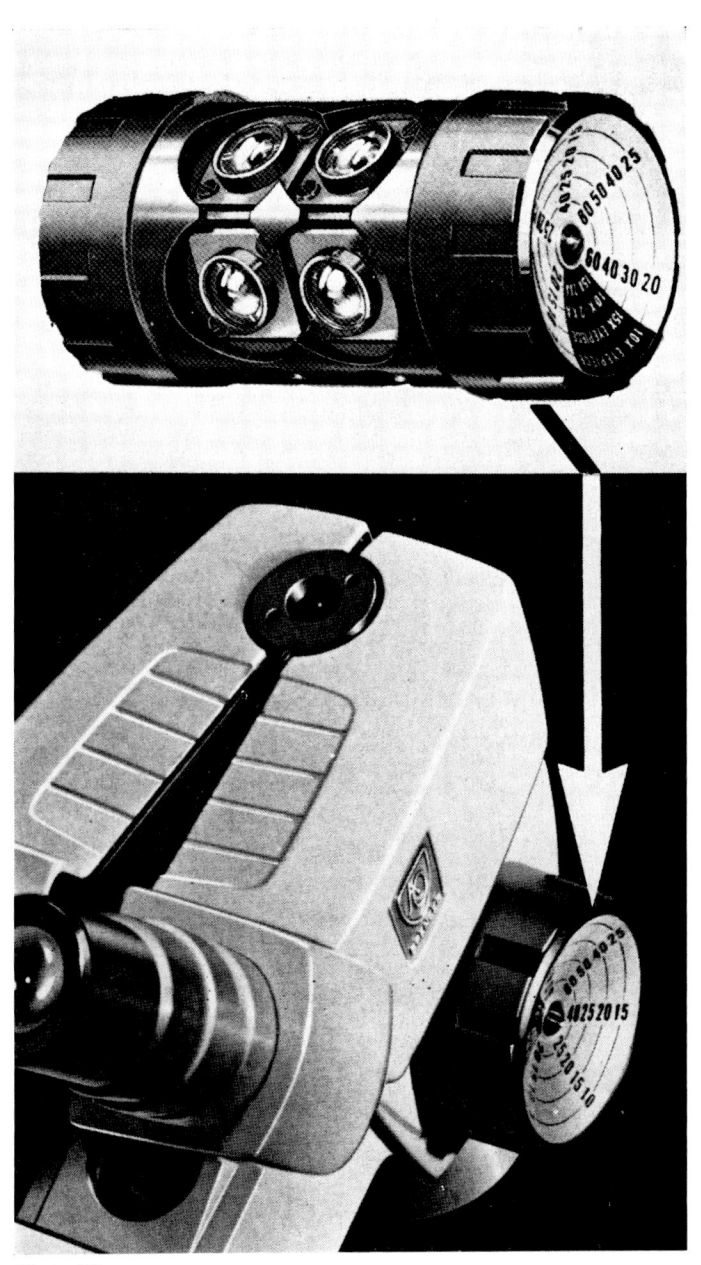

Plate 19.

Plate 20. Standard optical system for Bausch and Lomb laboratory microscope. This illustration shows how light from the Hi-Intensity illuminator travels through the optical system of a Dynazoom binocular laboratory microscope equipped with standard achromatic objectives and Huygen eyepieces. See also page 66. (Courtesy of Bausch and Lomb Optical Co.)

Plate 21. (a) The arrow points to a minute area of corrosion on the inside of a pressurized dispenser. So bulky a specimen produces problems when mounting and locating the field. An adjustable stage does enable the specimen to be moved by degrees.   (b) Photographed by near vertical incident light, $\times 8$.   (c) Recorded by grazed illumination revealing depth in the affected area. (b) does not demonstrate the corrosion as it really is.   (d) Egg of Pine hawk-moth on pine needle, $\times 16$. The egg is yellow but turns grey before hatching.

Plate 22.   Pollen of hazel, $\times 720$.   (b) Pollen of poppy, $\times 720$.   (c) Penicillin *Brevicompaction*, $\times 510$.   (d) Glass wool.   (e), (f) Colony of bacteria growing on agar, $\times 21$. (e) Was transmitted bright-field illumination 546 nm green, and (f) Exposed to UV radiations 265 nm. Note the increase in depth of field (depth of sharpness) and an overall improvement in quality. (See page 232.)

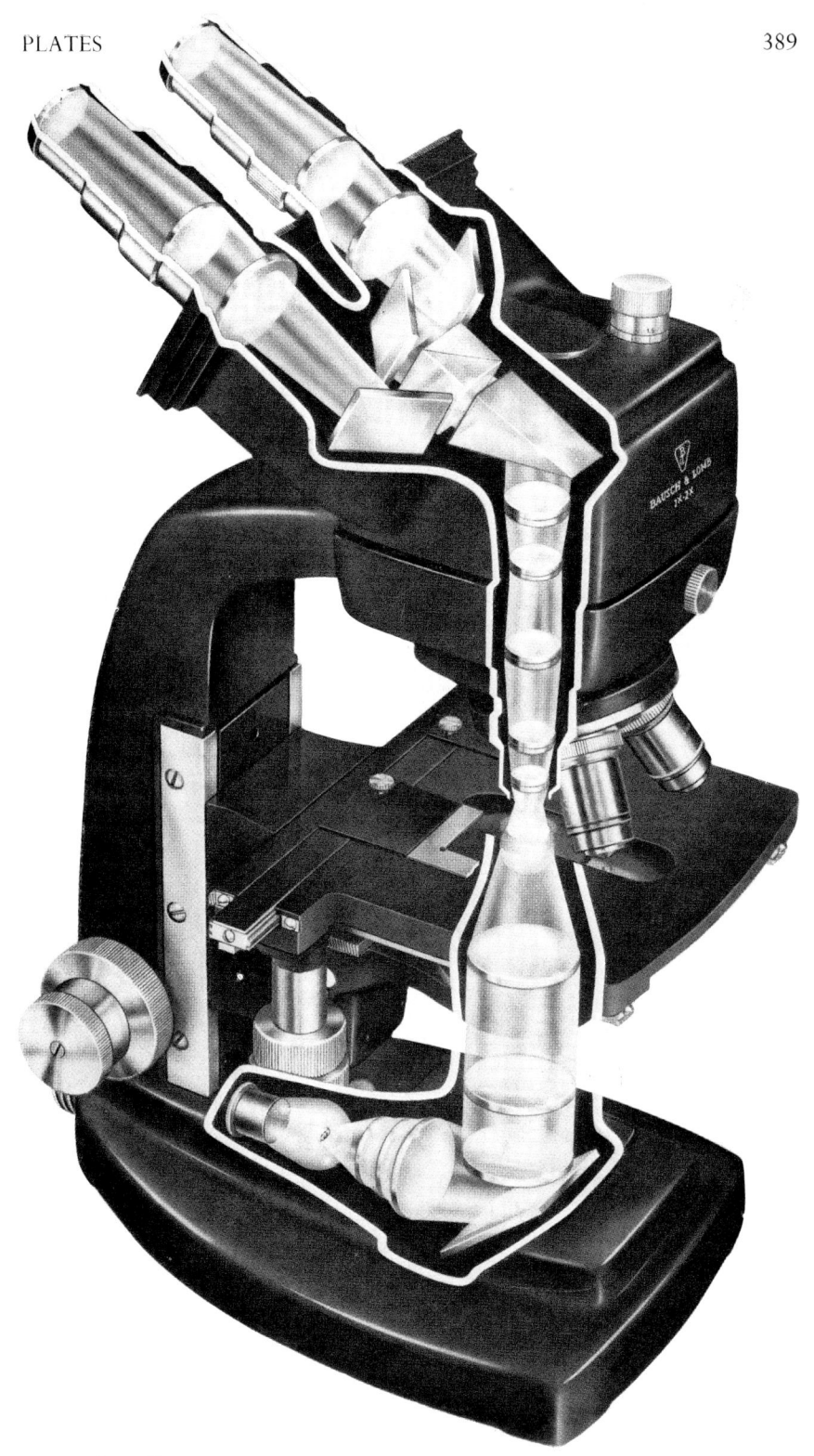

Plate 20.

Plate 21.

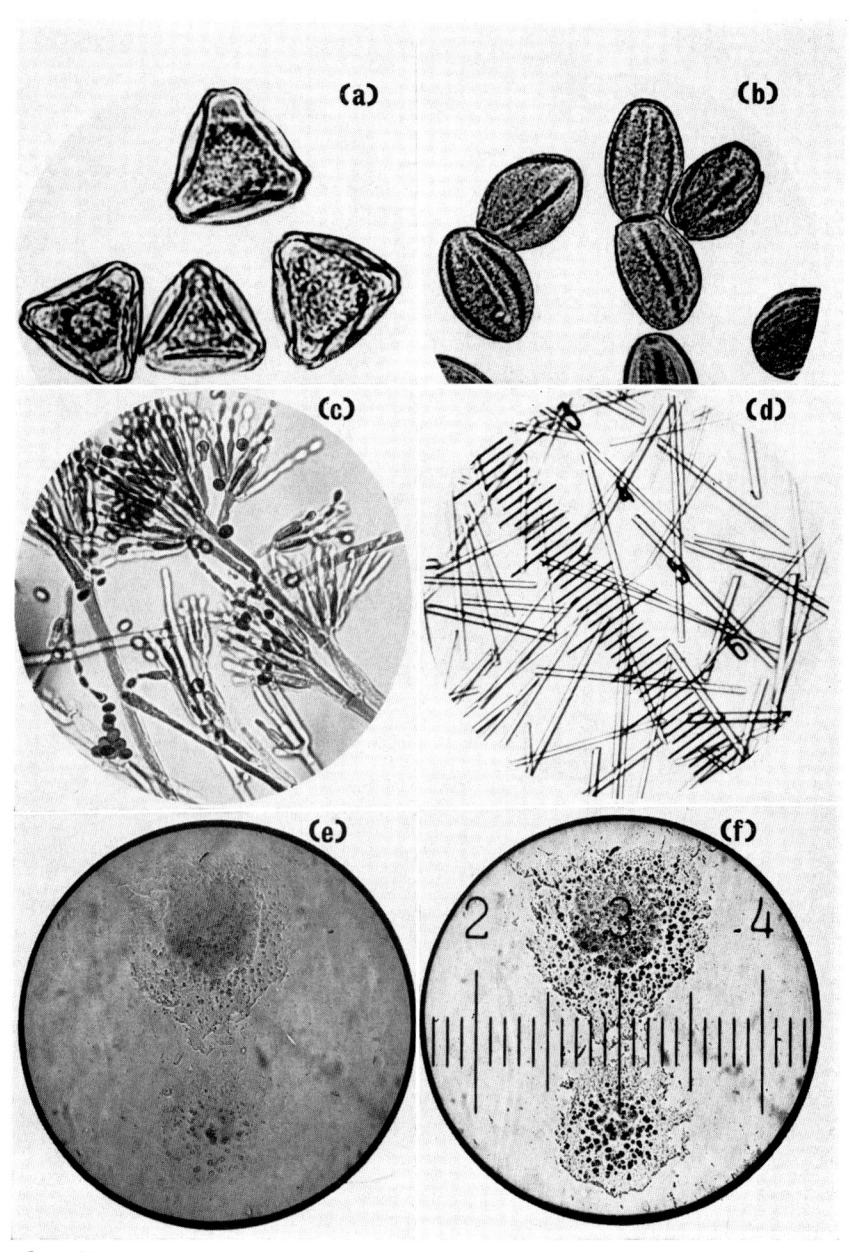

Plate 22.

Plate 23. The Vickers M15C microscope and 35 mm camera. There is built-in illumination and the reflector body contains a beam-splitting prism which allows 80% of all the light to form an image in the film plane. The remainder of the light is deflected to the focusing eyepiece or, by the rotating prism, to the photocell tube when the cadmium sulphide photometer is being used. The focusing eyepiece is fitted with a rectangular graticule which indicates the field of view to be photographed and a scroll mechanism allows the operator to focus onto the graticule. (Photo by Courtesy Vickers Ltd.)

Plate 24. Comparison eyepiece by Zeiss, Jena. *A* Projection and viewing eyepiece. *B* Connecting collar and support. See also page 82. (Courtesy of Zeiss, Jena.)

Plate 25. Field of view. Zeiss, Jena, comparison eyepiece.
Left: Pinewood, *Pinus silvestris*, ×27.
Right: Deciduous wood, *Aristolochia*, ×27. (By Zeiss, Jena.)

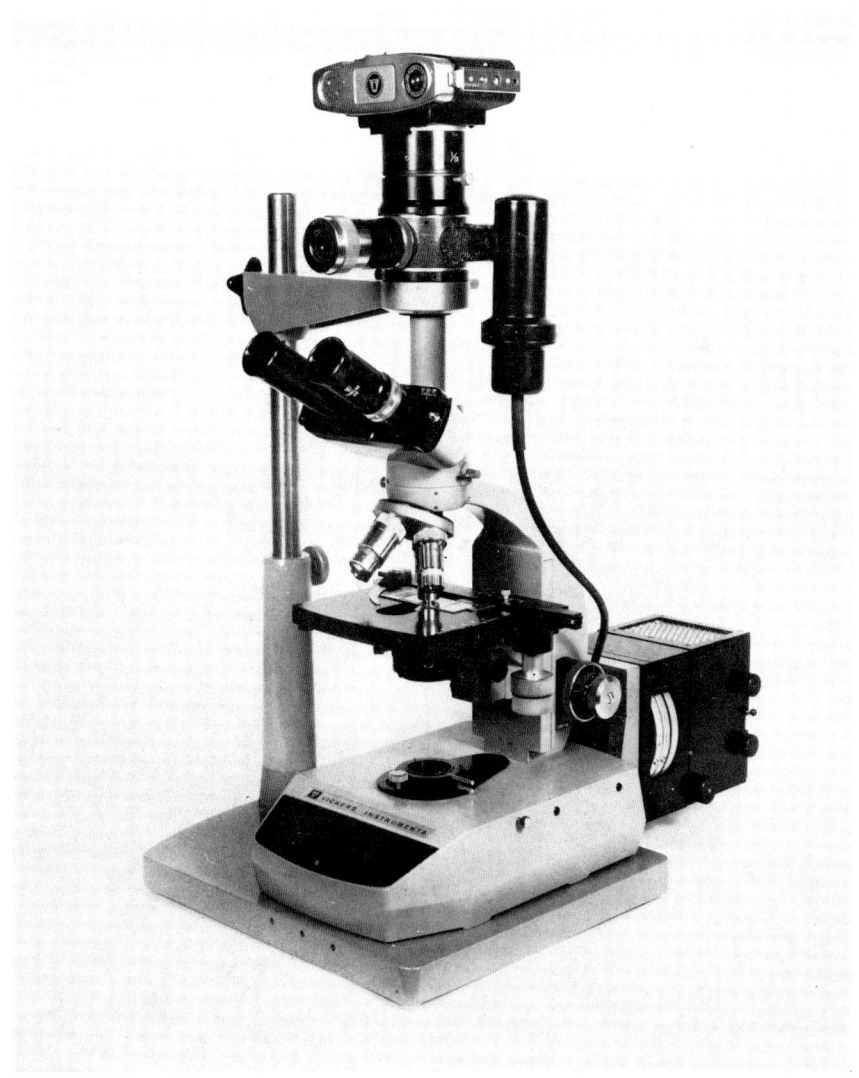

Plate 23.

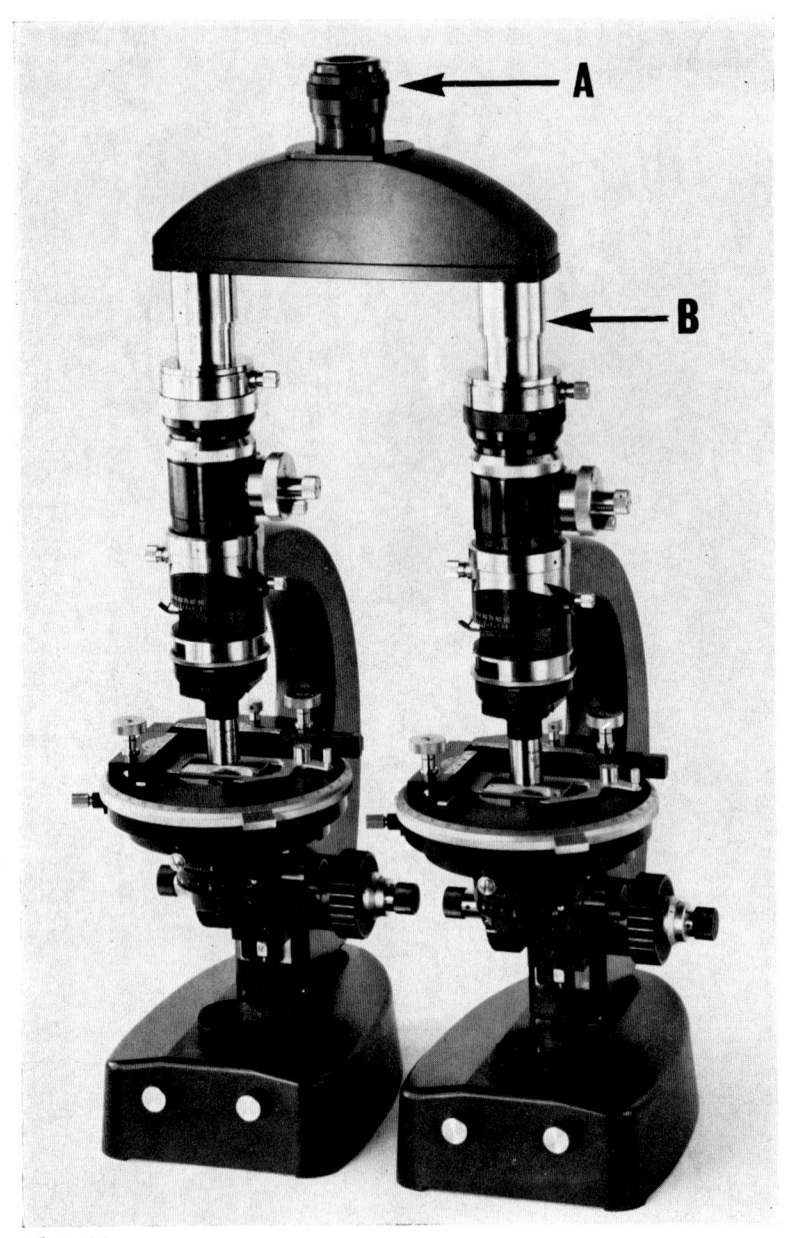

Plate 24.

PLATES

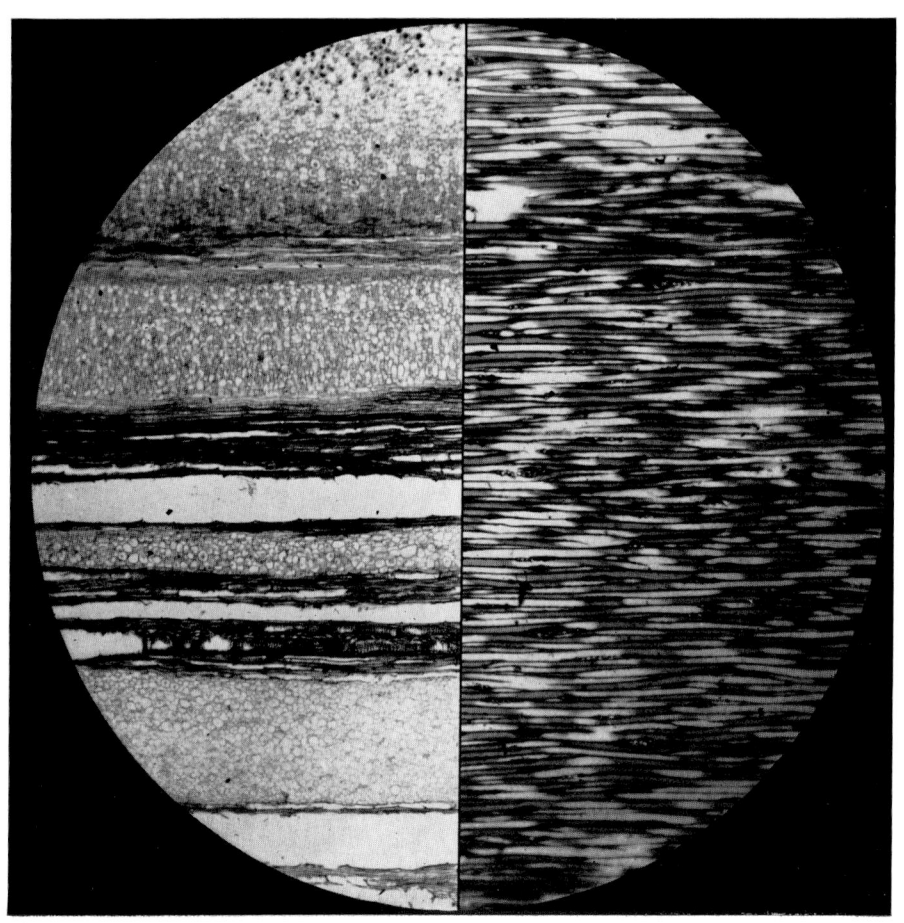

Plate 25.

Plate 26. (a) Vickers' micrometer eyepiece. (b) Vickers' screw micrometer eyepiece. (Courtesy of Vickers' Instruments.)

Plate 27. The Linhof Universal Accessory Stand for photomicrography. Left: camera swung aside for direct inspection purposes. Top left: showing the light tight connection between camera and eyepiece. (See page 84.)

Plate 28. (a) Damaged human hair, $\times 250$. Left of the damage the hair is dying, as indicated by the tone of the hair. Taken with a 25 mm objective and $\times 6$ eyepiece, initial magnification $\times 50$, after enlargement took the magnification to $\times 250$. (b) The same piece of hair but photographed with a change in the optical system. 8 mm objective and $\times 8$ eyepiece, contact print from the negative. (c) Polythene showing the fine structure, $\times 80$. Illuminated by one incident lamp at near vertical. (See also page 84.)

PLATES 397

(a)

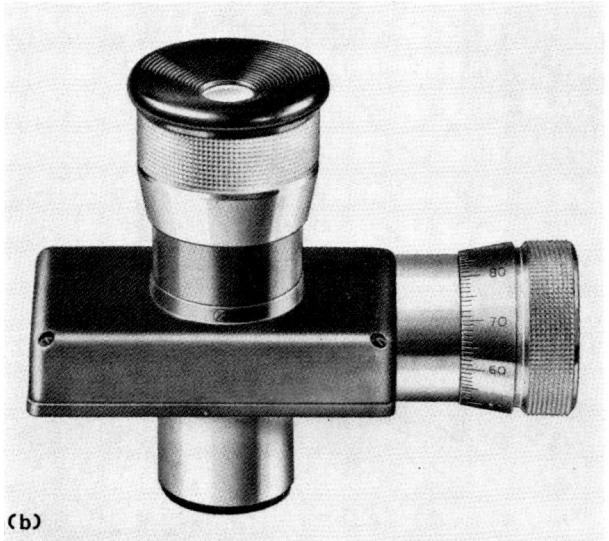

(b)

Plate 26.

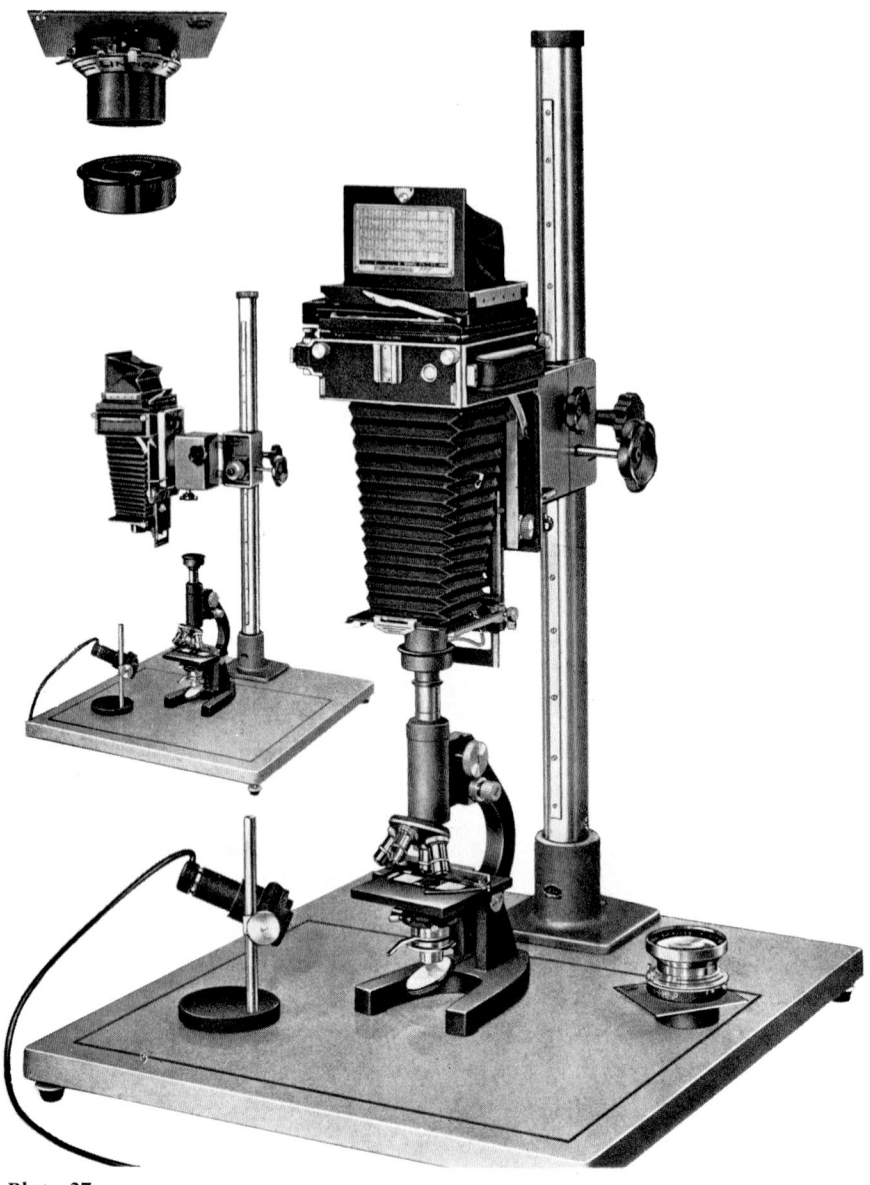

Plate 27.

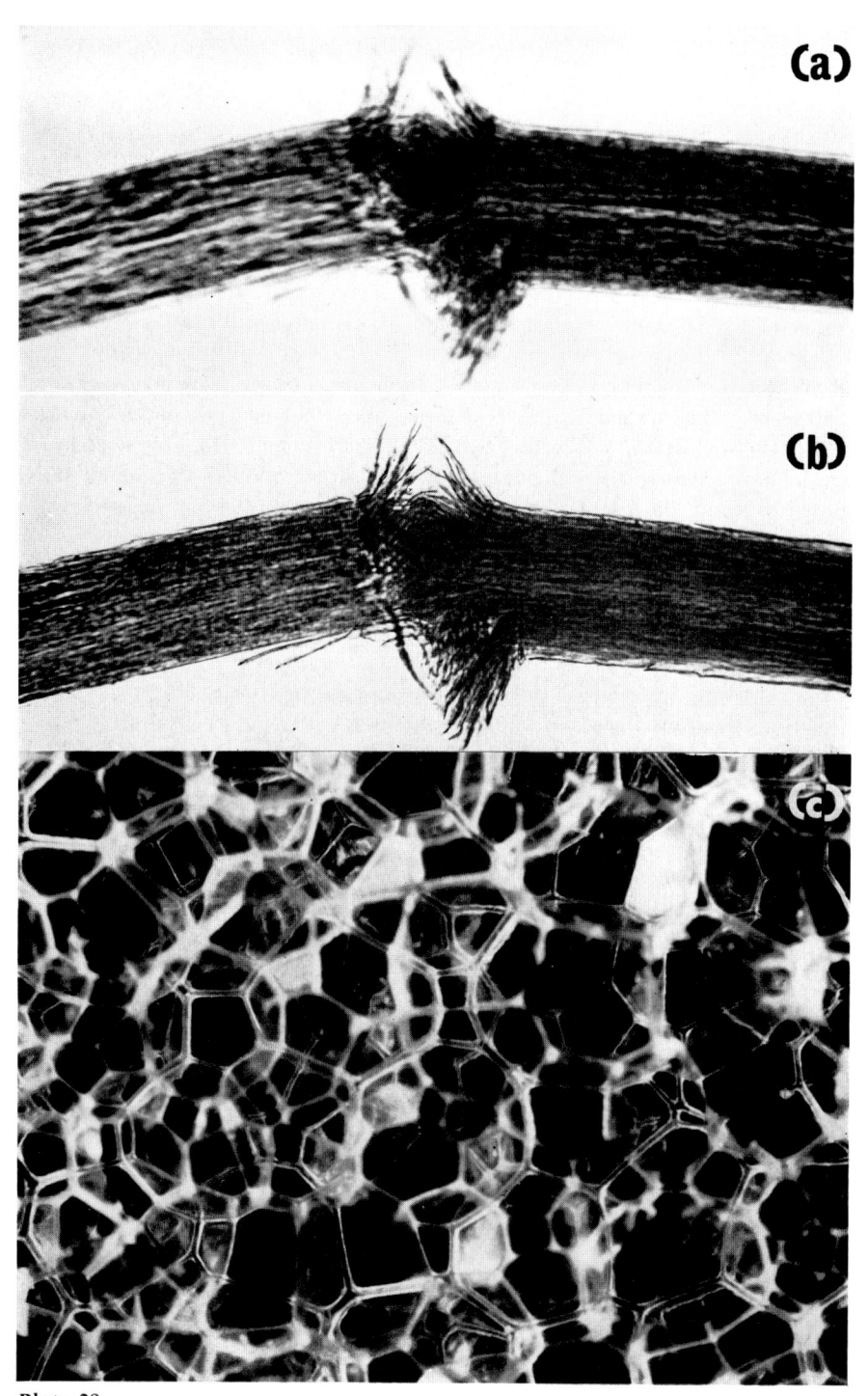

Plate 28.

Plate 29. (a) Palate of whelk, ×140. Transmitted light bright-field. Chromatic plate, green filter. (b) Phase contrast shows the chitinous teeth. Chromatic plate, green filter. (c) X-ray of left leg of mouse, 10 days after injection.

Plate 30. (a) Aluminium–tin alloy viewed with ordinary light microscope reveals only surface scratches and tin which occurs in some grain boundaries of aluminium, ×80. (Reproduction by Courtesy G.E.C.) (b) Photomicrograph of scratched chromium-plated surface, with shutter in closed position, ×144. (c) Aluminium–tin alloy clearly shows aluminium grain outlined by tin (dark lines). Specimen thickness 0·5 mm. X-ray (Fig. 205) exposure 10 min, ×80. (Reproduction by courtesy G.E.C.) (d) Interferogram of scratched chromium-plated surface, shutter in open position, ×144. The Watson 8mm and 16mm interference objectives are two-beam interferometers normally used with monochromatic light either from a sodium light (wavelength 0·59 microns) or, with a green light filter, a mercury lamp (wavelength 0·546 microns). A flat specimen will give straight line fringes and any angle may be measured by the lateral displacement of the fringes which are shifted one fringe spacing for each half-wavelength of height. ((b) and (d) by courtesy of Watson, the Optical Division of M.E.L. Equipment Co. Ltd.)

Plate 31. (a) Egg of Rare Swallowtail moth, ×48. A fine scale is attached to the egg. (b) Fossil promicroeras, ×6·4. The specimen has been preserved in iron sulphite and is claimed to be 140 to 170 million years old. (c) Ungraded sand, ×16. Here depth of field is very important.

(a)

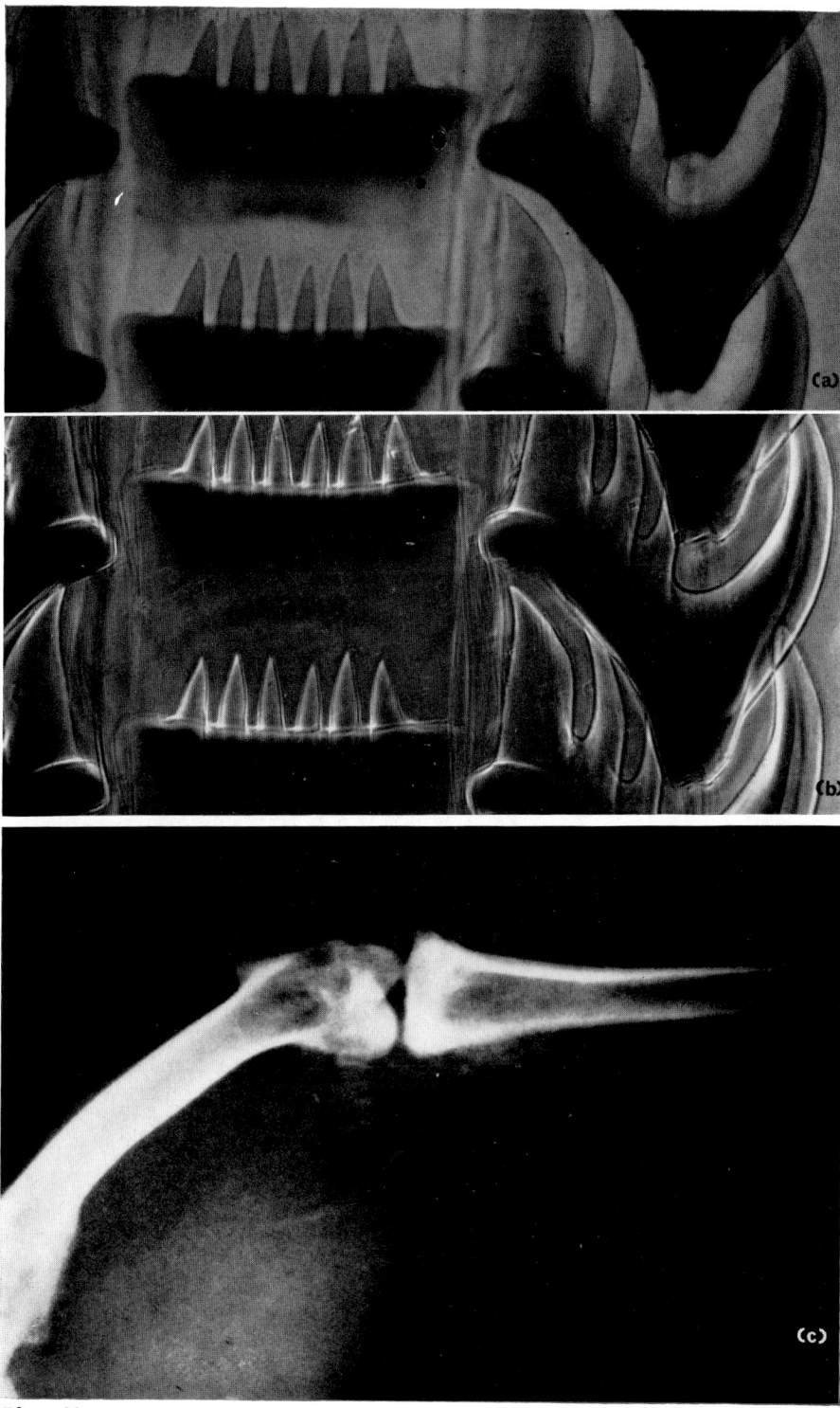

(b)

(c)

Plate 29.

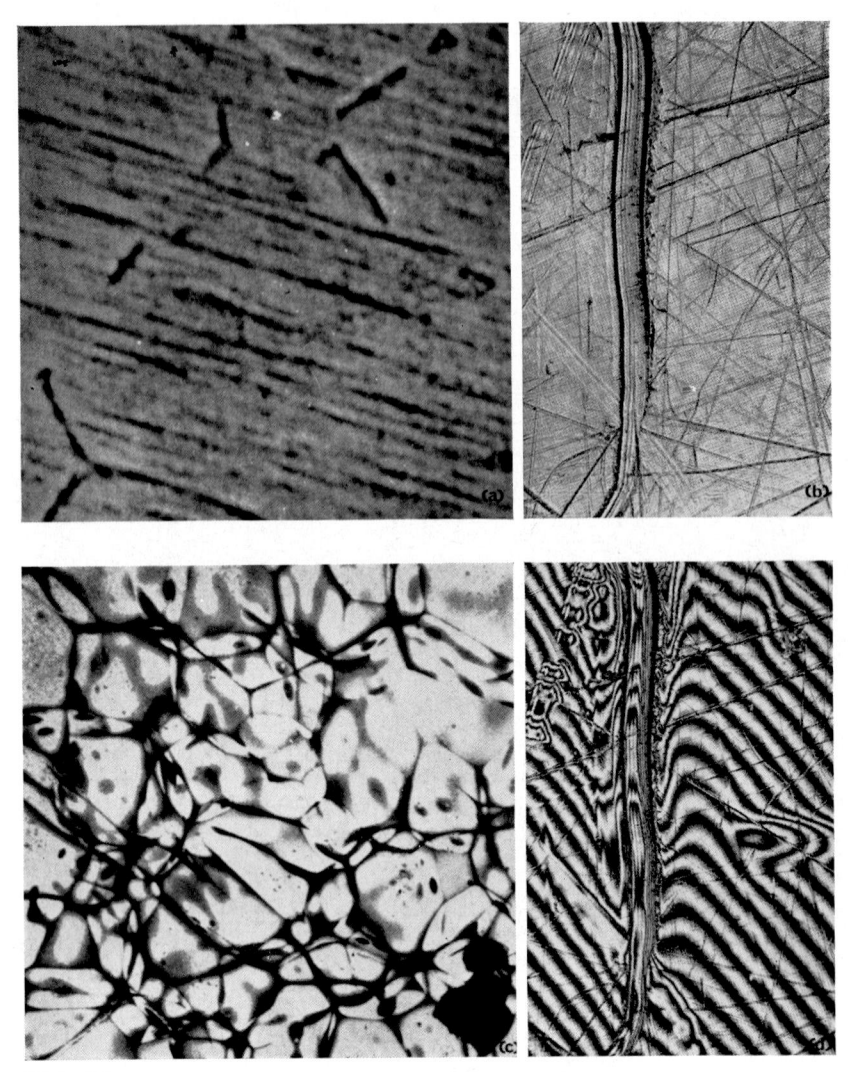

Plate 30.

PLATES

Plate 31.

Plate 32. Saw of Saw-fly. The small inset shows the saw life size. The whole saw was photographed with a low power objective less substage condenser. A small section covering about fourteen teeth was then photographed by a higher power objective. The teeth on either side of the image have been included to demonstrate the lack of coverage, producing an oblique effect in B and C as indicated. A was cut from the centre of the print where the subject was illuminated by the centre of the cone of light.

Plate 33. Guinea-pig siamese twins, shown life size in the insert. 100 mm objective plus bellows camera extension. Specimen submerged in glycerine. Two 12V 24W lamps used. Specimen taken from normal guinea-pig which had been fed on a normal diet.

Plate 34. (a) Section of skin of eel, $\times 320$. The transparency exhibits the natural colour of the skin, brought about by using oblique reflected light. Great care was given to the "play" of light used in the most advantageous manner to show the skin pattern with isolated spheroidal transparent bodies embedded in the skin. There is a rhythm about the flow of these scales, and the inclusion of those on the lower right and lower left give a base to this particular field. (b) *Licmophora flabellata*, $\times 480$. The entire marine plant presents us with a more-or-less complete flabella, or fan, upon the summit of the branches, with imperfect flabellae, or single frustules, scattered irregularly along the entire length of the curving stems. This beautiful plant is marine and is parasitic on seaweeds and zoophytes. It appears to be almost transparent and colourless at this magnification. Specimen wet mounted without pressure (Fig. 266). The aesthetic and pictorial application can be studied on page 290.

PLATES

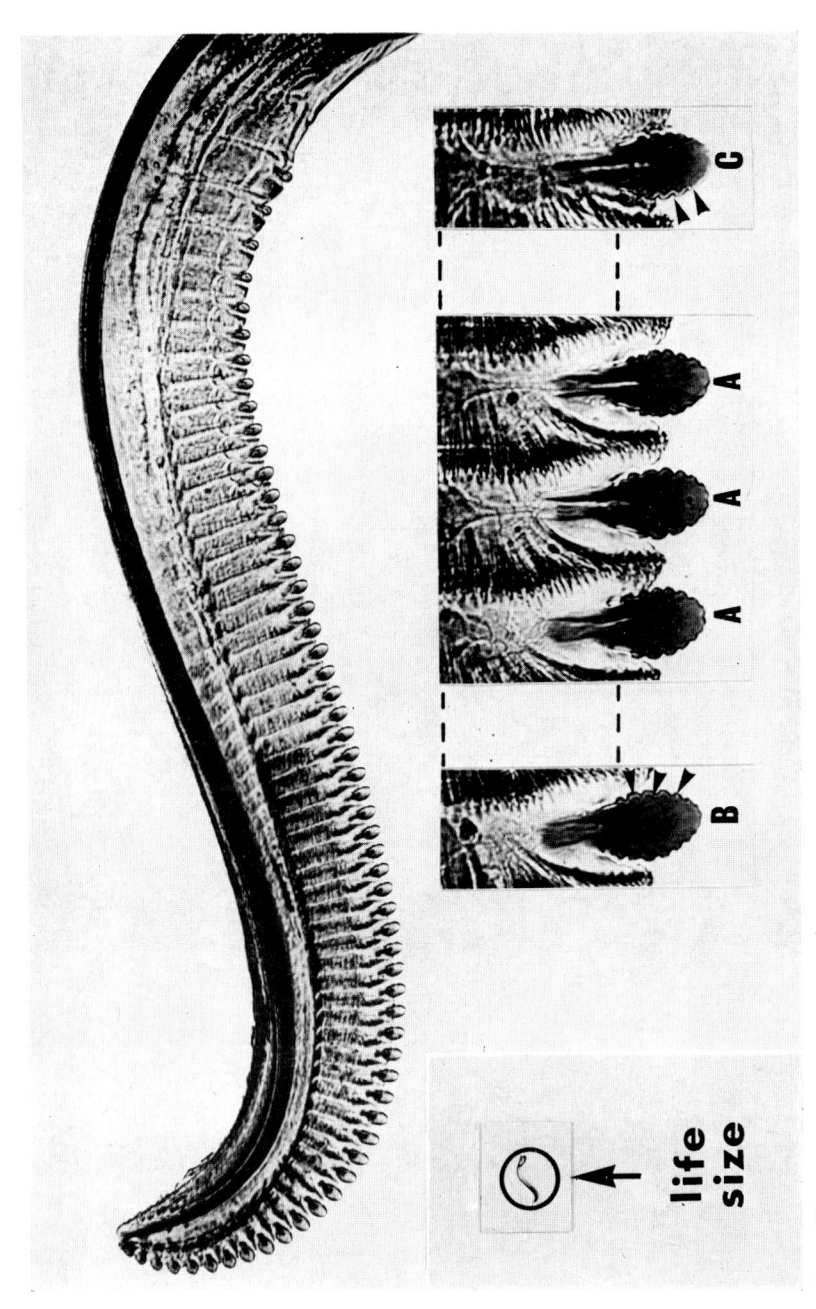

Plate 32.

Plate 33.

PLATES

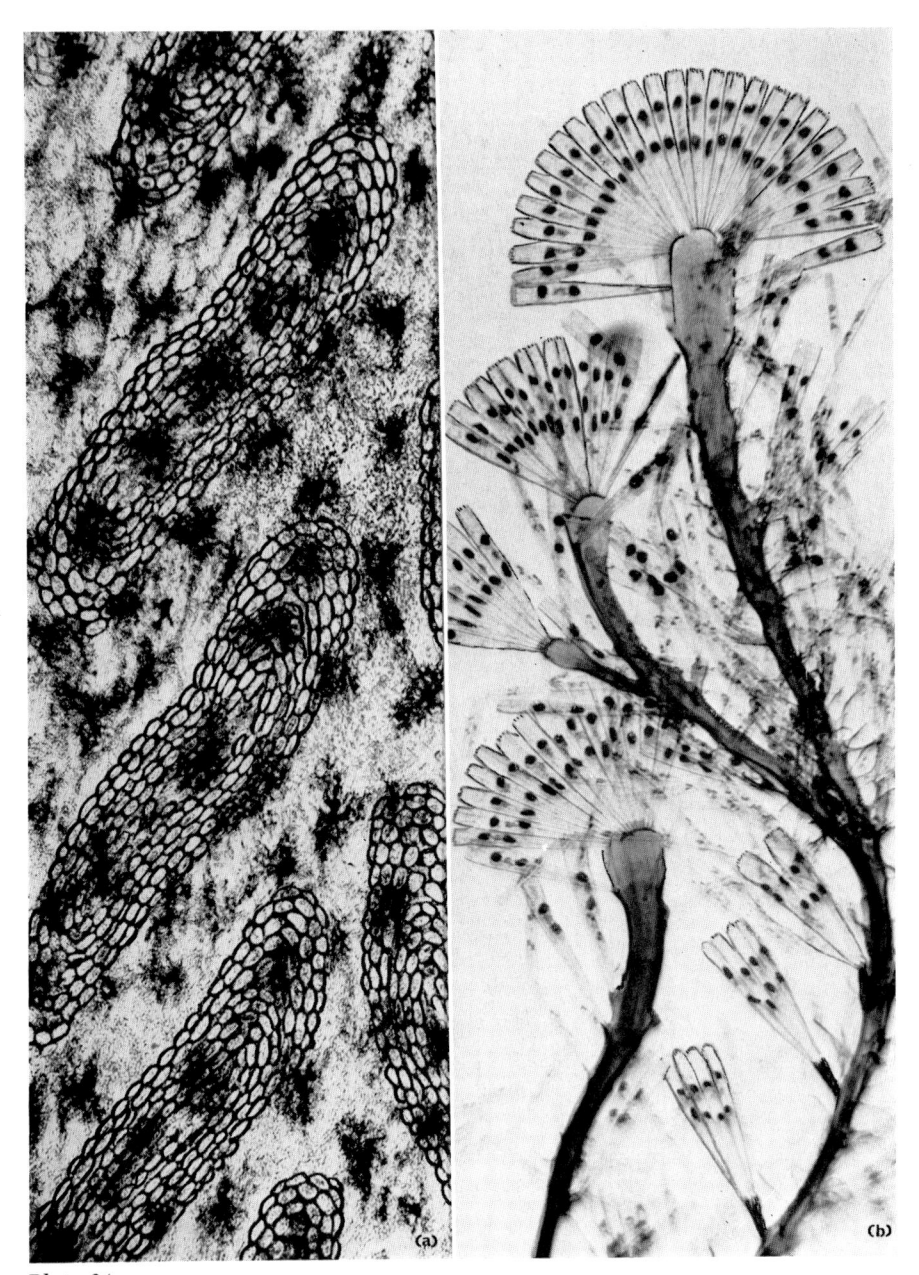

Plate 34.

Plate 35. (a) Tyndall beam, ×54, taken when using darkground illumination. (b) Foot of fly, ×48. (c) Rusted surface of metal after treatment with Deran, ×24. (d) Foot of sheep tick, ×54. (See arrow for fine detail.) Inset: actual size. (e) Scales of eel, ×18 (polarized light). (f) Surface of aluminium sheet after treatment with chromate-oxidation process.

Plate 36. (a) Beck versalite lamp. (b) Beck integral substage illuminator. (Courtesy of Ealing Beck.)

Plate 37. (a), (b) Wine crystals, ×161. (a) Taken with vertical transmitted light, producing a very poor quality image due to the material having the same refractive index as the mounting medium. (b) The same subject photographed by using oblique transmitted light (Fig. 98). Compare the two images. (c) Wheat Rust on wheat leaf, ×210. Showing group of uredospores and teleutospores and s germ spores. (d) Table salt, ×24·5. Vertical incident illumination.

PLATES 409

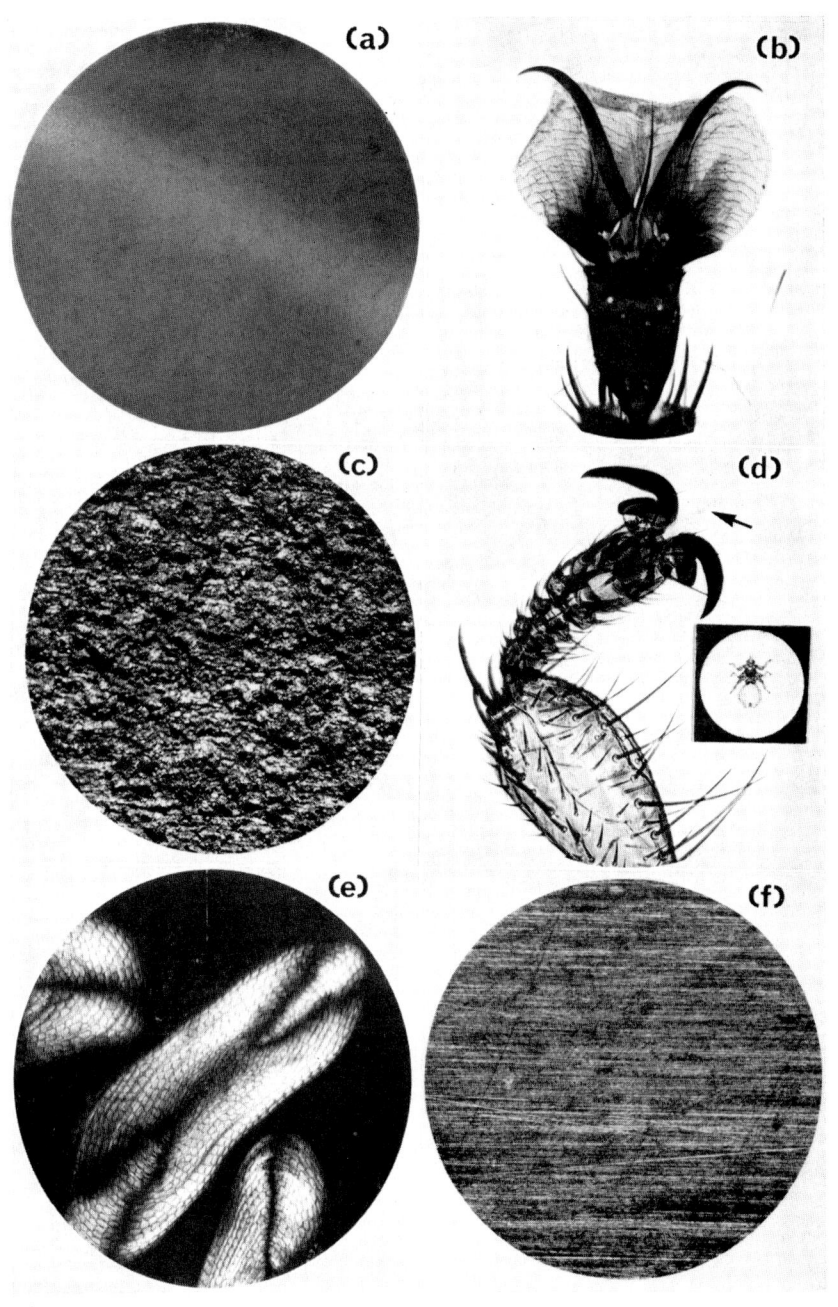

Plate 35.

(a)

(b)

Plate 36.

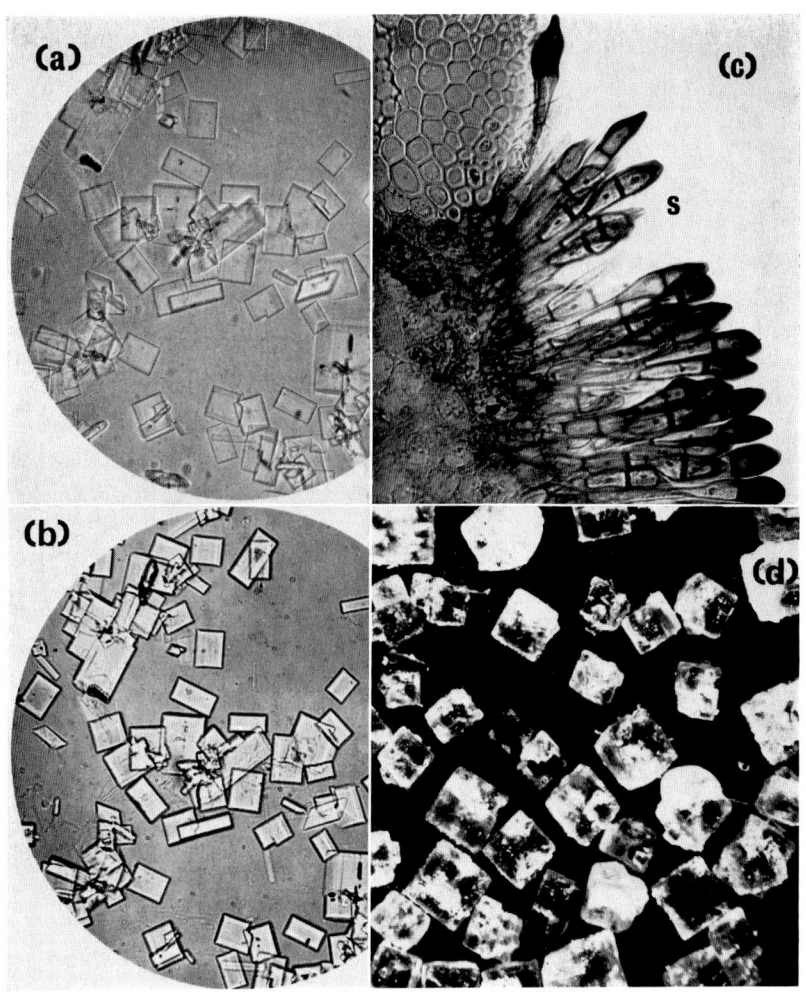

Plate 37.

Plate 38. Leitz "in-line" mirror monochromator (Fig. 198) and camera. (Courtesy of E. Leitz.)

Plate 39. Pin heads and one actual size, see arrow.

Plate 40. (a) Flea of human, *Pulex irritans* (male), ×49. (b) Daphnia Pulex, ×42, showing summer eggs in brood sac, above is the heart. Photographed in water in hollow ground slide.

Plate 40. (c) Flea of human, *Pulex irritans* (male), ×49. Illustrates how not to do it.

Plate 41. No eye strain. Minute detail can be studied collectively in a laboratory (Fig. 144).

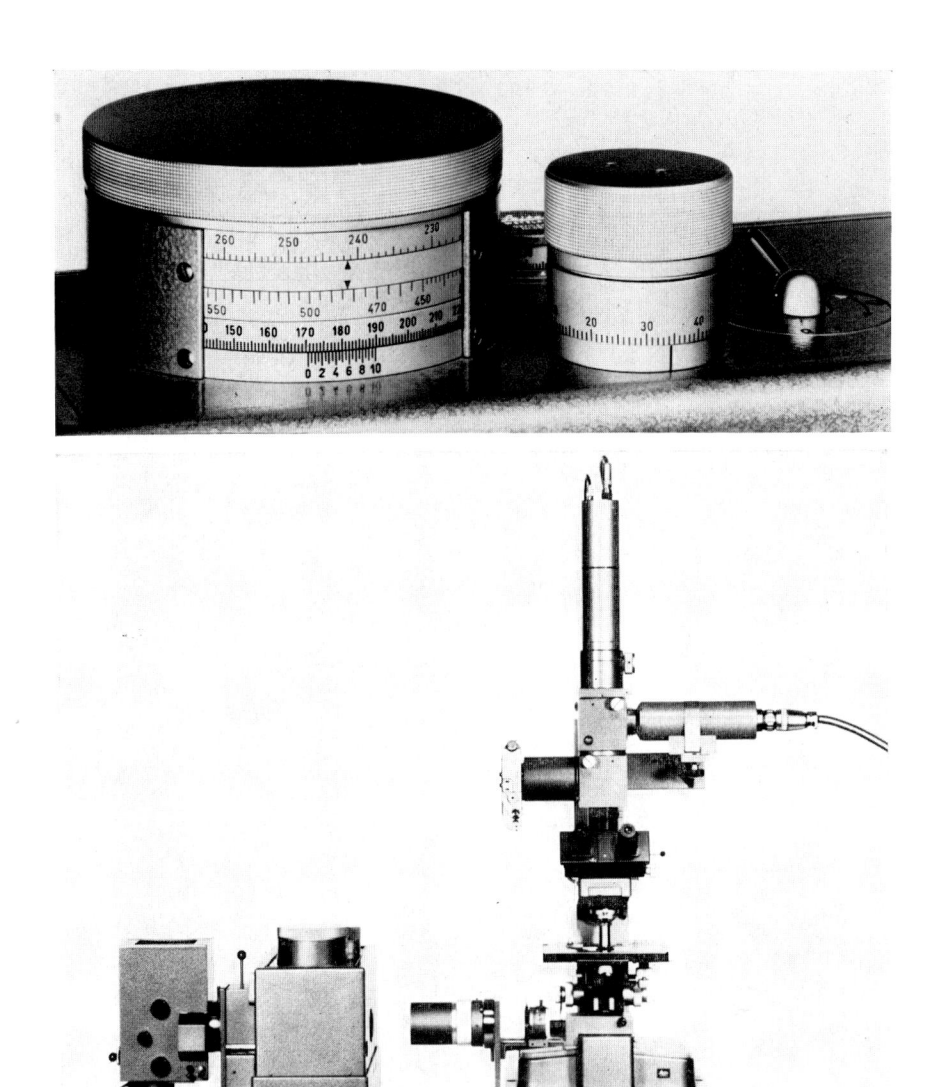

Plate 38.

Plate 39.

(a)

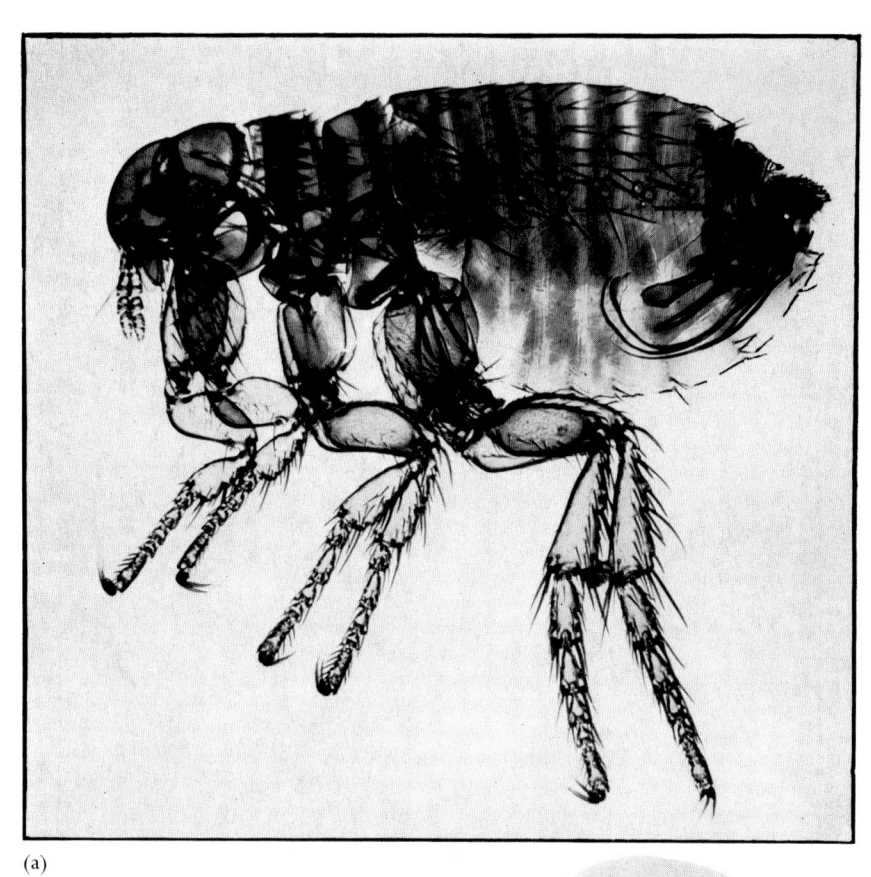

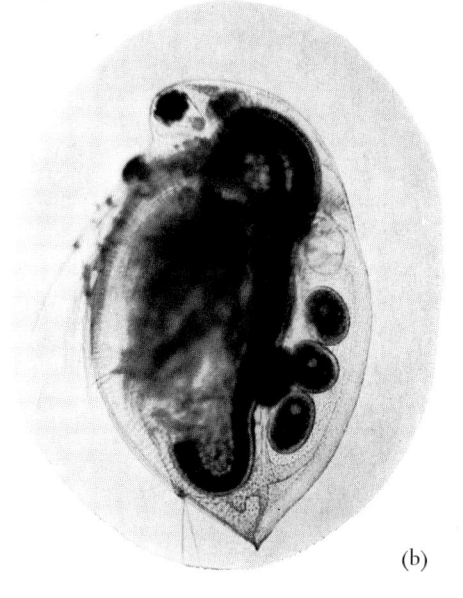

Plate 40.

(b)

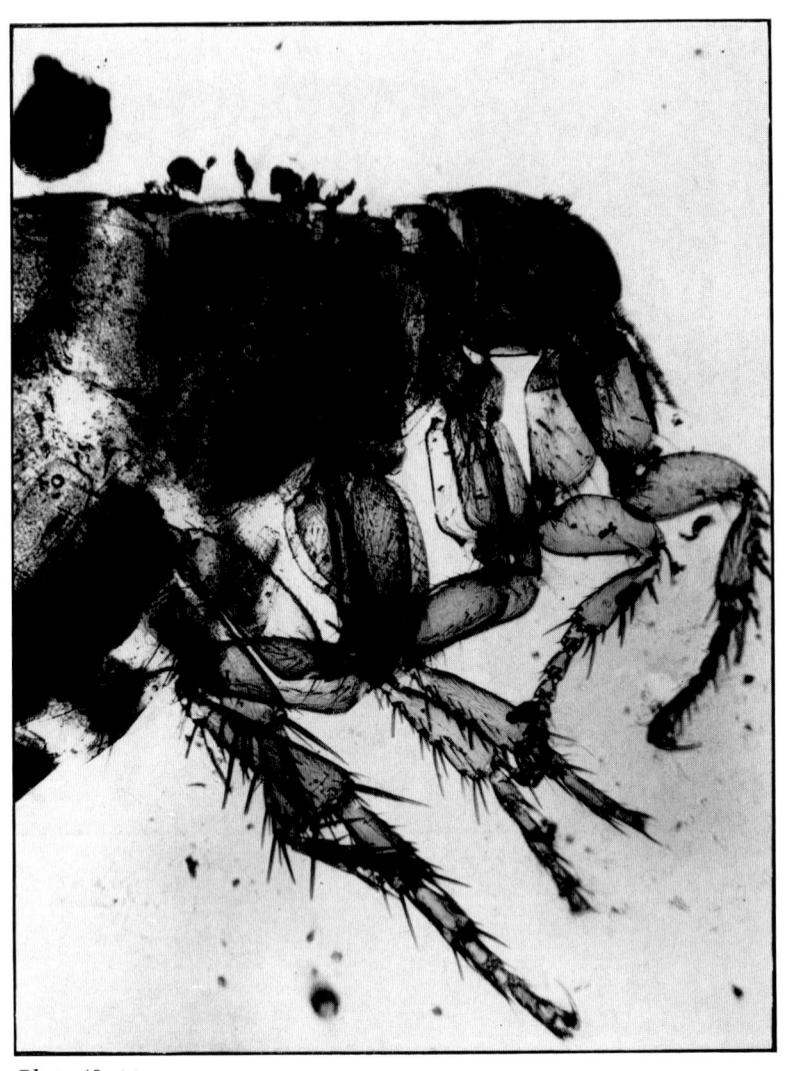

Plate 40. (c)

PLATES 417

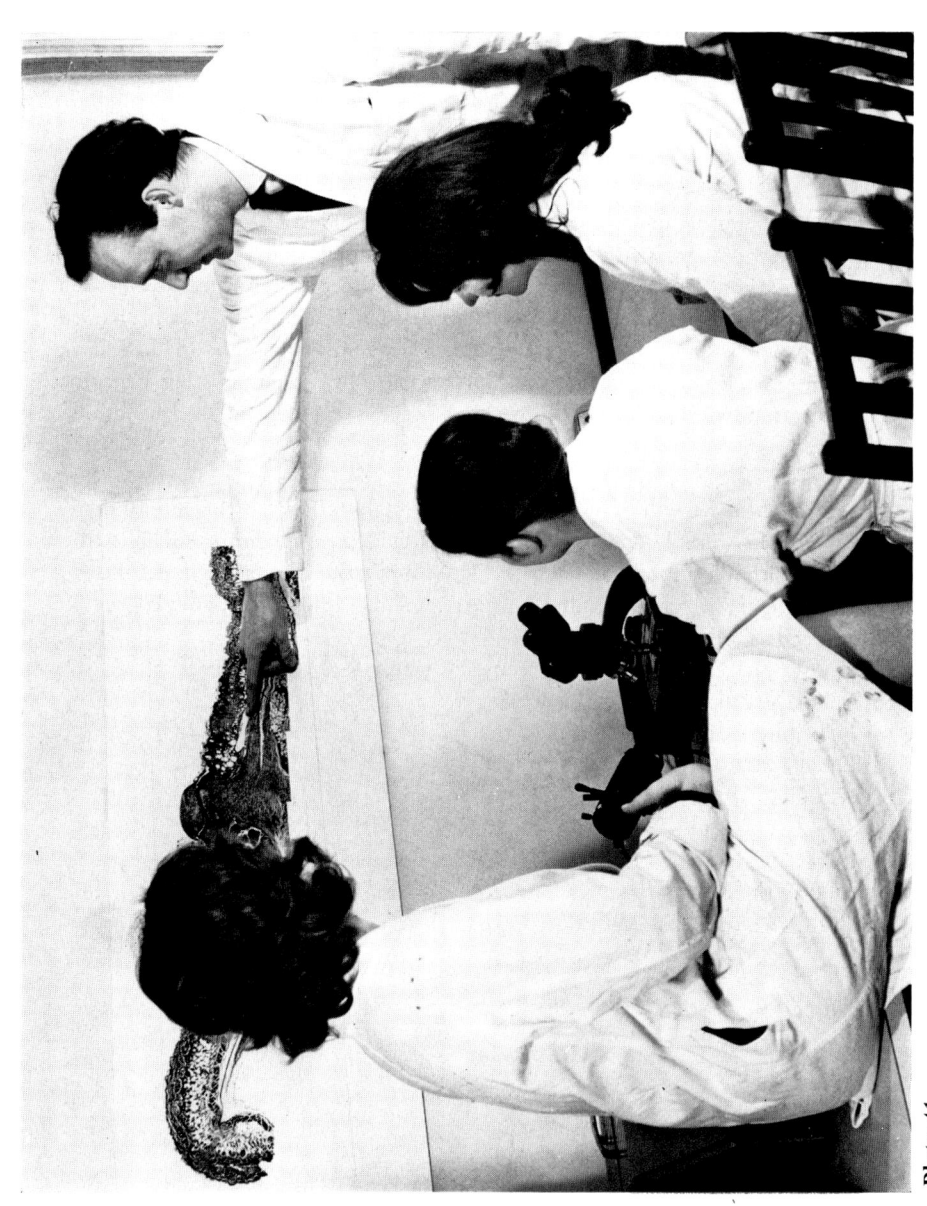

Plate 41.

Plate 42. Transverse section stem of Clematis, $\times 38\cdot 5$. Section stained red and brown. Top photographed with red filter in the light train. Centre without a filter. Lower photographed with a green filter in the light train.   (b) A succession of photomicrographs taken at intervals showing the dissolution of glycine and aspirin crystals, $\times 10\cdot 5$. The five crystals were placed on a piece of blackened blotting paper fastened to the microscope slide with Sellotape. The top illustration was photographed dry mounted, the blotting paper was then moistened and the six illustrations were photographed within 37 sec. Two independent light sources were used.

Plate 43. Molar tooth of human, part of vertical section showing site of decay (arrow), $\times 12$. Treble stained and photographed using transmitted light illumination without substage condenser. The preparation was stained as follows: enamel, pale green; dentine, pink/purple; cement, purple; site of decay, red. Filters used as follows:
(a) Yellow 110. factor $1\frac{3}{4}$. Deep green 404. factor 5.
(b) Yellow/green 402. factor 3.
(c) Light blue 301. factor $1\frac{1}{2}$ and 402.
(d) Orange 202. factor $2\frac{1}{2}$.
(e) Red 204. factor 4.

Plate 44. Carl Zeiss, Oberkochen, continuous running filter monochromator is attached to the high-performance microscope lamp by means of a special mount, page 143. (Courtesy of Carl Zeiss, Oberkochen.)

PLATES

(a)

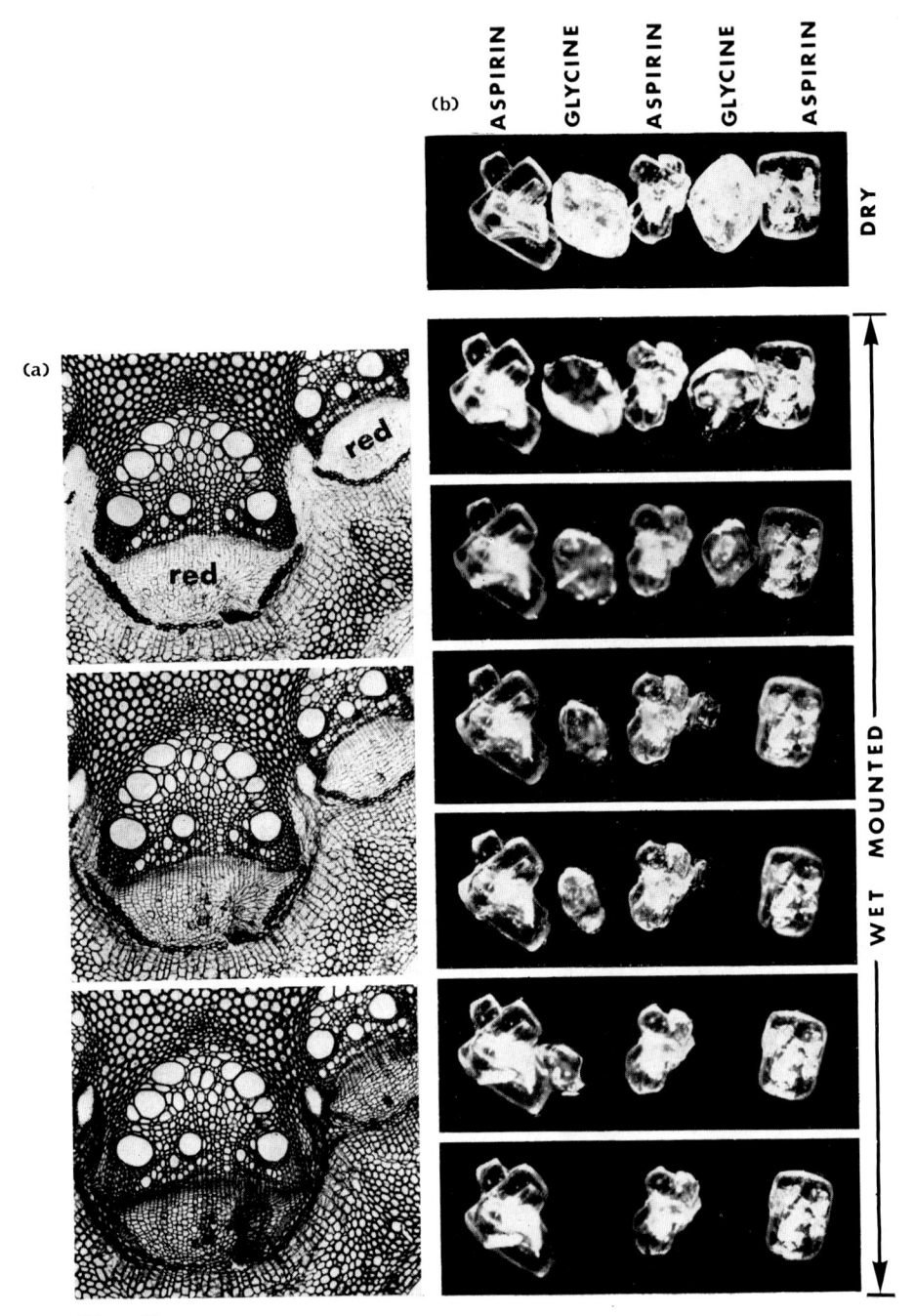

(b) ASPIRIN GLYCINE ASPIRIN GLYCINE ASPIRIN

DRY

WET MOUNTED

Plate 42.

420   PHOTOMICROGRAPHY

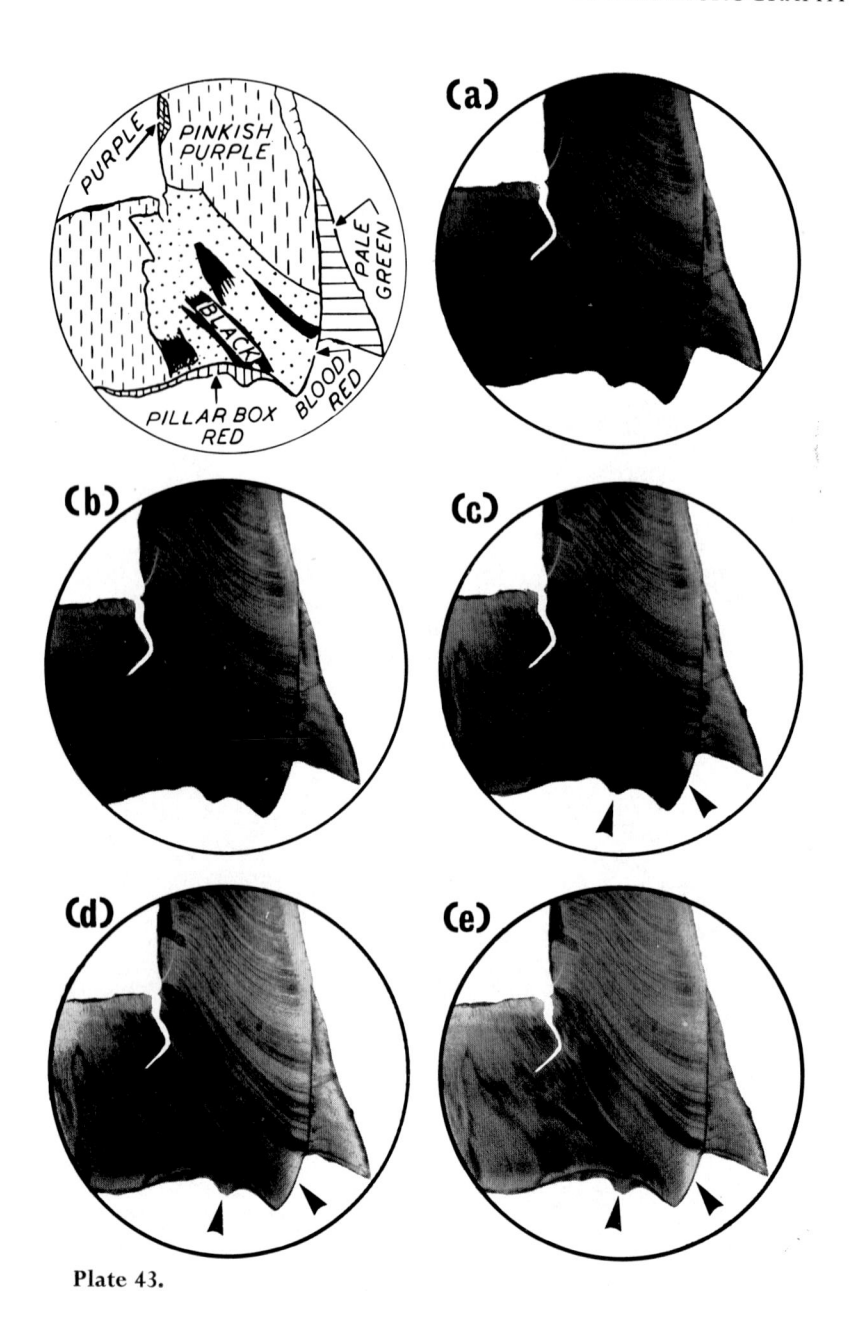

Plate 43.

PLATES

Plate 44.

Plate 45. Leitz Orthoplan with fully automatic microscope camera Orthomat and lamp housing 250. (Courtesy of Leitz.) (See colour Plates 75 and 76.)

Plate 46. Nikon Multiphot 4" × 5" bellows camera and Diascopic illuminator. The diascopic base, with low-voltage focusable illuminator, provides homogeneous illumination in the specimen plane. The mirror reflex housing mounted on the bellows carrier can be rotated more than 90° to right or left for image orientation. The camera can accommodate cut film or plates, Polaroid cut film holder and roll film (size 120) or sheet film. The diascopic illuminator provides bright, even illumination with good contrast for specimens up to 135 mm in diameter. Built-in mirror rotation permits oblique and darkground illumination. The light source is 6V 30W, with focusing collector lens. (By courtesy of The Projectina Company Ltd.)

Plate 47. Bausch and Lomb laboratory microscope cameras. (a) 35mm camera, held in position by a connecting collar. (b) Polaroid $\frac{1}{4}$ plate attachment. (c) 5 × 4 inch camera to take sheet film. Focusing is through the viewing eyepieces. (Photographs by courtesy of Bausch and Lomb.)

Plate 45.

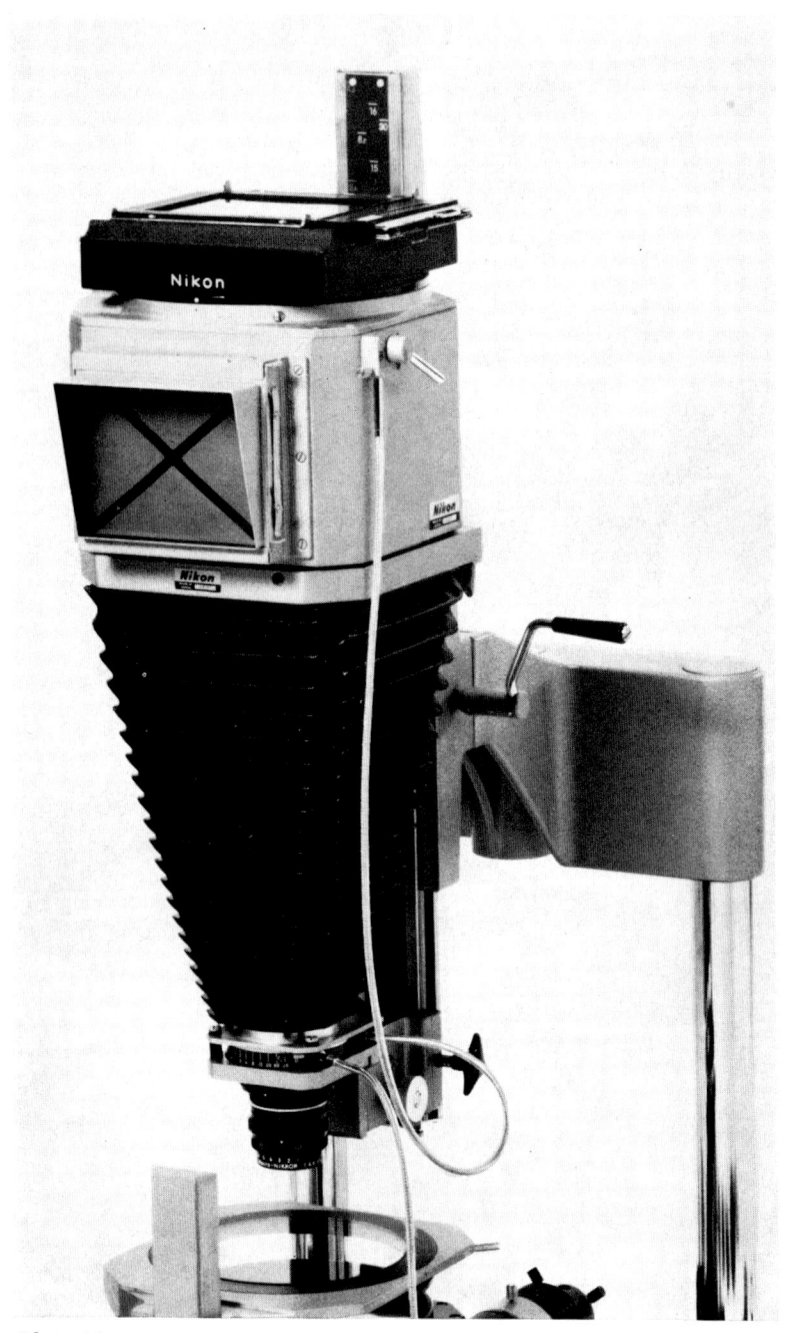

Plate 46.

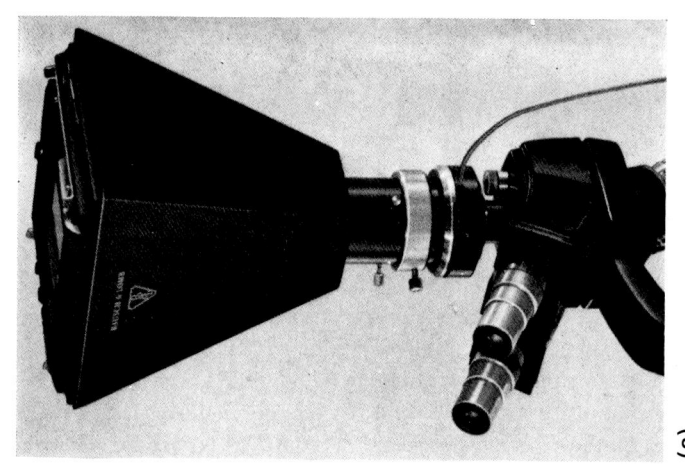

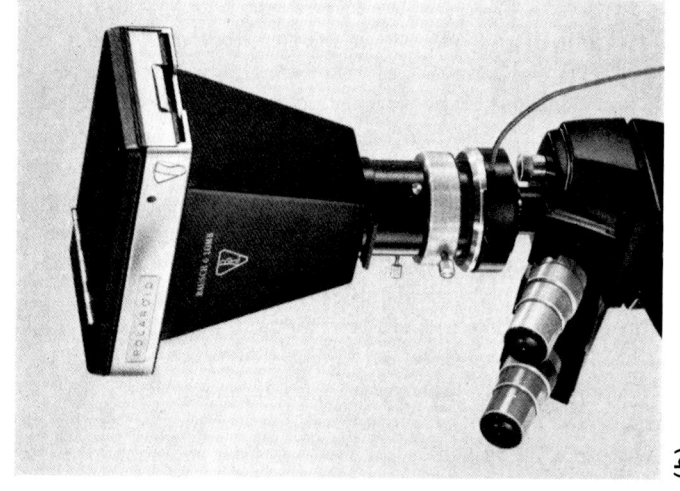

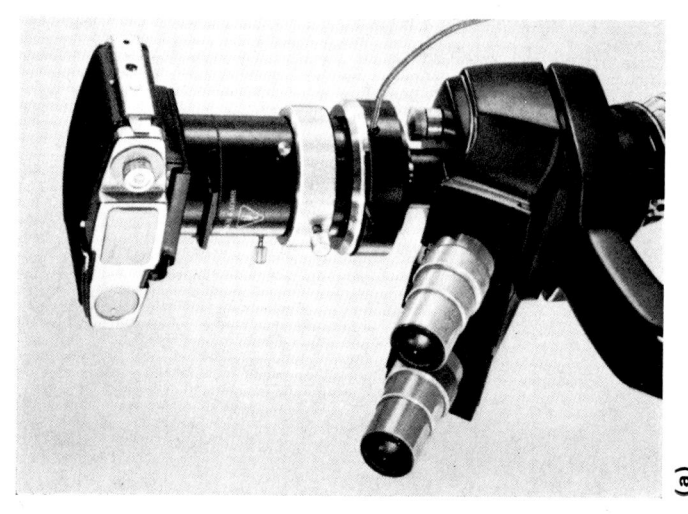

(a)

(b)

(c)

Plate 47.

Plate 48. Twin gall bladder in rabbit embryo. (a) Rabbit embryo life size. Arrow indicating the area of the gall bladder. (b) The minute twin gall bladder illuminated by two light sources, each operating at wide beam projection. The gall bladder was the same colour as the surrounding liver. (c) Wedge glass container (Fig. 150) used to photograph particles in liquid, page 171 and Plate 49. (d) Twin gall bladder illuminated for one minute by the source illustrated on the left of (e). The source on the right of the figure was switched on for approximately 5 sec.

Plate 49. Particles in broth, $\times 24$. A series taken, as set out in Fig. 149. The objective, 100 mm, N.A. 0·08, $\times 0·7$ used without an eyepiece produced an image size $\times 8$, enlarged here to $\times 24$. The series on the left were taken by transmitted light and those on the right by incident light. A series like these permits the close examination of pellets during the many days of antibiotic fermentation.

Plate 50. (a) L/S upper surface of precipitation band showing granular size $\times 150$. Wet mount, transmitted bright field. (b) T/S of precipitation band within the agar, showing granular structure $\times 150$. Wet mount, transmitted bright field. (c) Ouchterlony plate with simple precipitation band within agar gelatine, $\times 12$, illuminated according to the Lawson method, Fig. 151. (d) Oblique view of the same band pattern recorded as a luminous body, when in fact the band and surround are approximately the same refractive index. See (a) and (b) for the granular structure of a band. (e) Precipitation pattern, $\times 12$. In this instance the micro-gel was prepared on a microscope cover glass and submerged in distilled water for photography (Fig. 151). (f) The same subject photographed after staining with Ponceau stain and with a white base producing a bright field effect. Here a complicated number of bands are seen, some fine, which must be recorded on a negative without the bands uniting.

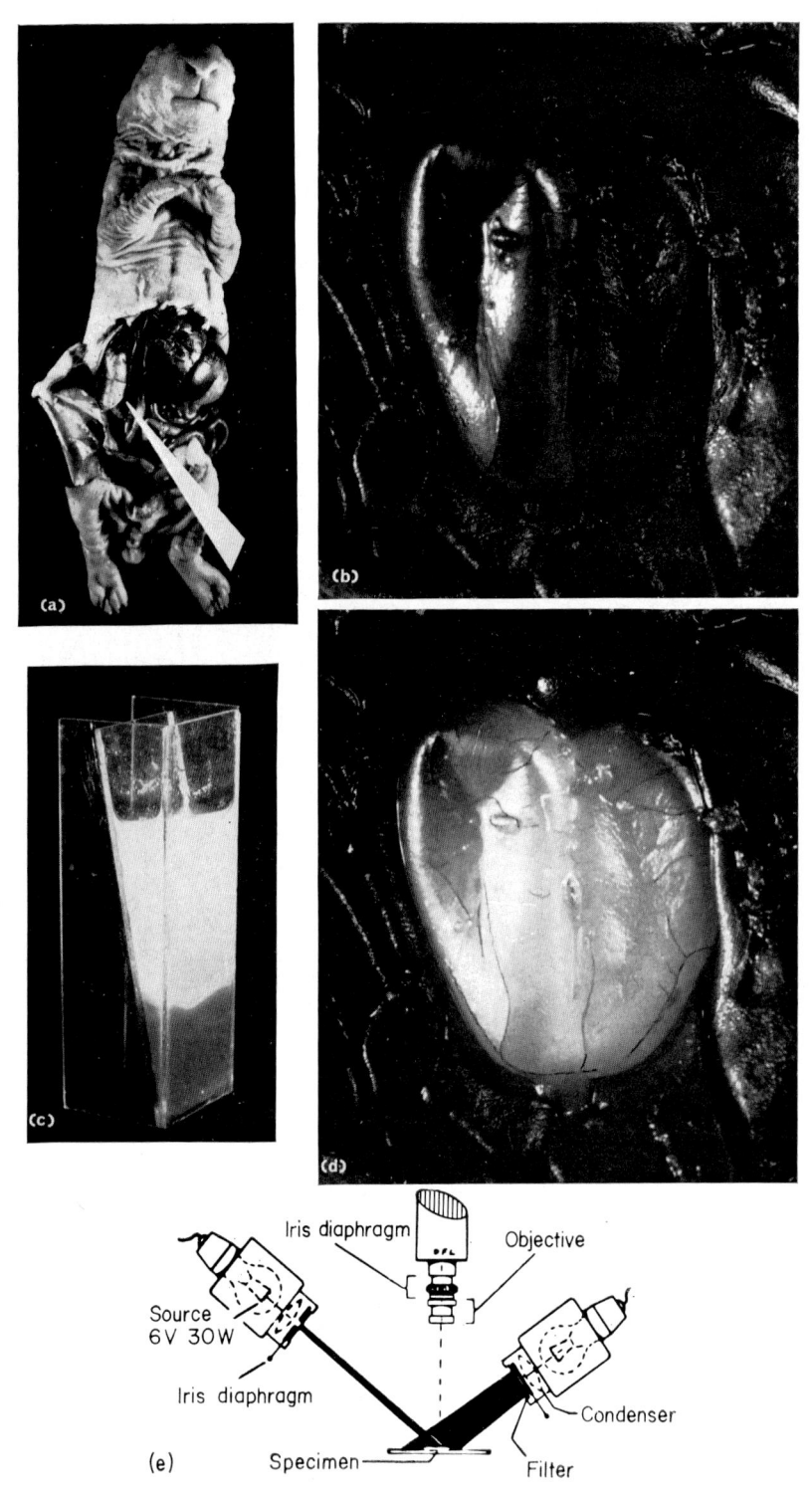

Plate 48.

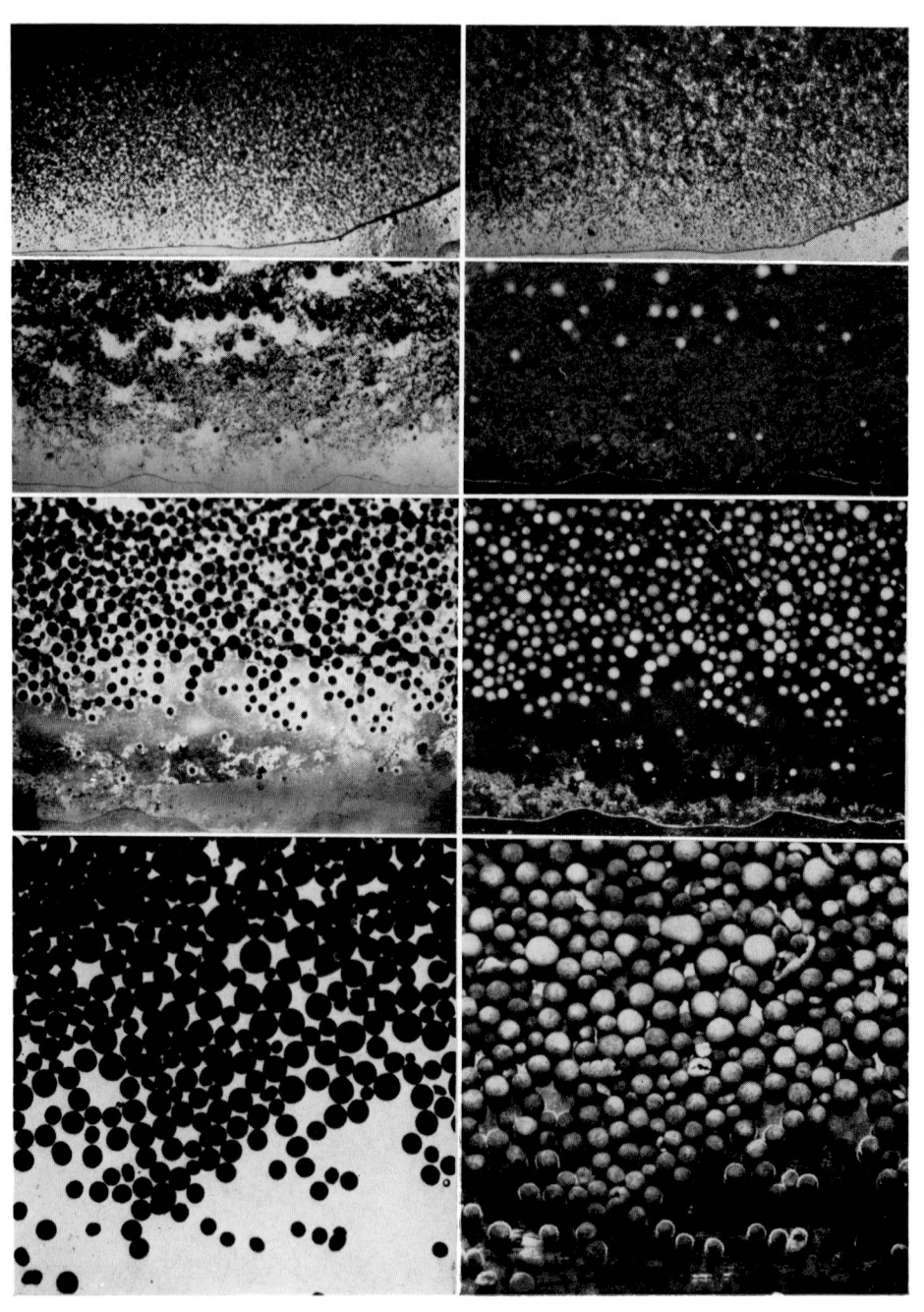

Plate 49.

PLATES 429

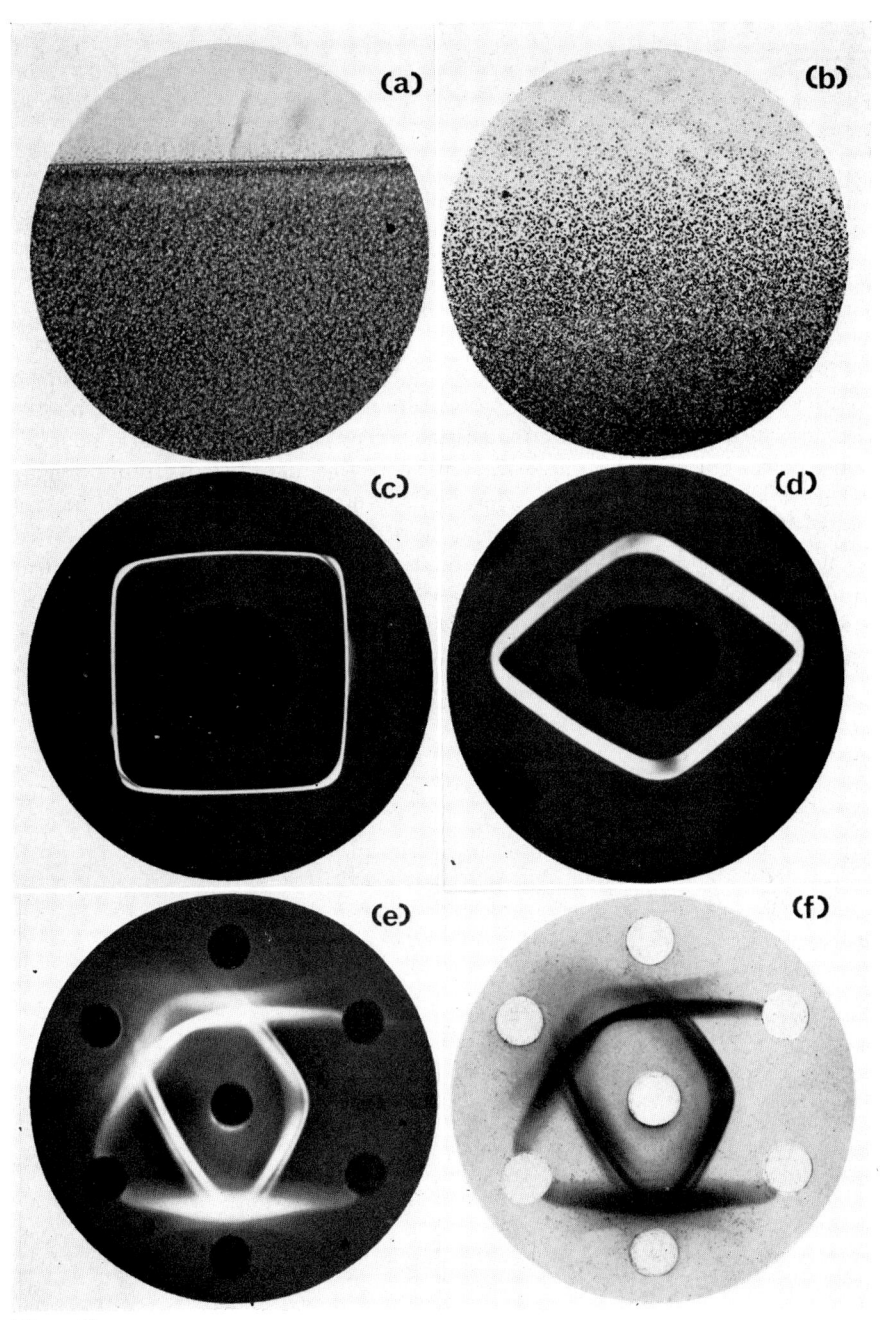

Plate 50.

Plate 51. Micro-gel precipitation bands in small glass capillaries. See inset and Fig. 152.

Plate 52. (a) Rabbit blastocysts, $\times 9 \cdot 9$, total magnification $\times 17$. (b) 14 day rat embryo, $\times 5 \cdot 4$, total magnification $\times 27$. See Fig. 151.

Plate 53. Deformed rabbit blastocysts, $\times 36$, total magnification $\times 124$. A white base was introduced into the apparatus with illumination provided by two photofloods (See Fig. 151). Initial magnification was achieved with a shallow mounted low-power objective, 25 mm $0 \cdot 15$ N.A., with a behind-the-lens iris diaphragm. The use of a small aperture, extensive lens to film distance, no direct light rays entering the lens and a slow emulsion, Ilford N40, meant a long exposure. After enlargement to the size seen in the plate. This method makes for more detail than in top left of Plate 52 (a). This magnification is about the highest one can achieve with the simple microscope yet the field covered and depth of field is far greater than that produced with a low-power compound microscope.

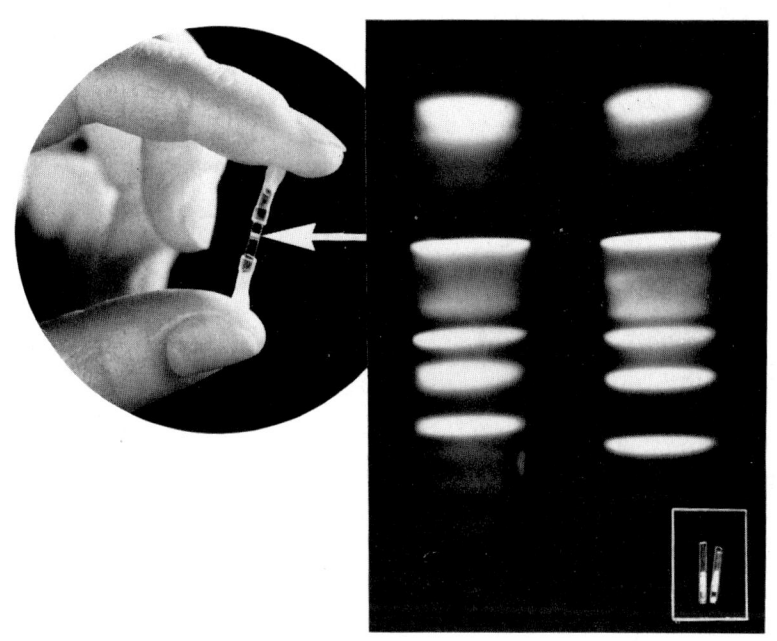

Plate 51.

432　　　　　　　　　　　　　　　　　　PHOTOMICROGRAPHY

(a)

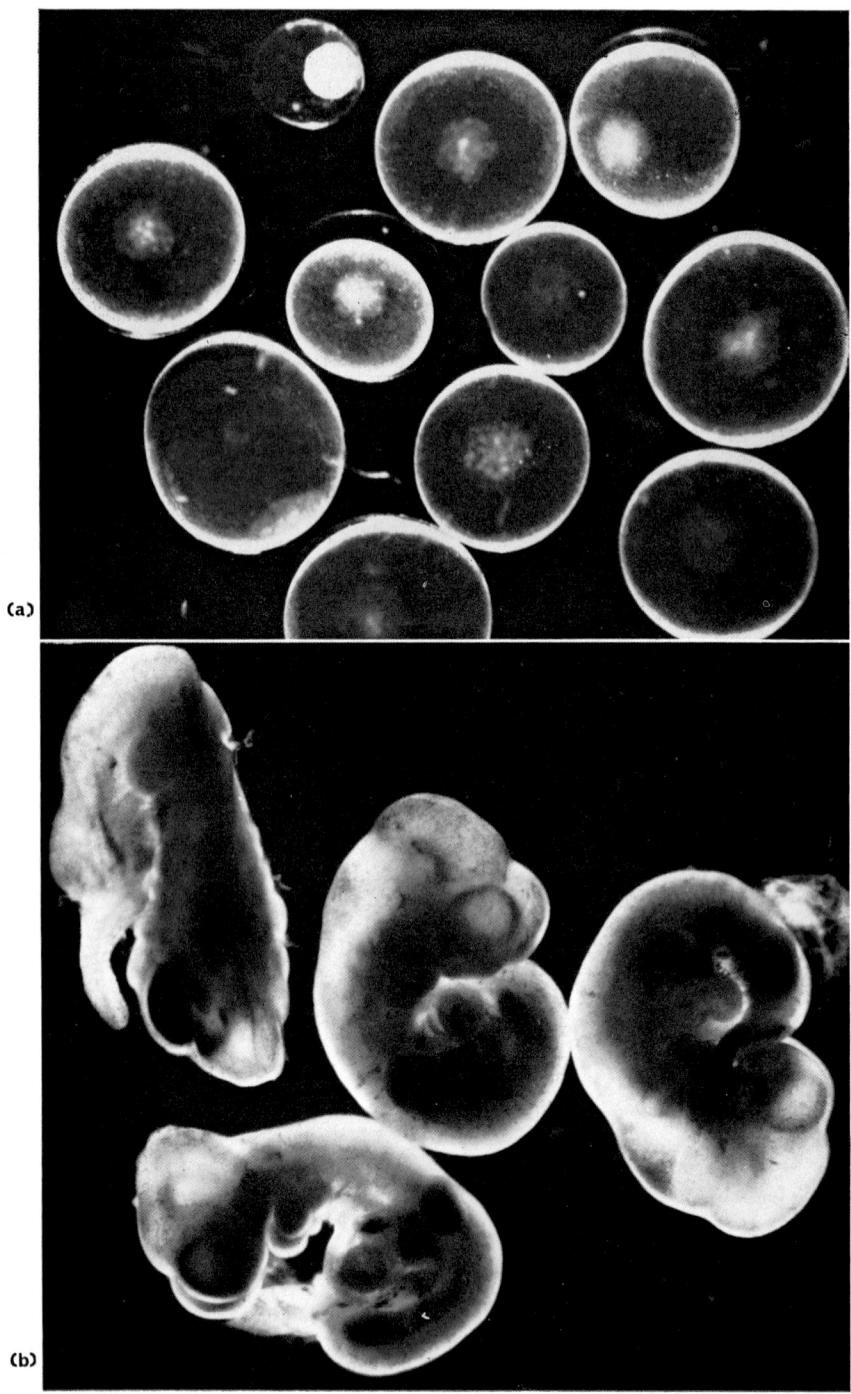

(b)

Plate 52.

Plate 53.

Plate 54. (a) ① Seen by the naked eye, recorded here on an MPP 18 inch bellows extension. ② Image projected by the eyepiece. ③ Transmitted illumination. (b) Interferogram of echeloned surface, ×160. Taken on Reichert Universal camera microscope MeF, with polarization interferometer attachment.

Plate 55. Gas bubbles at the solid–liquid interface, ×55. (a) Hydrophilic surface. Bubbles on the clean glass surface. (b) Hydrophobic surface. Bubbles clinging to a slightly greasy glass surface. (c) Bubble clinging to a heavily contaminated glass surface. (See Fig. 156.)

There was very little time in which to take the photomicrographs as the bubbles quickly became dislodged from the glass surface and floated to the top of the bottle.

Plate 56. Defective bottle tops (see Fig. 159).

PLATES 435

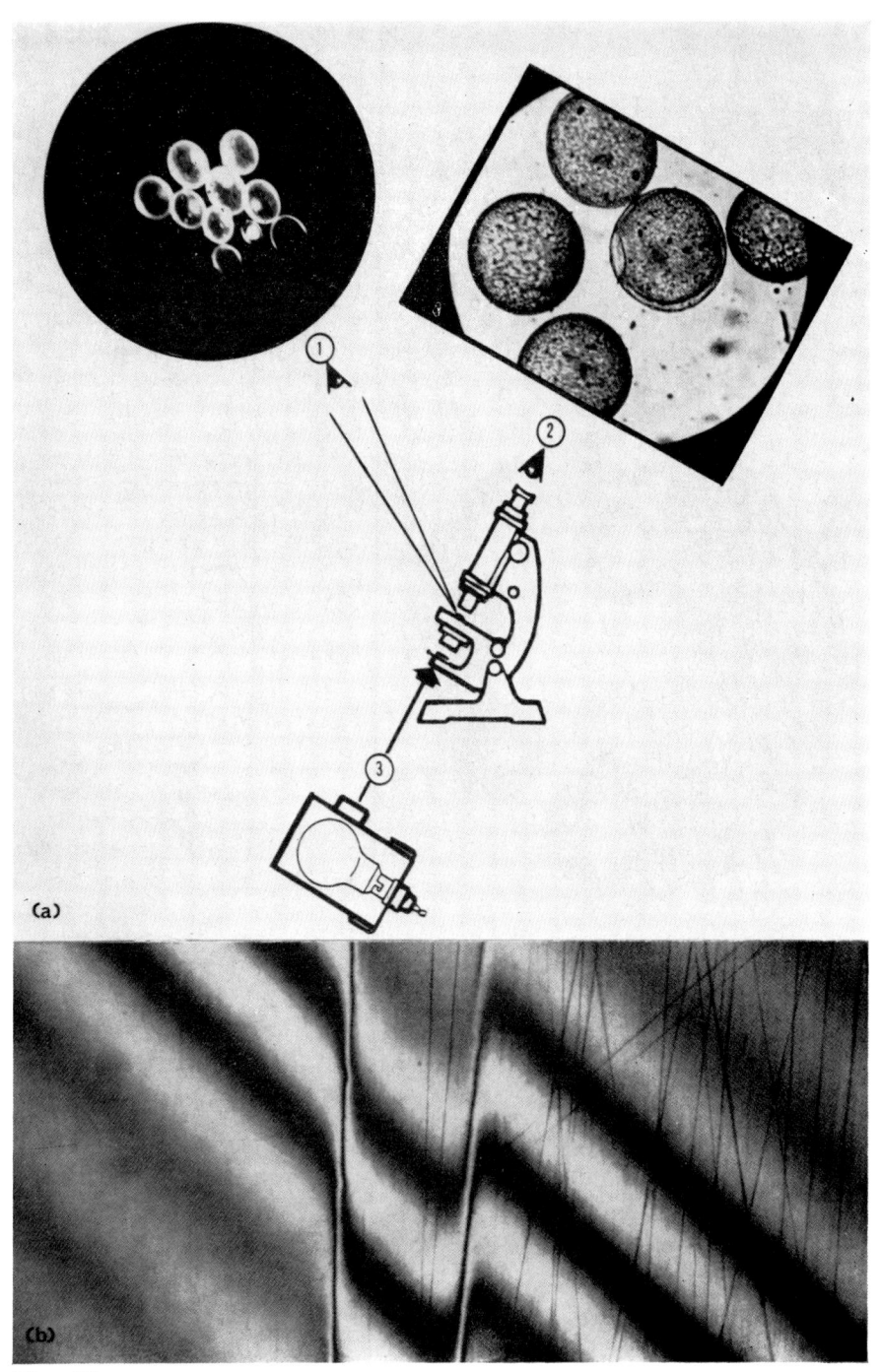

Plate 54.

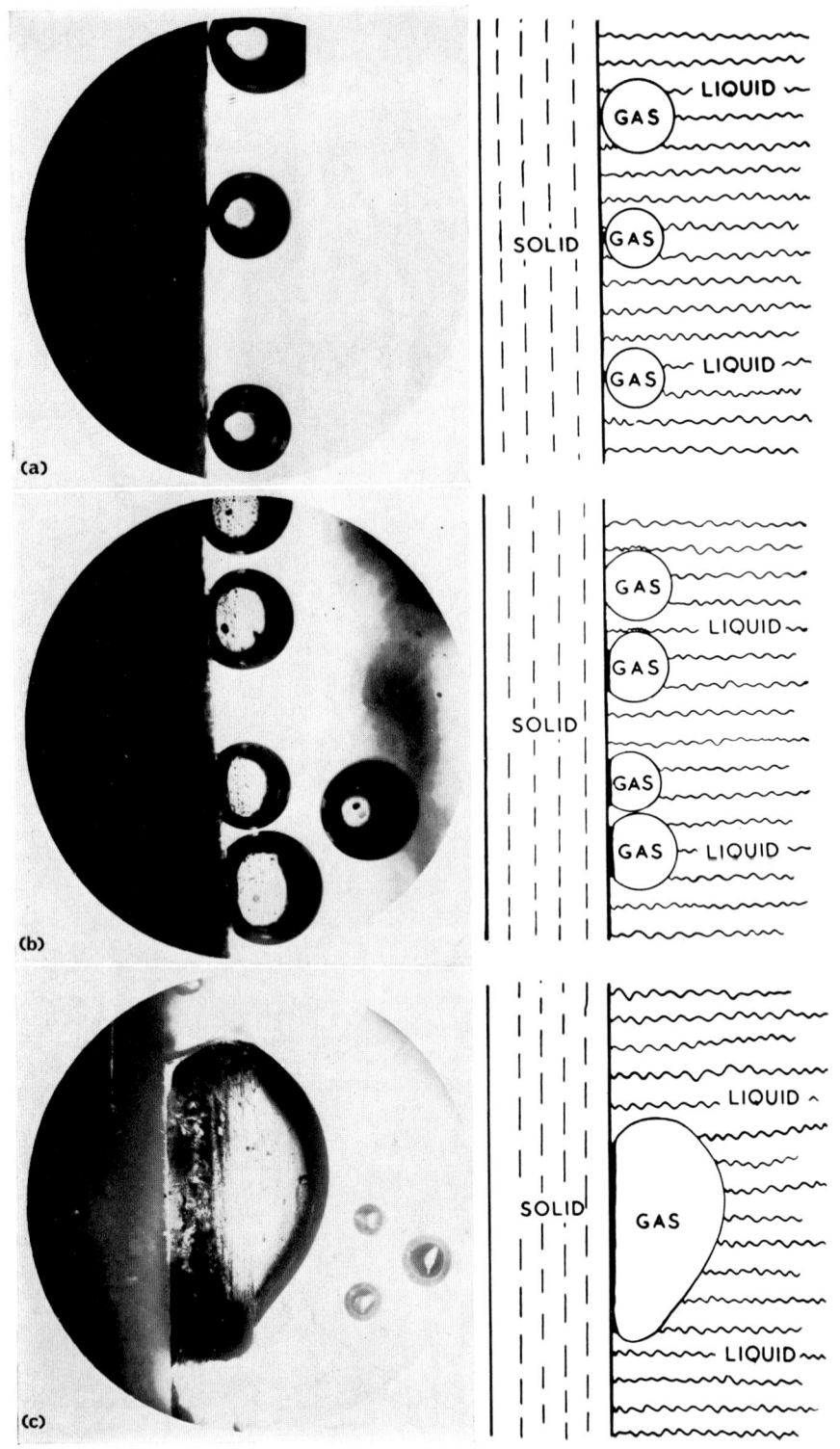

Plate 55.

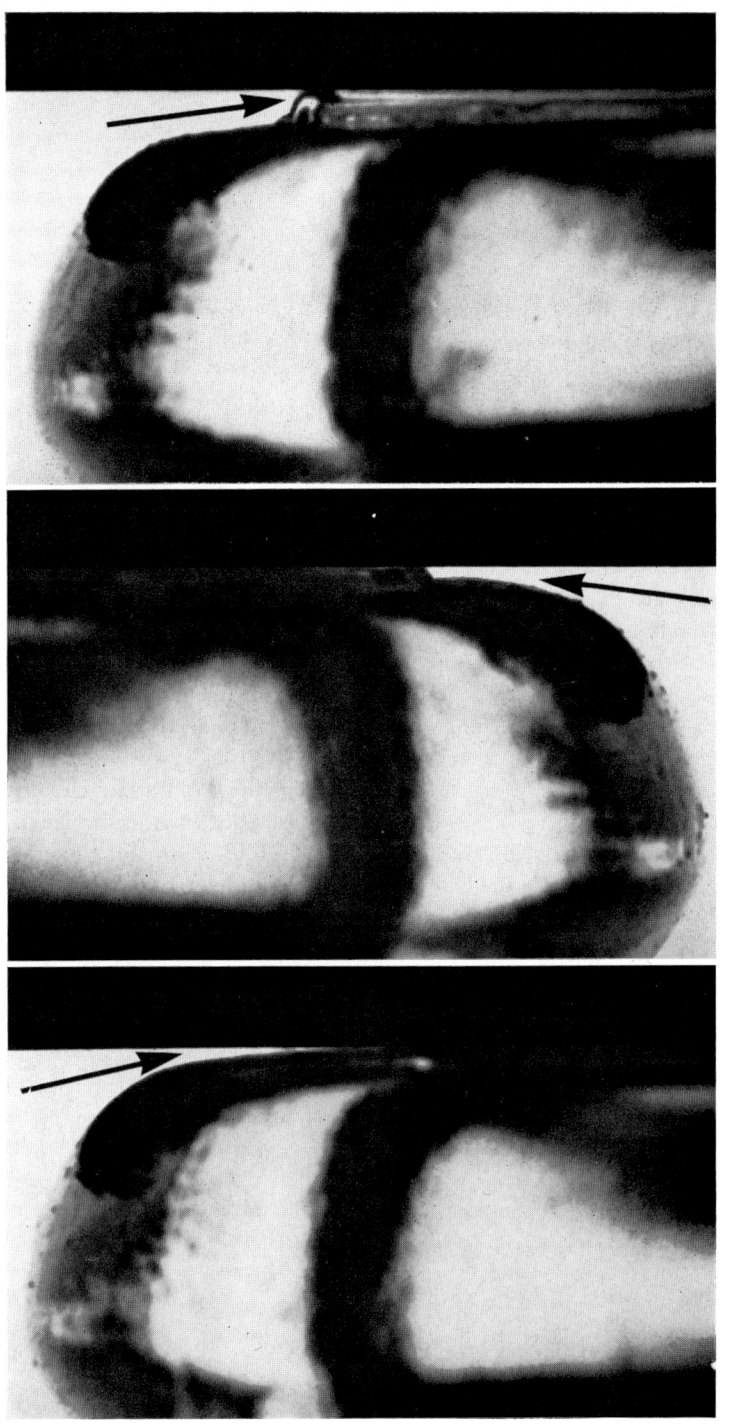

Plate 56.

Plate 57. Caterpillar of the Large Cabbage White butterfly in the act of hatching and eating. An opaque subject such as this is one of the most difficult to record. How to arrest movement when the means of illumination is limited; the lack of depth of field; point of focus, and deciding to generalise the focus; and exposure, which must be instantaneous, are all problems. Flash was used with a white reflection around one half of the field; note the light shadow area. The drawing is for guidance to what one is looking at.

Plate 58. Larva of *Dytiscus marginalis*, water beetle, ×2. The creature was photographed under water by transmitted light (Fig. 151), using a white base. Much detail can be extracted from such a subject and passed on in this style of layout. (Plate 65, 5) shows an air bubble attached to the rear of the larva, here in (j) the two air outlet valves can be seen and indicated by the small arrows. The complete breathing system can be followed and (i) and (k) illustrate two of the many air intake valves. (a) Mouth parts highly magnified. (b) Complete mouth parts. (c) Section of the antenna, base joint. (d) Section of leg. (e) Main breathing tube. (f) Leg joint and hairs, the base of which are as regular as a machine stitch. (g) Both breathing tubes coming together at the base. (h) Foot. (i) Breathing valve showing the inlet tube into the main trunking. (j) Base of larva, showing the two outlet valves. Note how small these are. (k) Breathing valve showing the external structure and inlet aperture on the surface of the outer lining. Of course a 20 × 16 inch display would detail the larva more clearly.

Plate 59a. Carl Zeiss, Oberkochen. Large fluorescence microscope with camera.

PLATES

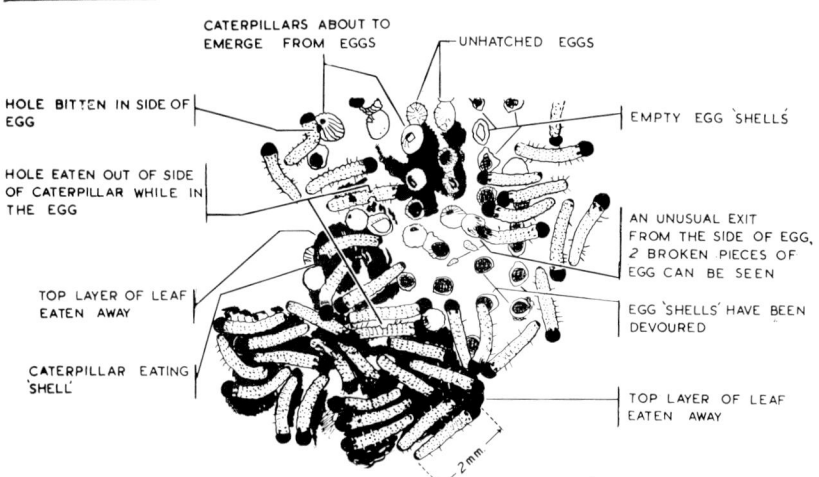

Plate 57.

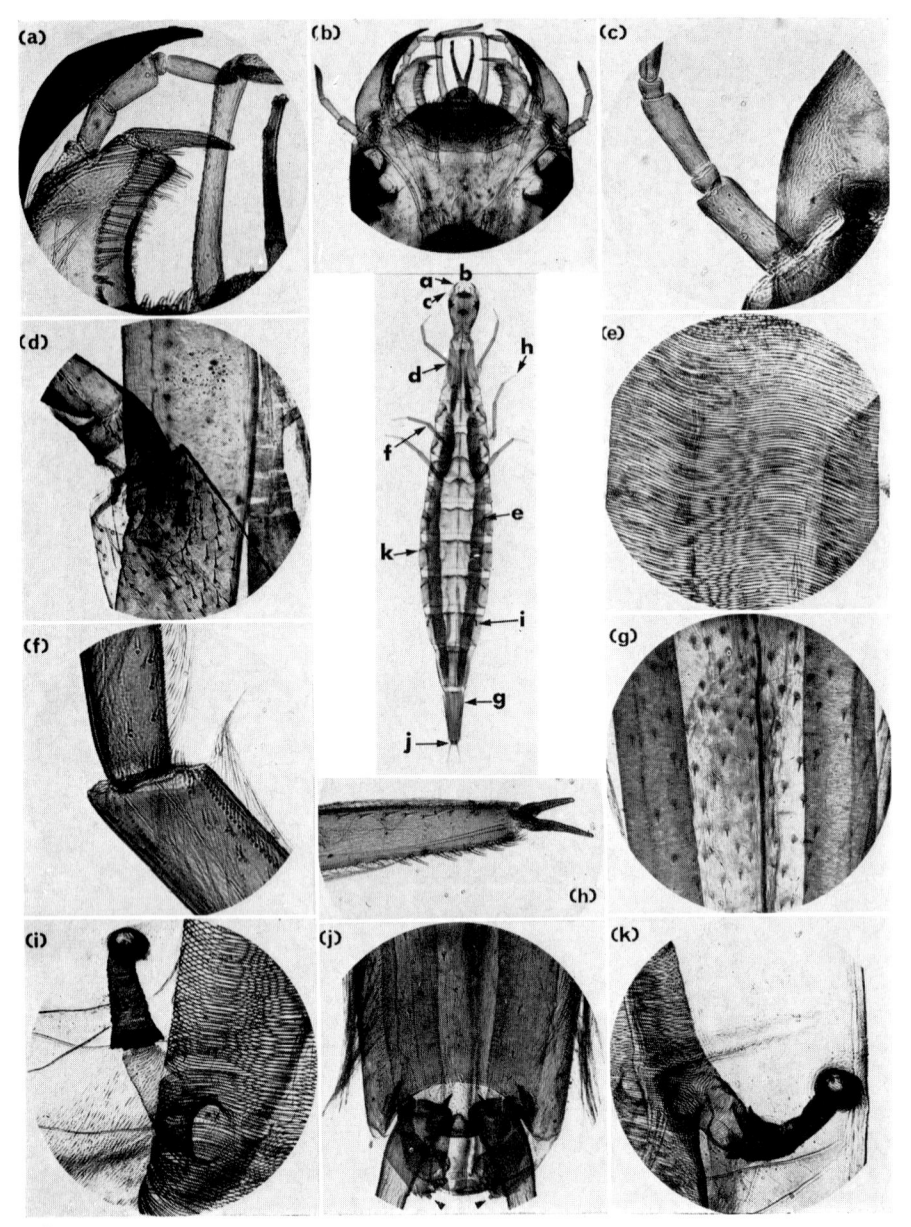

Plate 58.

(a)

Plate 59.

Plate 59b. Vickers' Autowind 35 mm camera, M15C microscope with projection base and J35 exposure unit. (Photo by courtesy of Vickers Instruments Ltd.)

(b)

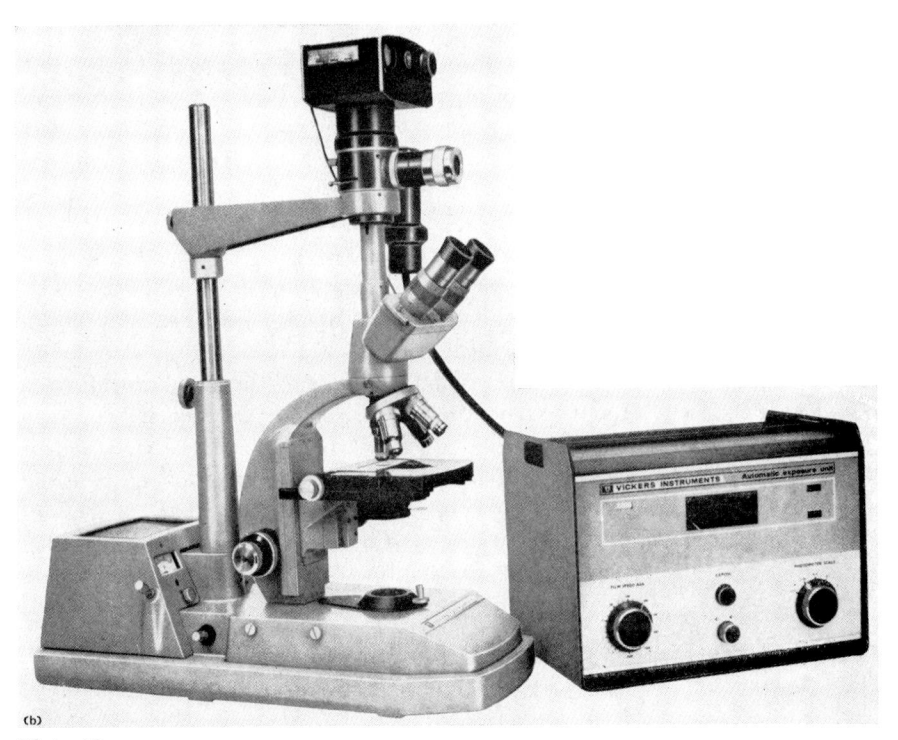

Plate 59.

Plate 60. Decaying rat teeth, ×15, total ×40. (a) Side view, (b) Plan view of the same teeth, (c) Side view and (d) Plan view of ground teeth. Inset: actual size. The five main ingredients of photomicrographs such as these, are: viewpoint, direction of light or lights, point of focus and magnification, plus negative material. In this case the jawbones were mounted on plasticine making the specimen rigid and providing a choice for angle of tilt. When tilting the specimen the point of focus must be considered. Two 6V 30W lamps were angled for each subject (see illustrations), according to the message being expressed. The point of focus was controlled by the angle of tilt, ensuring an overall sharp field. Focusing is critical and an iris diaphragm (as in Plate 3) is essential. This subject has considerable depth and so a small aperture was used. This is done when viewing the projected image in the ground-glass screen. The initial magnification is vitally important. If too low, it becomes excessive, after enlargement, revealing grain and a lack of definition. If too high, lack of depth of sharpness will be revealed and true perspective lost, limiting the degree of after enlargement. Of course fine grain emulsion, that records delicate half-tones when suitably developed, must be used. A middle course must be set and the initial magnification controlled by the required final magnification. Here a 50 mm focal length objective N.A. 0·11 was used with camera bellows extended to 25 in. (See Fig.)

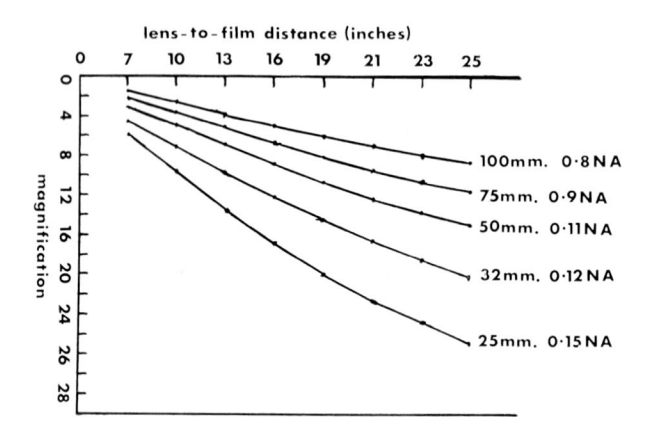

Plate 60.

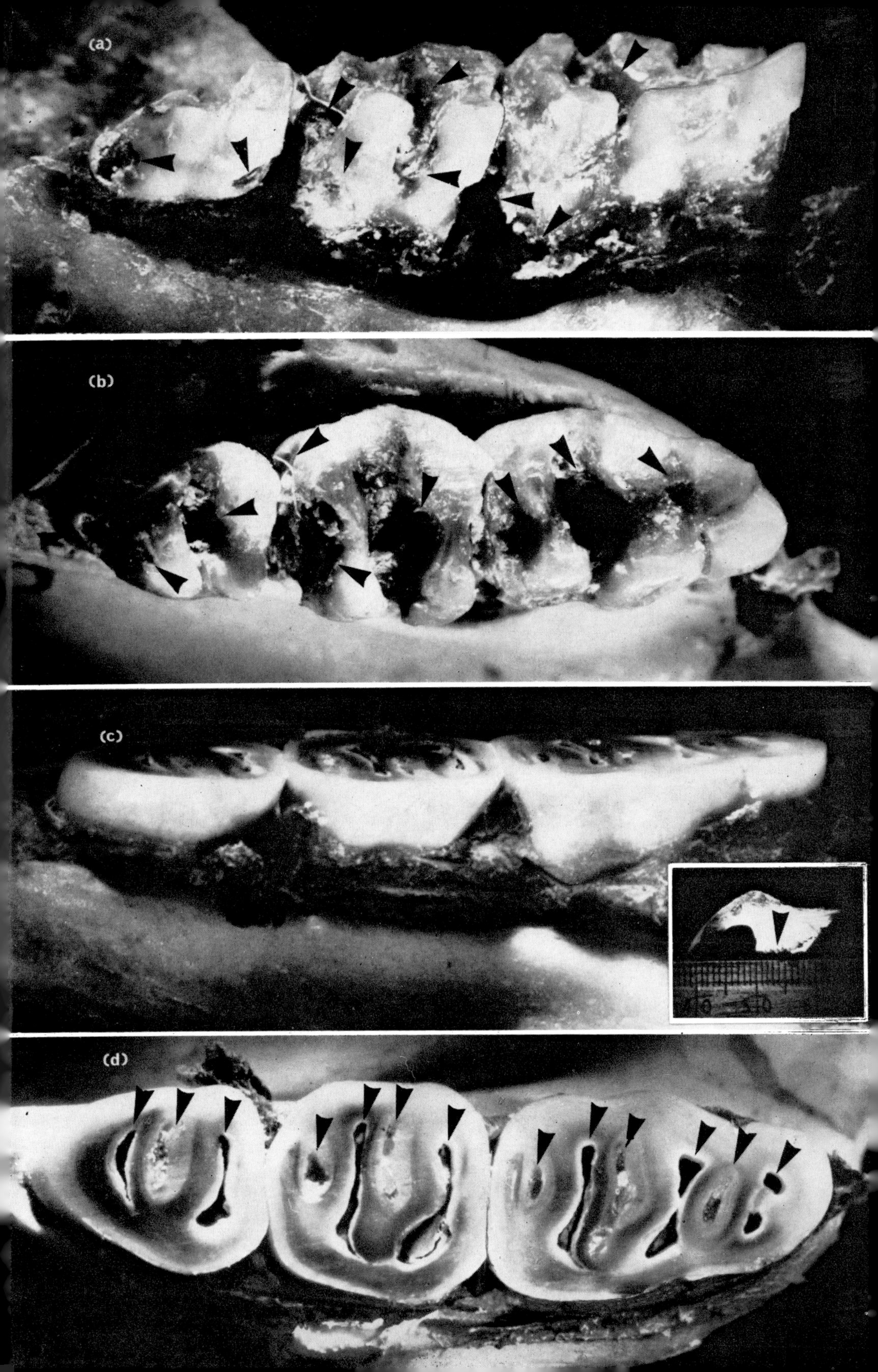
(a)
(b)
(c)
(d)

Plate 61. Wild M 20 binocular microscope, a representative modern research instrument with built-in Köhler illumination, coaxial coarse and fine adjustment. A wide range of interchangeable accessories allows the M 20 to be used for phase contrast, polarization techniques, fluorescence and interference techniques in incident light. A wide variety of light sources are available for the M 20 and camera can be fitted. (Courtesy of Wild Microscopes.)

Plate 62. (a), (b) Transverse section of testis of rat, $\times 117$. Specimen stained by haematoxylin and eosin. Transmitted light. (a) Photographed on orthochromatic material. (b) Photographed on chromatic material incorporating a green filter. (c) Crystal growth, $\times 180$. Oblique transmitted light. (Figs 98 and 99.)

Plate 63. (a) Carl Zeiss, Oberkochen, hot stage. (Courtesy of Carl Zeiss, Oberkochen.) (b) U.N. auto cleaner.

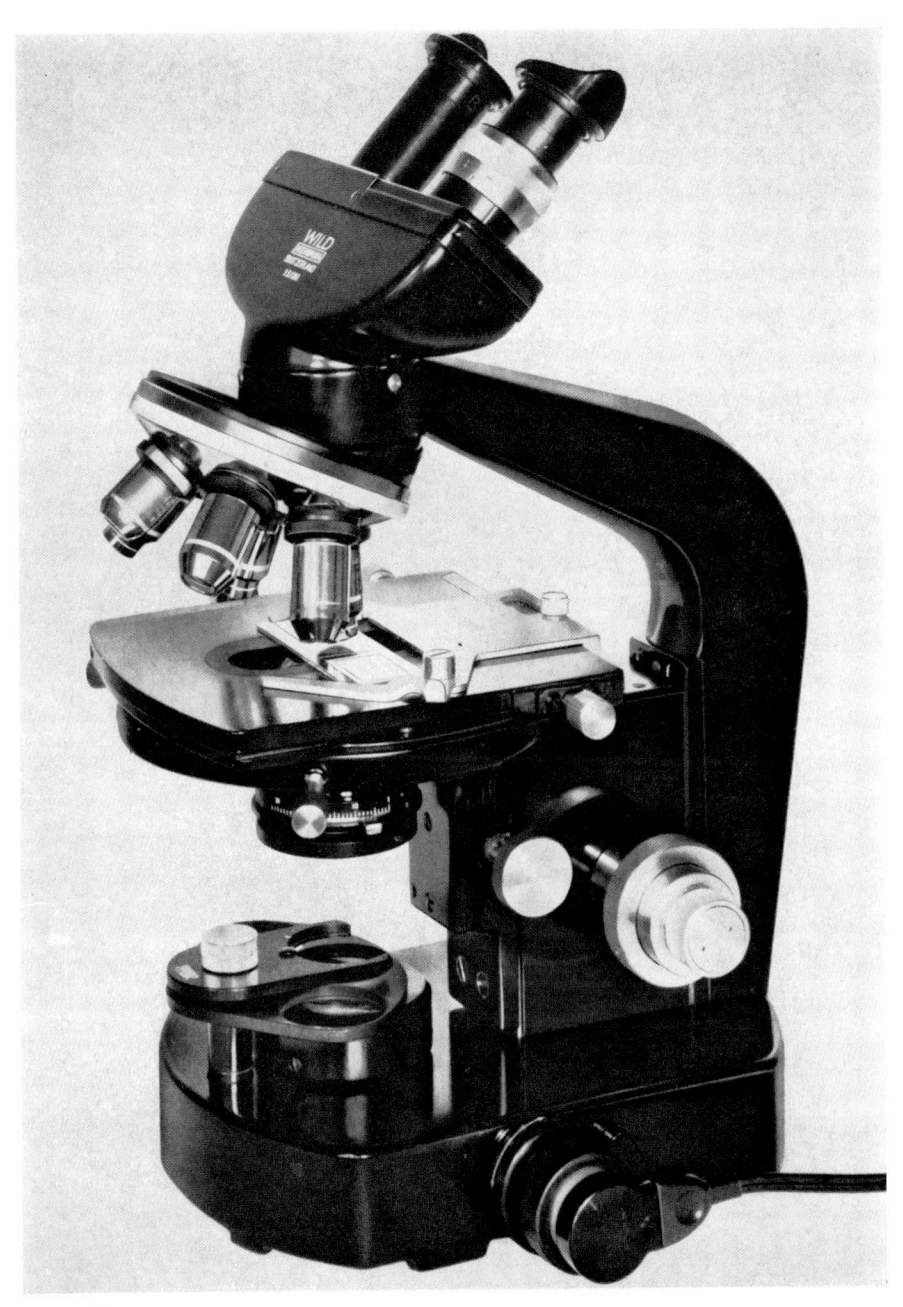

Plate 61.

448 PHOTOMICROGRAPHY

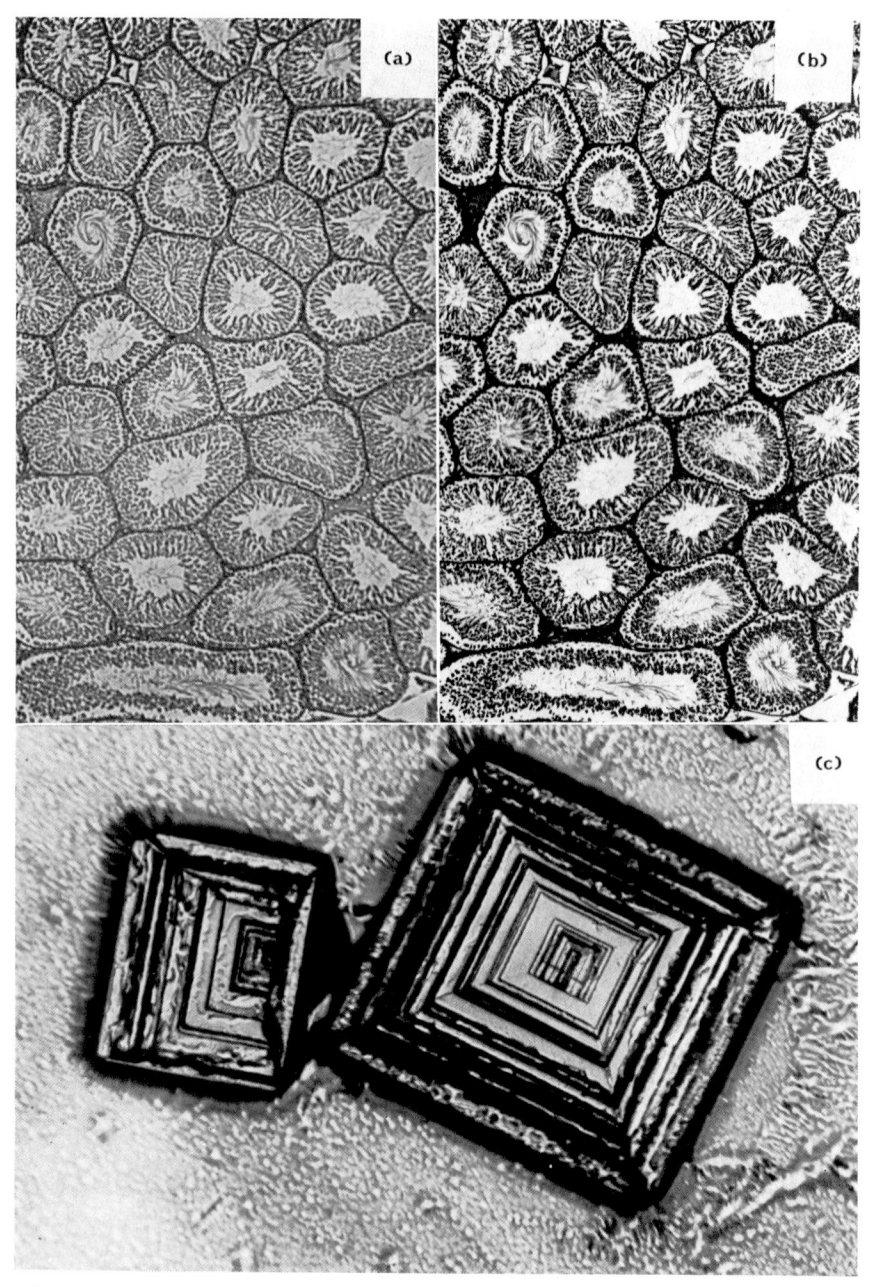

Plate 62.

PLATES

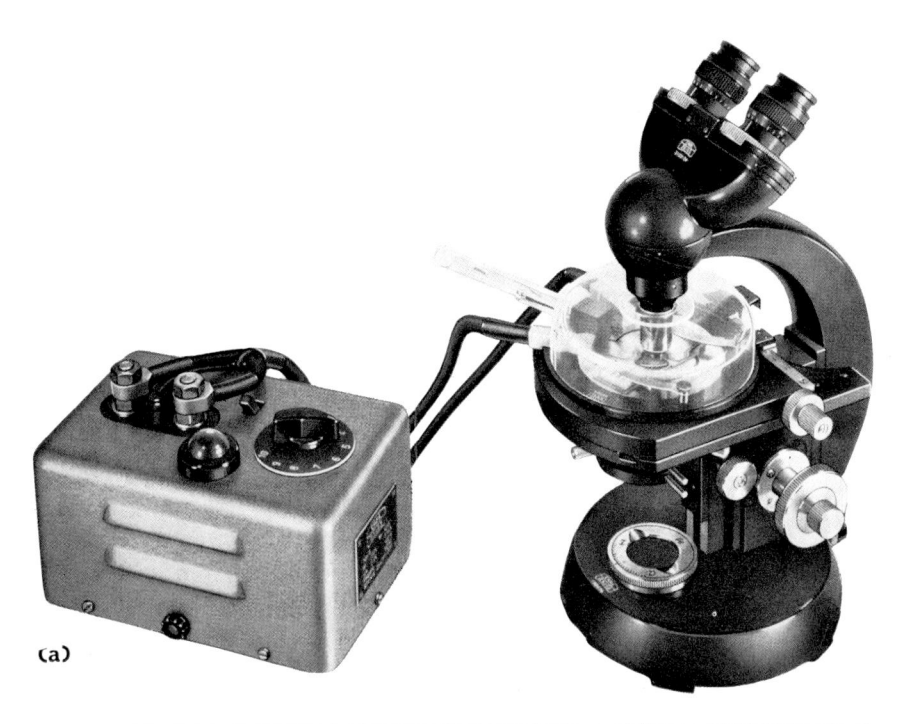

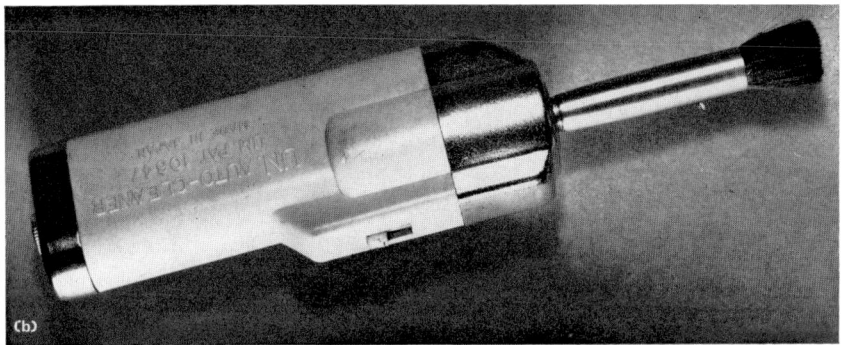

Plate 63.

Plate 64. Crystal growth, ×77. Four of a series of eight photomicrographs taken showing glycine crystals forming on and around pollen grains and complete in 3 min. The story is followed by studying the series of drawings.

Plate 65. Larva of Great water beetle *Dytiscus marginalis*, ×1·4. No. 1–10 shows this creature catching and devouring a tadpole, in 11 and 12 it has caught and eaten one of its own species. The prey is impaled by the mandibles and is not discarded until very little more than its empty skin remains. The larva is lighter than water and, for this reason, can only remain below the surface by holding on to weeds. When taking in air its hind-end is then pushed through the surface-film and the two hair-fringed cerci are spread out horizontally as seen in some of the pictures. Owing to their hydrophobe properties the water slides off these appendages and allows the large breathing pores to come in contact with the air. In this position the larva hangs head-downwards from the surface film and takes in air. See Plate 58. An air bubble can be seen escaping in upper 5.

Plate 66. (a) Complete cinémicrography equipment consisting of Wild microscopes M 20 KdEO with special attachment for cinémicrography, multi-purpose camera stand with anti-vibration plate and lamp-holder, universal lamp with ribbon filament lamp transformer, with additional low voltage lamp and regulating-transformer, complete with time lapse attachment and mirror galvanometer. (b) Vinten Mk. 1A ciné camera and Vinten special camera support and base. (c) Prior Cinémicroscope, a self-contained transportable unit (Fig. 240), fitted with an inverted microscope in an insulated cabinet, heated by forced circulation of warm air, thermostatically controlled.

PLATES 451

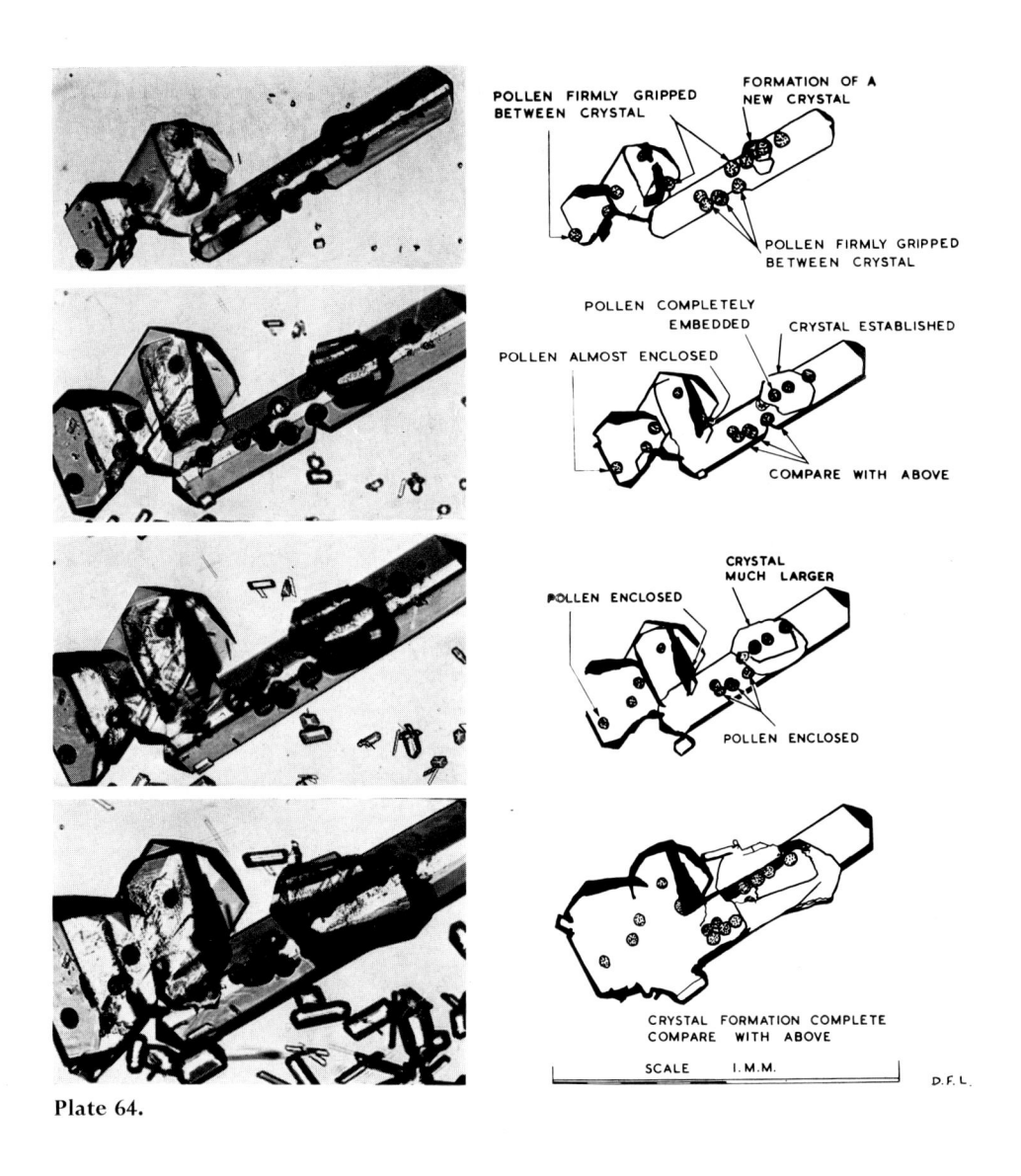

Plate 64.

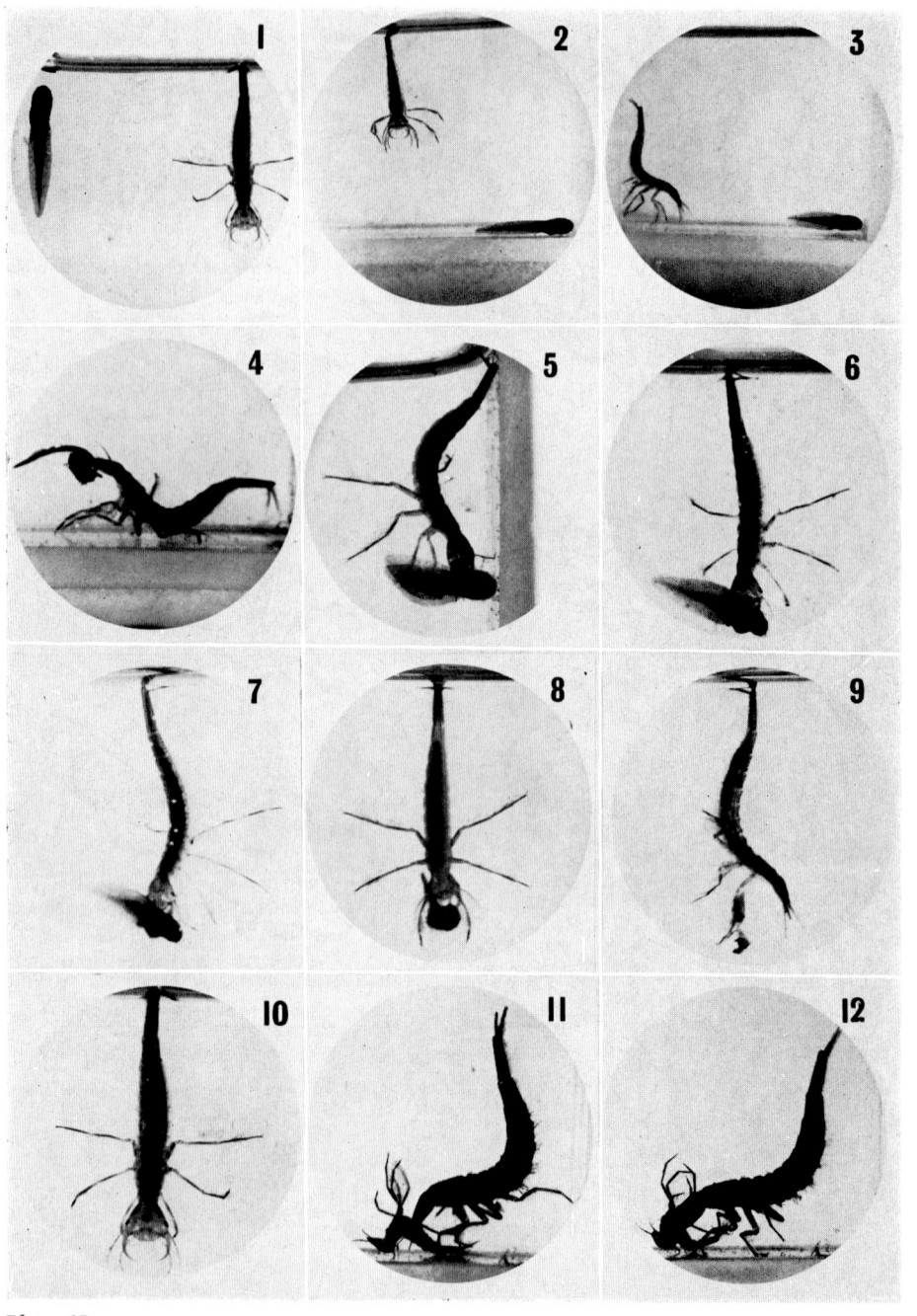

Plate 65.

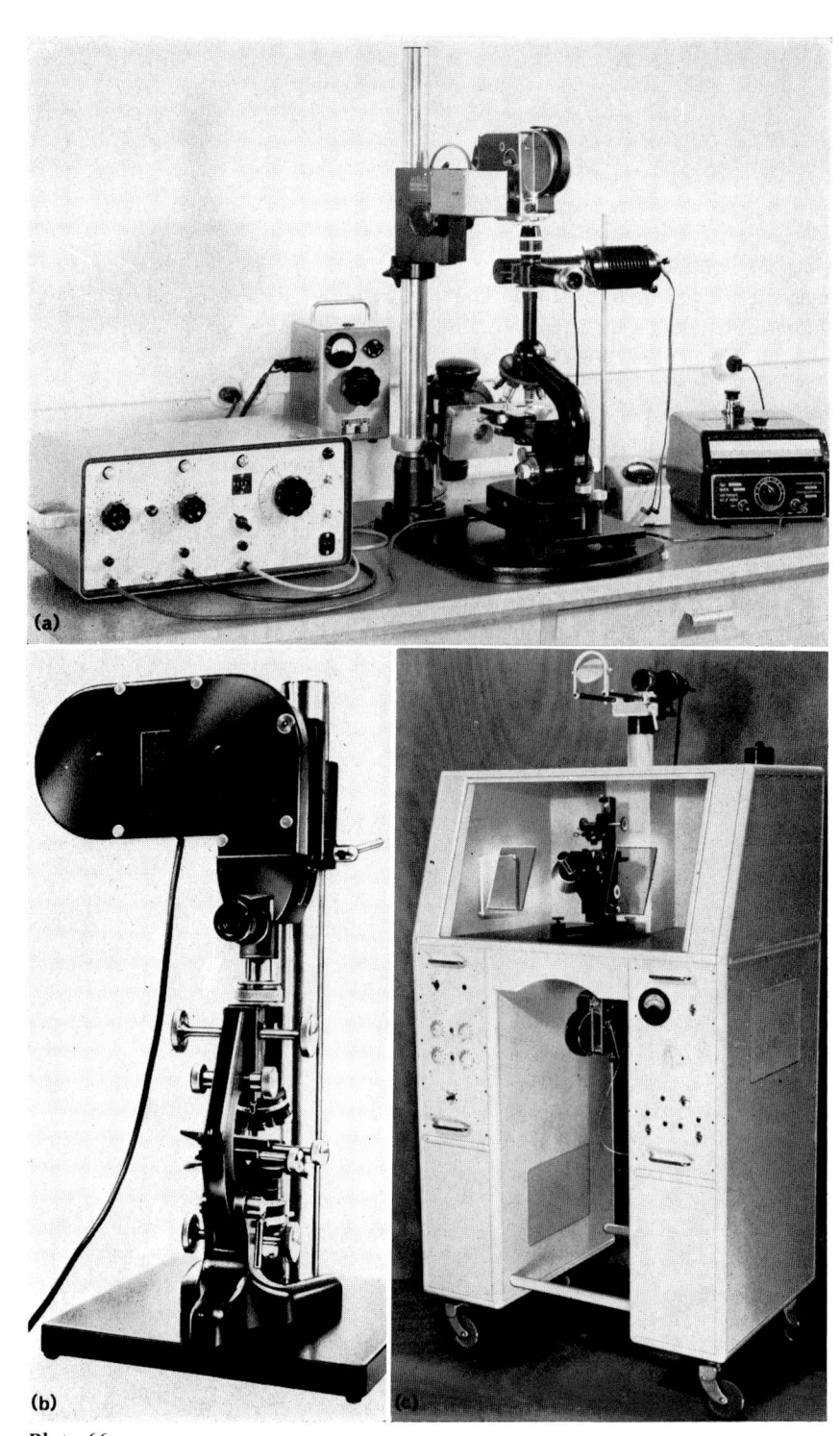

Plate 66.

Plate 67. Zeiss Stereomicroscope IV fitted with low voltage illuminator for transmitter light and diffuse-light illuminator with transformer for reflected light. Fluorescent tube can be fitted and camera, 6·5 × 9 cm, attachment with photo tube and sliding prism. (See page 267.)

Plate 68. Optical system with Bausch and Lomb objective; see page 67. (Photo by courtesy of Bausch and Lomb.)

Plate 69. Mast cell degranulation, ×800. (a) Taken on Kodak Gravure Positive Film 4135. The result appears to be unsharp and too soft. (b) Taken on Kodak Photomechanical Film Ortho 3. The focus is exactly the same as for (a) and yet the result gives the appearance that refocusing has taken place. It is important to use the correct film for each particular specimen, according to the colour of the stain. One film coating will not give good results for all subjects.

PLATES 455

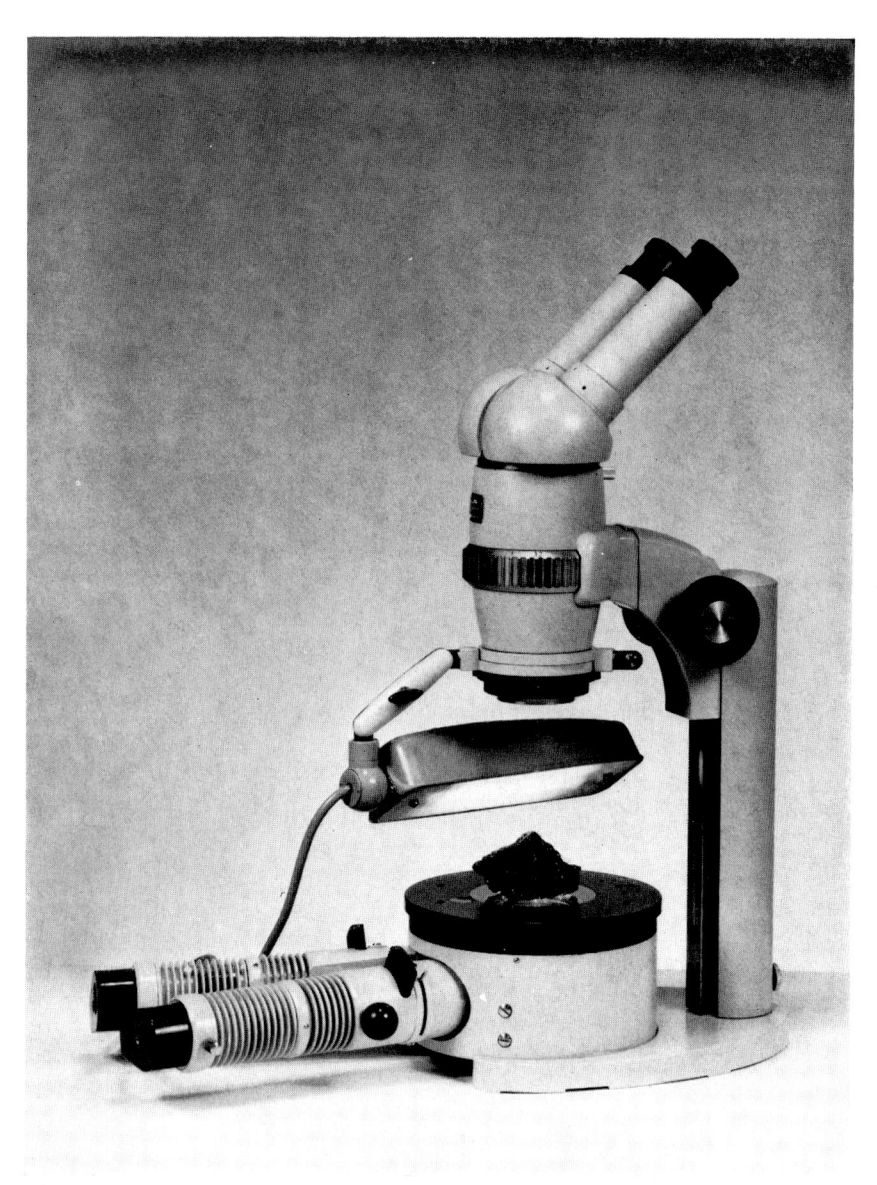

Plate 67.

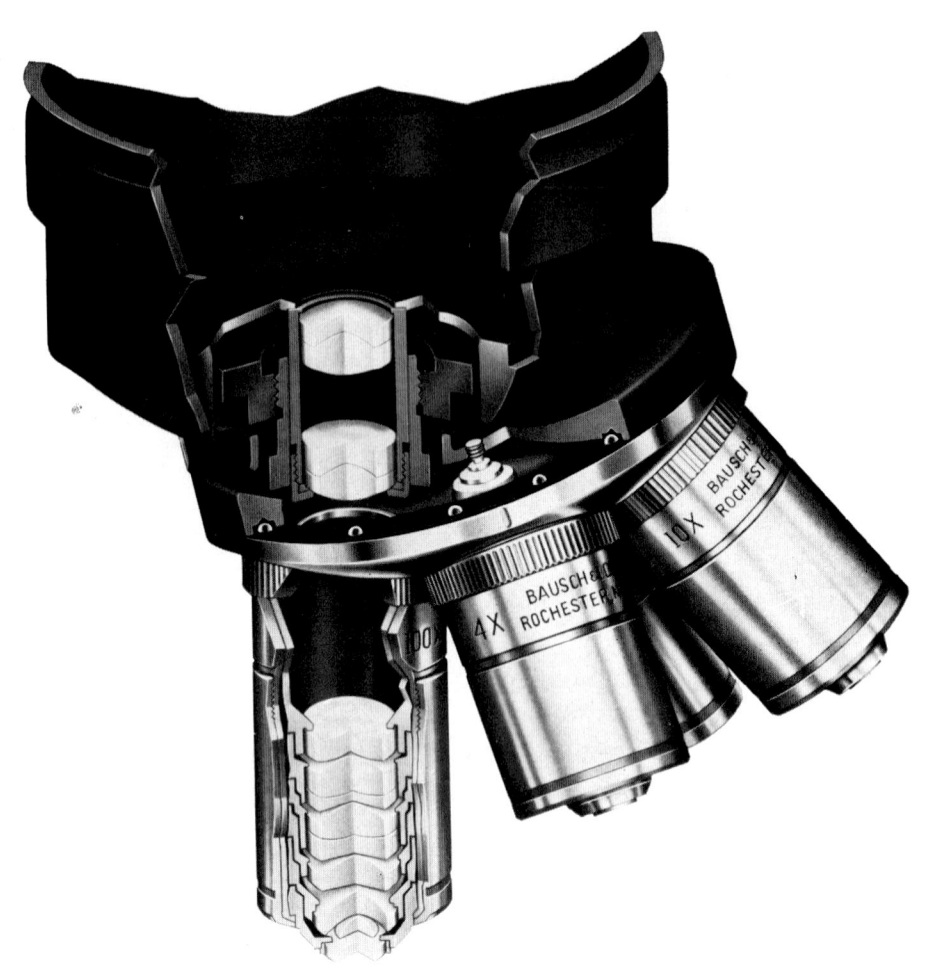

Plate 68.

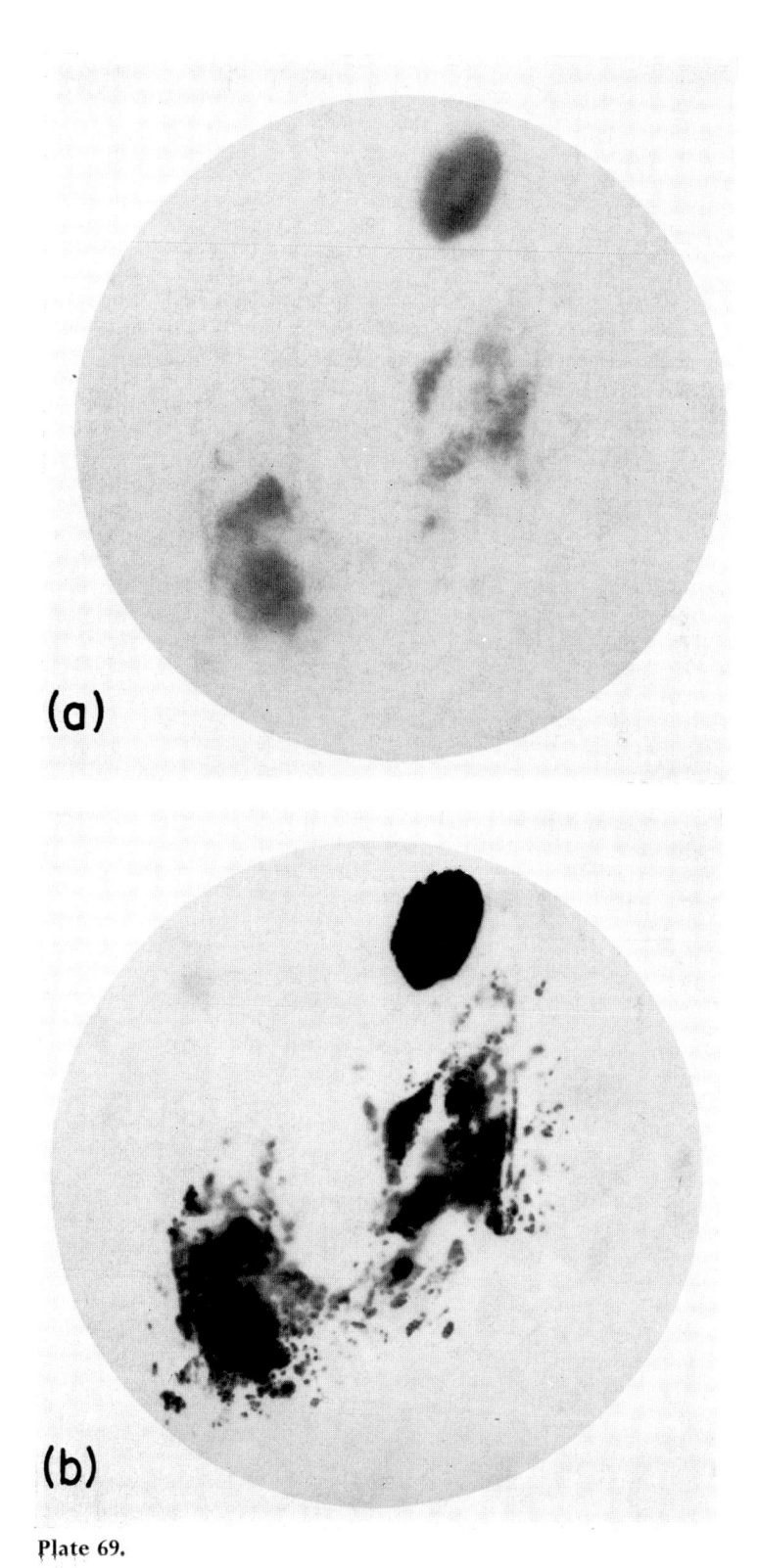

(a)

(b)

Plate 69.

Plate 70. *Upper.* Knobbed hairs on the edge of petal of Field spurrey *Spergula arvensis*, ×190. Transmitted and reflected light. A blue Wratten filter (82C) was placed in the transmitted light train which influenced the background colour. The petal was unstained. *Lower.* Raft of gnat eggs showing larvae inside, ×110. Specimen wet mounted. Transmitted light with blue Wratten filter (82A) in the light train to "stain" the background.

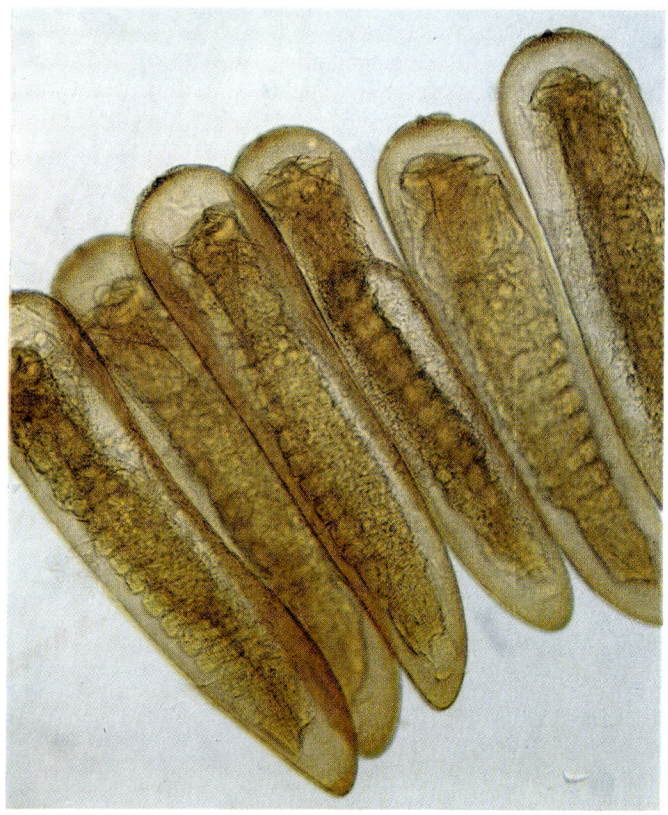

Plate 70.

Plate 71. *Upper*. Aspirin crystals, × 50. Demonstrating optical staining technique as set out on page 132. Direct transmitted light with Wratten filter 58 provided the background colour. Unfiltered indirect light illuminated the crystals. *Lower*. Aspirin crystals, × 50. Demonstrating optical staining. Direct transmitted light with Wratten filter 58 provided the background colour. Wratten filter 29 in the light train of the indirect light illuminated the crystals according to Fig. 131.

COLOUR PLATES 461

Plate 71.

Plate 72. *Upper.* Transverse section stem of clematis, ×25, illuminated by transmitted light with three Wratten filters (58 green, 38A blue, 21 orange), in the light train, provided the multicolour background. The section was chemically stained. *Lower.* Stained vertical section of human lip, ×10, illuminated by transmitted light with a light blue Wratten filter (82) in the path of the illuminating beam.

COLOUR PLATES 463

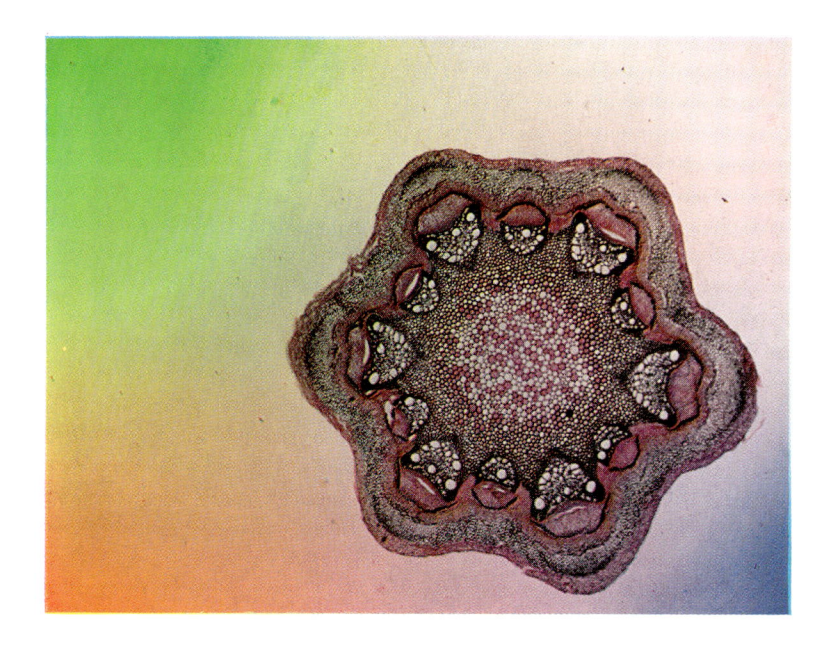

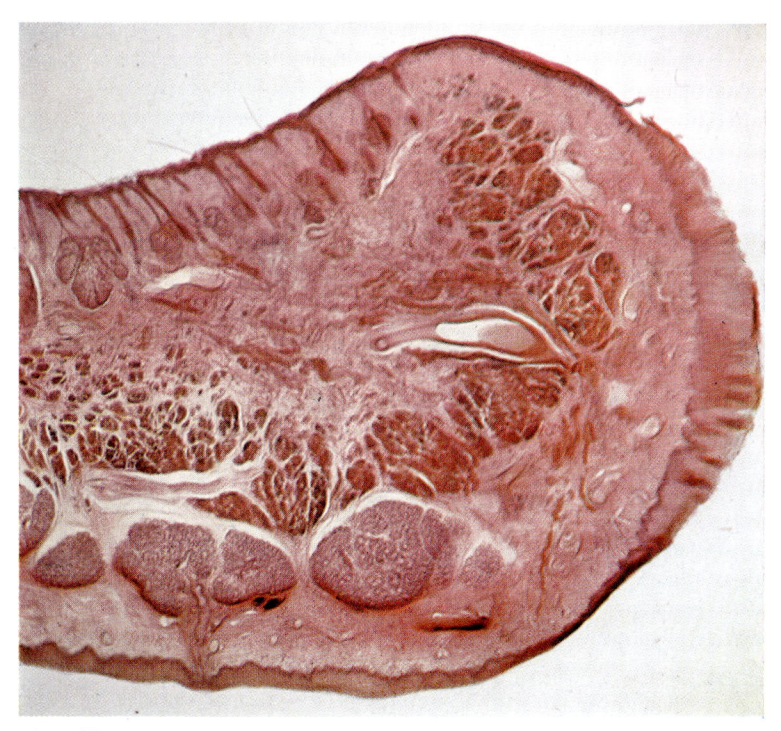

Plate 72.

Plate 73. *Upper.* Fused phloridzin crystals, ×100. Polarized transmitted light. *Lower.* Machine embroidery, ×16. Reflected light and reflector.

COLOUR PLATES

Plate 73.

Plate 74. *Upper.* Eggs of moth *Attacus edwardsii*, ×12. Illuminated by diffused sunlight. Ektachrome daylight sheet film was used here but plates 70–74 (lower) were taken on Ektachrome type B sheet film. *Lower. Amphioxus*, ×7. Unfiltered transmitted light, no substage condenser. This plate demonstrates the usefulness of an accompanying diagram.

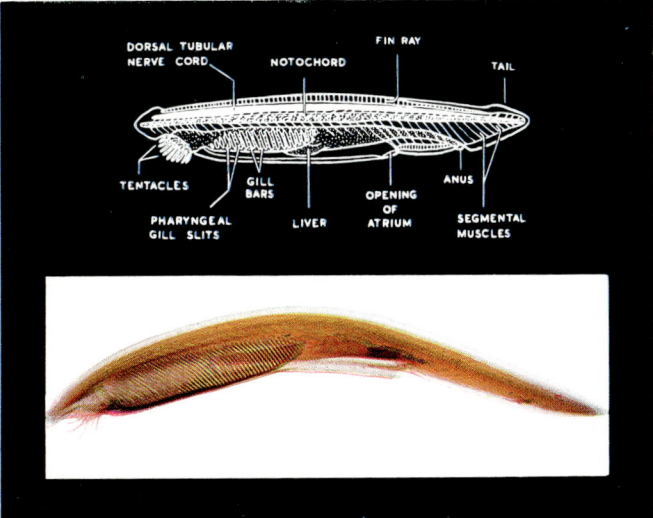

Plate 74.

Plate 75. *Upper.* Transverse section human jejunum, ×190, stained by Azane. Apochromatic objective, ×25; N.A. 0·65. The entire image field is used for measurement, the image fills the whole negative area. The fully automatic Orthomat 35mm microscope camera was used and the selector switch was set to 100%. *Lower.* Cancer pagurus L. (Common crab), chitinous tendon, stained by Congo red and photographed in polarized light using the Leitz fully automatic Orthomat microscope camera. Here the specimen fills about 65% of the negative area. Although this is a photomicrograph in polarized light with a dark background, integrating measurement was used because of the large ratio. Selector switch registered 65% in the area of the black wedge. A similar procedure is adopted when photographing specimens in brightfield. (Photomicrographs by H. Kornmann, courtesy of E. Leitz, Wetzlar.)

COLOUR PLATES 469

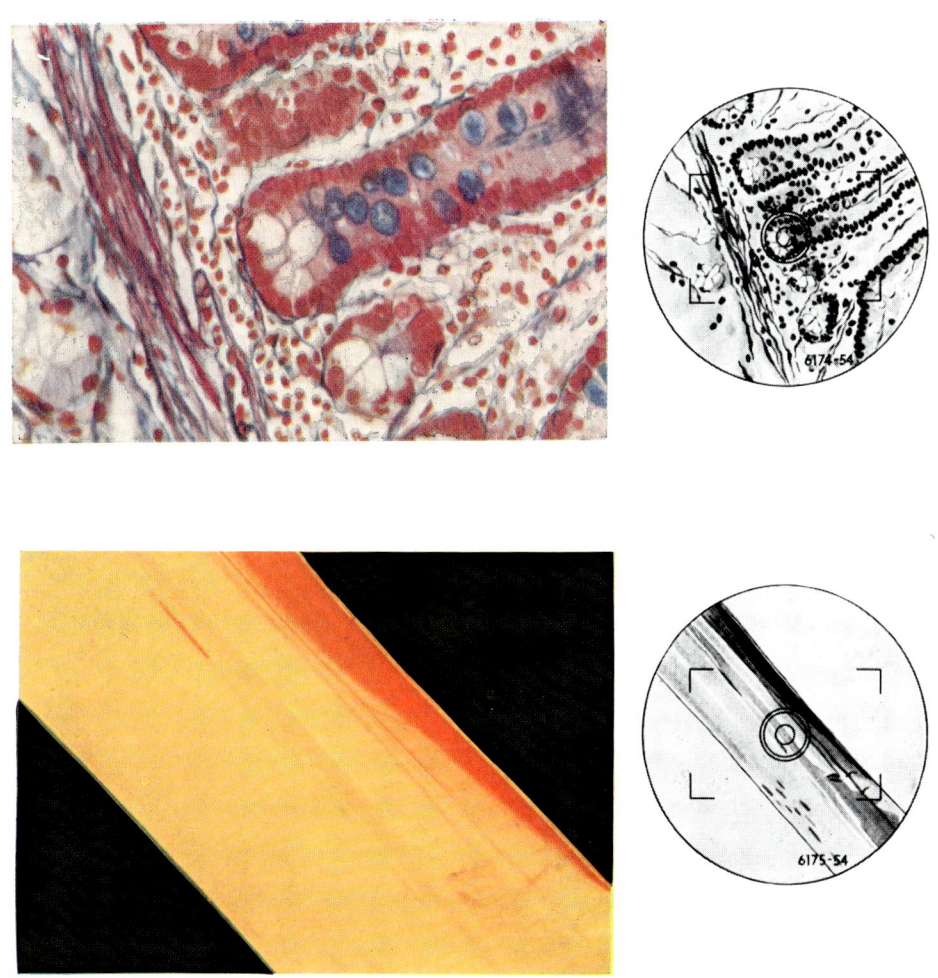

Plate 75.

Plate 76. *Upper.* Frog's blood, fluorochromed, ×700. Nuclei of the erythrocytes red, stroma green. Eosinophil granules of the leucocytes yellow. Compound stain with primuline, acridine orange and pyronine fluochromes. Leitz fluorescence equipment with Leitz fully automatic Orthomat microscope camera. Objective Fl. Oil immersion, ×95; N.A. 1·32, Agfa CT18, exposure time approximately 3 min. The small circle was used as a measuring field. A blood cell is brought to the inner circle by the mechanical stage and the area of the cell in proportion to this circle is estimated. As the objective fills the entire circle the specimen object proportion is 100%. *Lower.* Particles of asbestos in sputum, ×750; objective, ×63; N.A. 0·95. Detail measurement of small objects is also obtained, photographed on Leitz fully automatic Orthomat microscope camera. (Photomicrographs by Dr. E. Hain, courtesy of E. Leitz, Wetzlar.)

Asbestos dust attacks the lungs and makes breathing difficult. A large number of asbestosis cases develop lung cancer. Mesothelioma, another illness from asbestos dust, has been recently recognised. Photomicrographs show the dust acts on the pleural linings of the lungs and malignant tumours may develop from this.

COLOUR PLATES 471

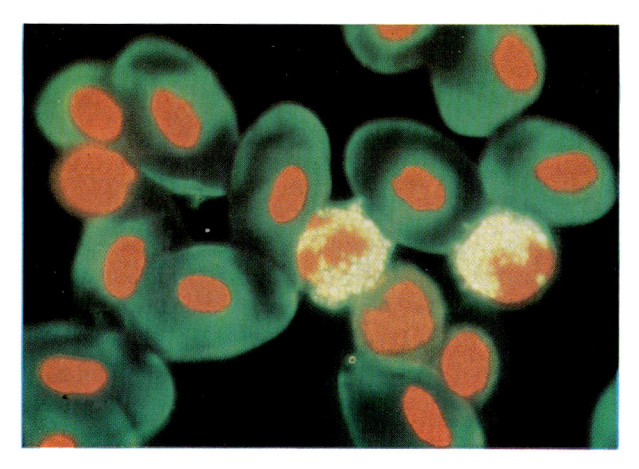

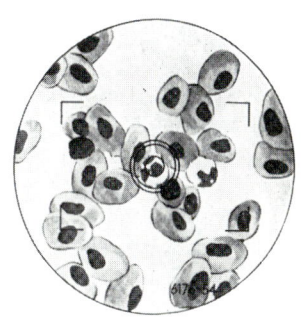

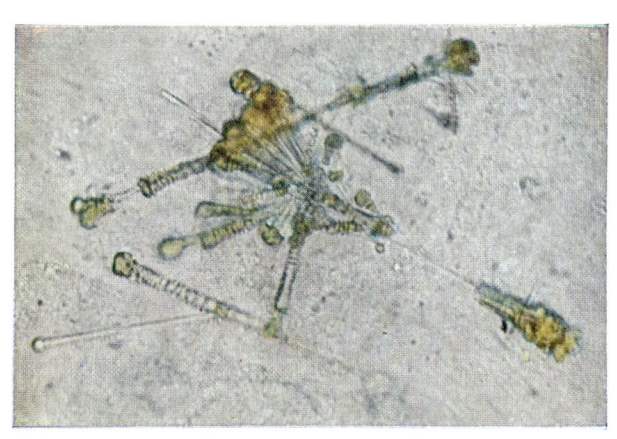

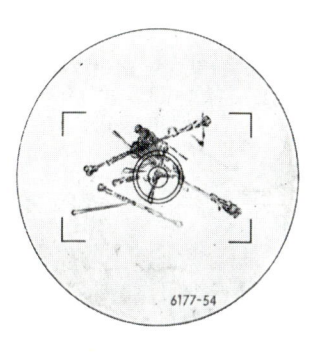

Plate 76.

Plate 77. Transmitted light interference photomicrography in cytological research. Living nerve cell from the ventral horn of the cervical enlargement of the spinal cord of the cat. These plates demonstrate three stages in the setting of the interference microscope. A 2 mm oil immersion objective was used with an immersion medium with refractive index 546 mμ of 1·3605 (a little less than the average refractive index of the cytoplasm). Calcite and $\lambda/2$ plates formed the substage beam-splitter which could be fitted along its principal crystalline axis. The beam-splitter was thus an asymmetrical compensator. Once the interference microscope is adjusted, the optical path difference between the measuring and reference beams can be regulated continuously, over a range of several wavelengths, while the specimen is observed orthoscopically.
Upper. This demonstrates "positive contrast" in which the immersion medium appears as a deep grey of about $\Gamma + 97$ mμ and parts of the network in deep red with blue centre. *Centre.* This is "positive contrast", in which the specimen appears a lower interference colour than the surrounding blue area. *Lower.* Here the microscope was adjusted to provide $\Gamma = 0$ for the immersion medium which is recorded as achromatic fringe, and results in a black background. (Photomicrographs by Dr. G. B. David, blocks by courtesy of Zeiss, Oberkochen.)

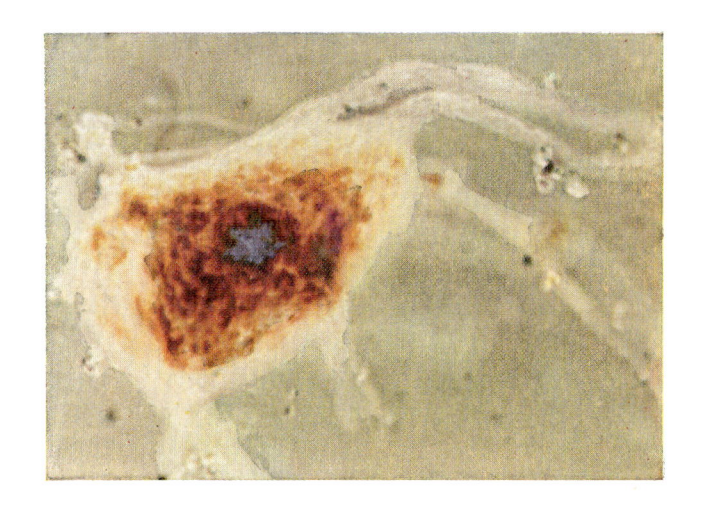

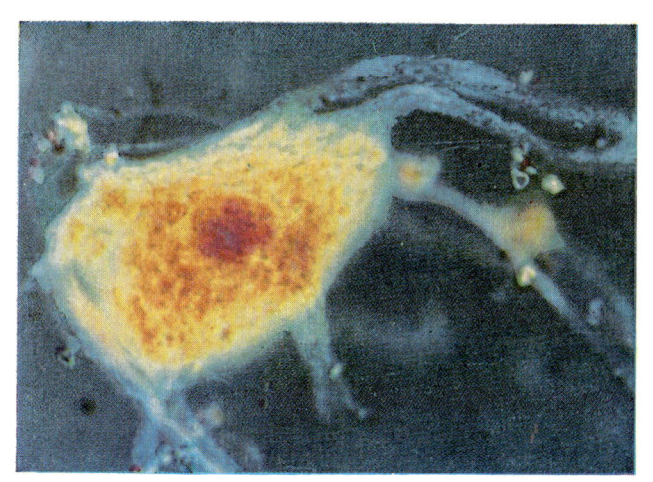

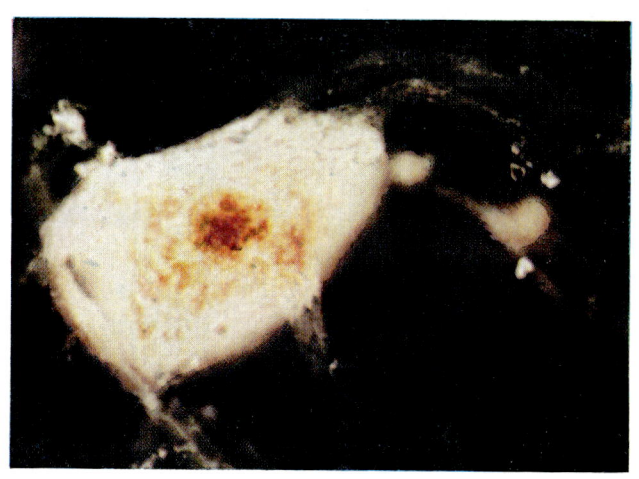

Plate 77.

Plate 78. Living nerve cell of the mesocerebral of the pulmonate gastropod *Helix pomatia* L. (the edible Roman snail) in a 100 μ deep micro-cuvette. A contrast to Plate 77: three interference photomicrographs of an isolated nerve cell, immersed in a medium of refractive 1·3605 and mounted, without compression, in a 100 μ micro-cuvette. There are numerous invaginations of the cell membrane, "trophospongium" of Holmgren (1899). Two small cells remain attached to invaginations in the lower part of each photomicrograph.
Upper. The microscope compensator was adjusted to provide "negative contrast". *Centre and lower.* These are adjusted to give "positive contrast". (Photomicrographs by Dr. G. B. David, blocks by courtesy of Zeiss, Oberkochen.)

Reference

Holmgren, E. (1899). *Anat. Anze.* **16**, 388.

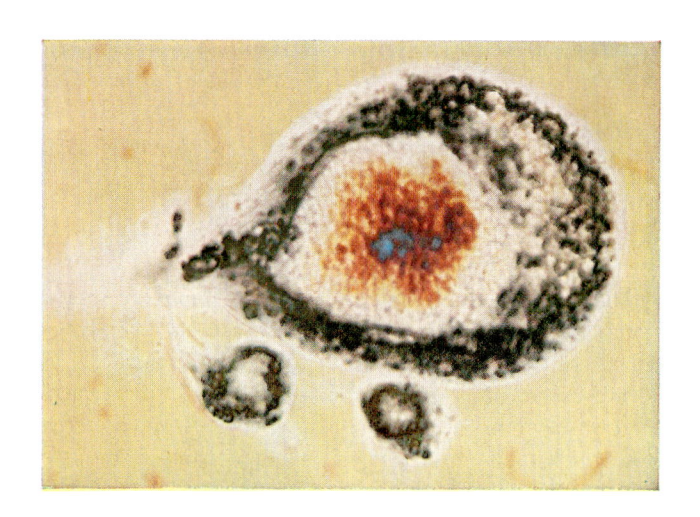

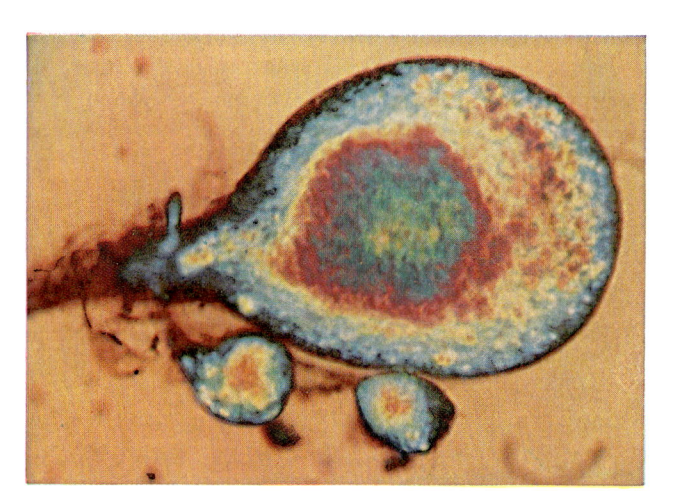

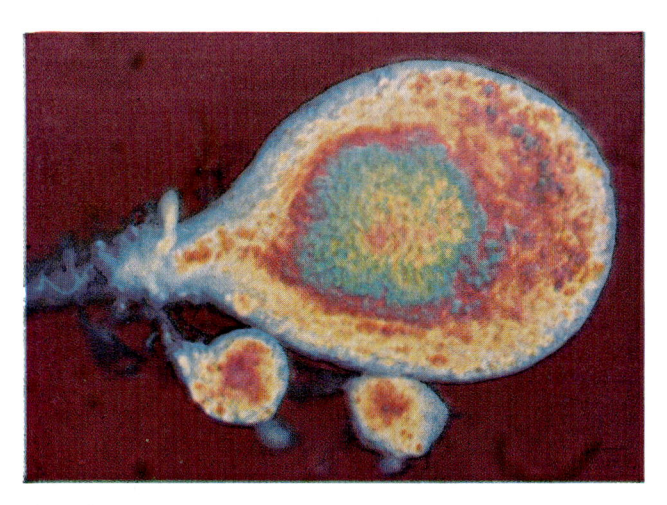

Plate 78.

Plate 79. Reflected light interference photomicrography in opaque material. *Upper.* The surface of a ground steel plate with sodium light fringes. Monochromatic light from a sodium lamp or a mercury lamp is normally used to give a fringe pattern over the whole area of the surface view. Where surface discontinuities several wavelengths deep are to be measured, a white light fringe system can be used in which the central fringe is distinguishable from all detail. The Watson 16 mm interference objective has a greater coverage and working distance. Here depth of surface finish and surface irregularities can be seen. *Lower.* A surface similar to that described above, but photographed with white light fringes. The central fringe can be seen. (By courtesy Watson, the Optical Division of M.E.L. Equipment Co. Ltd. Blocks kindly supplied by John Swain and Son Ltd.)

COLOUR PLATES 477

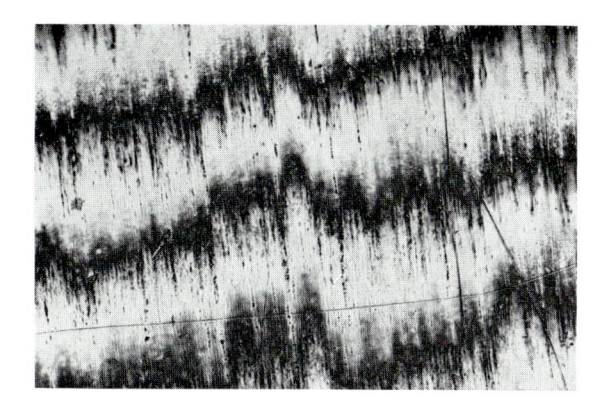

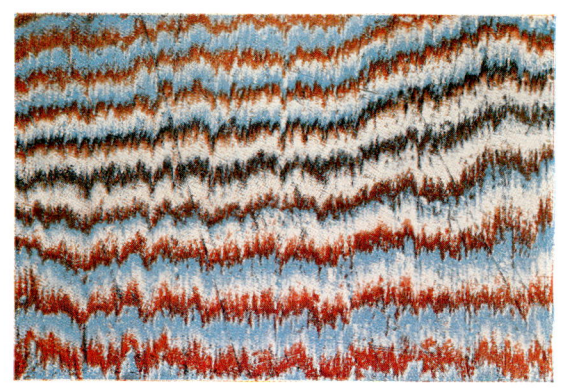

Plate 79.

Plate 80. Reflected light interference photomicrography in opaque material. *Upper.* Surface of germanium wafer polished to an angle of 2° to the top surface. The fringes, formed with sodium light, are perpendicular to the bevel angle. The depth of a diffused layer can be measured by counting the fringes down the bevel. *Lower.* An interferogram of the edge of a film of glue on a clear glass plate. By using white light fringes, the ambiguity in the film thickness measurement is eliminated. The thickness of this film is 0·75 wavelengths or 0·43 microns. Watsons 8 mm and 16 mm Interference objectives are made with built-in interferometer for examination of surfaces by reflected light. The objective has a built-in beam-splitter situated in the front lens, an illuminating tube with condenser lens system, and a reference surface with screws for adjusting the tilt in both planes and for focusing. There is a sliding shutter between the beam-splitter and the reference surface to cut off the interferometer when a return to ordinary bright-field incident illumination is desired. A detachable push-in green filter is available for use with the mercury lamp. (By courtesy Watson, the Optical Division of M.E.L. Equipment. Blocks by courtesy of John Swain and Son, Ltd.)

Specifications.
16 mm interference objective.
Focal length: 16 mm; Primary magnification at 160 tube length: ×10; N.A. 0·6; Field of view at ×100: 1·3 mm diameter; Working distance: 3·0 mm; Resolving power horizontally: 1·5 microns (approx.); Resolving power vertically: less than 1/10th fringe.
8 mm interference objective.
Focal length: 8 mm; Primary magnification at 160 mm tube length: ×20; N.A. 0·6; Field of view at ×200: 0·65 mm diameter; Working distance: 0·5 mm; Resolving power horizontally: 0·6 microns (approximately); Resolving power vertically: less than 1/10th fringe.

COLOUR PLATES 479

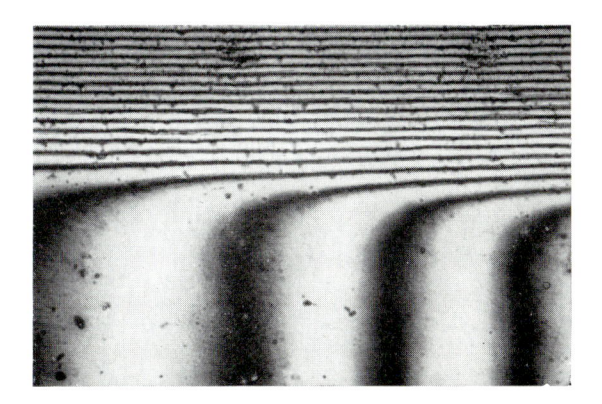

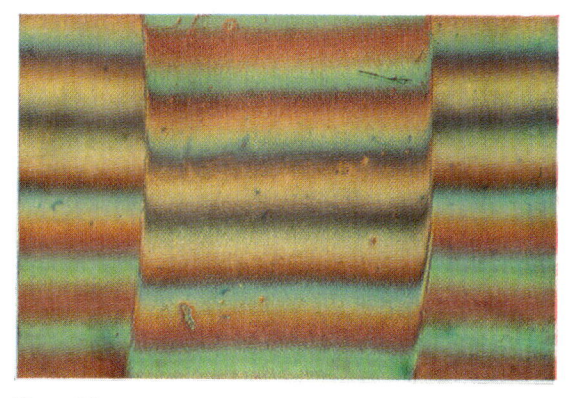

Plate 80.

Subject index

Aberration, 17, 22
　astigmatism, 26
　barrel distortion, 27
　chromatic, 24, 61, 64, 312, 317
　　correction, 24, 221
　coma, 25, 217, 313
　curvature of image, 26
　in X-ray, 250
　non chromatic and spherical, 222
　pin-cushion distortion, 27
　spherical, 23, 54, 57, 68, 347
　sphero-chromasy, 313
　spherical correction, 54, 68
　spherical in axis, 313
　　main ray, 313
　zonal 347
Abbe, 30
　test plate, 347
Absorption bands, 347
Acetic acid, 346
Achromatic, 9, 15, 23, 25, 62, 202, 266, 347
Acridine orange, 254
Aceto-carmine, 346
Adjustment coarse, 12
　fine, 13
Aesthetics, 291
Aesthetic and pictorial application, 290
Agar-czapek, 173, 339
Air, 54
　constant, 38
　bubbles, 336
　refractive index, 51
Airy disc, 35, plate 8
　radius, 35, 36
Axis, specimen, 48
Alcohol, 327
　acid, 328
　refractive index, 51
Aligarine, 331
Alignment, 347
Analyser, 196
Angle, cone of light, 85
Anisotropic material, 200

effect, 347
Animal tissues, 328
Anramine, 254
Antibodies, 173
Antigen, 173
Antiflex objective, 201
Aperture, angular, 30, 32, 347
　maximum 32
　numerical, 347
　narrow, 32, 35
Aplanatic, 347
　points, 23
Aptometer, 347
Araldite, 171
Arrangement, controlled, 291
Art, 290
Aspheric, 348
Aspherical, 348

Bacteria, staining, 332, 7, plate 2a
Balmer-lines, 243, 247
Beam-splitter, 207, 212, 215, 248
Bellows extension, 157, 176
Berberine sulphate, 254
Benzene, 332
Beryllium, 250
Bioluminescence, 176
Bismark brown, 344
Bleaching effect, 255
Blood cells, 339
　smear, 339
Blow pipe, 330
Body tube, 10, 154, 179
Borrodailes needle, 330
Botanical stains, 333
Brightness, 27, 161
　increasing and decreasing, 40
Bright spot, 35, 39
Brace-Köhler compensator, 209
Brownian movement, 275
Brunswick black, 338
Brush, camel-hair, 299, 336

481

SUBJECT INDEX

Cadium sulphite, 203
Calcite, 29, 107, 196, 197, 427
  transmission in IR, 226
Calcium oxalate ibioblast, 9
Camera, 21, 82, 84, 85, 86, 152, 156, 176, 180, plate 47
  bellows, 173, plate 46
  adaptor, 159, plate 47
  automatic, 162, 163, plate 59
  cassette, 303
  ciné, 162, 303, 306, plate 66
  Cook, Troughton and Simms ciné, 304
  Contarex, 162
  Exakta, 157, 297
  freestanding, 88, 154
  G45 ciné, 302
  height adjustment, 154
  hinged connector, 159
  horizontal, 180
  Leica bellows, 297
  Leitz fully automatic, plate 45
  lens, 156
  lucida, 8
  Nikon, 297, plate 46
  Pentax, 160
  projection distance, 159
  robot, 306
  stereo, 285
  strap support, 176
  support, 154
  television, 162
  time lapse, 64, 65, 306, 307, plates 42, 64, 65
  Vickers Autowind, plate 59b
  Vinten ciné 16 mm, 303, 304, 307
  Wild automatic, 306
Canada balsam, 50, 196, 203, 254, 332, 338, 343
Cane sugar refractive index, 236
Capillary tube, 173
Carbon dioxide, 175, 329
Carmine, 332, 346
Catoptric illuminators, 272
Cedarwood oil, 52, 54, 332
Cell, copper sulphate, 121
Cells, living, 300, plate 69
Cell ring, 335
Cellosolve dehydrant, 332
Cell ring, aluminium, 337
  Pyrex, 227

Teflon, 300, 337
Cement, Tolu balsam, 338
Chloroform, 338
Choroplast, 9
Chromic acid, 327
Chromatic emulsion, 25
  magnification, 313
Chromatin, 345
Ciné camera, 162, 304, 306, 453
Cinémicrography, 299, 304, 308
Circle of confusion, 36, 37
Cleaner, Auto, 299
Clove oil, 332
Cochineal, 331
Cohesion, 296
Collimator, 203
Colour cast, 152
  background, 245, 459, 463
  expression, 132
  cockpit drill, 152
  false, IR, 225
  film, IR, 230
  reproduction, 296
  transparency blue, 313
    green, 313
    yellow, 312, 313
  table, 147
  temperature, 126, 146, 149, 312
  pigmentary, 152
  versus black and white, 295
Colours, out-of-focus, 22
  interference between crossed Nicols, 201
Compensator, Brace and Köhler, 209
  Senarmonts, 209
Common faults, 310
Composition, 291, 292, 295
Condenser, 15, 39, 159, 348, 336
  Abbe, 70, 73, 94
  achromatic, 71, 73, 158
  adjustment, 116
    darkground, 271
    phase contrast, 193
  aperture, 69
  aplanatic anastigmatic, 269
  auxiliary, 158, 270
  bicentric double reflecting, 269
  bright-field, 68
  bull's eye, 348
  cardioid, 264

Condenser (contd)
  catoptric darkground, 269
  centering for Köhler illumination, 112
  chromatic, 267
  darkground, 71, 72, 266
  dust-free, 169
  eccentric stop, 284
  flare, 69
  flip-top, 70
  Heine, 190
  iris diaphragm, 41, 373
  ill adjusted, 312
  lamphouse focus, 193
  Leitz bicentric, 264
  Leuchtbild, 269
  marginal rays, 70
  oil immersion, 71
  paraboloid, 263
  phase contrast, 74, 158, 184, 186
  projected overlarge cone, 312
  quartz, 73, 254
  reflecting, 72
    darkground, 268
  Reichert toric darkground, 268, 269
  rotating turret, 185
  single lens, 73
  smear, 269
  substage, 348, plate 9
  UV, 232, 233
  Zeiss luminous spot, 264
Condensation, 366
Contrasts 295, 457
Converter tube, 226
Cooling chamber, 329
Coriphosphine, 254
Correction collar, 57
Corrosion in tube, 365
Cover glass, 31, 53, 54, 55, 56, 60, 107, 173, 216, 235, 300, 312, 334, 348, 375
  correction by tube length, 58
  breakdown in image, 220
  deviation in thickness, 56
  dispersion value, 60
  fluorescence, 261
  focus away, 192
  high power work, 336
  method of raising, 337
  over correction, 59
  properties, 55

  quartz, 233
  refractive index, 60
  thickness gauge, 60
  under correction, 59
  UV, 233
Crystal, biaxial, 52
  curved reflection radiography, 251
  isotropic, 52
  unixal, 52
Critical angle, 29, 348
Cryolite, 50
Cultures, 300
Curing agent, 346
Cutical wall, 9

Definition, 34, 55, 348
  critical, 168
Darkground illumination, 263
  incident light, 271, 272
Dark circular area-print, 312
Depth of field, 5, 35, 37, 43, 44, 45, 54, 101, 348
  control, 101
  increased, 49
  lack of, 311
  in print, 45
  shallow, 100
  simple microscope, 4, 5, 183, 365, 445
  ultra-violet, 232
Depth of focus, 40, 42, 43, 348
Dehydrant-cellosolve, 332
Detel, 339, 343
Diaphane, 343
Dichoric mirror, 260, 261
Diamond pencil, 330
Diethylene glycol mono-ethyl-ether, 338
Diffraction, 39, 349, 375
Diffusers, 121, 128
  opal, 146
Diffraction rings, 35, 36, 41, 311
  rotation-symmetrical, 39
Dispersion, 349
Dissecting microscope, 330
  scissors, 328
Dots and lines, 292
Drawings, 293
Draw-tube, 12, 349
Dust-free, 169

R

Dust, 121, 152, 298, 299, 311, 312, 336
Dyes, synthetic, 332

Edge zone, 50
Ehrlicks haematoxylin, 344
Electronmicrographs, 1, 117, 236
Electrophoresis plate, 173
Elliott IR objective, 227
Emulsion, 141
England finder, relocation, 318
Enlargement, 36, 361
Eosin, 332
Euparal, 343
Epidermis, 9
Equaral and sirax, 254
Equivalent focal length, 349
Essentials, 296
Ethyl alcohol, 332
Ethylene glycol mono-ethyl-ether, 338
Exhibition print, 91, 92
Exit pupil, 21, 349
Expanding stop, 266
Exposure, 27, 149, 152, 161, 162, 182, 420
Exposure increase factor, 183, 420
  control unit, 163
  determination, 160
  multi, 49
  table, 152
Extension tube, 160
Eyepiece, 2, 6, 8, 12, 14, 20, 45, 46, 74, 75, 349
  angle of view, 78
  Becke-Klein attachment, 215
  camera light trap, 82
  chromatised Ramsden, 75
  compensating, 38, 75, 78
  comparison, 80, 82, 88
  Epiplan, 325
  field lens, 75
    stop, 75
    of view, 77, 78
      number, 77, 78, 90
  focal plane, 81
  graticule, 82, 83, 314, 305
  holoscopic, 350
  Homal, 350
  Huygen, 75, 81, 350
    with graticule, 316
  incorrect, 313
  interferometer, 215
  Kellner, positive, 83, 316
  less, 162
  magnification, 78
  micro-disc, 79
  micrometer, 80, 82, 315, 316, 351
    calibration, 315
  negative type, 76, 221
  orthoscopic, 78
  parfocal, 79, 81
  periplanatic, 352
  periplan plano-compensating, 63, 78
  phase telescope, 183
  photometer, 215
  plano, 79
  pointer, 83
  positive type, 76
  projection, 75, 79
  quartz, 79
    Huygen, 221
  Ramsden, 75, 81, 352
    for measuring, 315
  screw-in, 7
  screw micrometer Vickers, 316
  telescope camera, 153
  telescopic eyepiece system, 62, 305
  ultra-violet, 238
    direct focusing, 80
  vertical image, 75
  viewing angle, 77
  visual for photomicrography, 80
  widefield, 353
  wide flat field, 79
    angle, 78
  Zeiss, Jena, micrometer, 316

Field, 349
  choice, 292
  curvature, 26
  diaphragm, 349
  of view, 38, 54, 349
  relocation finder, 318
    Zeiss, 326
Film fine grain, 47, plate 67
  holder, 162
  speeds, 163, 231
  transport, 163
Filming moving objects, 301

SUBJECT INDEX

Filter, 30, 152, 162, 142
  barrier, 120, 121, 257
    for fluorescence, 255, 256
  blue fluorescence, 254, 259
  Blazers IR transmission, 228
  Chance IR transmission, 228
  colour control, 159
    correction, 141
      background correction, 151
      balance, 150
  continuous running, 143
  diffusing, 121, 128, 145
  exciter, 119, 121
    and barrier, 257
    fluorescent, 256, 258
  factor, 420
  for use with mercury vapour, 216
  function, 140
  gelatine, 141
  Grey ND, 257
  heat absorbing, 121, 122, 128, 136, 145
    transmission curve, 225
  holder, 155, 159
  Hg, interference, 205
  Ilford IR transmission, 228
  infra-red, 143, 223
  interference, 110, 121, 124, 126, 127, 145, 158, 260
    transmission, 236, 237, 238
    for XBO 100 watt, 228
  Kodak IR transmission, 226
  light balancing, 150
  liquid, 141, 144
    IR transmission, 228
  maximum contrast, 140
  metal interference, 237
  multi-reflection, 145
  Muster Schmidt transmission, 228
  neutral density, 27, 30, 140, 145, 150, 162, 228, 255, 257
  optical, 221, 351
  polarizing, 151, 195, 201
  polaroid, 108, 215
  position and effect, 141
  reason for use, 138
  scatter, light, 245
  Schott and Gen. transmission, 228
  selective for, 255
  spectral curve Filtraflex, 144
  suppression, 121
  transmitting IR, 228
  transmission, 142, 224
  Turner, for iodine quartz, 259
  uses, 142
  ultra-violet, 144
    absorbing, 221, 226, 263
  Vario neutral density, 151
  Wratten 'M' series, 149
Finger marks, 312
Fixing, 327, 345
Flare spot, 15, 68, 69, 94, 152
Flatness of field, 54
Fluorescence, 119, 260
  illumination, 353
    incident light, 259
    phase contrast, 185
  impregnant, 122
  screen, lead glass, 251
  tubes, 261, 263
  with UV, 254
  blue light, 254
Fluorochromes, 346
Fluorite monochromator, 236
  transmission in IR, 226
Fluorspar, 254
Focal length, 6, 20, 39, 349
  plane, eyepiece, 21, 40, 43
  shutter, 349
Focus, above and below, 46
  integrated, 49
  out of, 298
  scale, 48
  shallow, 47
Focusing, 27, 48
  telescope, 306
Forceps, 330
Foreign marks, 311
Formaldehyde, 327
Formal-saline, 328

Galvanometer, 246
Ganglion cells, 226
Gillett and Sibert mirror objective, 96
Glacial acetic acid, 327
Glare, 350
Glass and air, 29
  cover correction, 55, 57, 58, 59
    thickness, 60
  container, 171

Glass and air (contd)
  crown, 24, 53
  flint, 24, 25
  IR transmission, 226
  pellets, 50
  refraction, 28
  tank, 53, 175
  unsuitable, 63
  ultra-violet absorption, 233
Glycerine, 174, 235, 254, 334, 337
  jelly, 337, 343
Gold size, 337
Graticules, 77, 79, 80, 87, 314, 316, 349
  concentric, 317
  eyepiece, 82, 83, 87
  Fairs series, 321
  globe and circle, 322
  net, 317
  Patterson, 320
  point counting, 323
  Portion, 319
Green damar, 344
Ground glass screen, 153

Haematoxylin, 331
Hanging-drop, 338
Hardening, 327
Hemispherical specimen, 47
Henry Green method, stain, 344
Homogeneity, 22
  abandoned, 341
Hot stage, 314, 299
Hydrax, 343

Iceland spar, 197, 209
Illuminator, Beck Chapman, 105
  bottle, 181
  built-in, 169
  diascopic, plate 46
  diffused-light, Zeiss, 261
  even, 152
  integral substage, 302
  metallurgical, 106
  polarizing, 107
  vertical, 105
    fluorescent, 160
    incident, 352
    transmitted, 352

Illumination, 46
  azimuth, 100, 108
  back-lighting, 181
  beam path for monochromator, 240
  bright-field, 348
  built-in, 95
  correct balance, 101
  critical, 350
  darkground, 92, 93, 114, 169, 263
    effect, 176
  even, line overlap, 166
  fluorescence, 119, 253
    tubes, 163
  grazed, 363
  Hauser, 96, 98
  incident-light, 104
  incident phase contrast, 191
  IR image converter, 227, 230
  IR transmission, 226
  interference, 201
    cyclic, 206
    round-the-square, 103, 199
  Köhler, 250 (see Köhler)
  oblique transmitted, 101, 103
    reflected, 93, 95
  parabolic mirror, 96
  polarization, 92, 194, 385
  phase contrast, 183
  phase-fluorescence, 185
  suitability, 98
  top, 94
  transmitted, 11, 92, 93, 114, 169, 170
  uneven, 31
  ultra-violet, 231
  vertical, 104
    via mirror, 104
    fluorescence, 260, 352
    incident, 93, 96
Image aerial, 42
  becomes diffraction, 40, 375
  curvature, 26, 313
  defects, 55, 457
  double, 137
  doubling, 178
  errors, 313
  formation of specimen, 21, 189
  iris diaphragm, 313, 348, 350, 373
  fuzzy, 47, 50, 457
  inverted, 7, 19, 21
  out-of-focus, 50, 457

SUBJECT INDEX

Image aerial (contd)
  quality, 23, 43, 399, 417
  real, 4, 20, 352
  refraction, 312
  secondary, 104, 433, 445
  TV intensifier, 205
  unsharp, 311, 457
  virtual, 17, 20, 353
Immersion liquid for UV, 235
  oil, Cargills, 254
  oil, 53, 54, 63, 68, 201, 338, 341, 343, 350
    non-drying, 54, 370
Incubating chamber, 300
Infra-red, 223
  colour film, 225
  converter, 229
  film, spectral sensitivity, 225
  sensitivity curve for IR film, 228
  spectrum, 63, 216
Ink-marker, 67
Institute of Incorporated Photographers, 15, 164
Interference compensator Brace and Köhler, 209, 210
  Ehringhaus, 209, 210
  Jarmin, 203
  Senarmounts, 209
  Soleil Babinet, 210
Interferometer plate, 212
Integrated focus, 49
Iodine solution, 332
Iris diaphragm, 41, 154, 373
  adjustment, 114, 136, 373
  behind-the-lens, 101, 103, 365
Isotropic material, 200

Jabus Green stain, 345

Kelvin (°K), 146, 312
Kentmere paper, 296
Kepler, 6
Köhler, 93, 103, 108, 112, 150, 152, 191, 270
  infra-red, 223

Lamp, microscope, 99
  adjustment, 110, 113, 152

candle flame, 147
carbon arc, 128
clear flash tube, 147
concentrated arc, 124, 125
colour temperature, 147, 149
CS 150 watt, 241, 259
deuterium spectral energy, 245
diffuser, 146
DL 100, deuterium, 241
electronic flash, 129, 147
flash tube, 131
fluorescent tube, 123, 147, 261, 263
halide, 122
Halogen arc, 125, 147, 149, 223
HBO 200 watt, 241, 259
Hg interference mic., 205
high pressure mercury, 119
house condenser, 154
HP 80 watt, 261
incandescent IR, 147, 223
independent, 110
intensity, 181
iodine quartz, 123, 147
  additive, 149
IR rays, 243
Leitz, 128, 250
low voltage filament, 128, 133, 147, 158, 171, 301
Lux "U", 132
mercury compact, 124, 215
  helium, 243, 245
  high pressure UV, 236, 256, 260
  vapour in IR, 223
micro-flash, 130, 162
photoflood, 147, 172, 180
projection, 147
ribbon filament, 111
special Reichert, 132
spectral energy distribution deuterium, 243
  tungsten, 147, 149
Tenslite, 110
tungsten filament, 146, 147, 148, 215, 221
  halide, 122
tungstic-iodine, 122
tubler 50 volt, 179
Versalite, 110
Wild universal, 111
xenon, 111, 147

SUBJECT INDEX

Lamp, xenon (contd)
   -arc flash, 126
   combination, 127
   high pressure, 126, 147
     XBO 150 watt, 223, 241
   Zeiss flash, 131
   Zirconium 100 watt, 223, 236
     arc, 124
Laser system, 247
Leeuwenhoek, 5
Leica camera, 156
Leitz 35 mm camera, 148, 153
Lens, concave, 24
   converging, 4, 9, 18, 24
   diameter, 5
   micro, 167
   negative, 17
   positive, 17
   various, 17
   wide angle, 22
Light, 350
   angle of, 53
   axial beam, 104
   beams interference, 190
   colour of, 146
   meter, 152, 161
   phase contrast, 187
   polarized, 194
   scatter, 39
   source, 119
Line, imaginary, 4
Line overlap, 165
Lines indistinct, 312
Lines resolved, 39
   UV, mercury arc, 232
Luminar head, 162
Luminous bodies, 173

Magnification, 3, 5, 19, 20, 32, 39, 56, 83, 86, 87, 88, 92, 159, 167, 350, 445
   empty, 92, 349
   primary, 39
   reproduction ratio, 183
   scale, 182
   secondary, 90, 298
   table, 90
   useful, 39

   void, 312
Magni-changer, 65
Magnolium, 216
Maintenance and care, 298
Malachite green, 344
Malformations, 47, 377, 406, 427, 432,
Mason's trichrome method, 344
Materials, bulk, 178
Material, inhomogeneous, 54
Measuring scale, 314
Mechanical tube length, 8, 14
Meniscus lens, 351
Mercuric chloride, 327
Mercury chloride, 345
Metal, polished, 216, 369
Metallurgical illuminator, 106
   microscope, 107, 221
Methoxybenzene, 254
Methylene iodine, 343
Microscope, 7, 154, 158, 423, 443
   adjustment, 154
   automatic head, 163
   base, 12, 179
   bench type, 11
   binocular body, 158
   biolaser system, 248
   body-tube adaptor, 215
   coarse adjustment, 12, 248
   comparison, 82
   compound, 348
   compound early, 6, 10, 14
   centering slide, 113
   ciné, 300, plate 66
   ciné, Wild, 304, 305
   cyclic interference, 205
   darkground, 263
   depth of field, 5
   dissecting, 330, 331
   electron, 1
   exposure control, 162
   fine adjustment, 13, 297, 349
   fluorescent, 253, 255, 259
   focusing scale, 48
   Greenough, 280
   handling, 298
   high temperature, 221, 222, 299
   limb, 12
   incubation chamber, 300
   infra-red, 223
   independent body tube, 179

SUBJECT INDEX

Microscope (contd)
  interference technique, 100, 203, 204, 206, 208, 211
    after Dyson, 212
  inverted Prior, 300, 309
  Leitz temperature control, 204
    interference, 204
  fluorescence, 120, 156, 259
    large fluorescence, 234
  maintenance, 298
  mirror, 14, 16
    UV, 235
  nosepiece, 351, 10, 11
  optical bench, 164
    system, Bausch and Lomb, 389, 456
  photomultiplier as detector, 164
  plane of focus, 46
  polarization, 194, 200
  reflecting, 56, 216
  Burch, working distance, 220
  Reichert polar, interference, 211
  round-the-square, 199, 204
  Schlieren, 275
  setting up for phase, 191
  simple, 4, 56, 162, 171, 173, 176, 180, 188, 365, 431, 432, 433
  slide, 54, 57, 58, 61, 216, 300, 334, 335, 337, 340, 342, 352
    cleaning, 265
    culture, 340
    hollow ground, 61
    ultra-violet, 232
    quartz, 254
  stage, 10, 352
  stand, 10
  stereo, 276, 282, 285
  strap support, 176
  shallow projection, 250
  substage condenser iris, 41
  tube, 2, 10, 13, 160
  tube diameter, 13
    head, 158
  ultra-violet, 232, 233
  water immersion, 175
  Watson zoom stereo, 281
  working distance, 179
  X-ray, 249
  Zeiss fluorescence, 441
  Zeiss research, 155, 156, 367
  zoom, 66

  zoom, interference, 207–210
Microtome, 328, 329, 330
Micrometer eyepiece, 351
  -disc turret, 324, 326
  stage, 351
Microradiograph, 253
Microphotograph, 1, 3, 351
Micrograph, 8, 45, 351
Micro-promar projector, 8
Millboard, 337
Mineral cryolite, 50
Mirror, 8, 15, 16, 136, 158, 212, 235
  dichroic, 260
    transmission, 262
  ellipsoidal, 222
  focal length, 15
  for reflection radiographs, 251
  galvanometer, 306
  plane, 330
  semi-aluminised, 212
  semi-permeable, 11
Mitochondria, 345
Monochromator accessories, 246
  applications, 246
  in-line, Leitz, 247
  in UV range, 247
  infra-red, 230
  Leitz beam path, 242
  Littrow, 220
Monochromatic light, 138, 351
Monobromonaphthaline, 202
Mounts temporary, 338
Mounting, 332
Mounting medium, 50, 60, 312, 336, 334
  old, 312
  corner post, 337
  turntable, 338

Nanometers, 63, 120
Narcotizing, 346
Navashin's fluid, 345
Negative coverage, 166
  development, 287
  evenly exposed, 166
  exposure, 287
  format, 297
  grain, 287

Negative coverage (contd)
  high quality, 168
  lines per mm, 286
  low contrast, 311
  material, 286
    for fluorescence, 263
    infra-red, 231
    minute, 3
    ultra-violet, 238
    unsharp, 311
    unsuitable, 312
    orthochromatic, 175
    over exposed, 49
    old stock, 313
    poor definition, 311
    quality, 153
    size, 287
Neutral wedge screens, 150
Newton colour scale, 201
Nickel, 216
Nomarski, 222
Nose-piece, 12, 13, 56, 154, 158, 351

Object marker, 67
Objects spacing, 296
Objective, 16, 21, 31, 61, 81
  achromatic, 9, 15, 23, 25, 32, 62, 202, 266, 347
  actual aperture, 33
  angular aperture, 33
  annotation, 55
  antiflex, 201
  aperture, 32
  aplanatically correct, 36
  apochromatic, 15, 25, 38, 60, 88, 91, 266, 348
    for fluorescent, 255
  aspheric reflecting mirror, 220
  back lens image, 114
  Berlyn Brixner optical for furnace, 222
  Burch aspheric surface mirror, 220
  catoptric (darkground), 236, 272, 273 (Vickers)
  colour range, 63, 64, 235
  correction collar, 56, 57, 348, 375
  damage, 298
  depth-of-field, 45, 298
  dioptric (darkground), 272
  dry, 30, 31, 349

Ealing-Beck, 219
eccentric stop, stereo, 285
electrostatic focusing, 250
Elliott IR, 227
epi, 195, 201
Epilan immersion, 201
first requirements, 32
flat field, 66, 67
  series, 68
fluorite, 64, 88
  for fluorescence, 255
focal length, 32
for opaque, 61
  phase contrast, 191, 235
  infra-red, 223
glycerol immersion, 63
half-reflecting, 97
heater, 222
high-power, 7, 50, 68
identification, 84
immersion, 68, 350
in glycerine, 52
interference, 211, 213
  cyclic, 206
  round-the-square, 199, 204
  Zeiss, 208
intermediate image, 21
Leitz, heating, 314
lines per cm, 140
Lister type, 24
long working, 56
low-power, 53
measuring interference, 204
methylene iodine immersion, 201
mirror, 95, 96
  after Nomarski, 221
Neofluars, 56, 61
N.A., 16, 29, 31, 32, 39, 45, 69, 85, 347, 351
object to image, 3
obscuration ratio, 218
oil, 31, 51, 52, 65, 99, 154, 351
parfocal, 64, 81, 352
phase contrast, 184
plano, 63
planochromat, 61, 62, 191, 266
primary magnification, 19, 32
protection cap, 53
quartz elements, 218
  fluorite and fluorite refracting, 221

SUBJECT INDEX

Objective, quarts elements (contd)
   for ultra-violet, 233, 236
   monochromatic, 233
  reflecting, 56, 65, 218, 219, 221
   long working distance, 219
   mirror, 64, 65, 95, 97, 104
    infra-red, 227
    Polaroid grey, 221
    Schwartzschild, 217
   solid quartz, 217, 218
  resolving power, 29, 38, 84, 168, 218, 220
  rotating sleeve, 56, 375
  semi-apochromatic, 25, 348
  sharp focus, 45
  short mounted, 181
  spherical reflecting, 230
  stopping down, 45
  transmission UV ultrafluars, 56, 235
  ultra-violet, 62, 63, 64
  water immersion, 31, 51, 53, 174
  working distance, 32, 61, 353
Oblique light, 351
Opaque subject, 47
Optics, 351
Optical alignment, 152
  axis, 15, 17, 53, 99, 154, 351
  bench, 351
  density, 28
  filters, 351
  glass, 351
   tank, 53
   tarnished, 298
  limitations, 297
  systems, 15, 158
  tube length, 14, 81

Palisade mesophyll, 9
Paper, tinted, 296
Parabolic mirror, 96
Particals in liquid, 171
Patch stop, 264, 266, 268, 274
Penicillin, 362, 374
Perspective, 169
Perspex, 180
Perspicillum, 6
Petri-dish, 177, 300
Phase contrast, 183, 352, 401

  adaptor, 191
  b-minus, 185
  bright contrast, 185
  dark contrast, 185
  plate, 184
  ring, 190
Phosphorescence, 254
Photocell, 11
Photometer, 246, 260, 393
Photomicrograph, 1, 352
  colour, 126
  common faults, 310
  data per subject, 382
  instant, 11
Photo emulsion examination of, 218
Photomultiplier, 246
Pictorial application,
Plane tissues, 328
Plates daguerrotype, 9
  silvered, 8
Pleurax, 343
Pleurosigma, 85
Plexiglas, 146, 314
Point of focus, 46
Polaroid, 198
  Land, 155
Polarizer, 196
Polarization, 50, 52, 194
  elliptically and circularly, 196
  filter, 195
  plane, 196, 198, 385
Polarizing vertical illuminator, 107
Polychromed methylene blue, 332
Potassium dichromate, 327, 345
  iodine, 333
Prints, 10, 168
  bright spot, 311
  exhibition, 289
  foreign marks, 312
  lacking detail, 311
   definition, 312
  light bar marks, 311
Printing, 288
Printing paper, 287, 289
Presentation, 168, 188, 313
Prism adjustment, 158
  Bergmann, 246
  deflecting, 153
  double image, 206
  eccentric, 104

Prism adjustment (contd)
  for interference, 203
  Nicol, 196
  Wollaston, 210

Quartz fluorite doublet element, 221
  fused, 231
  lenses, 218
  monochromat, 218
  plates, 190
  refractive index, 52
  transmission in IR, 226
    UV, 233
  wave plate, 200
  wedges, 200

Radiography by contact, 251
  reflection, 251
  shadow reflection, 252
  X, 251
Rabbit, 177
  eggs, 48, 432, 433
  embryos, 178
Rays axial, 26
Real image space, 20
  object space, 20
Realgar, 343
Record information, 152
Reflection, 31, 95
Reflecting objective Schwartzschild, 217
  surfaces, 216
Refraction, 28, 352
  double, 52
Refractive index, 24, 30, 50, 51, 196, 352
  measuring, 206
  various substances, 51
Reichert fluorescent microscope, 253
Replica technique, 346
Resolution, 84
  poor, 311
  oblique transmitted light, 103
Resolving power, 168, 218, 220, 232, 235
Respiratory cavity, 9
Rhodium, 203
Rheostat, 99
Ringer solution, 177
Rock salt transmission in IR, 226
Rodamine, 254

Safelights, 288
Saline solution, 334
Scale, 49
  line, 313
  measuring, 314
Scales, butterfly, 294, frontispiece
Scalpels, 330
Schlieren photomicrography, 275
Scissors, 330
Sectioning, 328
Section thin, 50
  lifter, 330
Selenium, evaporated, 228
Sellotape, 169
Shellac, 338
Shock transmission, 122
Shutter, 153
Silioset liquid, 346
Silver caesium oxide, 227
  chloride, 230
Simple microscope, 4, 365, 445
Sinks, 288
Slide, microscope, 54, 57, 58, 61, 216, 235, 300, 334, 340, 342, 352
  cabinet, 342
  centering, 113
  cultures, 330, 340
  darkground, 264
  form wedge, 171
  for fluids, 342
    fluorescence, 254
  hollow ground, 48, 61, 164
  labels, 342
  test, 58
  ultra-violet, 232
  wooden, 61, plate 16
  quartz, 61, 254
Snells law, 28
Sodium chloride, 328
  sulphate, 345
Source too bright, 312
Space real image, 20
Specimen, 2, 14, 48, 49, 55, 164, 165, 168, 174, 176, 181, 203, 260, 299, 301, 327, 328, 334, 338
  exhibit colour margins, 313
  feed, 329
  in water, 48
  multi-stained, 25
  off axis, 25

SUBJECT INDEX

Specimen (*contd*)
  opaque, 92, 297, 301, 339
  phase contrast, 188
  stain, fluorescent material, 254
  stained, 151
    by neoyanine, 226
      nitrate, 226
  stretcher, 329
  thickness, 60
  tilting, 25
  unmounted, 55
  wet mounted, 164
  bubbles in liquid, 180
  flat mounted, 177
  X-ray, 250
Spectrum, 63, 216, 352
  emission HBO 200 watt, 120
  colours separate, 313
Spectrosil, 240
Specular highlights, 108
Sphere, 5, 293
Sphero-chromasy, 313
Spongy mesophyll, 9
Spoon, mounting 336
Spring clip, 334
Stage auto-levelling, 158
  centering, 113
  cooling, 313
  hot, 299, 313, 449
  micrometer, 89, 351, 352, 391
  movement, 166
  rotating, 112, 129, 199, 202, 204
  square mechanical, 158, 351
  tilting, 117, 375
Stoma, 9
Stain, 140, 345
Stain botanical, 333, 345
  zoological, 333
Staining, 330
  blood, 345
  Gram, 332
  optical, 132
  plant chromosomes, 345
  reactions, 333
  reasons, 333
Semi-transparent, 137
Stereo eccentric stop, 282, 284
  effect, 117, 169
  cycloptic microscope, 282, 284
  half-lens method, 280

  pairs, mounting, 286
  photomicrography, 276
Stereoscopic microscope, 278
  prints, 278
  viewing prints, 279
  vision, 81, 277
Storage, 288
Styrax, 254
Substage, 10, 302, 348
  condenser, 15, 39, 159, 348, 336
  iris diaphragm, 39, 41, 47
Surface polished, 27
Sylvine transmission in IR, 226
Synclinal, 296

Teflon, 300
Telescope centering, 83
  focusing, 286
Temperatures from, 313
Thallium bromoiodide, 230
Thermometer, 313
Tissues stained, 203
Timber, 331
Time lapse, 419, 451
Tones in prints, 289
Transmission curves dichroic mirror, 262
  shock, 256
Transparent specimen, 137
Transparency faults, 312
Transverse section, 382
Translucent, 173
Tube, body 348
  diameter, 13
  length, 34, 220, 221, 342
    adjustment, 57, 58
    correction, 58
    corrector, 61
    mechanical, 13, 14, 16, 20, 61, 81, 352
    micro attachment, 153
    optical, 13, 14, 16, 20, 61, 81, 352
    practical, 15
Tweezers, 336
Tyndall effect, 274

Ultra-violet, 98, 103, 120, 216
  photomicrography, 231
  radiation, 121

Ultra-Violet (contd)
 transmission, 231

Vacuum grease, 337
Varnish, 337
Van Gieson's stain, 344
Velvet base,
Venetian turpentine, 254
Viewing points, 379
Virtual image space, 17, 20, 343
Visible spectrum, 63
Vision, 81, 277
Vision, angle of Volkensky's stain, 344

Watch glass, 89, 332, 334
Water, 48, 50, 334
 distilled, 53, 174, 334, 345
 refractive index, 53
Wave incident, 227
Wavefront, 30, 39

Wavelength, 24, 36, 40, 42, 120, 187
 drum-monochromator, 241
Whipple disc, 319
Willemite, 227
Wooden wedge, 171
Working distance, 5, 13, 31, 32, 37, 54, 56, 61, 64, 353

X-radiography, 251
X-ray, 249, 299, 401
 projection, 253
Xylene, 330, 332

Zernike, 183
Zeiss MPP, 156
 microscope, 158
 Ultrophot, 155
Zinc ortho-solicate, 227
Zirconium, 124
Zoom optical system, 65
 for stereos, 281